Plasma Diagnostics

Plasma Diagnostics

Editor

Bruno Soares Gonçalves

Basel • Beijing • Wuhan • Barcelona • Belgrade • Novi Sad • Cluj • Manchester

Editor
Bruno Soares Gonçalves
Instituto Superior Técnico,
Universidade de Lisboa
Lisbon
Portugal

Editorial Office
MDPI AG
Grosspeteranlage 5
4052 Basel, Switzerland

This is a reprint of articles from the Special Issue published online in the open access journal *Sensors* (ISSN 1424-8220) (available at: https://www.mdpi.com/journal/sensors/special_issues/Plasma_Diagnostics_Sensors).

For citation purposes, cite each article independently as indicated on the article page online and as indicated below:

Lastname, A.A.; Lastname, B.B. Article Title. *Journal Name* **Year**, *Volume Number*, Page Range.

ISBN 978-3-7258-1449-7 (Hbk)
ISBN 978-3-7258-1450-3 (PDF)
doi.org/10.3390/books978-3-7258-1450-3

© 2024 by the authors. Articles in this book are Open Access and distributed under the Creative Commons Attribution (CC BY) license. The book as a whole is distributed by MDPI under the terms and conditions of the Creative Commons Attribution-NonCommercial-NoDerivs (CC BY-NC-ND) license.

Contents

About the Editor vii

Bruno Gonçalves
Plasma Diagnostics
Reprinted from: *Sensors* 2024, 24, 3257, doi:10.3390/s24103257 1

Bruno Gonçalves, Paulo Varela, António Silva, Filipe Silva, Jorge Santos, Emanuel Ricardo, et al.
Advances, Challenges, and Future Perspectives of Microwave Reflectometry for Plasma Position and Shape Control on Future Nuclear Fusion Devices
Reprinted from: *Sensors* 2023, 23, 3926, doi:10.3390/s23083926 8

Raul Luís, Yohanes Nietiadi, Antonio Quercia, Alberto Vale, Jorge Belo, António Silva, et al.
Neutronics Simulations for DEMO Diagnostics
Reprinted from: *Sensors* 2023, 23, 5104, doi:10.3390/s23115104 40

Tommaso Patton, Alastair Shepherd, Basile Pouradier Duteil, Andrea Rigoni Garola, Matteo Brombin, Valeria Candeloro, et al.
Design and Development of a Diagnostic System for a Non-Intercepting Direct Measure of the SPIDER Ion Source Beamlet Current
Reprinted from: *Sensors* 2023, 23, 6211, doi:10.3390/s23136211 81

A. Malaquias, I. S. Nedzelskiy, R. Henriques and R. Sharma
The Heavy-Ion Beam Diagnostic of the ISTTOK Tokamak—Highlights and Recent Developments
Reprinted from: *Sensors* 2022, 22, 4038, doi:10.3390/s22114038 110

Igor Nedzelskiy, Artur Malaquias, Rafael Henriques and Ridhima Sharma
Affect of Secondary Beam Non-Uniformity on Plasma Potential Measurements by HIBD with Split-Plate Detector
Reprinted from: *Sensors* 2022, 22, 5135, doi:10.3390/s22145135 138

Anna Ponomarenko, Alexander Yashin, Gleb Kurskiev, Vladimir Minaev, Alexander Petrov, Yuri Petrov, et al.
First Results of the Implementation of the Doppler Backscattering Diagnostic for the Investigation of the Transition to H-Mode in the Spherical Tokamak Globus-M2
Reprinted from: *Sensors* 2023, 23, 830, doi:10.3390/s23020830 152

Alexander Yashin, Natalia Teplova, Georgiy Zadvitskiy and Anna Ponomarenko
Modelling of Backscattering off Filaments Using the Code IPF-FD3D for the Interpretation of Doppler Backscattering Data
Reprinted from: *Sensors* 2022, 22, 9441, doi:10.3390/s22239441 165

Ricardo Grosso Ferreira, Bernardo Brotas de Carvalho, Luís Lemos Alves, Bruno Gonçalves, Victor Fernandez Villace, Lionel Marraffa, et al.
VUV to IR Emission Spectroscopy and Interferometry Diagnostics for the European Shock Tube for High-Enthalpy Research
Reprinted from: *Sensors* 2023, 23, 6027, doi:10.3390/s23136027 191

Carlos Salgado-López, Jon Imanol Apiñaniz, José Luis Henares, José Antonio Pérez-Hernández, Diego de Luis, Luca Volpe and Giancarlo Gatti
Angular-Resolved Thomson Parabola Spectrometer for Laser-Driven Ion Accelerators
Reprinted from: *Sensors* 2022, 22, 3239, doi:10.3390/s22093239 222

Si-jun Kim, Sang-ho Lee, Ye-bin You, Young-seok Lee, In-ho Seong, Chul-hee Cho, et al.
Development of the Measurement of Lateral Electron Density (MOLE) Probe Applicable to Low-Pressure Plasma Diagnostics
Reprinted from: *Sensors* **2022**, *22*, 5487, doi:10.3390/s22155487 . . . . **235**

Si-jun Kim, In-ho Seong, Young-seok Lee, Chul-hee Cho, Won-nyoung Jeong, Ye-bin You, et al.
Development of a High-Linearity Voltage and Current Probe with a Floating Toroidal Coil: Principle, Demonstration, Design Optimization, and Evaluation
Reprinted from: *Sensors* **2022**, *22*, 5871, doi:10.3390/s22155871 . . . . **249**

Inho Seong, Sijun Kim, Youngseok Lee, Chulhee Cho, Jangjae Lee, Wonnyoung Jeong, et al.
Development of a Noninvasive Real-Time Ion Energy Distribution Monitoring System Applicable to Collisional Plasma Sheath
Reprinted from: *Sensors* **2022**, *22*, 6254, doi:10.3390/s22166254 . . . . **265**

Chulhee Cho, Sijun Kim, Youngseok Lee, Wonnyoung Jeong, Inho Seong, Jangjae Lee, et al.
Refined Appearance Potential Mass Spectrometry for High Precision Radical Density Quantification in Plasma
Reprinted from: *Sensors* **2022**, *22*, 6589, doi:10.3390/s22176589 . . . . **278**

Andrew Martusevich, Roman Kornev, Artur Ermakov, Igor Gornushkin, Vladimir Nazarov, Lyubov Shabarova, et al.
Spectroscopy of Laser-Induced Dielectric Breakdown Plasma in Mixtures of Air with Inert Gases Ar, He, Kr, and Xe
Reprinted from: *Sensors* **2023**, *23*, 932, doi:10.3390/s23020932 . . . . **289**

Youngseok Lee, Sijun Kim, Jangjae Lee, Chulhee Cho, Inho Seong and Shinjae You
Low-Temperature Plasma Diagnostics to Investigate the Process Window Shift in Plasma Etching of SiO_2
Reprinted from: *Sensors* **2022**, *22*, 6029, doi:10.3390/s22166029 . . . . **302**

Liyi Gu, Chintan Shah and Ruitian Zhang
Uncertainties in Atomic Data for Modeling Astrophysical Charge Exchange Plasmas
Reprinted from: *Sensors* **2022**, *22*, 752, doi:10.3390/s22030752 . . . . **314**

About the Editor

Bruno Soares Gonçalves

Bruno Soares Gonçalves holds a PhD in Physics and was awarded the "European Physical Society Plasma Division PhD Research Award" in 2005. He is a coordinator researcher at Instituto Superior Técnico, has worked as the President of Instituto de Plasmas e Fusão Nuclear (IPFN) since 2012 and leads the Systems Engineering and Integration Group at IPFN. He has spearheaded numerous research projects focusing on diagnostics development for nuclear fusion devices and control and data acquisition systems, and in 2022, he was awarded the ANACOM-URSI Portugal award. He serves on various committees of the European Fusion Programme, multiple scientific committees at international conferences and as an expert reviewer for different international organizations. He is co-inventor of four patents and co-author of over 130 articles published in international journals. He also engages in public outreach through seminars and discussion panels about nuclear energy, as well as energy and science, and through his book, "Nuclear Fusion in the Age of Climate Change".

Editorial

Plasma Diagnostics

Bruno Gonçalves

Instituto de Plasmas e Fusão Nuclear, Instituto Superior Técnico, Universidade de Lisboa, 1049-001 Lisbon, Portugal; bruno@ipfn.tecnico.ulisboa.pt

Citation: Gonçalves, B. Plasma Diagnostics. *Sensors* **2024**, *24*, 3257. https://doi.org/10.3390/s24103257

Received: 3 May 2024
Accepted: 6 May 2024
Published: 20 May 2024

Copyright: © 2024 by the author. Licensee MDPI, Basel, Switzerland. This article is an open access article distributed under the terms and conditions of the Creative Commons Attribution (CC BY) license (https://creativecommons.org/licenses/by/4.0/).

Plasma science and engineering is a multidisciplinary area encompassing some of the most exciting fundamental and applied research themes in today's scientific landscape, with an extraordinarily broad impact in science, technology, and industry. Although the mainstream areas of plasma research are readily identified as fusion (i.e., magnetic and inertial), laser–plasma interactions, and low-temperature plasma technologies, plasma science is often a key component of many other disciplines, including nanoscience, atomic and molecular physics, surface physics, biophysics, astrophysics, and space science.

The measurement of the parameters of plasmas, usually termed plasma diagnostics, is vital in the quest to harness the power of plasma. Its importance spans across various fields and applications, making it a cornerstone of plasma physics research. Measuring the parameters of plasmas is a key challenge in all these applications, providing essential data for understanding plasma behavior and the basic principles, validating theoretical models, and, in many cases, optimizing and controlling processes in plasma-based applications. In fusion energy research, accurate plasma diagnostics are key to optimizing the conditions for nuclear fusion. In space exploration, understanding plasma properties helps predict and mitigate the effects of space weather on satellites and spacecraft or optimize shielding for spacecraft reentry into the planet's atmosphere. In industrial applications, plasma diagnostics aid in improving processes like plasma etching and deposition in semiconductor manufacturing.

The range of different methods in plasma diagnostics is necessary to cover this diversity of applications. Each method offers unique insights into the complex behavior of plasmas under different conditions, contributing to the advancement of plasma physics and its applications. By analyzing parameters such as electron density, ion temperature, and electric potential, among many others, researchers can gain valuable insights into the complex phenomena occurring within plasma. Plasma diagnostics is based on a wide variety of characteristic plasma phenomena, and although most of the techniques used are already well established [1], plasma diagnostics is still a very challenging and vivid discipline. On one hand, there is a continuing effort to attain better spatial and temporal resolution, to reach higher accuracies, and to measure with more spatial channels. On the other hand, diagnostic techniques based on more subtle physical processes (compared to those used in routine diagnostics) are continuously being developed, and new tools are being added (e.g., machine learning techniques [2]). Furthermore, to obtain a better insight into the processes taking place in the plasma, it is a prerequisite that plasma parameters are diagnosed simultaneously, as much as possible, with multi-channel diagnostics, preferably with temporal and spatial resolutions smaller than the typical time and length scales of the instabilities.

In some areas, e.g., future fusion reactors [3], such as ITER and future power plant demonstrators, commonly known as DEMO [4–6], there will be a need to measure a wide range of plasma parameters in extreme conditions [7,8] of temperature, neutron, and gamma fluxes while providing inputs to control systems [9] with adequate reliability and long-term stability, enabling us to reach and sustain high levels of fusion power in a stationary manner. In such environments, the design of plasma diagnostics becomes a highly interdisciplinary endeavor [10]. The diagnostic design must adapt to the limited space restrictions for the integration of diagnostic components, and strong adverse effects acting on the diagnostic front-end components (neutron and gamma radiation, heat loads, erosion, and deposition) will degrade the diagnostic components over time. Moreover, the nuclear environment of DEMO imposes that any maintenance must be performed by remote handling, making it technically challenging to design diagnostic components for remote handling compatibility. Furthermore, all in-vessel components should be designed for a high degree of durability and reliability to minimize interventions for scheduled and non-scheduled maintenance. All these restrictions are highly demanding on the diagnostic design.

This Special Issue includes sixteen papers focused on the latest advancements in the field of plasma diagnostics, covering methods, instruments, and experimental techniques used to measure the properties of plasma, such as diagnostics for magnetic confinement fusion, beam plasmas and inertial fusion, low-temperature and industrial plasmas, reentry plasmas, and basic and astrophysical plasmas. Each of the sixteen original contributions accepted for publication has undergone a rigorous review process by a minimum of two expert reviewers across at least two rounds of revision and was judged by their technical merit and relevance. These studies published in the current collection are briefly summarized below.

In contribution 1, the authors address the development, design, and commissioning of a diagnostic system for a non-intercepting direct measure of the SPIDER ion source beamlet current, the so-called beamlet current monitor (BCM), aimed at directly measuring the electric current of a particle beam. Stable and uniform beams with low divergence are required in particle accelerators; therefore, beyond the accelerated current, measuring the beam current spatial uniformity and stability over time is necessary to assess the beam performance, since these parameters affect the perveance and thus the beam optics. For high-power beams operating with long pulses, it is convenient to directly measure these current parameters with a non-intercepting system due to the heat management requirement. Such a system needs to be capable of operating in a vacuum in the presence of strong electromagnetic fields and overvoltages due to electrical breakdowns in the accelerator.

In contribution 2, the authors examine the specifications and requirements for high-speed flow measurements for the plasmas produced in the European Shock Tube for High-Enthalpy Research (Portugal). This new state-of-the-art facility is tailored for the reproduction of spacecraft planetary entries in support of future European exploration missions. High-speed events, such as planetary entry shock waves, are very challenging to examine in shock tube facilities owing to their very short timescales (in the order of the µs), hence mandating the deployment of fast diagnostic techniques. The importance of examining the different spectral regions lies in characterizing the physical and chemical processes governing the behavior of the entry plasmas to perfect the numerical models. This paper discusses the design choices for the main diagnostics, with particular focus on VUV-to-IR emission spectroscopy and interferometry diagnostics. The spectroscopy setup covers a spectral window between 120 and 5000 nm, and the microwave interferometer can measure electron densities up to 1.5×10^{20} electrons/m^3.

Contribution 3 addresses the main challenges in the development of a plasma diagnostic and control system for a nuclear fusion DEMOnstration reactor (DEMO). The diagnostics need to cope with unprecedented radiation levels in a tokamak during long operation periods. This paper provides a broad overview of the radiation environment that diagnostics in DEMO are expected to face. Using the water-cooled lithium lead blanket configuration as a reference, neutronics simulations were performed for pre-conceptual designs of in-vessel, ex-vessel, and equatorial port diagnostics representative of each integration approach. Resorting to diagnostics representative of different integration approaches in DEMO—inner vessel, ex-vessel, and equatorial ports—neutronics simulations were performed to estimate the fluxes, heat loads, dose rates, and displacements per atom (dpa) in different sections of the tokamak using pre-conceptual CAD models of the diagnostics. Flux and nuclear load calculations are provided for several sub-systems, along with estimations of radiation streaming to the ex-vessel for alternative design configurations. The results can be used as a reference by diagnostic designers.

In contribution 4, the authors focused on the use of spectroscopy of laser-induced dielectric breakdown plasma in mixtures of air with inert gases Ar, He, Kr, and Xe. The generation of ozone and nitrogen oxides by laser-induced dielectric breakdown (LIDB) in mixtures of air with noble gases Ar, He, Kr, and Xe is investigated using OES and IR spectroscopy, mass spectrometry, and absorption spectrophotometry. It was shown that the addition of He to air does not fundamentally change the spectral pattern of air. In contrast, the addition of Ar suppresses the N+ band at 463.0 nm, while the other bands of nitrogen ions slightly decrease. The addition of Kr leads to even greater suppression of the line intensities of nitrogen ions as well as oxygen ions. It should be noted that when He, Ar, and Kr are added, the atomic oxygen line retains a low intensity. The addition of Xe results in complete suppression of the air component bands. It is expected that this method will be used on real systems, for example, for the treatment of abiogenic media obtained after irradiation with gas mixtures (for example, saline), with a further transition to blood treatment in order to increase its antioxidant potential.

Contribution 5 shows the first results of the implementation of the Doppler backscattering diagnostic (DBS) for the investigation of the transition to H-mode in the spherical tokamak Globus-M2. DBS allows the measurement of the poloidal rotation velocity and the turbulence amplitude of plasma. The multi-frequency DBS system installed on Globus-M2 can simultaneously collect data from different areas spanning from the separatrix to the plasma core, allowing the radial profiles of the rotation velocity and electric field to be calculated before and after the LH transition. DBS measurements of the poloidal plasma velocity and small plasma turbulence allowed for it to be investigated by observing the behavior of these parameters.

Contribution 6 also shows results obtained in Globus-M2. It has been observed that the DBS signal reacts to the backscattering from filaments, which are well known to strongly contribute to particle and energy losses both in L- and H-mode. However, the DBS data have proven difficult to analyze. The contribution focuses on modeling backscattering of filaments using two-dimensional full-wave simulations of backscattering off filaments with the code IPF-FD3D for the interpretation of Doppler backscattering data. This simulation enables us to understand what kind of information can be extrapolated from the signals.

Contribution 7 is focused on refined appearance potential mass spectrometry (APMS) for high-precision radical density quantification in plasma to improve the precision of plasma diagnosis and help elucidate the plasma etching process. As the analysis of complicated reaction chemistry in bulk plasma has become more important, especially in plasma processing, quantifying radical density is now in focus. With a simple modification of the original APMS approach, the fitting process was eliminated, and the He density was obtained over the entire electron energy range. A comparison of the neutral densities in He plasma between the conventional method and the new method, along with the real neutral density obtained using the ideal gas equation, confirmed that the proposed quantification approach can provide more accurate results.

As the importance of ion-assisted surface processing based on low-temperature plasma increases, the monitoring of ion energy impinging on wafer surfaces becomes more important. Monitoring methods that are noninvasive, real-time, and comprise ion collisions in the sheath have received much research attention. Contribution 8 is focused on the development of a noninvasive real-time ion energy distribution (IED) monitoring system based on an ion trajectory simulation where the Monte Carlo collision method and an electrical model are adopted to describe collisions in sheaths. In previous works, IED monitoring systems had the limitations of, for example, neglecting collisions, measuring invasively, and assuming the plasma potential to be a sine wave. To overcome these limitations, the authors investigated the IED measurement with the proposed method and compared it with the results of IEDs measured via a quadrupole mass spectrometer under various conditions.

A noninvasive and real-time IED monitoring system was proposed and validated in an asymmetric RF CCP discharge in Ar plasma. The comparison results show that there was no major change in the IEDs as radio-frequency power increased or as the IED gradually became broad as gas pressure increased, which was in good agreement with the results of the mass spectrometer.

Contribution 9 addresses low-temperature plasma diagnostics to investigate the process window shift in plasma etching of SiO_2. As low-temperature plasma plays an important role in semiconductor manufacturing, plasma diagnostics have been widely employed to understand changes in plasma according to external control parameters, which has led to the achievement of appropriate plasma conditions, normally termed the process window. During plasma etching, shifts in the plasma conditions both within and outside the process window can be observed. In this contribution, the authors utilized various plasma diagnostic tools to investigate the causes of these shifts. Cutoff and emissive probes were used to measure the electron density and plasma potential as indicators of the ion density and energy, respectively, that represent the ion energy flux. Quadrupole mass spectrometry was also used to show real-time changes in plasma chemistry during the etching process. The obtained diagnostic results were able to sufficiently explain the process window shift and, in addition, were in good agreement with the etch model prediction. By extending the SiO_2 etch model with rigorous diagnostic measurements (or numerous diagnostic methods), more intricate plasma processing conditions can be characterized, which will be beneficial in applications and industries where different input powers and gas flows can make notable differences to the results.

Contribution 10 is focused on the development of a high-linearity voltage and current probe with a floating toroidal coil (FTC). As the conventional voltage and current (VI) probes widely used in plasma diagnostics have separate voltage and current sensors, crosstalk between the sensors leads to degradation of measurement linearity, which is related to practical accuracy. The authors propose a VI probe with a floating toroidal coil that plays both roles as a voltage and current sensor and is thus free from crosstalk. In this paper, the operation principle of the FTC was demonstrated, and its optimum design was established through 3D electromagnetic wave simulation. It is expected that the proposed VI probe could be applicable to plasma diagnostics as well as process monitoring with higher accuracy.

Contribution 11 proposes the measurement of a microwave probe, the measurement of lateral electron density (MOLE) probe, applicable to low-pressure plasma diagnostics. The basic properties of the MOLE probe are analyzed via three-dimensional electromagnetic wave simulation, with simulation results showing that the probe estimates electron density by measuring the surface wave resonance frequency from the reflection microwave frequency spectrum (S11). An experimental demonstration on a chamber wall measuring lateral electron density is conducted by comparing the developed probe with the cutoff probe, a precise electron density measurement tool. Experimental demonstrations, including a cutoff probe for comparison, exhibit that the MOLE probe represents good linearity with the cutoff probe in bulk as well as on the chamber wall, which means that the MOLE probe can measure the lateral electron density. Based on both simulation and experiment results, the MOLE probe is shown to be a useful instrument to monitor lateral electron density.

Contribution 12 is focused on the heavy ion beam diagnostic (HIBD), the only tool for direct measurements of plasma potential in magnetically confined fusion plasmas. The measurements of plasma potential fluctuations are of special interest and importance in investigations of turbulent transport. In a heavy ion beam diagnostic (HIBD), the plasma potential is obtained by measuring the energy of the secondary ions resulting from beam–plasma collisions by an electrostatic energy analyzer with a split-plate detector (SPD), which relates the secondary ion beam energy variation to its position determined by the difference in currents between the split plates. This paper considers the possible influence of the secondary beam non-uniformity on plasma potential and its fluctuation measurements using the SPD technique. The results are supported by experimental data from the tokamak ISTTOK HIBD.

Contribution 13 goes further in addressing the highlights and recent developments obtained in ISTTOK HIBD. The heavy-ion beam diagnostic installed on the ISTTOK tokamak (Lisbon, Portugal) has been conceptualized to provide simultaneously the plasma radial profile evolution of the plasma temperature, electron density, plasma poloidal magnetic field, and plasma potential. In fact, this diagnostic has the capability to scan the plasma in 2D, although in ISTTOK, it is limited in its coverage by geometrical factors. In practice, it can provide a 1D full-diameter profile of each of the above parameters. This paper describes the capabilities that have been developed in this diagnostic and includes a more detailed description of the physics basis for the measurement of the plasma poloidal magnetic field profile in ISTTOK. ISTTOK HIBD is based on a unique configuration that allows the collection in a multiple-cell array detector of the probing secondary beams generated from the whole plasma diameter. The wealth of information obtained allows for accounting for the path integral effects and retrieving the local values of the plasma parameters at the ionization volume in the plasma. The method allowed for the obtaining of plasma-like pressure profiles with good spectral resolution. Exploitation of the plasma pressure-like measurements in AC discharges allowed for the identification and characterization of MHD activity and turbulent transport in edge polarization experiments. Interesting results are also shown on the real-time determination of the plasma column center, on an innovative method to determine the plasma current poloidal magnetic field, and on the most recent developments in measuring the plasma potential using a cylindrical energy analyzer.

Contribution 14 reports the development, construction, and experimental test of an angle-resolved Thomson parabola (TP) spectrometer for laser-accelerated multi-MeV ion beams that can measure the spectra of discretized beamlets with different emission angles from a laser–plasma interaction experiment at a high repetition rate, simultaneously sorting the ion species by their charge-to-mass ratio. The work presents a multi-pinhole Thomson parabola spectrometer, which combines sharp spectral/angular precision with ionic species sorting capability. A novel analysis method, which can examine the crossing parabolic traces on the detector plane, grants access to several variables simultaneously. The contribution also describes the first test of the spectrometer at the 1PW VEGA 3 laser facility at CLPU, Salamanca (Spain), where up to 15 MeV protons and carbon ions from a 3 μm laser-irradiated Al foil are detected. Such kinds of detectors open interesting prospects for beam analysis of novel acceleration mechanisms, such as collisionless shock acceleration (CSA) or radiation–pressure–acceleration (RPA), as well as for measurements of transported ion beamlines.

Relevant uncertainties in theoretical atomic data are vital to determining the accuracy of plasma diagnostics in several areas, including, in particular, the astrophysical study. Contribution 15 presents a new calculation of the uncertainties on the present theoretical ion-impact charge exchange atomic data and X-ray spectra, based on a set of comparisons with the existing laboratory data obtained in historical merged-beam, cold-target recoil-ion momentum spectroscopy, and electron beam ion traps experiments.

To conclude, contribution 16 addresses the advances, challenges, and future perspectives of microwave reflectometry for plasma position and shape control on future nuclear fusion devices. Microwave reflectometry is a radar-based technique that can be used to determine the radial distribution of plasma density in fusion experiments. This technique has been proposed as an alternative to magnetic measurements for plasma position control. This contribution presents the multiple engineering and physics challenges addressed while designing reflectometry diagnostics using radio science techniques. Specifically, short-range dedicated radars for plasma position and shape control in future fusion experiments, the advances enabled by the designs for ITER and DEMO, and future perspectives. One key development is also in electronics, aiming at an advanced, compact, coherent, fast-frequency sweeping RF back-end (23–100 GHz in a few μs) that is being developed at IPFN-IST using commercial monolithic microwave integrated circuits (MMIC).

The sixteen contributions in this Special Issue offer invaluable updates and insights into various aspects of plasma diagnostics development, addressing the challenges in these research domains. These findings can serve as a catalyst for future endeavors aimed at enhancing and optimizing next-generation plasma diagnostics. Each contribution reflects the collective ingenuity and relentless efforts of researchers committed to pushing the boundaries of knowledge, underscoring the multidisciplinary nature of the field.

To conclude, I would like to add a word of heartfelt gratitude to all the authors for their invaluable contributions to this Special Issue, as well as to the reviewers who generously devoted their time and expertise to ensure the quality of the published papers, thus contributing to the success of this endeavor.

Conflicts of Interest: The author declares no conflicts of interest.

List of Contributions:

1. Patton, T.; Shepherd, A.; Duteil, B.P.; Garola, A.R.; Brombin, M.; Candeloro, V.; Manduchi, G.; Pavei, M.; Pasqualotto, R.; Pimazzoni, A.; et al. Design and Development of a Diagnostic System for a Non-Intercepting Direct Measure of the SPIDER Ion Source Beamlet Current. *Sensors* **2023**, *23*, 6211. https://doi.org/10.3390/s23136211.
2. Ferreira, R.G.; Carvalho, B.B.; Alves, L.L.; Gonçalves, B.; Villace, V.F.; Marraffa, L.; da Silva, M.L. VUV to IR Emission Spectroscopy and Interferometry Diagnostics for the European Shock Tube for High-Enthalpy Research. *Sensors* **2023**, *23*, 6027. https://doi.org/10.3390/s23136027.
3. Luís, R.; Nietiadi, Y.; Quercia, A.; Vale, A.; Belo, J.; Silva, A.; Gonçalves, B.; Malaquias, A.; Gusarov, A.; Caruggi, F.; et al. Neutronics Simulations for DEMO Diagnostics. *Sensors* **2023**, *23*, 5104. https://doi.org/10.3390/s23115104.
4. Martusevich, A.; Kornev, R.; Ermakov, A.; Gornushkin, I.; Nazarov, V.; Shabarova, L.; Shkrunin, V. Spectroscopy of Laser-Induced Dielectric Breakdown Plasma in Mixtures of Air with Inert Gases Ar, He, Kr, and Xe. *Sensors* **2023**, *23*, 932. https://doi.org/10.3390/s23020932.
5. Ponomarenko, A.; Yashin, A.; Kurskiev, G.; Minaev, V.; Petrov, A.; Petrov, Y.; Sakharov, N.; Zhiltsov, N. First Results of the Implementation of the Doppler Backscattering Diagnostic for the Investigation of the Transition to H-Mode in the Spherical Tokamak Globus-M2. *Sensors* **2023**, *23*, 830. https://doi.org/10.3390/s23020830.
6. Yashin, A.; Teplova, N.; Zadvitskiy, G.; Ponomarenko, A. Modelling of Backscattering off Filaments Using the Code IPF-FD3D for the Interpretation of Doppler Backscattering Data. *Sensors* **2022**, *22*, 9441. https://doi.org/10.3390/s22239441.
7. Cho, C.; Kim, S.; Lee, Y.; Jeong, W.; Seong, I.; Lee, J.; Choi, M.; You, Y.; Lee, S.; Lee, J.; et al. Refined Appearance Potential Mass Spectrometry for High Precision Radical Density Quantification in Plasma. *Sensors* **2022**, *22*, 6589. https://doi.org/10.3390/s22176589.
8. Seong, I.; Kim, S.; Lee, Y.; Cho, C.; Lee, J.; Jeong, W.; You, Y.; You, S. Development of a Noninvasive Real-Time Ion Energy Distribution Monitoring System Applicable to Collisional Plasma Sheath. *Sensors* **2022**, *22*, 6254. https://doi.org/10.3390/s22166254.
9. Lee, Y.; Kim, S.; Lee, J.; Cho, C.; Seong, I.; You, S. Low-Temperature Plasma Diagnostics to Investigate the Process Window Shift in Plasma Etching of SiO_2. *Sensors* **2022**, *22*, 6029. https://doi.org/10.3390/s22166029.
10. Kim, S.-j.; Seong, I.-h.; Lee, Y.-s.; Cho, C.-h.; Jeong, W.-n.; You, Y.-b.; Lee, J.-j.; You, S.-j. Development of a High-Linearity Voltage and Current Probe with a Floating Toroidal Coil: Principle, Demonstration, Design Optimization, and Evaluation. *Sensors* **2022**, *22*, 5871. https://doi.org/10.3390/s22155871.
11. Kim, S.-j.; Lee, S.-h.; You, Y.-b.; Lee, Y.-s.; Seong, I.-h.; Cho, C.-h.; Lee, J.-j.; You, S.-j. Development of the Measurement of Lateral Electron Density (MOLE) Probe Applicable to Low-Pressure Plasma Diagnostics. *Sensors* **2022**, *22*, 5487. https://doi.org/10.3390/s22155487.
12. Nedzelskiy, I.; Malaquias, A.; Henriques, R.; Sharma, R. Affect of Secondary Beam Non-Uniformity on Plasma Potential Measurements by HIBD with Split-Plate Detector. *Sensors* **2022**, *22*, 5135. https://doi.org/10.3390/s22145135.
13. Malaquias, A.; Nedzelskiy, I.S.; Henriques, R.; Sharma, R. The Heavy-Ion Beam Diagnostic of the ISTTOK Tokamak—Highlights and Recent Developments. *Sensors* **2022**, *22*, 4038. https://doi.org/10.3390/s22114038.

14. Salgado-López, C.; Apiñaniz, J.I.; Henares, J.L.; Pérez-Hernández, J.A.; de Luis, D.; Volpe, L.; Gatti, G. Angular-Resolved Thomson Parabola Spectrometer for Laser-Driven Ion Accelerators. *Sensors* **2022**, *22*, 3239. https://doi.org/10.3390/s22093239.
15. Gu, L.; Shah, C.; Zhang, R. Uncertainties in Atomic Data for Modeling Astrophysical Charge Exchange Plasmas. *Sensors* **2022**, *22*, 752. https://doi.org/10.3390/s22030752.
16. Gonçalves, B.; Varela, P.; Silva, A.; Silva, F.; Santos, J.; Ricardo, E.; Vale, A.; Luís, R.; Nietiadi, Y.; Malaquias, A.; et al. Advances, Challenges, and Future Perspectives of Microwave Reflectometry for Plasma Position and Shape Control on Future Nuclear Fusion Devices. *Sensors* **2023**, *23*, 3926. https://doi.org/10.3390/s23083926.

References

1. Hutchinson, I.H. Principles of plasma diagnostics. *Plasma Phys. Control. Fusion* **2002**, *44*, 2603. [CrossRef]
2. Humphreys, D.; Kupresanin, A.; Boyer, M.D.; Canik, J.; Chang, C.S.; Cyr, E.C.; Granetz, R.; Hittinger, J.; Kolemen, E.; Lawrence, E.; et al. Advancing fusion with machine learning research needs workshop report. *J. Fusion Energy* **2020**, *39*, 123–155. [CrossRef]
3. Neilson, G.H.; Basile, A.; Cohen, A.; Cometa, F.; de Looz, M.A.; Fair, R.; Gattuso, A.; Jariwala, A.; Muscatello, C.; Pablant, N.; et al. Diagnostics for Burning Plasmas. *IEEE Trans. Plasma Sci.* **2022**, *50*, 4144–4149. [CrossRef]
4. Biel, W.; Albanese, R.; Ambrosino, R.; Ariola, M.; Berkel, M.V.; Bolshakova, I.; Brunner, K.J.; Cavazzana, R.; Cecconello, M.; Conroy, S.; et al. Diagnostics for plasma control–From ITER to DEMO. *Fusion Eng. Des.* **2019**, *146*, 465–472. [CrossRef]
5. Biel, W.; Ariola, M.; Bolshakova, I.; Brunner, K.J.; Cecconello, M.; Duran, I.; Franke, T.; Giacomelli, L.; Giannone, L.; Janky, F.; et al. Development of a concept and basis for the DEMO diagnostic and control system. *Fusion Eng. Des.* **2022**, *179*, 113122. [CrossRef]
6. Hu, L.Q.; Liu, Y. Progress of engineering design of CFETR diagnostics. *Fusion Eng. Des.* **2020**, *155*, 111731. [CrossRef]
7. Costley, A.E.; Sugie, T.; Vayakis, G.; Walker, C.I. Technological challenges of ITER diagnostics. *Fusion Eng. Des.* **2005**, *74*, 109–119. [CrossRef]
8. Donné, A.J.H. Plasma diagnostics in view of ITER. *Fusion Sci. Technol.* **2012**, *61*, 357–364. [CrossRef]
9. Perek, P.; Makowski, D.; Kadziela, M.; Lee, W.R.; Zagar, A.; Simrock, S.; Abadie, L.; Lee, J.H.; Lee, S.J.; Kim, H.J. Evaluation of ITER Real-Time Framework in plasma diagnostics applications. *Fusion Eng. Des.* **2023**, *192*, 113623. [CrossRef]
10. Belo, J.H.; Nietiadi, Y.; Luís, R.; Silva, A.; Vale, A.; Gonçalves, B.; Franke, T.; Krimmer, A.; Biel, W. Design and integration studies of a diagnostics slim cassette concept for DEMO. *Nucl. Fusion* **2021**, *61*, 116046. [CrossRef]

Disclaimer/Publisher's Note: The statements, opinions and data contained in all publications are solely those of the individual author(s) and contributor(s) and not of MDPI and/or the editor(s). MDPI and/or the editor(s) disclaim responsibility for any injury to people or property resulting from any ideas, methods, instructions or products referred to in the content.

Review

Advances, Challenges, and Future Perspectives of Microwave Reflectometry for Plasma Position and Shape Control on Future Nuclear Fusion Devices

Bruno Gonçalves [1,*], Paulo Varela [1], António Silva [1], Filipe Silva [1], Jorge Santos [1], Emanuel Ricardo [1], Alberto Vale [1], Raúl Luís [1], Yohanes Nietiadi [1], Artur Malaquias [1], Jorge Belo [1], José Dias [1], Jorge Ferreira [1], Thomas Franke [2], Wolfgang Biel [3], Stéphane Heuraux [4], Tiago Ribeiro [2], Gianluca De Masi [5], Onofrio Tudisco [6], Roberto Cavazzana [5], Giuseppe Marchiori [5] and Ocleto D'Arcangelo [6]

1. Instituto de Plasmas e Fusão Nuclear, Instituto Superior Técnico, Universidade de Lisboa, 1049-001 Lisboa, Portugal
2. Max-Planck-Institut für Plasmaphysik, Boltzmannstr. 2, D-85748 Garching, Germany
3. Institut für Energie- und Klimaforschung, Forschungszentrum Jülich GmbH, D-52425 Jülich, Germany
4. Institut Jean Lamour, UMR 7198 CNRS-Université de Lorraine, BP 50840, F-54011 Nancy, France
5. Consorzio RFX, 35127 Padova, Italy
6. ENEA, Fusion and Technologies for Nuclear Safety Department, C.R. Frascati, Via E. Fermi 45, 00044 Frascati, Italy
* Correspondence: bruno@ipfn.tecnico.ulisboa.pt

Abstract: Providing energy from fusion and finding ways to scale up the fusion process to commercial proportions in an efficient, economical, and environmentally benign way is one of the grand challenges for engineering. Controlling the burning plasma in real-time is one of the critical issues that need to be addressed. Plasma Position Reflectometry (PPR) is expected to have an important role in next-generation fusion machines, such as DEMO, as a diagnostic to monitor the position and shape of the plasma continuously, complementing magnetic diagnostics. The reflectometry diagnostic uses radar science methods in the microwave and millimetre wave frequency ranges and is envisaged to measure the radial edge density profile at several poloidal angles providing data for the feedback control of the plasma position and shape. While significant steps have already been given to accomplish that goal, with proof of concept tested first in ASDEX-Upgrade and afterward in COMPASS, important, ground-breaking work is still ongoing. The Divertor Test Tokamak (DTT) facility presents itself as the appropriate future fusion device to implement, develop, and test a PPR system, thus contributing to building a knowledge database in plasma position reflectometry required for its application in DEMO. At DEMO, the PPR diagnostic's in-vessel antennas and waveguides, as well as the magnetic diagnostics, may be exposed to neutron irradiation fluences 5 to 50 times greater than those experienced by ITER. In the event of failure of either the magnetic or microwave diagnostics, the equilibrium control of the DEMO plasma may be jeopardized. It is, therefore, imperative to ensure that these systems are designed in such a way that they can be replaced if necessary. To perform reflectometry measurements at the 16 envisaged poloidal locations in DEMO, plasma-facing antennas and waveguides are needed to route the microwaves between the plasma through the DEMO upper ports (UPs) to the diagnostic hall. The main integration approach for this diagnostic is to incorporate these groups of antennas and waveguides into a diagnostics slim cassette (DSC), which is a dedicated complete poloidal segment specifically designed to be integrated with the water-cooled lithium lead (WCLL) breeding blanket system. This contribution presents the multiple engineering and physics challenges addressed while designing reflectometry diagnostics using radio science techniques. Namely, short-range dedicated radars for plasma position and shape control in future fusion experiments, the advances enabled by the designs for ITER and DEMO, and the future perspectives. One key development is in electronics, aiming at an advanced compact coherent fast frequency sweeping RF back-end [23–100 GHz in few μs] that is being developed at IPFN-IST using commercial Monolithic Microwave Integrated Circuits (MMIC). The compactness of this back-end design is crucial for the successful integration of many measurement channels in the reduced space

Citation: Gonçalves, B.; Varela, P.; Silva, A.; Silva, F.; Santos, J.; Ricardo, E.; Vale, A.; Luís, R.; Nietiadi, Y.; Malaquias, A.; et al. Advances, Challenges, and Future Perspectives of Microwave Reflectometry for Plasma Position and Shape Control on Future Nuclear Fusion Devices. *Sensors* 2023, *23*, 3926. https://doi.org/10.3390/s23083926

Academic Editor: Hossam A. Gabbar

Received: 4 March 2023
Revised: 30 March 2023
Accepted: 6 April 2023
Published: 12 April 2023

Copyright: © 2023 by the authors. Licensee MDPI, Basel, Switzerland. This article is an open access article distributed under the terms and conditions of the Creative Commons Attribution (CC BY) license (https:// creativecommons.org/licenses/by/ 4.0/).

available in future fusion machines. Prototype tests of these devices are foreseen to be performed in current nuclear fusion machines.

Keywords: microwave antennas; microwave propagation; millimetre wave propagation; microwave circuitry; millimetre wave circuitry; microwave measurements; plasma diagnostic; fusion plasma

1. Introduction

The International Thermonuclear Experimental Reactor (ITER) in France and the conceptual design of a demonstration power plant (DEMO) are advancing the development of nuclear fusion as a viable solution for large-scale energy production. In these reactors, nuclear fusion reactions occur when a deuterium and tritium (D-T) plasma is heated to temperatures ten times higher than at the core of the Sun. The position control of this extremely hot plasma inside the reactor's fusion chamber is one of the most critical issues in the operation of these power-generating devices. As such, real-time feedback control of the plasma position plays a vital role in machine protection and disruption avoidance, being crucial for successful reactor operation. However, controlling plasma parameters in future reactor-grade fusion tokamaks, such as ITER and DEMO, presents significant challenges. During the ramp-up phase, it is crucial to prevent the plasma from impinging the inner vessel walls and to avoid destructive disruptions during steady-state operation. Presently, magnetic measurements are used for plasma control, but in future long pulse tokamak devices, such as ITER and DEMO, these measurements may be impacted by drifting integrators, radiation-induced voltages, or radiation damage to the magnetic pickup coils. These effects could compromise the magnetic equilibrium reconstruction, leading to premature discharge termination or damage to plasma-facing components.

Several modern measurement diagnostics used in fusion rely on Radiofrequency (RF) techniques to probe the plasma. The low amplitude of the probing waves introduces negligible perturbations to the plasma. However, the interaction between the electromagnetic (EM) field of the propagating waves and the magnetized plasma can cause changes in the amplitude, phase, polarization state, and spectrum of the waves due to propagation close to the plasma layers where cut-off reflection occurs. The location of the reflecting layers can be deduced by measuring the delay caused by the round-trip of the reflected waves. This information is used to reconstruct the electron density profile, which enables the real-time location of the separatrix—the last closed magnetic surface—provided a good estimation for its density is known.

Microwave reflectometry is a radar-based technique that can be used to determine the radial distribution of plasma density in fusion experiments. This technique has been proposed as an alternative to magnetic measurements for plasma position control. The first proof of principle for reflectometry-based plasma position feedback control was successfully demonstrated on ASDEX Upgrade [1]. Microwave reflectometry is a well-proven technique in the plasma fusion community and is used for a variety of physics measurements. One application is the study of turbulence characteristics in fixed electron density layers by continuously probing the plasma with a fixed frequency wave. Other uses for microwave reflectometry can also be found in the literature [2].

2. Probing Plasmas with Microwave Reflectometry

The Maxwell equations and the motion of plasma particles induced by the EM waves can be used to describe the propagation of EM waves in an inhomogeneous medium, such as a fusion plasma. This description is possible if we assume [3] that: First, when there are no EM perturbations, the plasma particles remain in their equilibrium positions (known as the cold plasma approximation). Second, the plasma is treated as a two-fluid medium consisting of ions and electrons that are coupled through the EM field of the wave (known as the fluid approximation). Third, it is considered that only the electrons contribute to the medium polarisation because the frequency of the EM waves is much higher than the eigen ion plasma frequencies, cyclotron f_{ci} and ion plasma f_{pi} frequencies (high-frequency approximation). Lastly, the plasma is considered homogeneous in all directions except for the direction of wave propagation (known as the slab approximation). This approach allows for a comprehensive understanding of the behaviour of EM waves in fusion plasma.

Suppose the former approximations are assumed for a wave propagation direction perpendicular to the plasma magnetic field ($k \perp B$), where k is the propagation direction and B is the plasma magnetic field, two propagation modes can be identified [4]: the ordinary mode (O-mode), where the wave's electric field is parallel to the plasma magnetic field ($E \parallel B$), as illustrated in Figure 1, and the extraordinary mode (X-mode), where the electric field of the wave is perpendicular to the magnetic field of the plasma ($E \perp B$). For the case of a tokamak plasma, the poloidal component of the magnetic field is negligible when compared to the toroidal component ($B_\varphi \gg B_\theta$), and therefore, only the toroidal component of the magnetic field is considered, i.e., $B \simeq B_\varphi$.

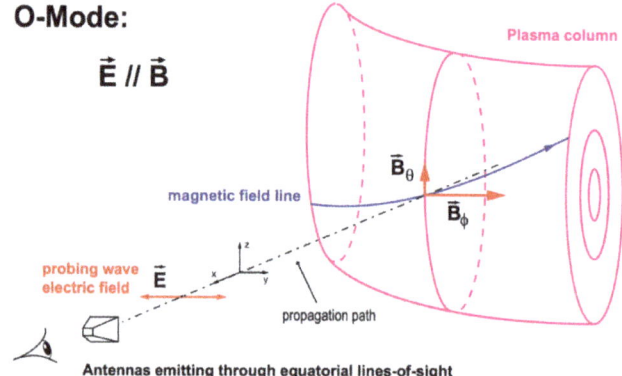

Figure 1. Schematic view of the O-mode reflectometry set-up where the wave electric field is not perfectly aligned with the tokamak magnetic field.

The refractive index, N_O, for an O-mode wave with frequency f is described by [5]:

$$N_O = \left[1 - (f_{pe}/f)^2\right]^{1/2} = [1 - n_e/n_{co}]^{1/2} \qquad (1)$$

where n_e is the electron density and $f_{pe} = (1/2\pi)(n_e e^2/\epsilon_0 m_e)^{1/2}$ is the electron plasma frequency. Wave propagation can occur for $f > f_{pe}$ or for $n_e < n_{co}$, where n_{co} is the cut-off density given by:

$$n_{co} = (2\pi f/e)^2 \epsilon_0 m_e \qquad (2)$$

where e is the electron charge, m_e is the electron mass and ϵ_0 is the permittivity of free space.

A microwave reflectometer is a short-range radar dedicated to the plasma measurement: a wave with frequency f is launched at normal incidence into the plasma, propagating through it until it reaches the cut-off layer with density $n_{co}(f)$, for O-mode polarization, where the refractive index becomes zero and the wave is reflected. The position of the density layers, at which the corresponding reflections occur, can be obtained by sweeping the frequency of the incident probing waves and measuring the waves' group delay due to the round trip from the launching antenna to the plasma and back to the receiving antenna. This information, plus an initial cut-off position for the lowest probing frequency, can then be used to calculate the plasma density profile along the reflectometer's line-of-sight (LOS). Frequency-modulated continuous wave (FMCW) reflectometry using ordinary (O-mode) plasma waves can measure the radial electron density profile independently of any parameters other than the plasma density (n_e) since O-mode propagation depends solely on the electron density distribution. For that reason, reflectometry measurements have been proposed on ITER to complement the standard magnetic measurements that are used in present machines for plasma position control [6] during the steady-state operation periods. In addition, the reflectometry diagnostic displays significant advantages, namely its minimum access requirements and compatibility with the harsh environment of a fusion device, plus the capability to perform density profile measurements in a short time scale (on the order of a few μs) inherent to a radar technique. Although for control purposes, measurements of the edge density profile are not an identical replacement for magnetic measurements, under certain conditions, they can provide similar capabilities [7]. In current fusion devices, position control is accomplished by changing the currents in the poloidal field coils based on the real-time estimation of the plasma shape and, in particular, the estimation of the plasma magnetic separatrix. Because the separatrix results exclusively from the reconstruction of the magnetic flux distribution, there is no direct relation with the electron density. Thus, a good estimation for the density just inside this flux surface is needed to track the gap between the separatrix and the vessel's first wall using a radial density profile given by reflectometry. A fusion plasma is a turbulent medium, and obtaining reliable measurements from waves propagating in it has proven to be a difficult and complex task. Before considering this technique as a reliable alternative for measuring plasma position, multiple challenges must be addressed. This article outlines the necessary advancements required to integrate this radar-like technique into the control systems of a fusion machine.

3. Demonstration of the PPR Principle at ASDEX-Upgrade

The use of microwave reflectometry for position control was successfully demonstrated for the first time on the ASDEX Upgrade (AUG) tokamak [1]. To control the plasma column position, the location of the outer plasma boundary was tracked in real-time (RT)—1 ms measurement rate/1 ms control loop cycle duration—using a custom-built very high bandwidth streaming data acquisition system [8], an optimized digital signal processing analysis and a dedicated neural-network based profile reconstruction algorithm [9]. The approach combined the RT reflectometry edge profile and an RT scaled line integrated density measurement from interferometry. After the first successful demonstration of plasma position control using reflectometry, the diagnostic's hardware was updated to acquire a higher number of signals and to improve its RT data-acquisition and data-processing capabilities. The improvements were aimed at producing a second demonstration of a plasma position control using both AUG's equatorial reflectometers, probing the tokamak high (HFS) and low field sides (LFS) simultaneously. This upgrade also aimed to achieve a four-fold increase of the system RT measurement rate (250 μs measurement rate/1 ms control loop cycle duration) and to integrate the RT reflectometry diagnostic in the AUG's Discharge Control System (DCS), a solution closer to ITER's PPR foreseen operation mode [10]. Both successful position control demonstrations were performed in ITER-relevant plasma regimes and configurations, providing a decisive contribution to the design of future PPR control systems.

The AUG broadband reflectometry diagnostic started operation at the beginning of the 1990's. The system was designed by IPFN/IST researchers and is also operated by researchers from the research unit in close collaboration with the AUG Team of the Max-Planck Institut für Plasmaphysik (IPP). For the plasma position control experiments, two FMCW O-mode reflectometry systems [7] were used to probe the plasma, installed at the tokamak equatorial plane, from the high (inner) and low (outer) magnetic field sides (HFS and LFS, respectively) of the device (see Figure 2). The HFS reflectometers operate in the 18–75 GHz frequency range, using bands K (18–26.5 GHz), Ka (26.5–40 GHz), Q (33–50 GHz), and V (50–75 GHz). In the LFS, this range is extended, up to ~110 GHz, with an additional reflectometer operating in the W band (75–110 GHz). In the LFS, two additional channels operating in X-mode are used to cover the 35–75 GHz range (using bands Q and V). The FMCW microwave reflectometry system on ASDEX Upgrade uses fundamental rectangular waveguides connected to hog-horns with elliptical mirrors. Their orientation sets the electrical field of the probing beam in the toroidal direction. The ASDEX Upgrade magnetic pitch angle is in the order of 10° [11], which implies that 97% of the probing signal is sent in O-mode. The 3% injected in X-mode are attenuated in the waveguide. Therefore, no calibration procedure is required to set the correct polarization. If circular waveguides were used, the correct polarization would require calibration using a mirror with a grid. The original system design was heavily influenced by the limited space available within the tokamak to route the waveguides and position the antennas behind the vessel heat shield. This shield, consisting of heat-resistant tiles, protects the inner vessel walls and plasma-facing components. The constraints, along with the experience with reflectometers built to the old ASDEX tokamak, resulted in an optimized design using mono-static focused hog-horn antennas (emission and reception are made using a single antenna), where the reception is always optimized with moderated directivity. This configuration allows for a more compact solution, especially in the HFS region, where space is more restricted. The system proved to cope quite well with rather large vertical displacements in the order of 20 cm [12]. The higher frequency bands, V and W, which probe plasma layers further away from the antennas, were later improved to increase the system's Signal to Noise Ratio (SNR) by implementing a heterodyne detection scheme with a phase-locked-loop (PLL) and a second oscillator to drive a harmonic mixer to down-convert the reflected signals to an intermediate frequency (IF). In these channels, the IF signal is still detected with a Schottky-diode detector at approximately 1 GHz. O-mode full-wave 1D simulations were performed to assess measurement sensitivity to plasma turbulence and initialization of the non-probed plasma density range ($0 \leq n_e \leq 0.36 \times 10^{19}$ m^{-3}) [13]. These simulations demonstrated that the spatial resolution always remains below 5 mm. Due to its design, the AUG reflectometry system remains the only broadband diagnostic worldwide capable of directly probing the plasma from the machine's HFS.

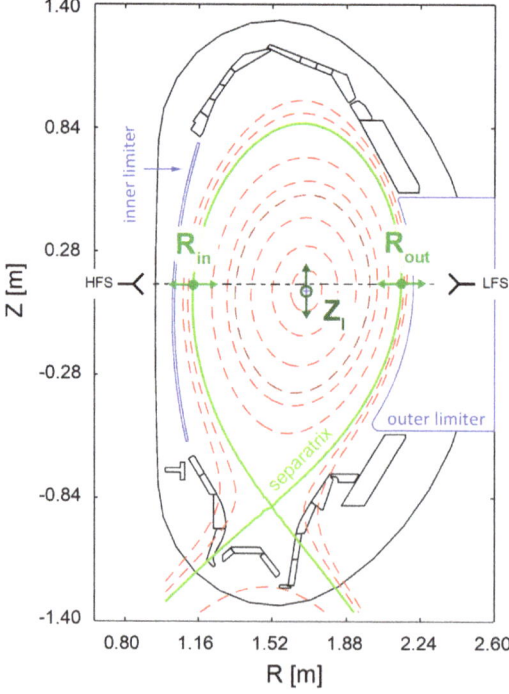

Figure 2. Simplified view of an ASDEX Upgrade poloidal cross-section, showing the magnetic separatrix position, the points used (R_{in}, R_{out}, Z_I) for the plasma position, and the lines-of-sight of the HFS and LFS microwave reflectometers.

The usual method for controlling the position of the plasma volume is to monitor the location of a few specific control points of the magnetic separatrix (green line on the plot of Figure 2). Then, the position controller changes the configuration of the magnetic fields that confines the plasma on the estimation of their locations (R_{in}, R_{out}, Z_I). For reflectometry to serve as a replacement for magnetic diagnostics, it must provide the controller with the same input information accurately, within an 1 cm error bar. In the position feedback control demonstrations performed at AUG, reflectometry provided RT estimates for R_{in} and R_{out}, while Z_I remained magnetics based. The developed systems and real-time algorithms' performance was evaluated by conducting dedicated discharges, where the magnetic and reflectometry-based controllers were switched. The controller demonstrated its ability to handle transitions between the two input sources (magnetic and reflectometric) and maintained the controlled position close to the programmed trajectories throughout the process. During the demonstration of plasma position control using both HFS and LFS reflectometers, the estimates of the inner, R_{in}, and outer, R_{out}, separatrix positions were combined to estimate the geometric plasma radius, $R_{geo} = (R_{in} + R_{out})/2$, which replaced the corresponding input magnetic measurement in the position controller.

The experiment was performed during the high-confinement mode ('H-mode') in the presence of Edge Localized Modes (ELMs) instabilities which results in bursts of energy and particles at the plasma edge. Electron density profiles measured by reflectometry can be severely affected by the ELMs, and it was required to develop algorithms to automatically validate reflectometry measurements to cope with the effects of these cyclic transitory periods. Figure 3 shows the main time evolution of the measured parameters of one of the four discharges used in the demonstration. The top plot shows the line integrated density (H-1) at the equatorial plane, Deuterium (D) fuelling, neutral beam power (NBI), ECRH heating power, and plasma current (I_{pa}). During the ELMy H-mode flat-top phase, the position control switched to the reflectometry-based controller from $t \sim 2.6$ s until $t \sim 7.6$ s. After that moment, the ramp down of the plasma is performed by the magnetic controller. For the period where reflectometry was used to control the plasma, the geometric radius of the plasma column was programmed (CTRL trace on the bottom plot of Figure 3) to swing 1.5 cm (non-symmetrically) around its original position. The reflectometry-based controller maintained the reflectometry's R_{geo} within $\sim\pm 0.5$ cm of the target trajectory. The reflectometry estimates for R_{in}, R_{out}, and R_{geo} are coherent with their magnetic counterparts, demonstrating very good precision although with improvable accuracy: ~ 1.5 cm and ~ 2 cm offsets to the magnetics at the LFS and HFS, respectively. The corresponding R_{geo} input offset was successfully handled by the position controller when switching to and from reflectometry input at $t \sim 2.6$ s and $t \sim 7.6$ s. Similarly, to the original LFS-only demonstration, in these experiments, the system operated flawlessly during the programmed reflectometry control phases, and its performance was reproducible in all four discharges.

The success of these experiments proved that it would be possible to fulfil the ITER requirements for plasma control using microwave reflectometry as an alternative to standard magnetic measurements. The use of a reflectometry-based plasma position control can contribute to achieving the high availability and reliability levels required for the operation of ITER, helping to pave the way toward the realization of future fusion power plants.

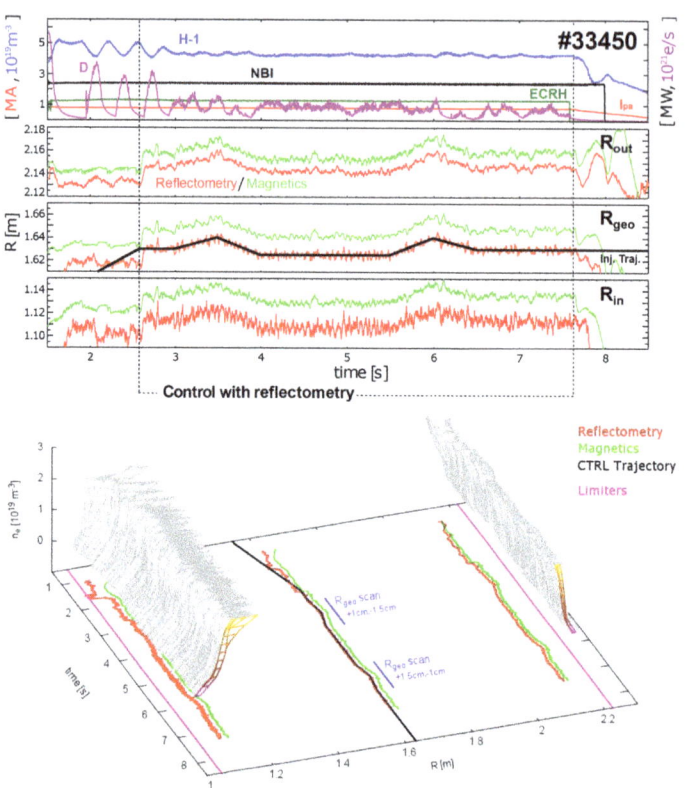

Figure 3. Time traces during ASDEX Upgrade discharge #33450 showing the position controller target trajectory and the magnetic and reflectometric separatrix positions (the reflectometry-based control of the plasma was performed during the highlighted period, ~2.6–7.6 s).

4. On the Integration of a PPR System at COMPASS

The COMPASS tokamak, located in IPP-CAS Prague, is a compact divertor device that features an ITER-relevant geometry and can operate in H-mode, which makes it an ideal platform for conducting pedestal and edge physics studies relevant to ITER. COMPASS has installed a microwave reflectometer, operating in O-mode, and has a real-time control system based on the Multi-Threaded Application Real-Time executor (MARTe). These two features create optimal conditions for the further development and advanced demonstration of a reflectometry-based PPR system. To implement the PPR system at COMPASS [14], it was necessary to seamlessly integrate the microwave reflectometry diagnostic with the Real-Time Control system (RTCS) [15,16], which is responsible for the control of the plasma position based on standard magnetic signals. Although the COMPASS reflectometry channels can be swept in <10 μs, the limited upload streaming capabilities of the system's ATCA acquisition system and the consequent reflectometry RT measurement latency conditioned the implemented PPR to be used in COMPASS's slow 500 μs control loop cycle.

COMPASS O-mode reflectometry diagnostic [17] probes the plasma through a quartz vacuum window located at a midplane LFS port. K, Ka, and U microwave bands are used to probe plasma electron densities in the $0.8 - 4.3 \times 10^{19}$ m^{-3} range. Two separate crates are used to enclose the microwave hardware and the control electronics (one for the K and Ka bands and the other for the U band). The system is also ready for an E band upgrade in the future. Each crate has a controller board consisting of a PC/104 embedded computer with additional electronics to interface and drive the microwave oscillators. Its digital electronics are synchronized with a local high-precision 10 MHz time-base synchronization clock. The diagnostic can operate at fixed, hopping, or swept frequency modes. The controllers allow for the remote operation of the diagnostic via the TCP/IP protocol over Ethernet, with a set of proprietary commands. The microwave electronics of each band provide the single-sideband (SSB) modulation, the in-phase (I), and the quadrature (Q) detection where the SSB signal has the highest bandwidth of the set, corresponding to an IF of 48 MHz for the K and Ka bands and 96 MHz for the U band. The profile measurements used by the PPR system are obtained by sweeping all bands simultaneously using a configurable sweep duration; an external trigger source sets the sweep repetition rate.

Figure 4a shows the real-time reflectometry profiles for PPR control demonstration discharge 19691 (circular L-mode discharge). Figure 4b shows the separatrix location estimated from the reflectometry profiles (green) from which the PPR magnetic radius controller input was derived. During the period highlighted in grey, ~140 ms, the estimated magnetic radius was used for the radial position controller input (red), replacing the actual magnetic measurement (blue) in the slow controller (500 µs control loop cycle time). In the case of COMPASS, the separatrix location had to be inferred from the profiles using a pre-validated fixed density, representative of the plasma regime, unlike in AUG, where a fraction of the average density measured in real-time was used.

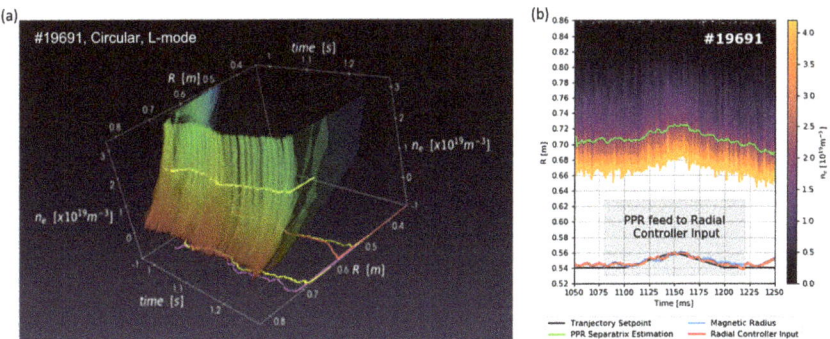

Figure 4. (a) Real-time reflectometry profiles obtained during COMPASS discharge #19691; and (b) estimation of the separatrix location based on reflectometry profiles (green) and controller input magnetic radius (red). In the shaded region, the controller input was the magnetic radius derived from the reflectometry separatrix position estimation and not the actual magnetic measurement (blue).

The O-mode microwave reflectometer was successfully integrated into the existing real-time diagnostic network system. This integration enabled the use of the diagnostic for position feedback experiments by providing the MARTe main controller with a PPR-based radial position estimation. The well-defined software block structure of the MARTe framework largely expedited the design, development, and commissioning of the real-time application. The implementation of PPR at COMPASS made it possible to extend its applicability limits by demonstrating that it can cope with twice as fast control dynamics as in the original demonstrations performed in AUG.

5. Plasma Position Reflectometry at ITER

The ITER Plasma Position Reflectometry (PPR) system was planned to play a supplementary role to magnetic diagnostics in providing information to the ITER Plasma Control System (PCS) about the gap between the plasma and the first wall. Although the project was descoped in 2020, a considerable amount of work has been accomplished that paved the way to solve many of the similar (and worse) problems faced by the implementation of plasma position reflectometry in future fusion devices such as the Divertor Test Tokamak (DTT) and DEMO.

On ITER, the PPR system consisted of four FMCW O-mode reflectometers in full bi-static configuration covering densities up to $\sim 7 \times 10^{19}$ m^{-3} (15–75 GHz). These reflectometers were to be installed at four different locations, known as gaps 3, 4, 5, and 6, both in-vessel (gaps 4 and 6) and inside port plugs (gaps 3 and 5). The main goal of the diagnostic was to measure the edge density profile in real time with high spatial (<1 cm) and temporal (100 µs) resolutions. To reduce transmission losses, the system used oversized 20 × 12 mm rectangular waveguides inside the vacuum vessel and port plugs.

The antennas and part of the transmission lines of gaps 4 and 6 were to be installed inside the vacuum vessel, as depicted in Figure 5. Due to direct exposure to the plasma, these antennas and the 90° bends connecting to the transmission lines were expected to suffer significant loads that could compromise their integrity.

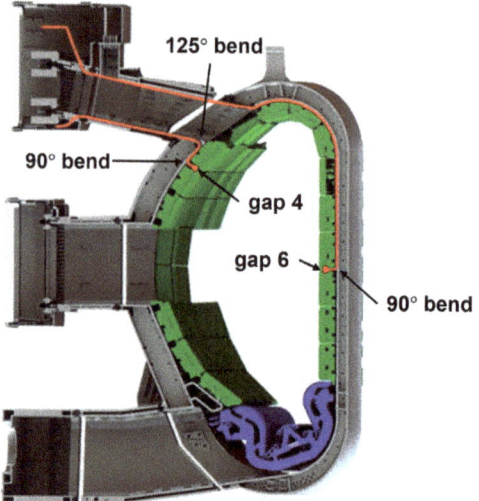

Figure 5. Location (red) of the ITER PPR in-vessel components of gaps 4 and 6, indicating some of the parts that required a detailed design, namely the 125° bend, the 90° bends, and the antennas. The in-vessel waveguides are routed behind the blanket modules (coloured green) near the vacuum vessel wall. The divertor structure is represented in blue.

The system of gap 4 had the antennas located in the low-field side (LFS) of sector 9, between the blanket modules (BMs) in rows #11 and #12. The support of the 90° bend was bolted to ITER standard in-vessel attachments welded to the wall [18]. From the 90° bends, the microwaves were routed the feed-outs at the bottom of upper port 1 using straight and curved sections of the waveguide.

In the case of gap 6, the antennas were to be installed at the high-field side (HFS) of sector 7, between the BMs in rows #3 and #4. The support of the 90° bend would be attached to the vessel in the same way as in gap 4. The transmission line of gap 6 was then routed along the inner wall of the vacuum vessel between the bend and one of the feed-outs on the top of upper port 14. Further details about the design of the PPR system are provided in [19,20].

5.1. Thermal Loads and Material Testing

As mentioned, the in-vessel components of gaps 4 and 6, namely the antennas and 90° bends, would be subjected to significant thermal loads that could compromise their integrity. Figure 6 shows the expected temperature distributions in these two components for gap 6, which were estimated using ANSYS Mechanical by applying the load specifications for the PPR in-vessel components as per the ITER guidelines for thermal analyses. As shown, the maximum operating temperatures would be well above the limit of 450 °C for the selected material (ITER-grade stainless steel) under neutron irradiation. Thus, it was proposed that these components were manufactured from other materials, such as tungsten or a nickel-based superalloy [21].

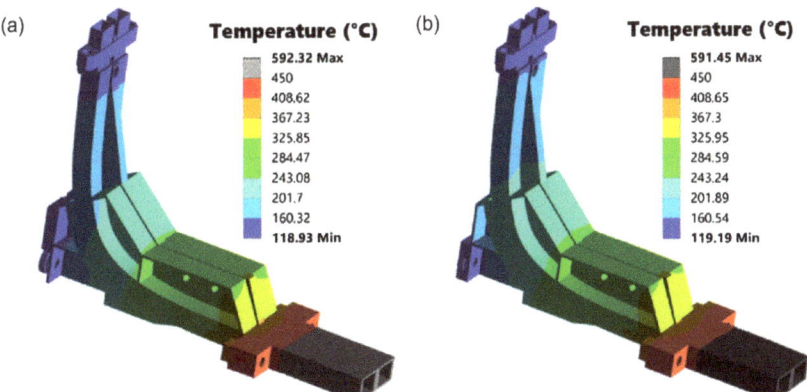

Figure 6. Temperature distributions of the front-end components of gap 6 after extending the design volume in the poloidal direction (**a**) and in the toroidal direction (**b**).

To minimise the EM forces induced during plasma disruptions, the in-vessel waveguides were made from ITER grade stainless steel (SS 316(LN)-IG), but coated inside with a thin (15–25 µm) Copper layer to keep ohmic losses at a minimum. Prototypes of these waveguides were tested at the IPFN-IST microwave laboratory. The results obtained (see Figure 7) have shown that the stainless-steel, Cu-coated waveguides had similar performance to the theoretical predictions and to Cu-only waveguides, both in O-mode and X-mode-like polarizations. These results were reproducible for all the waveguide prototype samples tested.

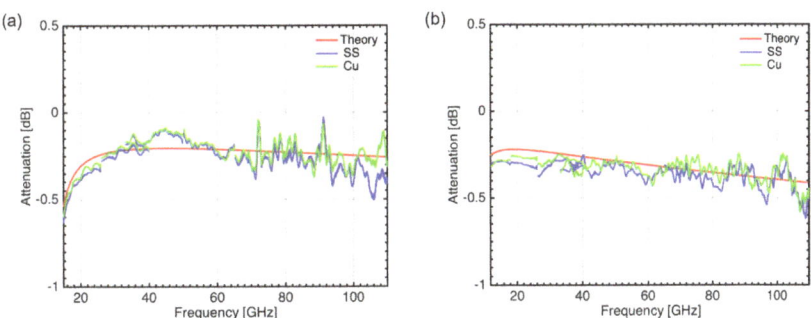

Figure 7. Performance of the stainless-steel copper-coated waveguides (blue) against the copper-only waveguides (green) and theoretical predictions (red) both for O-mode (**a**) and X-mode- (**b**) like polarizations.

5.2. Performance Assessment of Waveguides Bends

As expected, the performance impact of the in-vessel oversized waveguide bends of gaps 4 and 6 was a critical aspect of the PPR design, namely the 90° bends just behind the antennas and the 125° bend of gap 4. It is well known that these bends are prone to excite higher-order modes and create resonances, which can adversely affect performance if not designed carefully. The 90° and 125° bends of the in-vessel PPR systems were first studied via 3D EM simulations using both the frequency-domain solver of ANSYS HFSS and the time-domain solver of CST Microwave Studio (MWS).

Figure 8a shows the results obtained with HFSS for the optimization of the 125° bend, which was split into two identical 62.5° bends with hyperbolic secant geometry. The optimized shape (C9) clearly improved performance with respect to the constant radius bend (Baseline) across the 15–75 GHz frequency range (with just a small degradation < 0.2 dB visible above 70 GHz) while complying with the in-vessel space restrictions [22]. Figure 8b depicts the results obtained with HFSS for the optimized bend against the ones obtained using MWS, which exhibit a perfect match between both tools. The electric field distributions inside the optimized 125° bend (C9), as obtained with HFSS for 15 GHz, 45 GHz, and 75 GHz, are shown in Figure 9.

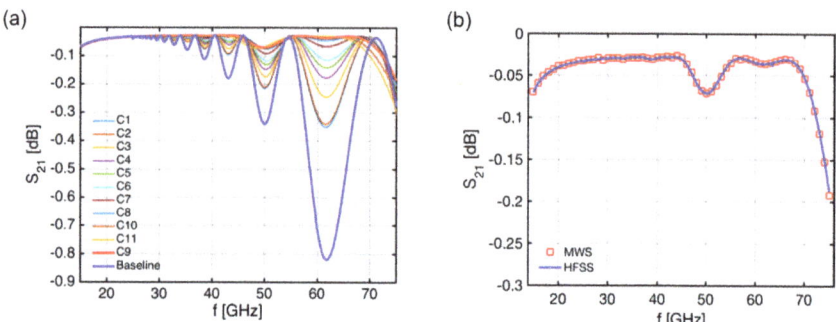

Figure 8. Performance of the constant-radius bend (baseline design) and of various hyperbolic secant 125° bends (C1 to C9) (**a**) and comparison between the results obtained with HFSS and MWS for the optimized design (C9) (**b**).

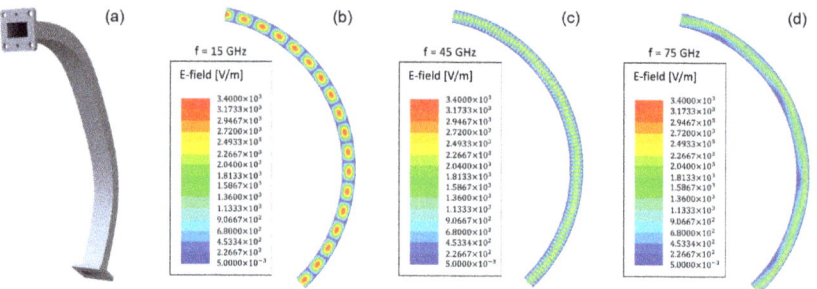

Figure 9. Electric field distribution inside the optimized (C9) 125° bend (**a**) obtained with HFSS for (**b**) 15 GHz, (**c**) 45 GHz, and (**d**) 75 GHz.

The 125° bend was prototyped to assess the EM performance of the optimized design and the manufacturing process, namely the copper-coating and the bending to the specified geometry. During the site acceptance inspection of the prototype samples, some problems were readily identified:

- All samples showed a lack of (copper) coating near the edge of the flanges, as shown in Figure 10a, as well as in some spots near the flanges, as observed in Figure 10b. The lack of copper was also visible inside the waveguides along the corners formed by the inner walls and in the walls themselves.
- Marks were visible in the inner surface of the bend, probably caused by some tool inserted in the bend during bending, as depicted in Figure 10c,d.
- local shape deformation was clearly visible in all samples around the middle of the bend, as shown in Figure 10e, where, per specification, there should be a straight section due to the connection between the two 62.5° hyperbolic-secant bends—as is

visible in Figure 10f, the deformation extends to the inner wall of the waveguide, which anticipates performance issues.
- Figure 10e clearly shows that the received samples were not symmetric components as specified, and judging by the relative position of its flanges, it seems the samples were all slightly twisted around the propagation axis.
- Finally, measurements taken as part of the site acceptance tests have shown that the internal section of the received samples was about 0.2 mm (both in height and width) smaller than the specified dimensions.

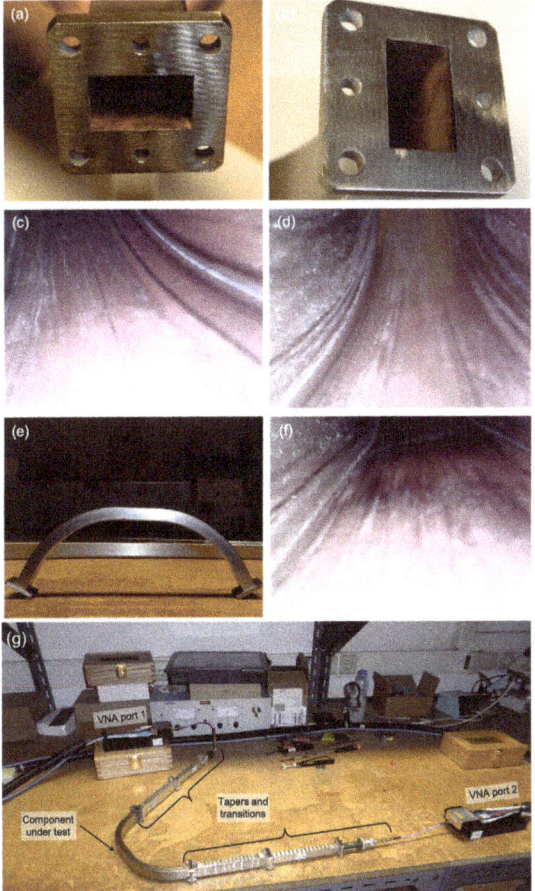

Figure 10. Prototype of the 125° bend (**a**–**f**) and experimental setup at IPFN-IST microwave laboratory (**g**) used to measure the EM performance of the prototype samples.

The main goal of the tests was to measure the performance of the samples in the frequency range 15–75 GHz for the TE_{01} waveguide mode (O-mode-like polarization), as used by the PPR system. The testing frequency range was covered using the standard frequency bands K (extended down to 15 GHz), Ka, Q, and V. Waveguide tapers were used to enable the connection between the non-standard dimensions of the bend and the standard dimensions of the hardware equipment used in the tests. For each band, the measurements were preceded by a calibration for that particular band of the Vector Network Analyzer (VNA) used in the tests. Due to the small attenuations of the bend in certain frequency ranges, the difference to the reference measurements was within the resolution limit of the VNA (±0.1 dB). Thus, the measurements presented here were performed by activating the Secondary Match Correction (SMC) feature of the VNA, which improves the resolution of the measurements by slowing down the frequency sweep.

For each tested sample, the measured attenuation was compared with what can be expected from the analytical conductivity losses for the TE_{01} mode propagating through a straight rectangular copper waveguide, α_c, given by

$$\alpha_c[\text{dB/m}] = 8.686 \frac{2R_m}{bZ_0\sqrt{1-k_c^2/k_0^2}} \left[\left(1 + \frac{b}{a}\right)\frac{k_c^2}{k_0^2} + \frac{b}{a}\left(\frac{1}{2} - \frac{k_c^2}{k_0^2}\right) \right], \qquad (3)$$

where, a and b, with $a > b$, are the waveguide cross-section dimensions, k_c is the cut-off wave number, k_0 is the wave number corresponding to frequency $\omega = 2\pi f$, Z_0 represents the free-space impedance, and $R_m = (\omega \mu_0 / 2\sigma_{Cu})^{1/2}$, where μ_0 is the free-space permittivity and σ_{Cu} is the electric conductivity of copper.

The results obtained from the tests of the prototype samples (numbered 18110703 to 18110706 and 18110708) are depicted in Figure 11, along with the analytical attenuation of a straight 0.33 m-long rectangular waveguide with the same cross-section, as obtained from Equation (3). As can be observed for all bends, there are two frequency regions where the experimental attenuation deviates significantly from the analytical losses: between 43 GHz and 54 GHz, where the bends exhibit moderate attenuation (up to 2 dB), and above 62 GHz, where the attenuation rapidly increases from ~0.5 dB up to ~10 dB at 75 GHz. Out of these regions, all samples exhibited good performance, with the attenuation remaining below 0.5 dB.

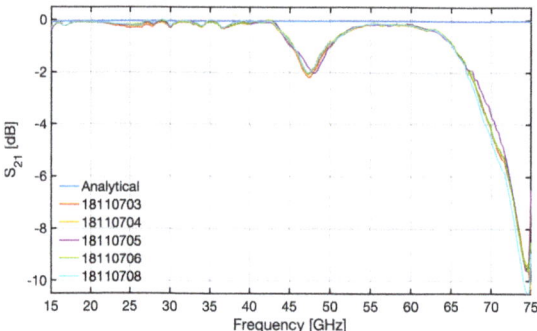

Figure 11. Analytical losses and experimental attenuation for the various prototype samples (numbered 18110703 to 18110706 and 18110708) of the 125° hyperbolic secant rectangular (20 × 12 mm) stainless-steel copper-coated waveguide bend. The analytical losses were calculated using Equation (3) for a straight rectangular waveguide with the same dimensions and equivalent length of 0.33 m.

These results have shown that the tested bends were inadequate for the PPR system and were not a surprise given the observations made during the acceptance of the prototypes, which indicated that the (bending) process used to manufacture this component, although reproducible as demonstrated by the results of the tests, is clearly unable to provide the precision and accuracy required by the specified hyperbolic secant geometry, whose main goal is to mitigate the excitation of higher-order modes that are known to deteriorate performance. In fact, although not exactly coincident with the ones of the tested samples, the attenuations estimated with HFSS and MWS also exhibit two frequency regions where the performance of the bend deteriorates slightly with respect to the theoretical losses, which are shown to be mainly due to the excitation of the TE_{02} mode. This similarity indicated that the bad performance of the prototype could be due to mode conversion losses into higher order modes, which were not mitigated as expected due to the local deformations and overall poor shape accuracy of the manufactured prototype. One of the lessons learned for future developments was that the prototyping required to validate the

EM design should be separated from the prototyping required to validate manufacturing processes to allow for a better understanding of the root causes of performance issues.

5.3. Avoiding Crosstalk in the In-Vessel Waveguide Flanges

The ITER in-vessel waveguides consisted of multiple sections of parallel waveguides connected using flanges. Therefore, crosstalk at the flanges was a concern. The envisaged solution to prevent crosstalk consisted of matching male-female flanges with a centre pole, as depicted in Figure 12a. A prototype flange was tested using the setup illustrated in (Figure 12b). First, a reference isolation measurement was performed with both ports of the VNA directly terminated by a matched load. Secondly, the crosstalk measurement (A→D) was performed with the flanges inserted, and B and C terminated using matched loads. The results of those tests are illustrated in Figure 13. As shown, the envisaged solution effectively reduced the flange crosstalk to negligible levels across the entire 15–75 GHz frequency range.

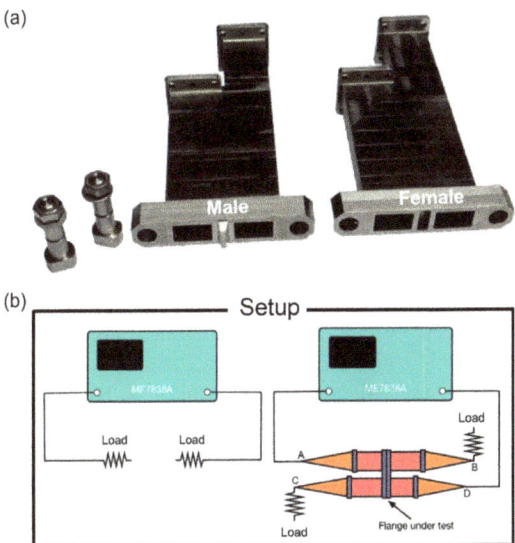

Figure 12. Prototype flanges designed to avoid crosstalk (**a**) and experimental setup to measure the crosstalk at the flanges (**b**).

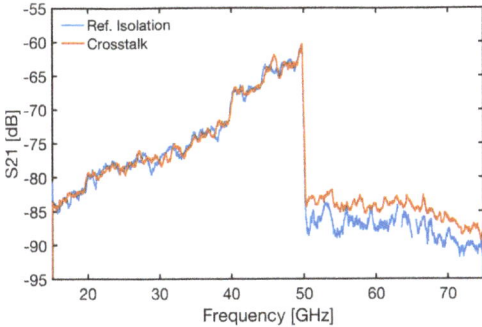

Figure 13. Crosstalk measurements on the prototype flange versus frequency. The discontinuity at 50 GHz is due to the different sensitivity of the VNA in the 50–75 GHz band.

5.4. Testing the Integration of the Antennas of Gaps 4 and 6 between Blanket Modules

To optimize the design of the antennas for gaps 4 and 6, extensive simulations were performed using the REFMUL Finite-Difference Time-Domain (FDTD) code [23–25], which is able to consider the geometric characteristics of the space between BMs where the antennas would be installed. From these simulations, some concerns were raised regarding the effect of the blankets on the signals due to reflections and diffraction. Thus, prototypes of the antenna assembly, consisting of the antennas and feeding waveguides, were built and tested, both isolated and integrated into a mock-up of the blanket modules to simulate the effect of these structures. The prototypes of the antennas and the mock-up of the BMs for gap 6 are shown in Figure 14.

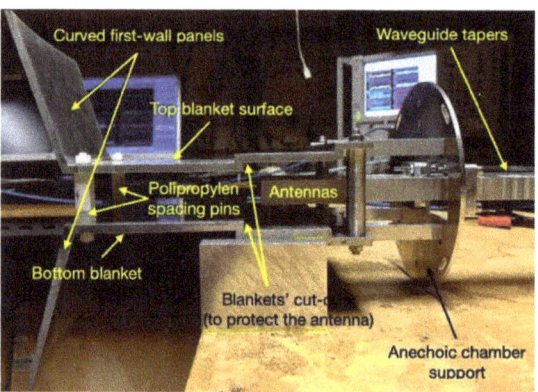

Figure 14. Prototype antennas and mock-up of the BMs for gap 6.

The radiation patterns of the antenna assembly were measured in an $3.8 \times 2.5 \times 2$ m anechoic chamber with and without the BMs. Figure 15 shows the antenna inside the anechoic chamber during these measurements. Two different antennas were prototyped and tested: antenna #1 (baseline), consisting of two parallel pyramidal horns with a length of 115 mm, a toroidal flare of ± 2 mm, and poloidal flare of ± 1 mm, and antenna #2, consisting of two parallel pyramidal horns with a length of 115 mm, a toroidal flare of ± 2 mm, and poloidal flare of ± 4 mm.

Figure 15. Prototype antenna assembly for gap 6 inside the anechoic chamber during the measurement of the radiation patterns.

To reduce the time required to complete the measurements, instead of full 360° scans, the radiation patterns were only measured in the range $-120°$ to $+120°$ (240° scans). For each frequency band, the radiation pattern of a standard calibrated horn antenna was measured to absolutely calibrate the radiation patterns of the Antenna Under Test (AUT). Before each scan, the phase centre of the AUT was aligned with the Probe Antenna (PA). In the measurements, both the AUT and the PA could be rotated in their vertical plane; the AUT was also able to be rotated in the horizontal plane. The measurement plane (E or H) was selected by rotating the AUT and PA antennas in the vertical plane as desired. For cross-polarization measurements, the AUT and PA antennas were rotated such that they were in complementary planes.

For the different bands, the radiation patterns were measured at the following discrete frequencies:

- Ku band: 15 GHz, 16.5 GHz.
- Ka band: 18 GHz, 19.7 GHz, 22.25 GHz, and 24.8 GHz.
- Ka band: 29.2 GHz, 33.25 GHz, and 37.3 GHz.
- Q band: 42 GHz, 45 GHz, and 48 GHz.
- V band: 55 GHz, 62.5 GHz, 70 GHz, and 75 GHz.

Depending on the frequency range, different setups were used in the anechoic chamber due to the need to keep the dynamic range of the measurements above 25 dB for all bands (the loss of dynamic range occurs because the output power from the Agilent Technologies N8361A VNA used in the measurements decreases as the frequency increases with the simultaneous increase of the losses on the high-frequency cables).

We expected the measurements to show the effects of the blanket modules to be most noticeable in the poloidal plane, which for the TE_{01} mode corresponds to the antenna's H-plane, due to the fact that, with respect to the dimensions of the antennas, the aperture between blankets is sufficiently large in the toroidal plane (antenna's E-plane) so that it does not significantly affect the radiation patterns in this plane. The E-plane co-polar radiation patterns measured for antennas #1 and #2 with and without the blanket modules in place confirmed those expectations showing that the modifications in the radiation patterns of both antennas were, in fact, not significant. For both antennas, the H-plane co-polar radiation patterns measured with and without the BMs are illustrated in Figure 16.

As can be observed, the most noticeable effect of the blanket modules in the lower end of the frequency range (15 GHz, 18 GHz, 22.25 GHz, and 29.2 GHz) was a narrowing of the radiated beam and the appearance of side-lobes (radiation angles $< 90°$) and back-lobes (radiation angles $> 90°$). The beam narrowing effect had to do with the larger directivity of the radiating structure formed by the blanket surfaces and first-wall panels, into which the emission antenna radiates. The side- and back-lobes had to do with the beam truncation effect induced by the BMs' cut-outs located near the antennas' apertures (see Figure 14), which created reflections that, because of the broader antenna beam, were more significant at the lower frequencies. For the higher frequencies (37.3 GHz, 45 GHz, 55 GHz, 62.5 GHz, and 75 GHz), the dominant effect was the secondary lobes that appeared near the main lobe. With respect to antenna #1, the modifications induced by the blankets in the radiation patterns of antenna #2 were less noticeable, again due to the larger directivity of this antenna, which made it more insensitive to the surrounding BM structures.

The results obtained during the tests of the baseline antenna of gap 6 indicated that the BMs below and above the antennas shaped the radiation patterns and, because of the reflections and resonance effects originated in the cut-outs of the first-wall surfaces, which had a negative effect on the measurement performance below approximately 35 GHz. These effects were partially mitigated by increasing the poloidal flare of the antenna to increase its poloidal directivity. This design change significantly reduced the effects of the reflections in the first-wall cut-outs and directly improved the system's performance.

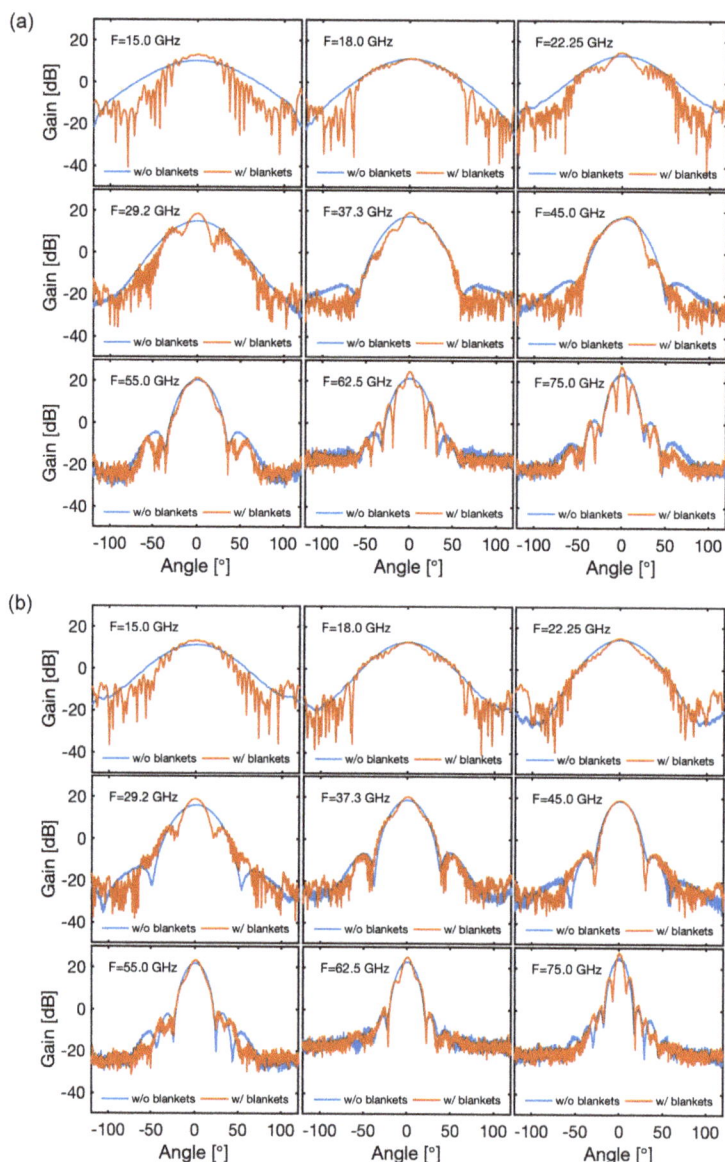

Figure 16. H-plane co-polar radiation patterns for antenna #1 (**a**) and antenna #2 (**b**), with (red) and without (blue) the BMs.

Another critical design parameter in terms of the system performance was the gap between the antennas' apertures and the cut-outs, in particular, at the lower end of the frequency range, where the radiation patterns are very broad. The tests have shown that if this distance was reduced from the nominal 15 mm to 10 mm the effect of the reflections in the first-wall cut-outs was reduced, improving the performance of the system.

If the above design changes were applied simultaneously, the effects of the structures surrounding the antenna were restricted to the K band (15–26.5 GHz) [20], and the measurement performance of the system was significantly improved with respect to the baseline design. Furthermore, the tests have shown that these modifications also helped in mitigating the impact of the blanket modules if, after installation, the antenna does not end up exactly centred between the blanket surfaces due to the tolerances of the different components involved.

In the case of gap 4, the in-vessel waveguides would run through the bottom of upper port 01, behind the BMs, close to the inner shell of the vacuum vessel, and down to the antennas. The waveguides would be attached to the bottom of the upper port and to the vessel's inner shell and could not be maintained or replaced during the lifetime of ITER. The antennas, which as for gap 6 consisted of a bi-static array of parallel small-flare pyramidal horns, would be installed at the low-field side (LFS), above the midplane, in the small space between two BMs consisting of two main components: (i) the shield block, a large stainless-steel block directly attached to the vacuum vessel, whose main purpose would be to reduce the neutron flux, and (ii) the first-wall, a relatively thin structure attached to the shield block and directly facing the plasma, whose main purpose would be to sustain the high thermal and nuclear loads generated during the plasma discharges. Near the antennas, the first-wall panels are slanted both in the poloidal and toroidal directions. The 3D EM simulations have shown that the BMs would modify the antenna radiation pattern at the lower frequencies and that the geometry and location of the cut-outs required to accommodate the antennas would have a significant negative impact on the EM performance. Moreover, 2D full-wave plasma simulations confirmed the modification of the antenna radiation patterns by the blanket modules and have shown, for gap 6, that the slanted geometry of the first wall in front of the antennas could also have a strong negative effect on the measurement performance of the system. In fact, the results of the simulations indicated that there would be situations in which the system of gap 6 could not be able to meet its measurement requirements. The 2D full-wave plasma simulations also indicated potential issues with the performance of gap 4 due to the specific geometry of the enclosure, in particular, the fact that the first-wall surfaces above and below the antennas would be oblique to each other, forming an aperture that narrowed toward the plasma. Therefore, as for gap 6, it was decided to build a mock-up of gap 4 and its environment to test the performance of this system. This mock-up has been procured and provided by the ITER Organization (IO).

Figure 17 depicts the mock-up and the PPR antennas of gap 4 during the tests, which were performed with the two prototypes corresponding to the baseline and optimized antennas developed for gap 6. The results obtained with the baseline antenna have shown that the measurement performance of gap 4 was very good, with the system being able to recover all the mirror distances with position errors below the target ± 1 cm margin, except in the K band, where the errors exceeded that margin. For these lower frequencies, the very broad beam radiated by the baseline antennas, due to the small poloidal flare (1 mm) was deflected and reflected in the oblique first-wall surfaces, shattering the beam and creating relevant side lobes, which produced multiple lines-of-sight that reduced the measurement performance.

With the optimized antenna, the results show that the measurement performance of gap 4 improved, mainly due to the higher poloidal directivity of the antennas, which contributed to a reduction of the level of deflections and reflections in the first-wall surfaces. Indeed, the results have shown that the performance of the system would be compliant with the target error margin across the entire operating frequency range of the PPR system (15–75 GHz), if it were not for the results obtained in the K band for the closest distance of the mirror (25 cm), where the position error was shown to exceed that margin for some frequencies, if only by a few millimeters. Finally, the results also suggested that the performance of gap 4 could be further improved to achieve full compliance by increasing the poloidal flare of the antennas, which in this gap, contrary to gap 6, would be possible due to the relatively large distance between the antennas and the surrounding components.

Figure 17. Mock-up of gap 4 provided by the IO featuring the relevant geometry of the first-wall surfaces and BMs surrounding the PPR antennas.

Concerning the measurement difficulties of gap 4 anticipated by the 2D full-wave plasma simulations, the tests with the baseline antenna confirmed, although less severe, the estimated difficulties in the lowest frequency band, K, but have not shown any problems in the highest frequency band, V. This discrepancy could be due to the 2D nature of the simulations, which, for the 3D geometry of the mock-up tests, collapses all energy in the poloidal plane and decreases the rate at which that energy dissipates, leading to an enhancement of the non-primary contributions to the measured signal [23].

5.5. Lessons Learned in ITER PPR

Although the ITER PPR system was descoped in 2020, the work performed until then to comply with the many interfaces and restrictions posed by its integration in ITER has revealed crucial to the development of many of the design methodologies that are currently being used in the design of a PPR system for DTT and in the conceptual design of the PPR for DEMO, both of which face similar challenges as the design for ITER. The extensive use of both EM and thermo-mechanical simulations to expedite the prototyping phase was also a key takeaway from the development of the ITER PPR system.

6. Microwave Reflectometry as a Plasma Position Diagnostic for DEMO

The objective of MW reflectometry in DEMO is to determine the radial edge density profile and to provide data for real-time control of plasma position and shape [26–28]. In order to fulfil its requirements, reflectometry shall be able to measure the plasma electron density profile in the pedestal region, where the high-density gradient occurs, with a temporal resolution of 1 ms, maximum precision error of 5%, the maximum noise level of 2%, and maximum latency of 0.01 s [29]. To be able to provide shape information, the system will use several antennas distributed at several poloidal locations, with direct access to the plasma [30,31]. Depending on the location, each reflectometer will access the plasma along a specific line of sight. Therefore, raytracing and 2D FDTD simulations were conducted to assess the expected performance of the reflectometers at the various locations. The results showed that close to the equatorial plane, a single pair of antennas can provide good spatial resolution. In regions of higher plasma curvature, however, clusters of 3–5 antennas (1 for emission and 2–4 for reception) may be needed to fully capture the reflected waves [32,33]. Considering measurements in 16 gaps on a poloidal plane around the plasma, up to 80 antennas might be required to fulfil the measurement requirements. Figure 18 shows possible poloidal locations for these antennas (100 antennas were simulated) and the simulation results for three of those locations, illustrating the need for clusters of antennas at the more extreme positions. The dimensions of the antennas and feeding waveguides are driven by the lowest frequency (15 GHz) used by the system, which corresponds to a density of $n_e = 0.3 \times 10^{19}$ m^{-3}.

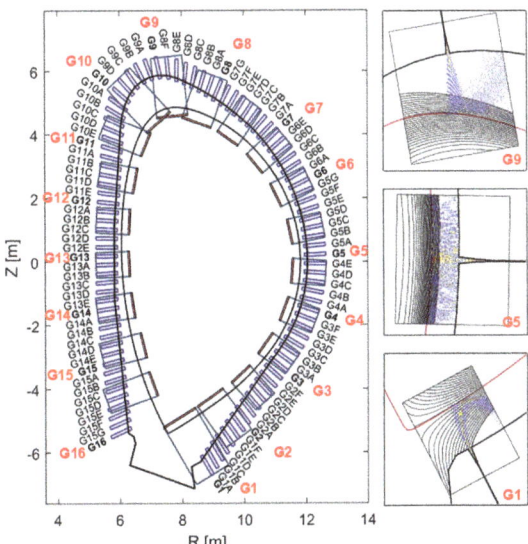

Figure 18. Poloidal distribution of the antennas aligned perpendicularly to the separatrix tangent line (**left**) and their expected performance at three locations (**right**) for DEMO when the LOS is set perpendicular to the wall. The smaller blue boxes correspond to the region to place the antenna setup, and the larger blue rectangles correspond to some examples of the selected region of interest for the tested poloidal positions [31].

The integration of such a high number of antennas and waveguides with the breeding blanket (BB) is by itself an extremely challenging task. In addition, these components need to endure radiation loads far exceeding those expected for ITER during long operation periods, and their design must be compatible with the DEMO Remote Maintenance (RM) systems such that they can be replaced in case of failure. The Diagnostics Slim Cassette (DSC) concept, explained in detail in Section 6.1, aims to address this requirement while providing a feasible solution for the integration of all the antennas and waveguides with the BB [29,34].

The plasma-facing antennas are foreseen to be made of tungsten, which is the material used in the BB first wall. To keep erosion below 10 μm per year, it has been recommended that the antennas are retracted from the BB surface by about 100 mm—erosion leads to higher surface roughness, which increases the reflection coefficient and the voltage standing wave ratio (VSWR), and therefore the losses. The waveguides are proposed to be made of EUROFER with copper coating for increased conductivity and must connect the antennas to the ex-vessel, crossing the primary vacuum boundary.

As with the ITER PPR system, the integration of MW reflectometry in DEMO is a multidisciplinary endeavour. EM simulations are required to optimize the location and design of the antennas, and Computer-Aided Design (CAD) and integration studies, such as space occupation, RH compatibility, and definition of interfaces with other systems, are essential. Additionally, the integration of the antennas and feeding waveguides inside the vacuum vessel will have an impact on the waveguide routing and may impose bends that, if not carefully designed, may affect the SNR. Neutronics simulations (using MCNP) [34,35] and thermo-structural analyses [36,37] play an important role not only in ensuring the reliability of the diagnostic but also in providing information about any deformations that may be induced on the components under the prescribed loads and the impact on the diagnostic measurements permitting to establish a correction to apply to minimize deformation effects if the thermal load is known.

Similar to the ITER PPR system, the additional components to implement MW reflectometry in DEMO may not be classified as requiring scheduled remote maintenance but must be prepared to an opportunistic operation of maintenance. An integrated and holistic approach is used to design the DSC considering the other systems, the expected remote maintenance operations, such as cutting/welding, alignment, grasping, manoeuvring, and transportation, the Technology Readiness Levels (TRL) of the available tools and their availability during the lifetime of the reactor. In [34] the DSC was updated to be compatible with the hybrid kinematic mechanism designed to transport the BB [38].

The maintenance addresses nominal operations, i.e., operations that are planned and expected to be performed. A failure is considered when a nominal operation is interrupted. Risk of failure may always occur no matter how well-designed and integrated the reflectometry system is in DEMO. Failure mode, effects, and criticality analyses (FMECA) are used to detect, analyse, and find solutions to avoid failures, or to mitigate their effects on the performance of the system. An initial FMECA of the DSC RH operations was performed, and mitigation actions were proposed [34]. In case of failure, the operation shall be interrupted, and the following actions may be needed: (i) start a recovery operation, where no additional equipment is required, or (ii) start a rescue operation, involving the support of additional remote maintenance systems. The recovery and rescue operations are planned to consider all possible failure situations.

6.1. The DSC Concept

The proposal to integrate MW reflectometry in DEMO is the Diagnostics Slim Cassette (DSC), a slim, solid EUROFER structure with 20–25 cm of toroidal thickness to be attached to or integrated with the blankets [39]. The DSC will host the antennas and their feeding waveguides, which are routed to the upper ports, as detailed in Figure 19. The plasma-facing antennas are grouped around 16 locations, known as gaps G1–G16. The gaps are planned to be arranged in three toroidal planes to avoid clashing between the waveguides (each with an inner cross-section of 19 mm × 9.5 mm).

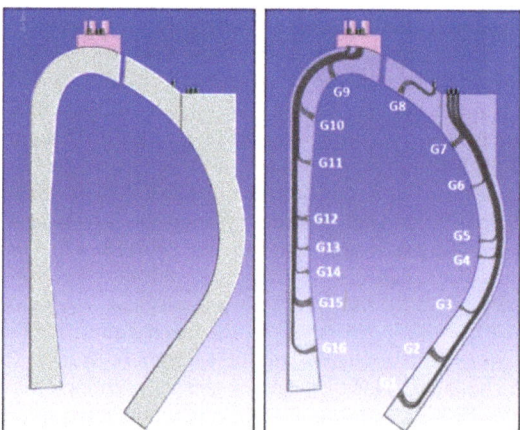

Figure 19. Poloidal view of the DSC with (**left**) and without (**right**) applying transparency showing the waveguide routing.

The number of antennas in each location was derived from ray-tracing simulations [32]. However, their arrangement is also dependent on space restrictions in the upper ports, where waveguide extensions will be used to connect the DSC to the primary vacuum boundary. Due to space restrictions, the waveguides must be grouped in sockets and routed side-by-side with the blanket pipes [34]. The antenna arrangement is then influenced by two main constraints: the need to avoid clashing between waveguides belonging to different gaps and the requirement of grouping the waveguides in a small number of sockets.

To assess the possibility of symmetrical waveguide distribution, a CAD model was developed for the antenna clusters in the first wall of the DSC, aligned perpendicular to the separatrix. The options are illustrated in Figure 20 and consist of (a) two antennas aligned vertically; (b) 2 antennas aligned horizontally; (c) a cluster of 3 antennas aligned horizontally; (d) a T-shape alignment of 4 antennas, with the emitting antenna placed below the receiving antennas (suitable for the divertor region); and (e) 5 antennas, in '+' and '×' shapes, respectively. These two shapes, being complementary, can be used together to maximize the number of waveguides that can be routed without clashes.

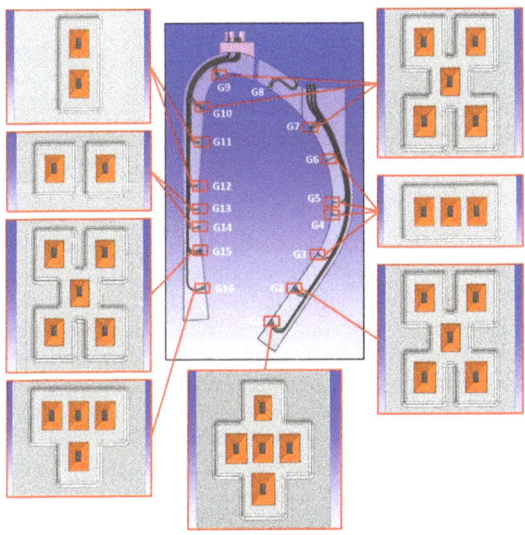

Figure 20. Possible distribution of antennas in the first-wall DSC openings [37].

The arrangement presented in Figure 20 was optimized to maximize the number of reception antennas while minimizing the number of sockets needed to route the waveguides in the upper port. It must be noticed, however, that the waveguide bends were not yet optimized from the point of view of EM performance. With all reflectometers running simultaneously in FMCW mode, the cross-interference will be challenging due to low gain horn, non-normal incident angle, wall reflection, etc. This in-vessel global microwave pollution has been clearly observed in DIII-D [40] and other facilities. Two approaches can be followed: (a) synchronized sweeping of all reflectometers with a time shift to ensure that there are no reflectometers at the same frequency at a given time; (b) Use Direct Digital Synthesis (DDS) but instead of using a linear sweep, use an encoded sweep with a different code for each reflectometer and only the received signals that show high correlation with the transmitted one will be selected. A lot of these technics are being developed for the automotive industry to overcome a similar problem of interference between the radars of different vehicles that share the same road.

One important question to address in an extreme environment such as DEMO is whether the shapes of the waveguides in the DSC are preserved to minimize power losses. Thermal analyses can be used to assess the waveguide deformation and its impact on the propagation of the EM waves. The full analysis performed for the DSC outboard segment is presented in [37,39]. The expected deformation of the waveguides in the DSC is illustrated in Figure 21, with the maximum values obtained in the antennas and waveguides located closer to the equatorial plane.

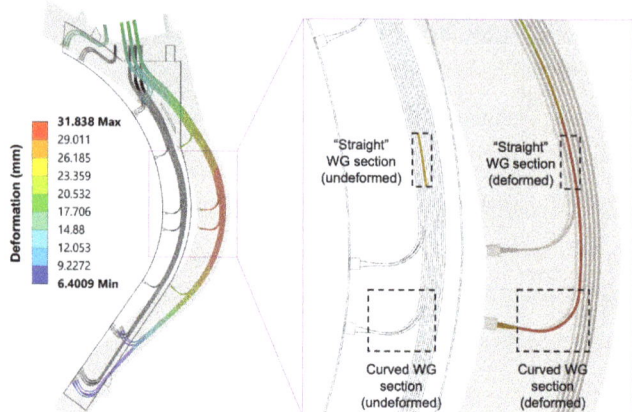

Figure 21. Preliminary results for the expected deformation of the waveguides inside the DSC (exaggerated representation). The waveguide sections selected for the preliminary EM performance assessment are highlighted on the inset.

Once the expected deformations of the waveguides are known, the performance of the reflectometry system can be assessed by conducting EM analyses that include the entire transmission lines from the source up to the antennas. However, considering that the length of the in-vessel transmission lines alone is approximately 21 m (including the waveguide extensions) and the span of frequency bands, simulating the complete transmission lines would require very extensive computational resources. Therefore, critical segments of the transmission lines containing the larger expected deformations were selected for a preliminary assessment using ANSYS HFSS. The input for the simulations was the deformed geometries presented in Figure 21, divided into smaller components and then chosen for the simulations according to the deformation values. Two important waveguide sections (highlighted in Figure 21) were thus selected: a curved section just before one of the antennas and a straight section between the antennas and the upper port.

The EM analyses aimed to estimate the power losses and compare them between the undeformed and deformed waveguides, not only for the fundamental mode, TE_{10} but also for other relevant higher-order modes. Symmetry boundary conditions were applied to the waveguides to increase the simulation speed. The frequency was swept between 15 GHz and 75 GHz, with meshes optimized for each band (Ku, K, Ka, U, and V) to further optimize the simulation times. Nevertheless, all these simplifications were not sufficient within the available computational resources to complete the analysis for frequencies above 75 GHz.

Comparing the losses obtained for the two selected waveguide sections, presented in Figure 22, one can observe that the straight section exhibits, at most, a very small power loss of less than 0.12 dB, at 35 GHz. This has a minor impact on the overall system performance. On the other hand, the curved waveguide section has much higher losses, mostly for frequencies close to 50 GHz. In this case, losses above 1 dB may compromise the measurements since they will be propagated by the number of curves in each transmission line. These losses were expected, as the shape of the waveguide bends, which is critical to avoid losses due to mode conversion [41,42], is not yet optimized at this stage. Previous simulations for the 90° bends of the ITER PPR system have shown that curves with optimized hyperbolic secant shapes can eliminate losses of this magnitude [42]. These optimizations will be implemented at a later design stage when the number of antennas and waveguides and their positions within the system are fully determined.

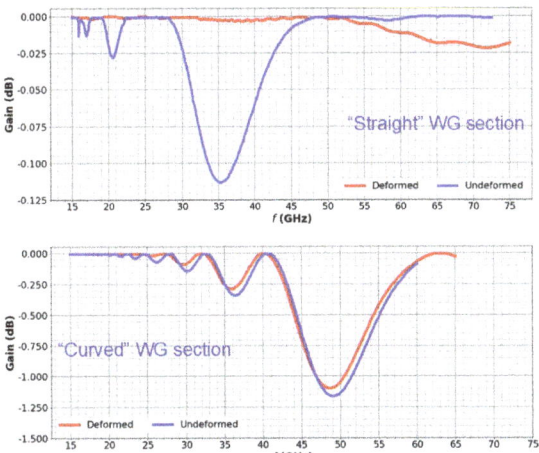

Figure 22. Attenuation of the TE_{10} mode for the two selected waveguide sections.

Compared to the losses due to lack of optimization of the waveguide bends, waveguide deformation due to thermal loads has a small impact on the microwave propagation (in fact, slightly lower losses were obtained with the deformed geometries). Nevertheless, it can be concluded from this study that for optimal performance, the optimization studies must consider the operating conditions.

Multiple aspects must be considered in the design of diagnostics for DEMO, and those were taken into consideration in the design of the DSC. Space restrictions must be considered in the BBs and in the equatorial and upper ports regions. The harsh conditions of operation (radiation and heat loads) impose restrictions on the design and the selection of materials, such as in the BB region, where only metallic components are allowed, and cooling systems are needed. Additionally, plasma-facing components need to be retracted in protected locations. Finally, RH compatibility must be considered early in the design of components that will be installed in areas where hands-on maintenance will not be possible.

6.2. Main Achievements and Future Work on the Development of MW Reflectometry for DEMO

Fusion reactors are highly complex devices. The harsh environment and high number of interfaces between their multiple systems impose many challenges to diagnostic integration and call for an iterative design process. The MW reflectometry system for DEMO is in the conceptual design phase, and a considerable amount of work has already been accomplished concerning its integration inside the vacuum vessel. The use of high-performance codes such as REFMUL enabled the development of synthetic reflectometry diagnostics that have been used to optimize the location and distribution of diagnostic antennas. The design of the system strongly benefited from the work performed for the ITER PPR system, which has provided design solutions that have been tested in relevant prototyping activities. However, much more work is still required to address the many issues that persist:

- the interface between the DSC and the BB, including the respective cooling services.
- the detailed definition of the antenna configurations in the first wall and the waveguide routing inside the DSC.
- the interface for attaching/detaching the waveguide extensions to/from the BB chimneys.
- the limited space available in the upper port (especially at the inboard) and the need to avoid toroidal bending of the waveguides.
- the displacements between the vacuum vessel and the blankets, which must be accommodated by the waveguide extensions, and the design of the in-vessel/ex-vessel waveguide transitions.

The listed issues are not only dependent on the design of the reflectometry system itself but also on the outcomes of the R&D being performed for other DEMO systems, such as the BBs and the RH system. Addressing these issues requires extensive simulation work, which involves CAD design and modelling, nuclear, thermo-mechanical, seismic, EM analyses, and EM simulations. A comprehensive structural analysis will also be required to demonstrate, considering the applicable codes and standards, that the system is capable of sustaining the prescribed loads and loading conditions, thus, qualified for its intended use.

7. Plasma Position and Shape Reflectometry in DTT—A Test Bed for DEMO

The Divertor Test Tokamak (DTT) facility is a good test bed to test and validate relevant non-magnetic control diagnostics contributing to improving their design and implementation in DEMO. For this reason, plasma position and shape reflectometry (PP&SR) is being investigated for installation on DTT with the aim of testing its capabilities in view of its application in DEMO as a complementary/backup system for magnetic sensors. In DTT, the PP&SR system should use four lines of sight: one in the LFS midplane, one in the upper vertical port, and two in the HFS region. Depending on the physical space available, a full band configuration (K, Ka, U, V, W, and D band) or only a subset of it will be used by the reflectometers at each line of sight to monitor the full radial density profile or the regions around the last closed flux surface. The utilization of FDTD codes enables a detailed description of the reflectometers, which includes aspects such as propagation in realistic plasmas, the specific location within the vacuum vessel, and access to the plasma. Additionally, the results of these simulations can be used to test new signal-processing techniques. To predict the behaviour and capabilities of the proposed reflectometry system for DTT, three synthetic diagnostics (SD) were prepared. These SDs correspond to three O-mode reflectometers located on the LFS, with one on the midplane and two others located away from the midplane [43]. The simulations were performed using the 2D FDTD full-wave code REFMULF in two plasma scenarios envisioned for DTT, namely the 5 MA single null (SN) and double null (DN) configurations during the start-of-flat (SOF) phase of the discharge. One of these simulated SD, located at the upper section of the LFS, is depicted in Figure 23, showing its position on a cross-section of the DTT vacuum vessel. This is an example of non-standard PPR, which may play a key role in next-generation machines. Figure 24 depicts snapshots of the full system 2D simulations for this gap during an SN plasma at four frequencies corresponding to the K ($f = 26.5$ GHz), Ka ($f = 40$ GHz), V ($f = 75$ GHz), and W ($f = 95$ GHz) bands. Note that the snapshot for the K band is the one represented in machine coordinates in Figure 23. Although it is not shown here, the Q band was also simulated. A simulated FMCW signal is excited at the antenna with frequencies ranging from the minimum to the maximum frequency of each band. The probing beam propagates to the plasma and is reflected back to the antennas, where it is decoupled from the emission through the use of a Unidirectional Transparent Source (UTS) [44]. These simulations include the plasma curvature and the full structure surrounding the plasma resulting in a visible distortion of the field patterns with the appearance of resonances in the lower K and Ka bands passing to a regime of multi-reflections at higher frequency bands. Ancillary simulations performed with slab plasmas were used to obtain reference cases against which the performance of the different gaps is evaluated, in particular, the measurement error. Results from the simulations of reflectometers at LFS, in equatorial, upper, and lower sections of the machine demonstrate good performance with measurement errors within the required ±1 cm and small deviations, of the order of some millimeters, at the separatrix. These simulations consider the plasma and surrounding structures, as well as their interplay, demonstrating that the proposed reflectometer positions (or any small variations) are suitable for DTT. Since 3D simulations are computationally intensive and require access to high-performance computer (HPC) facilities, the use of 2D codes is more common. However, it is important to evaluate if a 2D model is sufficient to describe the problem at hand and what compromises may be involved in using such a model or if a 3D approach is necessary, especially since the received power can be a concern on larger devices. To address this, simulations were carried out using two full-wave codes, namely REFMULF (2D) and REFMUL3 (3D), to make a comparison between the two [45].

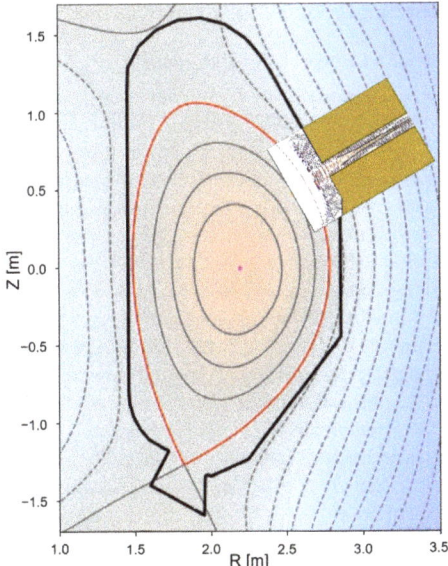

Figure 23. Cross-section of the DTT vacuum vessel showing an LFS synthetic reflectometer placed at the upper part of the machine. The line in red represents the Last Closed Flux Surface (LCFS).

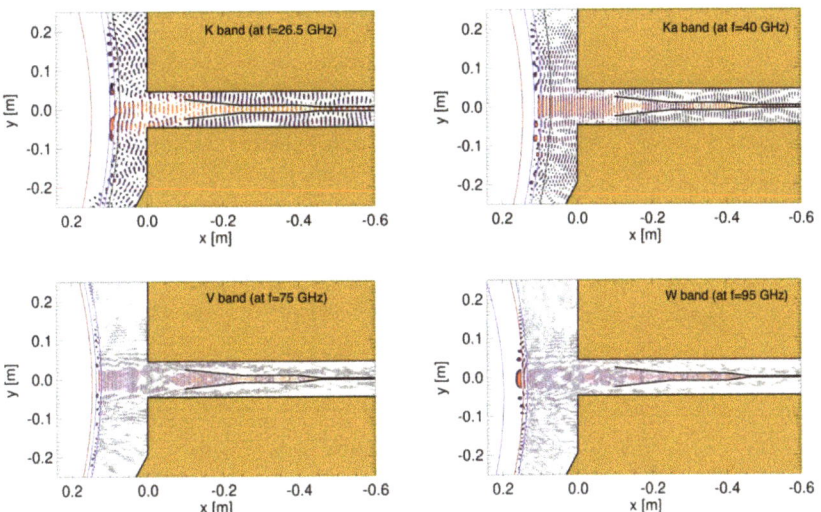

Figure 24. Snapshot of electric field E_z for LFS SD located at the upper part of the machine (see Figure 23) in the DTT $f = 26.5$ GHz (top left), Ka band at $f = 40$ GHz (top right), V band at $f = 75$ GHz (bottom left) and W band at $f = 95$ GHz (bottom right). The vertical blue lines mark the iso-density region corresponding to the limits of the band; the red line shows the position of the separatrix.

The overall results are encouraging for the implementation of reflectometry systems at DTT and represent an important milestone for reflectometry in D-shaped tokamaks. Even a reduced system with an equatorial gap and only one of the other two gaps (either on the upper or lower part of the machine) would be a significant advancement for reflectometry in general and for DTT and DEMO, as reflectometry has never been used before to measure the density profile from a location away from the midplane in a D-shaped tokamak.

8. Development of a Compact Reflectometer

The number of channels needed to achieve good poloidal and toroidal plasma coverage in fusion reactors such as DEMO is larger than in current machines, which could take advantage of the use of compact and modular millimetre wave back-ends. For that reason, Instituto de Plasmas e Fusão Nuclear at Instituto Superior Técnico (IPFN-IST) in Lisbon is developing and prototyping a coherent fast frequency sweeping RF back-end using commercial Monolithic Microwave Integrated Circuits (MMICs) leveraging the large-scale availability of high-performance MMIC at affordable prices.

The new back-end system was designed to be able to accommodate different frequency ranges in a flexible way. Thus, the core back-end covers the NATO J-band (10 GHz to 20 GHz) and is capable of driving external full-band frequency multipliers to achieve ultra-wideband coverage up to 160 GHz. Although developed for fusion plasma diagnostics, this new compact reflectometer may also be used in other FMCW radar applications, such as space re-entry experiments. The details of its implementation are described in [46].

The system architecture is shown in Figure 25. Heterodyne radars use a microwave sweep oscillator to generate a signal, which is then used for different purposes in the circuitry: (i) for target probing through the transmission channel (Ax); (ii) to drive a frequency translator. The frequency-translated signal is used to drive two mixers: the front-end mixer of the receiver channel (Bx), which collects the reflecting signal from the target and generates the beat signal around the Intermediate Frequency (IF) carrier, and the local reference mixer that generates the reference signal used on the quadrature detector to demodulate the reflected signal. The fact that the IF is generated by the frequency translation of the probing signal ensures the coherence of the system [47]. This approach improves the radar dynamic range with respect to homodyne detection. The developed system uses two high-frequency PCB boards: (i) the back-end board, which is responsible for generating all the signals (shaded green in Figure 25), and the quadrature detector board, which is used to demodulate the reflected signal, providing amplitude and phase information (shaded pink in Figure 25). The front-end part of the system (yellow-shaded area in Figure 25) includes the final frequency multipliers, mixers, transmission lines, and antennas. The compact reflectometer will have to face two major challenges, (i) high performance with excellent shielding against X-ray, Gamma, and neutron; (ii) affordable price. The prototype was developed with commercial circuit components on a customized PCB base. It is one possible solution, but some may consider it outdated. The most promising solution for high-level integration microwave diagnostics is system-on-chip, which develops all microwave circuits on a 10 mm^2 semi-conductor chip (CMOS, InP, GaAs, and GaN). This technology has been successfully developed on DIII-D since 2019 [48]. However, the reflectometers on a chip on DIII-D were applied to imaging reflectometry where fixed frequency or very limited frequency sweep is used. The presented prototype could have been produced with a larger level of integration, but in this way, it was possible to better optimize the different stages of the reflectometer. Radar on a chip exists for massive applications in the automotive industry but is bandwidth limited. It should also be noted that the US Domestic Agency, which is responsible for the development of the main plasma low field side ITER reflectometers, did not propose the "on a chip" solution for those systems. However, it is expectable that the future will lead to a large integration system on a chip or a couple of chips, and those advances will be considered for future developments as they will require a larger number of reflectometers.

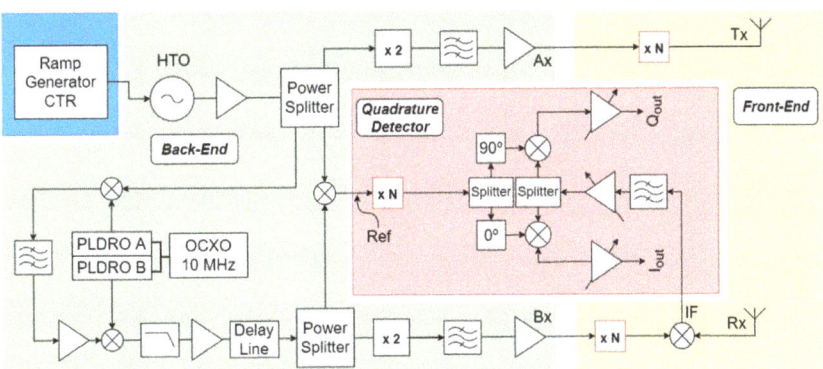

Figure 25. Block diagram of the compact reflectometer.

Figure 26 shows a prototype of the compact reflectometer being tested. To be capable of efficiently driving external frequency multipliers, the system generates full band signals exceeding 8 dBm. To keep all undesirable harmonics at least 15 dB below the desired output across the entire frequency range of interest, all signal paths were fine-tuned using appropriate in-line attenuators. The system also proved capable of coping with low-level reflected signals.

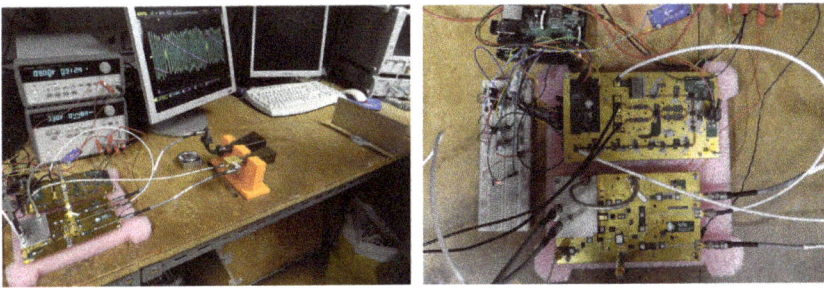

Figure 26. Compact reflectometer prototype being tested with a metallic mirror, setup (**left**), and top view on the electronic hardware (**right**).

The back-end and in-phase and quadrature (I/Q) boards included EM shielding and were designed for easy installation on a standard 3U 19″ rack.

The precision of FMCW radars is known to be strongly dependent on the linearity of the frequency sweep. Modern fully integrated Direct Digital Synthesis (DDS) chips can generate signals with very high frequency, phase precision, and resolution, improving the linearity of the radar probing signals and allowing very linear and agile frequency chirps. The direct digitalization of the IF signals makes it possible to use highly flexible data processing techniques. A new prototype using these two techniques is being developed. The system is expected to be ready for deployment into experimental devices by early 2024.

9. Conclusions

In 2017, the US National Academy of Engineering considered that providing energy from fusion was one of the grand challenges for engineering. The engineering community faces the challenge of scaling up the fusion process to commercial proportions in an efficient, cost-effective, and environmentally friendly manner. Controlling the burning plasma is one of the critical issues that need to be addressed. Plasma Position Reflectometry is expected to have an important role in next-generation fusion machines, such as DEMO, as a diagnostic to monitor the position and shape of the plasma. Proofs of concept focused on the ability to use the diagnostic for plasma control were performed in the ASDEX-Upgrade and COMPASS tokamaks. The success of these experiments has shown that it is possible to fulfil the ITER and DEMO requirements for plasma control using microwave reflectometry as an alternative to standard magnetic measurements. Namely, it can achieve the high availability and reliability of plasma control required for the operation of ITER and DEMO, helping to pave the way towards the realization of future fusion power plants.

Due to the nature of microwave reflectometry measurements, the front end of the reflectometers (antennas and feeding waveguides) is directly exposed to the plasma, thus subjected to fluxes of high-energy neutrons (14 MeV) and high thermal loads. Therefore, it is indispensable to ensure that these components can operate in the harsh environment of a fusion reactor, where they will face fusion powers of 500 MW for ITER and 2000 MW for DEMO (heat fluxes up to 1–2 MW/m^2 [49]). For that purpose, complex design studies (involving neutronics, thermo-mechanical studies, and EM analyses) are crucial to validate the design of these components and to ensure that they can sustain the expected loads without damages that could seriously compromise their performance and low aging to ensure their time life to be compatible with a fusion power plant. The neutronics and thermo-mechanical studies provide an insight into what needs to be optimized in the design to meet certain requirements, while the EM studies provide an understanding of the impact of the thermo-mechanical loads on the ability of the systems to perform measurements.

Some of the developments achieved so far have strongly benefited from the progress made for the REFMUL suite of codes developed at IPFN for reflectometry simulation. The suite includes REFMUL, a 2D O-mode simulation code, and REFMULX/REFMULXp, a 2D X-mode (serial/parallel) simulation code. Recently, two new codes were added: REFMULF and REFMUL3. REFMULF is a 2D code able to cope with full polarisation waves, treating all components of the electric and magnetic fields of the wave [50]. REFMUL3, which recently entered production, is a 3D full-wave code. Additional tools were developed to automate the simulation and data analysis of a generic PPR system and make it adaptable to different devices. These techniques can be used to optimize a given PPR system for a given equilibrium scenario and study its stability. Furthermore, an effort has been devoted to EM modelling to further understand how edge turbulence impacts density profile reconstruction using O-mode reflectometry [51].

The ITER PPR system would allow testing the use of this diagnostic for control of burning plasma, assess its performance under harsh conditions, and evaluate the design decisions taken and how those would affect the performance of the diagnostic. The R&D and design of the in-vessel components progressed according to schedule. Unfortunately, ITER terminated the PPR project due to concerns that captive components (waveguides and supports needed to install the system between the ports and the diagnostic hall) would not be procured and delivered in time. This would have resulted in unacceptable delays in installing the remaining captive components from other systems that could only be installed after the PPR system components were fully in place. As a result, ITER opted to rely on alternative technologies for measuring plasma position. Although descoped, the design of the ITER PPR system allowed the development and testing of some important aspects of using reflectometry as a plasma position system and the identification of critical issues that still require further development or prototyping.

The MW reflectometry system being proposed for DEMO is at an early stage of development, and a significant amount of work is still required. The work performed so far illustrates the level of multidisciplinary required to develop and integrate microwave reflectometry diagnostics on future fusion devices such as DTT and DEMO. Such developments require an integrated approach that encompasses, for instance, microwave reflectometry simulations, CAD design and integration studies (space occupation, RH compatibility, definition of interfaces with other systems, etc.), neutronics simulations (MCNP), thermo-mechanical analyses, EM full-wave simulations and multiple design iterations, incorporating constraints from each area.

Many of the developed concepts will have to be further tested in the laboratory and in fusion experiments. DTT will provide a unique opportunity to test further the use of reflectometry for plasma control with the aim of testing its capabilities in view of a possible reactor application as a supplementary/backup system for the magnetics.

The recent developments in the design of a compact reflectometer will also be crucial for the successful integration of many measurement channels in the reduced space available in future fusion machines.

Author Contributions: Conceptualization, A.M., A.S., F.S., A.V., T.R., E.R. and J.D.; methodology, R.L., A.S., F.S., Y.N., E.R., S.H., T.R. and J.F.; software, P.V., A.S., F.S., E.R., J.B., J.F., T.R., J.D., G.D.M. and G.M.; validation, A.S., F.S., J.F., E.R. and T.R.; formal analysis, P.V., A.S., A.V., E.R., S.H. and G.M.; investigation, P.V., J.S., R.L., A.S., Y.N., J.B., S.H., J.D., A.V., O.T., R.C., O.D. and G.D.M.; resources, B.G., F.S., E.R. and T.R.; data curation, P.V., A.S., F.S., E.R. and J.F.; writing—original draft preparation, B.G., P.V., R.L., F.S. and Y.N.; writing—review and editing, All; visualization, A.S., R.L., F.S., Y.N., J.B., E.R., J.F. and J.D.; supervision, A.S., J.S., A.M., S.H., T.F. and W.B.; project administration, B.G., P.V., J.S., T.F. and W.B.; funding acquisition, B.G., P.V., F.S. and W.B. All authors have read and agreed to the published version of the manuscript.

Funding: IPFN activities received financial support from "Fundação para a Ciência e Tecnologia" through projects UIDB/50010/2020 and UIDP/50010/2020. This work has been carried out within the framework of the EUROfusion Consortium, funded by the European Union via the Euratom Research and Training Programme (Grant Agreement No. 101052200—EUROfusion). The compact reflectometer development was supported by the European Regional Development Fund (FEDER) through the Competitiveness and Internationalization Operational Program (COMPETE 2020) of the Portugal 2020 framework, Project, RETIOT, POCI 01 0145 FEDER 016432. Views and opinions expressed are, however, those of the author(s) only and do not necessarily reflect those of the European Union or the European Commission. Neither the European Union nor the European Commission can be held responsible for them.

Institutional Review Board Statement: Not applicable.

Informed Consent Statement: Not applicable.

Data Availability Statement: Not applicable.

Conflicts of Interest: The authors declare no conflict of interest.

References

1. Santos, J.; Guimarãis, L.; Zilker, M.; Treutterer, W.; Manso, M.; ASDEX Upgrade Team. Reflectometry-based plasma position feedback control demonstration at ASDEX Upgrade. *Nucl. Fusion* **2012**, *52*, 032003. [CrossRef]
2. Sabot, R.; Sirinelli, A.; Chareau, J.-M.; Giaccalone, J.-C. A dual source D-band reflectometer for density profile and fluctuations measurements in Tore-Supra. *Nucl. Fusion* **2006**, *46*, S685–S692. [CrossRef]
3. Yeh, K.C.; Liu, C.H. *Theory of Ionospheric Waves*; Academic Press: Cambridge, MA, USA, 1972.
4. Ginzburg, V.L. *The Propagation of Electromagnetic Waves in Plasmas*, 2nd ed.; Pergamon Press: Oxford, UK, 1970.
5. Hutchinson, I.H. *Principles of Plasma Diagnostics*, 2nd ed.; Cambridge University Press: Cambridge, UK, 2002. [CrossRef]
6. Bretz, N.L.; Kessel, C.; Doyle, E.J.; Vayakis, G. ITER Position Control Reflectometry—Conceptual Design. Diagnostics for Experimental Thermonuclear Fusion Reactors 2; Stott, P.E., Gorini, G., Prandoni, P., Sindoni, E., Eds.; Plenum: New York, NY, USA, 1998; pp. 129–137. [CrossRef]
7. Santos, J.; Manso, M.; Varela, P. Plasma position measurements from ordinary FM-CW reflectometry on ASDEX Upgrade. *Rev. Sci. Instrum.* **2003**, *74*, 1489–1492. [CrossRef]
8. Santos, J.; Zilker, M.; Guimarãis, L.; Treutterer, W.; Amador, C.; Manso, M. COTS-Based High-Data-Throughput Acquisition System for a Real-Time Reflectometry Diagnostic. *IEEE Trans. Nucl. Sci.* **2011**, *58*, 1751–1758. [CrossRef]
9. Santos, J. Fast Reconstruction of Reflectometry Density Profiles on ASDEX Upgrade for Plasma Position Feedback Purposes. Ph.D. Thesis, Instituto Superior Técnico, Universidade de Lisboa, Lisbon, Portugal, 2022. [CrossRef]
10. Santos, J.M.; Guimarãis, L.; Rapson, C.; Santos, G.; Silva, A.; Treutterer, W.; Zilker, M.; ASDEX Upgrade Team; EUROfusion MST1 Team. Real-time reflectometry—An ASDEX Upgrade DCS plugin App for plasma position and shape feedback control. *Fusion Eng. Des.* **2017**, *123*, 593–596. [CrossRef]
11. Prisiazhniuk, D.; Krämer-Flecken, A.; Conway, G.D.; Happel, T.; Lebschy, A.; Manz, P.; Nikolaeva, V.; Stroth, U.; ASDEX Upgrade Team. Magnetic field pitch angle and perpendicular velocity measurements from multi-point time-delay estimation of poloidal correlation reflectometry. *Plasma Phys. Control Fusion* **2017**, *59*, 025013. [CrossRef]

12. Silva, A. The ASDEX Upgrade Broadband Microwave Reflectometry System. Ph.D. Thesis, Instituto Superior Técnico, Universidade Técnica de Lisboa, Lisbon, Portugal, 2006.
13. Santos, J.; Hacquin, S.; Manso, M. Frequency modulation continuous wave reflectometry measurements of plasma position in ASDEX Upgrade ELMy H-mode regimes. *Rev. Sci. Instrum.* **2004**, *75*, 3855–3858. [CrossRef]
14. Lourenço, P.D.; Santos, J.M.; Havránek, A.; Bogár, O.; Havlicek, J.; Zaja, J.; Silva, A.; Batista, A.J.N.; Hron, M.; Pánek, R.; et al. Real-time plasma position reflectometry system development and integration on COMPASS tokamak. *Fusion Eng. Des.* **2020**, *160*, 112017. [CrossRef]
15. Janky, F.; Havlicek, J.; Batista, A.J.N.; Kudlacek, O.; Seidl, J.; Neto, A.C.; Pipek, J.; Hron, M.; Mikulin, O.; Duarte, A.S.; et al. Upgrade of the COMPASS tokamak real-time control system. *Fusion Eng. Des.* **2014**, *89*, 186–194. [CrossRef]
16. Neto, A.C.; Alves, D.; Boncagni, L.; Carvalho, P.J.; Valcarcel, D.F.; Barbalace, A.; De Tommasi, G.; Fernandes, H.; Sartori, F.; Vitale, E.; et al. A Survey of Recent MARTe Based Systems. *IEEE Trans. Nucl. Sci.* **2011**, *58*, 1482–1489. [CrossRef]
17. Zajac, J.; Bogár, O.; Varanin, M.; Zacek, F.; Hron, M.; Pánek, R.; Nanobashvili, S.; Silva, A. Upgrade of the COMPASS tokamak microwave reflectometry system with I/Q modulation and detection. *Fusion Eng. Des.* **2017**, *123*, 911–914. [CrossRef]
18. Vidal, C.; Luís, R.; Nietiadi, R.; Velez, N.; Varela, P. Thermal analysis of a waveguide section of the ITER plasma-position reflectometry system on the high-field side. *Fusion Eng. Des.* **2019**, *146*, 2389–2392. [CrossRef]
19. Udintsev, V. *System Design Description (DDD) 55.F3 Plasma Position Reflectometry*; ITER Report 76MW26 v1.9; ITER Organization: Saint-Paul-lez-Durance, France, 2012.
20. Varela, P.; Belo, J.H.; Silva, A.; da Silva, F. Design status of the in-vessel subsystem of the ITER Plasma Position Reflectometry system. *J. Instrum.* **2019**, *14*, C09002. [CrossRef]
21. Nietiadi, Y.; Vidal, C.; Luís, R.; Varela, P. Thermal analyses of the in-vessel frontends of the ITER plasma position reflectometry system. *Fusion Eng. Des.* **2020**, *156*, 111599. [CrossRef]
22. Belo, J.; Varela, P.; Ricardo, E.; Silva, A.; Quental, P. Performance assessment of critical waveguide bends for the ITER in-vessel plasma position reflectometry systems. *Fusion Eng. Des.* **2017**, *123*, 773–777. [CrossRef]
23. Kane, Y. Numerical solution of initial boundary value problems involving maxwell's equations in isotropic media. *IEEE Trans. Antennas Propag.* **1996**, *14*, 302–307. [CrossRef]
24. Taflove, A.; Hagness, S.C. *Computational Electrodynamics: The Finite-Difference Time-Domain Method*, 3rd ed.; Artech House: Norwood, MA, USA, 2005.
25. da Silva, F.; Pinto, M.C.; Després, B.; Heuraux, S. Stable explicit coupling of the Yee scheme with a linear current model in fluctuating magnetized plasmas. *J. Comput. Phys.* **2015**, *295*, 24–45. [CrossRef]
26. Biel, W.; Albanese, R.; Ambrosino, R.; Ariola, M.; Berkel, M.V.; Bolshakova, I.; Brunner, K.J.; Cavazzana, R.; Cecconello, M.; Conroy, S.; et al. Diagnostics for plasma control—From ITER to DEMO. *Fusion Eng. Des.* **2019**, *146*, 465–472. [CrossRef]
27. Biel, W.; Ariola, M.; Bolshakova, I.; Brunner, K.J.; Cecconello, M.; Duran, I.; Franke, T.; Giacomelli, L.; Giannone, L.; Janky, F.; et al. Development of a concept and basis for the DEMO diagnostic and control system. *Fusion Eng. Des.* **2022**, *179*, 113122. [CrossRef]
28. Malaquias, A.; Silva, A.; Moutinho, R.; Luís, R.; Lopes, A.; Quental, P.B.; Prior, L.; Velez, N.; Policarpo, H.; Vale, A.; et al. Integration Concept of the Reflectometry Diagnostic for the Main Plasma in DEMO. *IEEE Trans. Plasma Sci.* **2018**, *46*, 451–457. [CrossRef]
29. Jesenko, A.; Mazon, D.; Ariola, M.; Luís, R.; Treutterer, W. *Diagnostics and Control (DC) System Requirements Document (SRD)*; EURO Fusion Report IDM: EFDA_D_2MNK4R_v3.0; European Space Agency: Paris, France, 2020.
30. Marchiori, G.; De Masi, G.; Cavazzana, R.; Cenedese, A.; Marconato, N.; Moutinho, R.; Silva, A. Study of a Plasma Boundary Reconstruction Method Based on Reflectometric Measurements for Control Purposes. *IEEE Trans. Plasma Sci.* **2018**, *46*, 1285. [CrossRef]
31. Ricardo, E.; da Silva, F.; Heuraux, S.; Silva, A. Assessment of a multi-reflectometers positioning system for DEMO plasmas. *J. Instrum.* **2019**, *14*, C08010. [CrossRef]
32. Silva, A. *Advanced Simulation Studies on Microwave Diagnostics under DEMO Conditions, Advanced Concept on in Vessel and Ex Vessel Components, Complete Set of Expected Performance Data of All MW Diagnostics*; EURO Fusion Report IDM: EFDA_D_2P5AU9 v1.0; European Space Agency: Paris, France, 2020.
33. Belo, J.; Nietiadi, Y.; Luís, R.; Silva, A.; Vale, A.; Gonçalves, B.; Franke, T.; Krimmer, A.; Biel, W. Design and integration studies of a diagnostics slim cassette concept for DEMO. *Nucl. Fusion* **2021**, *61*, 116046. [CrossRef]
34. Luís, R.; Moutinho, R.; Prior, L.; Quental, P.B.; Lopes, A.; Policarpo, H.; Velez, N.; Vale, A.; Silva, A.; Malaquias, A. Nuclear and Thermal Analysis of a Reflectometry Diagnostics Concept for DEMO. *IEEE Trans. Plasma Sci.* **2018**, *46*, 1247–1253. [CrossRef]
35. Luís, R.; Nietiadi, Y.; Belo, J.H.; Silva, A.; Vale, A.; Malaquias, A.; Gonçalves, B.; da Silva, F.; Santos, J.; Ricardo, E.; et al. A diagnostics slim cassette for reflectometry measurements in DEMO: Design and simulation studies. *Fusion Eng. Des.* **2023**, *190*, 113512. [CrossRef]
36. Nietiadi, Y.; Luís, R.; Silva, A.; Ricardo, E.; Gonçalves, B.; Franke, T.; Biel, W. Nuclear and thermal analysis of a multi-reflectometer system for DEMO. *Fusion Eng. Des.* **2021**, *167*, 112349. [CrossRef]
37. Nietiadi, Y.; Luís, R.; Silva, A.; Belo, J.H.; Vale, A.; Malaquias, A.; Gonçalves, B.; da Silva, F.; Santos, J.; Ricardo, E.; et al. Thermomechanical analysis of a multi-reflectometer system for DEMO. *Fusion Eng. Des.* **2023**, *190*, 113530. [CrossRef]
38. Keep, J.; Wood, S.; Gupta, N.; Coleman, M.; Loving, A. Remote handling of DEMO breeder blanket segments: Blanket transporter conceptual studies. *Fusion Eng. Des.* **2017**, *124*, 420–425. [CrossRef]

39. Nietiadi, Y. Nuclear Technology and Engineering Studies on Reflectometry Systems for ITER and DEMO. Ph.D. Thesis, Instituto Superior Técnico, Universidade de Lisboa, Lisbon, Portugal, 2022.
40. Zhu, Y.; Yu, J.-H.; Chen, M.; Tobias, B.; Luhmann, N.C. New Trends in Microwave Imaging Diagnostics and Application to Burning Plasma. *IEEE Trans. Plasma Sci.* **2019**, *47*, 2110–2130. [CrossRef]
41. Meschino, S.; Ceccuzzi, S.; Mirizzi, F.; Pajewski, L.; Schettini, G.; Artaud, J.F.; Bae, Y.S.; Belo, J.H.; Berger-By, G.; Bernard, J.M.; et al. Bends in oversized rectangular waveguide. *Fusion Eng. Des.* **2011**, *86*, 746–749. [CrossRef]
42. Ricardo, E.; Varela, P.; Silva, A.; Gonçalves, B. Assessment and performance optimization of the ITER plasma position reflectometry in-vessel oversized waveguide bends. *Fusion Eng. Des.* **2015**, *98–99*, 1593–1596. [CrossRef]
43. da Silva, F.; Ferreira, J.; Santos, J.; Heuraux, S.; Ricardo, E.; De Masi, G.; Tudisco, O.; Cavazzana, R.; D'Arcangelo, O. Assessment of measurement performance for a low field side IDTT plasma position reflectometry system. *Fusion Eng. Des.* **2021**, *168*, 112405. [CrossRef]
44. da Silva, F.; Heuraux, S.; Hacquin, S.; Manso, M.E. Unidirectional transparent signal injection in finite-difference time-domain electromagnetic codes –application to reflectometry simulations. *J. Comput. Phys.* **2005**, *203*, 467. [CrossRef]
45. da Silva, F.; Ricardo, E.; Ferreira, J.; Santos, J.; Heuraux, S.; Silva, A.; Ribeiro, T.; De Masi, G.; Tudisco, O.; Cavazzana, R.; et al. Benchmarking 2D against 3D FDTD codes for the assessment of the measurement performance of a low field side plasma position reflectometer applicable to IDTT. *J. Instrum.* **2022**, *17*, C01017. [CrossRef]
46. Silva, A.; Dias, J.; Santos, J.; da Silva, F.; Gonçalves, B. FM-CW compact reflectometer using DDS signal generation. *J. Instrum.* **2021**, *16*, C11005. [CrossRef]
47. Meneses, L.; Cupido, L.; Manso, M.E.; JET-EFDA Contributors. New frequency translation technique for FM-CW reflectometry. *Rev. Sci. Instrum.* **2010**, *81*, 10D924. [CrossRef]
48. Zhu, Y.; Chen, Y.; Yu, J.-H.; Domier, C.; Yu, G.; Liu, X.; Kramer, G.; Ren, Y.; Diallo, A.; Luhmann, N.C.; et al. System-on-chip approach microwave imaging reflectometer on DIII-D tokamak. *Rev. Sci. Instrum.* **2022**, *93*, 113509. [CrossRef]
49. Vizvary, Z.; Arter, W.; Bachmann, C.; Barret, T.R.; Chuilon, B.; Cooper, P.; Flynn, E.; Firdaouss, M.; Franke, T.; Gerardin, J.; et al. European DEMO first wall shaping and limiters design and analysis status. *Fusion Eng. Des.* **2020**, *158*, 111676. [CrossRef]
50. da Silva, F.; Heuraux, S.; Ribeiro, T. Introducing REFMULF, a 2D full polarization code and REFMUL3, a 3D parallel full wave Maxwell code. In Proceedings of the 13th International Reflectometry Workshop for Fusion Plasma Diagnostics—IRW13, Daejeon, Republic of Korea, 10–12 May 2017.
51. Ricardo, E.; da Silva, F.; Heuraux, S.; Silva, A. On the edge turbulence effects on the density profile reconstruction using O-mode reflectometers. *Fusion Eng. Des.* **2021**, *171*, 112652. [CrossRef]

Disclaimer/Publisher's Note: The statements, opinions and data contained in all publications are solely those of the individual author(s) and contributor(s) and not of MDPI and/or the editor(s). MDPI and/or the editor(s) disclaim responsibility for any injury to people or property resulting from any ideas, methods, instructions or products referred to in the content.

Article

Neutronics Simulations for DEMO Diagnostics

Raul Luís [1,*], Yohanes Nietiadi [1], Antonio Quercia [2], Alberto Vale [1], Jorge Belo [1], António Silva [1], Bruno Gonçalves [1], Artur Malaquias [1], Andrei Gusarov [3], Federico Caruggi [4], Enrico Perelli Cippo [4], Maryna Chernyshova [5], Barbara Bienkowska [5] and Wolfgang Biel [6]

1 Instituto de Plasmas e Fusão Nuclear, Instituto Superior Técnico, Universidade de Lisboa, Av. Rovisco Pais 1, 1049-001 Lisbon, Portugal; ynietiadi@ipfn.tecnico.ulisboa.pt (Y.N.); avale@ipfn.tecnico.ulisboa.pt (A.V.); jbelo@ipfn.tecnico.ulisboa.pt (J.B.); silva@ipfn.tecnico.ulisboa.pt (A.S.); bruno@ipfn.tecnico.ulisboa.pt (B.G.); artur.malaquias@ipfn.tecnico.ulisboa.pt (A.M.)
2 DIETI/Consorzio CREATE, Università Federico II, Via Claudio 21, 80125 Napoli, Italy; antonio.quercia@unina.it
3 SCK CEN Belgian Nuclear Research Center, 2400 Mol, Belgium; andrei.goussarov@sckcen.be
4 Institute for Plasma Science and Technology, National Research Council, 20125 Milan, Italy; f.caruggi@campus.unimib.it (F.C.); enrico.perellicippo@istp.cnr.it (E.P.C.)
5 Institute of Plasma Physics and Laser Microfusion, Hery 23, 01-497 Warsaw, Poland; maryna.chernyshova@ifpilm.pl (M.C.); barbara.bienkowska@ifpilm.pl (B.B.)
6 Institute of Energy and Climate Research, Forschungszentrum Jülich GmbH, 52428 Jülich, Germany; w.biel@fz-juelich.de
* Correspondence: rluis@ipfn.tecnico.ulisboa.pt; Tel.: +351-916529266

Abstract: One of the main challenges in the development of a plasma diagnostic and control system for DEMO is the need to cope with unprecedented radiation levels in a tokamak during long operation periods. A list of diagnostics required for plasma control has been developed during the pre-conceptual design phase. Different approaches are proposed for the integration of these diagnostics in DEMO: in equatorial and upper ports, in the divertor cassette, on the inner and outer surfaces of the vacuum vessel and in diagnostic slim cassettes, a modular approach developed for diagnostics requiring access to the plasma from several poloidal positions. According to each integration approach, diagnostics will be exposed to different radiation levels, with a considerable impact on their design. This paper provides a broad overview of the radiation environment that diagnostics in DEMO are expected to face. Using the water-cooled lithium lead blanket configuration as a reference, neutronics simulations were performed for pre-conceptual designs of in-vessel, ex-vessel and equatorial port diagnostics representative of each integration approach. Flux and nuclear load calculations are provided for several sub-systems, along with estimations of radiation streaming to the ex-vessel for alternative design configurations. The results can be used as a reference by diagnostic designers.

Keywords: neutronics; diagnostics; tokamaks; DEMO; nuclear fusion; MCNP

Citation: Luís, R.; Nietiadi, Y.; Quercia, A.; Vale, A.; Belo, J.; Silva, A.; Gonçalves, B.; Malaquias, A.; Gusarov, A.; Caruggi, F.; et al. Neutronics Simulations for DEMO Diagnostics. *Sensors* **2023**, *23*, 5104. https://doi.org/10.3390/s23115104

Academic Editor: Hossam A. Gabbar

Received: 28 April 2023
Revised: 24 May 2023
Accepted: 25 May 2023
Published: 26 May 2023

Copyright: © 2023 by the authors. Licensee MDPI, Basel, Switzerland. This article is an open access article distributed under the terms and conditions of the Creative Commons Attribution (CC BY) license (https:// creativecommons.org/licenses/by/ 4.0/).

1. Introduction

One of the main challenges in the development of a plasma diagnostic and control (D&C) system for the demonstration fusion reactor (DEMO) is the need to cope with unprecedented radiation levels in a tokamak during long operation periods. Projected to operate with a fusion power of 2 GW, the DEMO plasma will produce 14 MeV neutrons from deuterium–tritium (D-T) reactions at an approximate rate of 7×10^{20} n s^{-1} [1]. Although this results in neutron fluxes in the first wall that are not significantly increased when compared to ITER, the longer pulses in DEMO will lead to higher fluences and displacements per atom (dpa) in the plasma-facing materials [2]. Presently, DEMO operation is scheduled in two phases: a first phase with a "starter" blanket, designed to withstand up to 20 dpa in the first wall steel, and a second phase after blanket replacement, with blankets designed for a higher limit of 50 dpa. ITER plasma-facing components, for comparison, will remain below 4 dpa [3]. The development of materials that can cope with loads one order of magnitude higher than those expected for ITER is one of the main challenges towards the realization of DEMO [4].

In DEMO, the design of a D&C system is the task of the Work Package Diagnostic and Control (WPDC). The aim of this project is to design a D&C system with high reliability and accuracy that allows safe operation of the plasma near its operational limits, to maximize the power output [5]. Based on constraints which go far beyond the case of ITER, including the harsher radiation environment, the need for compatibility with remote maintenance operations and space limitations dictated by the requirements of first wall integrity and tritium breeding, a list of diagnostics required for plasma control has been developed within WPDC during the pre-conceptual design phase. This list includes [6]:

- Magnetic diagnostics (pickup coils, saddle loops, full-flux loops, diamagnetic loops, Rogowski coils, Hall sensors);
- Faraday sensors;
- Infrared (IR) polarimetry/interferometry;
- Neutron and gamma cameras;
- Microwave (MW) reflectometry;
- Electron cyclotron emission (ECE);
- Divertor thermocurrent measurements;
- Radiated power and soft X-ray intensity;
- X-ray spectroscopy;
- Vacuum ultraviolet spectroscopy (VUV) spectroscopy;
- IR/visible (VIS)/near-UV divertor spectroscopy;
- VIS spectroscopy and thermography of limiters;
- Pellet monitoring;
- Collective Thomson scattering (CTS).

Figure 1 illustrates the different approaches followed for the integration of these diagnostics in DEMO (collapsed in a single DEMO sector for easier visualization). Most sub-systems are designed to be integrated in equatorial port (EP) plugs dedicated to diagnostics (five or six EPs are foreseen), in some cases with additional lines of sight in the upper ports (UPs) if there is space reserved for diagnostics in the UPs. These include spectroscopy diagnostics [7], neutron/gamma cameras [8,9], radiated power and soft X-ray intensity [10] and IR polarimetry/interferometry [11], with the eventual addition of collective Thomson scattering [12], still under study. For diagnostics that require access to the plasma from several poloidal positions, such as MW reflectometry [13] and ECE [14], the diagnostics slim cassette (DSC) concept has been developed [15–17] as a modular approach compatible with the remote handling operations of the breeding blanket (BB). Thermocurrent measurements are planned to be integrated within the divertor cassette [18], while Faraday sensors are distributed poloidally on the outer surface of the vacuum vessel (VV) [19]. Finally, magnetic sensors are distributed on the inner and outer surfaces of the VV [20–22].

According to each integration approach, diagnostics in DEMO will be exposed to different radiation levels. This will have a considerable impact on their design. Moreover, radiation streaming to the ex-vessel, either through diagnostic ducts in the ports or due to inadequate shielding from the DSC or other diagnostic components, shall be minimized, in order to comply with the radiation limits defined for the DEMO plant. These include 0.3–0.5 W/cm^3 and 2.75 dpa in the VV stainless steel, 50 W/m^3 in the winding packs of the superconductor coils and 100 µSv/h of dose rate in the port cells 12 days after shutdown [1].

This paper aims to provide estimations of the fluxes and nuclear loads in preliminary designs of diagnostics representative of each integration approach, to be used as a reference for diagnostic designers. Although compliance with all the radiation limits set out for DEMO is beyond the scope of this work, such an evaluation is presented when possible, namely with regard to the nuclear heat loads and dpa in the VV. Previous studies have been published with neutronics simulations for the DSC concept [17,23,24], but with limited results for the full DSC and its impact on the neutron and gamma fluxes in the VV. A work focused on an early design of the divertor survey visible high-resolution spectrometer has also been published, which aimed mainly to assess the loads in the first mirrors and the impact of the number of doglegs in the EP ducts on the radiation streaming to the port cells [25]. The objective now is to extend those simulations to more complex geometries and to include additional diagnostics that were not studied before.

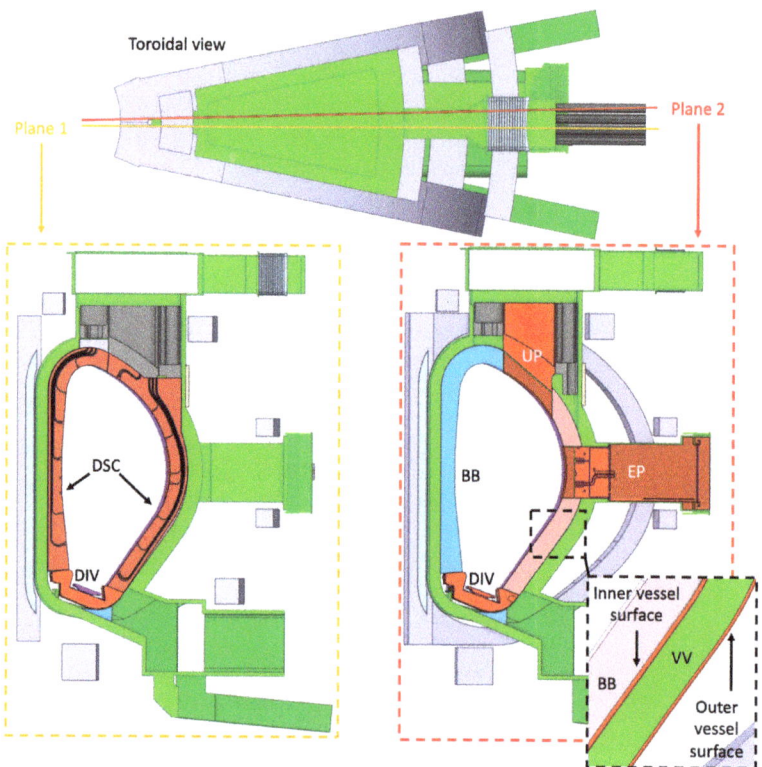

Figure 1. Representation of one DEMO sector and foreseen locations for diagnostics, represented in red (for visualization only—diagnostics will be distributed in different sectors). DSC: Diagnostics Slim Cassette (red). EP: Equatorial Port (red). UP: Upper Port (red). BB: Breeding Blanket (blue—inboard and pink—outboard). DIV: Divertor (red). VV: Vacuum Vessel (green).

Section 2 provides a description of the simulation methods common to all the analyses presented in the paper. Section 3 is the main body of the paper, presenting the models and results obtained for each set of diagnostics: inner-vessel diagnostics, ex-vessel diagnostics (Faraday sensors) and equatorial port diagnostics. Finally, a summary and discussion of the results are provided in Section 4.

2. Simulation Methods

2.1. Simulation Workflow

All simulations were performed with the Monte Carlo simulation program MCNP6 [26,27], approved for neutronics simulations in the DEMO project [1]. The JEFF-3.3 [28] and FENDL-3.1d [29] neutron cross-section libraries were used in the simulations. The CAD models were produced or edited with CATIA V5 [30] and simplified for conversion to the MCNP input format using ANSYS SpaceClaim 2021R2 [31]. Conversion was carried out with SuperMC 3.3.0 [32,33] and/or McCad v0.5 [34], depending on the requirements of the conversion, and the MCNP simulations were run in the MARCONI-FUSION high-performance computing cluster [35].

The results were processed using Mathematica 13.0 [36], Python 3.9 [37] and Paraview 5.9 [38]. Neutron fluxes, when presented, follow the VITAMIN-J 175 group structure [39].

2.2. DEMO Reference Models

The simulations are grouped into three sections, according to each integration approach:

- Inner-vessel diagnostics (excluding port diagnostics);
  - In-vessel magnetics sensors;
  - Diagnostics slim cassette (reflectometry);
- Ex-vessel diagnostics;
  - Faraday sensors;
- Equatorial port diagnostics;
  - Spectroscopy diagnostics;
  - Neutron/gamma cameras;
  - Radiated power and soft X-ray intensity.

The first group includes the in-vessel diagnostics that are not integrated in the ports. Although the divertor thermocurrent diagnostic was not simulated, the results obtained for magnetics sensors below the divertor allow a first estimate of the fluxes and loads in this region, with the important caveat that the integration studies for the divertor in DEMO are still in a very preliminary phase, with ongoing studies and experiments to define the best configurations and strategies to deal with the very high thermal loads [3,40]. Similarly, since the ECE diagnostic is expected to be integrated in a DSC, the results obtained for reflectometry are representative as a first estimation for ECE components. The second group contains simulations for Faraday sensors, which also allow for a first estimate of the fluxes in the ex-vessel magnetics sensors, to be studied in more detail at a later stage. The third group contains several equatorial port diagnostics. Combined with the results presented in reference [41], it provides a broad perspective for most of the diagnostics projected for the ports, with the exception of IR polarimetry/interferometry, not yet simulated, and collective Thomson scattering, for which no design has been proposed for DEMO yet. Due to the lack of a consolidated design for the upper ports, including updated designs of the blanket pipe modules, shielding materials and other systems that will impact the design of diagnostics, no simulations are presented here for upper port systems, which are expected to contain additional lines of sight for the equatorial port diagnostics listed above. This shortcoming shall be addressed in future works.

The blanket configuration assumed in all the simulations was the water-cooled lithium lead (WCLL) BB. The alternative configuration, helium-cooled pebble bed (HCPB), has not been studied yet. Based on comparisons between the two blankets, it can be anticipated that the fluxes behind the blankets with the HCPB configuration would exceed those obtained here by up to one order of magnitude, or even more [1]. This would mostly impact the results obtained for diagnostics located in the inner and outer surfaces of the VV (magnetics and Faraday sensors). It would also imply a redesign of the DSC with helium cooling, with an obvious impact on the shielding and thermomechanical performance of the DSC.

2.2.1. In-Vessel Diagnostics

The MCNP reference model used for the first group of diagnostics, including the in-vessel diagnostics not integrated in the ports, is represented in Figure 2 [41]. A 22.5-degree model was used, corresponding to a full sector out of the 16 into which DEMO is divided. The blankets of this model were filled with a mixture representative of the WCLL BB, composed of tungsten, EUROFER, water and PbLi [17]. Using a homogenized material in the blankets reduces the complexity of the model, improving the simulation time and allowing for a reasonable first estimate of the fluxes and loads behind the blankets.

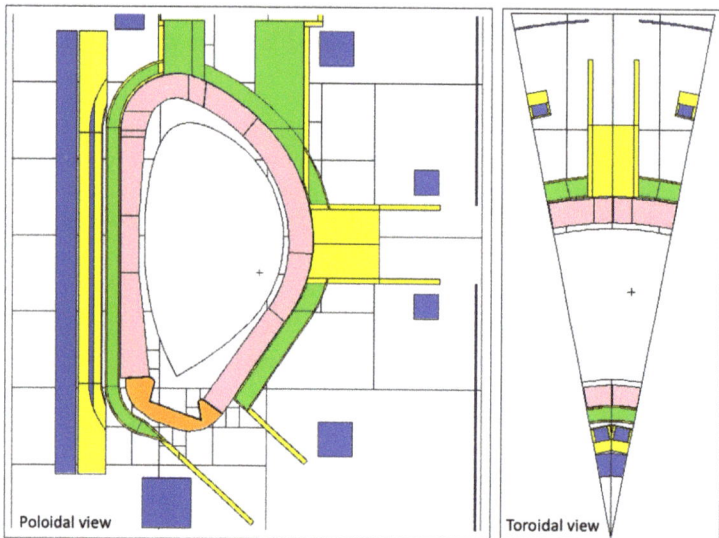

Figure 2. DEMO MCNP reference model used for the simulation of in-vessel diagnostics (excluding port diagnostics). **Left**: Plane y = 10.5 cm. **Right**: Plane z = 0.

2.2.2. Ex-Vessel Diagnostics

The reference model used for ex-vessel diagnostics (Faraday sensors) is represented in Figure 3 [42]. This is a smaller model when compared to the previous one (11.25° instead of 22.5°), containing a semi-heterogeneous representation of the WCLL blanket which provides good accuracy for ex-vessel simulations while decreasing the simulation times [43].

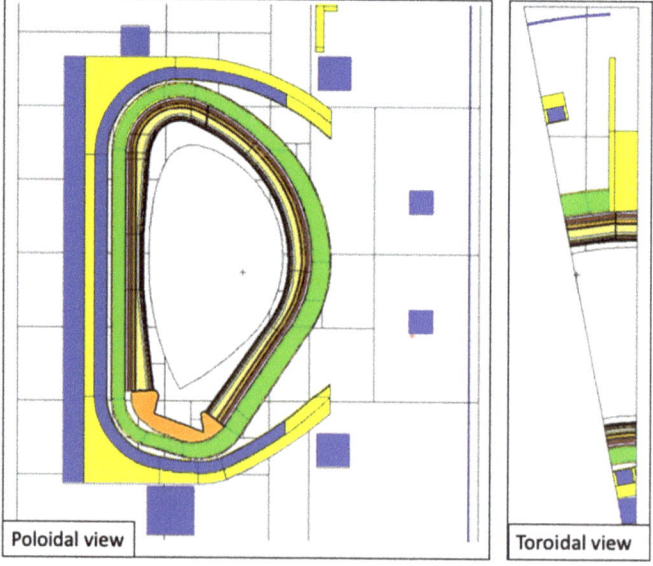

Figure 3. Reference model used in the simulations of the ex-vessel Faraday sensors.

2.2.3. Equatorial Port Diagnostics

The reference MCNP model used for equatorial port diagnostics is illustrated in Figure 4 [44]. The model features an upper port, equatorial limiter port and lower pumping port and includes layered representations of two HCPB BBs and one layered representation of the WCLL BB, as described with more detail in reference [45]. For these studies, the WCLL option was chosen for the BB (MCNP universe u = 882). All the remaining geometry definitions were kept unchanged except for the equatorial port (universe u = 210 in the MCNP model) and part of the bioshield and cryostat (u = 900), which were adapted to include the models created for the equatorial port, defined in Section 3.3.

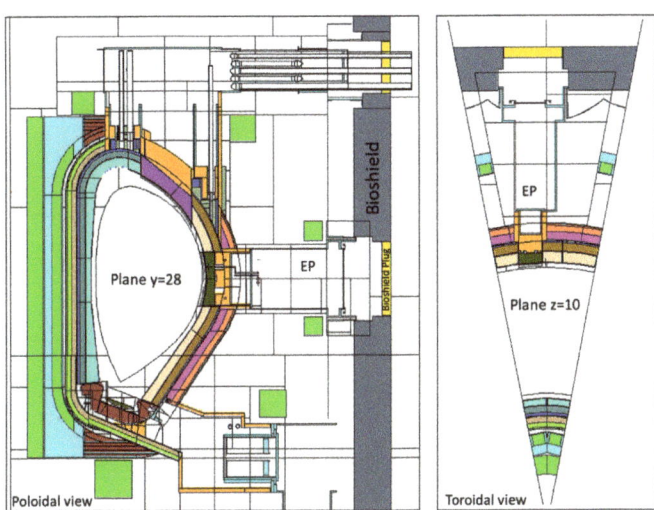

Figure 4. MCNP reference model used in the simulations of equatorial port diagnostics. **Left**: Plane y = 28 cm. **Right**: Plane z = 10 cm.

The equatorial port of the reference model is shown with more detail in Figure 5. The bioshield is made of concrete and has a thickness of 2 m, while the bioshield plug is a 50 cm thick slab of heavy concrete with a density of 3.6 g/cm^3. As there were no CAD models available with details of the DEMO bioshield plugs at the start of this work, the design available in the reference neutronics model was used, with the addition of the diagnostic duct openings in the bioshield plug.

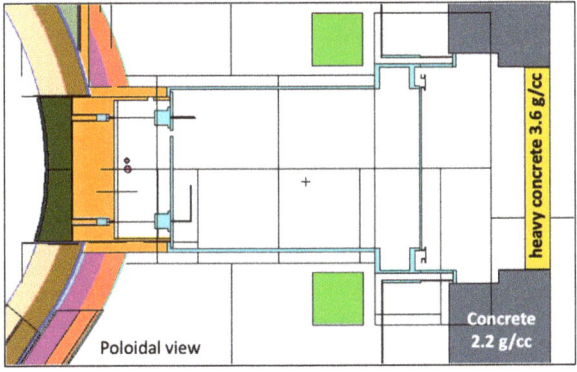

Figure 5. Detail of the MCNP reference model used in the simulations of equatorial port diagnostics, showing the EP and the bioshield plug.

2.3. Weight Window Generation

Common to all cases studied in this paper was the use of weight windows for variance reduction, to reduce the statistical errors of the simulation results. This is a crucial point when the aim is to calculate fluxes or loads at large distances from the plasma, as is the case with most simulations presented here. Two examples are provided in Figure 6. On the left, the weight windows were tuned to calculate the fluxes and loads in the Faraday sensors, using the reference model of Figure 3. In this case, the weight window generator of MCNP was used, after multiplying the density of all materials in the model by 1/10. In subsequent simulations, the material densities were progressively increased (1/5, 1/2 and finally 1), optimizing the weight windows at each step. This allowed us to bias the simulations towards the outer surface of the VV (from red to blue), where the sensors are located. For gammas, this weight window mesh was duplicated with the iWW-GVR [46] code and multiplied by 0.1.

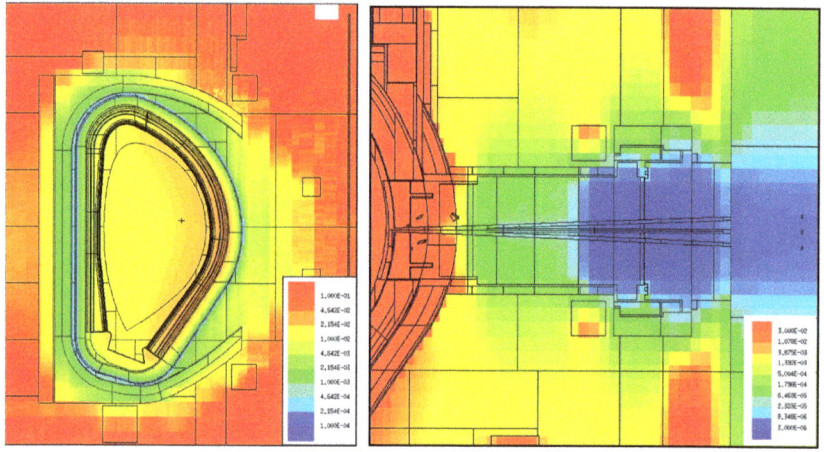

Figure 6. Weight windows used in the simulations. **Left**: Faraday sensors, installed on the outer layer of the VV. **Right**: EP diagnostics.

Another example is provided on the right side of Figure 6, for equatorial port diagnostics (reference model of Figure 4). In this case the aim was to obtain fluxes in the port cell, more than 12 m away from the plasma. The weight windows were generated with the ADVANTG code [47] and further manipulated with the iWW-GVR tool. They were tuned, in each simulation, to bias the propagation of neutrons and gammas towards the bioshield and the mirrors in the port cell.

3. Results

3.1. Inner-Vessel Diagnostics

The main objective of this analysis was to estimate the heat loads and the dose rates in the positions where magnetics sensors are expected to be placed, within the inner surface of the VV. Such an estimation, although preliminary, is important to assess the requirements for magnetics sensors. Since one possibility for the integration of magnetics sensors would be to attach them to the back of a blanket or even to a DSC, a comparison is made in this section between results obtained on the back of the blanket and on the back of a DSC. This comparison allows us to simultaneously assess the effect that the introduction of the DSC, the proposed integration approach for reflectometry and ECE in DEMO, would have on the fluxes and loads in the VV. What follows is a first study of neutron fluxes, nuclear heat loads, dose rates and dpa around the plasma with focus on the 60 poloidal locations where magnetics sensors are expected to be installed.

3.1.1. MCNP Models

The model of the DSC is presented in Figures 7 and 8, integrated in the reference model of Figure 2. As described in more detail in [16], the DSC has a thickness of 25 cm in the toroidal direction and the same poloidal shape as the blankets, ~12 m in height and approximately 52 tons in weight (similar density to the WCLL blankets). A homogenized mixture of EUROFER and water was assumed for the DSC, with volume fractions of 83.6% and 16.4%, respectively, representative of a small module with a cooling system studied in a previous work [24]. This water volume fraction is similar to the one in the blanket (15.9%). The MW antennas and waveguides were kept in the geometry, to provide a more conservative estimation and because they have been shown to have a small impact on the neutron fluxes and nuclear loads in the VV.

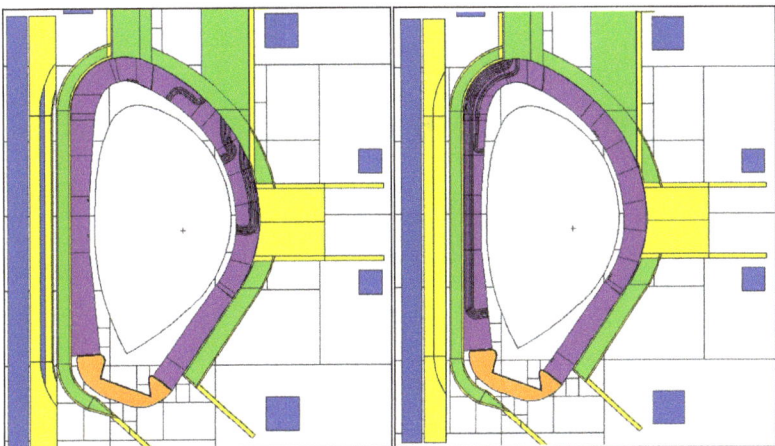

Figure 7. MCNP model of the DSC, containing the antennas and WGs designed for reflectometry. **Left**: Plane y = 10.5 cm. **Right**: Plane y = 6 cm.

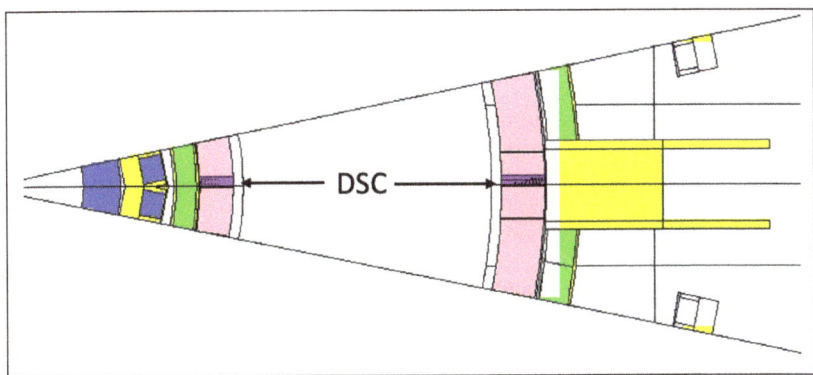

Figure 8. MCNP model of the DSC (plane z = 0).

For the first simulations, three tally cells (spheres with 1.5 cm diameters) were created at three toroidal positions (y = 0, y = 10.5 cm and y = 45 cm) in plane z = 0, as illustrated on the left side of Figure 9. These three positions correspond to the point between blanket modules, the middle of the DSC and the middle of the blanket, and were used to evaluate how the distance to the 2 cm gap between blanket segments affects the neutron fluxes and nuclear loads in the magnetics sensors. Afterwards, the fluxes, heat loads, dose rates and dpa in the blankets and DSC were estimated using FMESH tallies with multiplication factors. Using Paraview, the values for each of these quantities were obtained at the 60 poloidal positions where the magnetics sensors are expected to be placed, for the two studied configurations: with the DSC (sensors located behind the DSC) and without the DSC but considering a BB with the full (toroidal) width and having the sensors in the same positions (sensors located behind the blanket). These positions are illustrated on the right side of Figure 9.

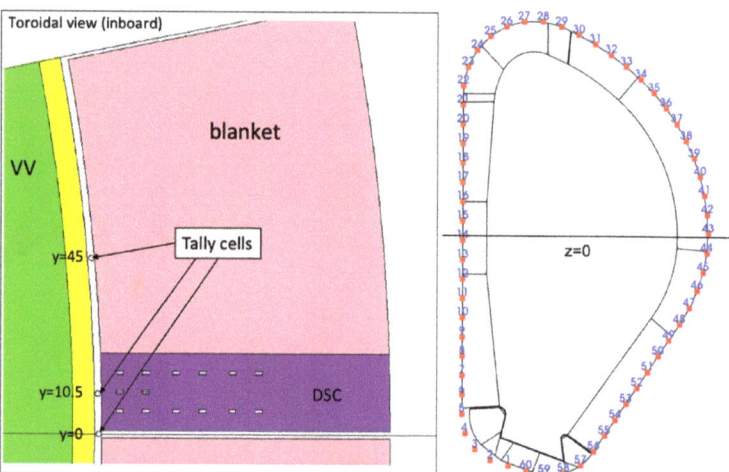

Figure 9. *Left*: Detail of the DSC, highlighting the cells used to tally fluxes, heat loads and dose rates (plane z = 0). *Right*: The 60 poloidal positions around the plasma foreseen for magnetics sensors.

The heat loads and dose rates were estimated for the following candidate materials:
- Alumina (Al_2O_3)—3.95 g cm^{-3};
- DuPont-951 ceramic (43%at. Al, 31%at. Si, 20%at. O, 6%at. Ca)—3.1 g cm^{-3};
- Aluminum nitride (AlN)—3.26 g cm^{-3};
- Magnesia (MgO)—3.58 g cm^{-3};
- Silicon dioxide (SiO_2)—2.65 g cm^{-3};
- Silicon nitride (Si_3N_4)—3.17 g cm^{-3}.

3.1.2. Fluxes, Nuclear Heat Loads and Dose Rates in the Tally Cells

The neutron fluxes, gamma fluxes, nuclear heat loads and dose rates at the three toroidal locations presented in Figure 9 (left) are summarized in Table 1, for the two studied cases (with and without DSC, in which case the DSC portion in Figure 9 is replaced by the blanket). As expected, the neutron and gamma fluxes at the center of the 2 cm gap between BB modules are three to four times higher than the fluxes behind the DSC or the blanket (at y = 10.5 cm). In the middle of the blanket (y = 45 cm), the neutron fluxes are 65% (with DSC) to 74% (without DSC) lower than at y = 10.5 cm. Without the DSC, the gamma fluxes at y = 10.5 cm are a factor of 2.6 higher than at y = 45 cm, due to the increased production of gammas by uncollided neutrons at y = 0 (between blankets). With the DSC, this factor increases to 6.4, due to the increased production of gammas in the DSC. Overall, between the three locations and the two configurations, the neutron fluxes vary between 7×10^{11} and 5×10^{12} n cm^{-2} s^{-1} and the gamma fluxes vary between 7×10^{10} and 1.4×10^{11} γ cm^{-2} s^{-1}.

Table 1. Neutron fluxes, nuclear heat loads and dose rates at 3 toroidal positions in the equatorial plane IB. For the flux calculations, the statistical errors are also presented. For the heat loads and dose rates, the neutron contribution to the total is included (remaining contribution by gammas).

Position			Between BBs (y = 0)	Behind DSC/BB (y = 10.5 cm)	Behind WCLL BB (y = 45 cm)
with DSC	Neutron flux (n cm^{-2} s^{-1})		4.56×10^{12} (1.8%)	1.22×10^{12} (3.6%)	7.38×10^{11} (4.7%)
	Gamma Flux (γ cm^{-2} s^{-1})		1.35×10^{12} (3.1%)	4.72×10^{11} (5.3%)	7.34×10^{10} (13.9%)
	Nuclear heat load (mW cm^{-3})	Al$_2$O$_3$	82.84 (66%n)	11.60 (17%n)	3.14 (44%n)
		DP-951	65.79 (66%n)	8.86 (12%n)	2.17 (35%n)
		AlN	73.96 (69%n)	9.96 (21%n)	2.66 (46%n)
		MgO	82.86 (69%n)	10.89 (20%n)	3.04 (47%n)
		SiO$_2$	62.55 (70%n)	8.13 (21%n)	2.33 (49%n)
		Si$_3$N$_4$	90.12 (70%n)	10.16 (23%n)	2.75 (47%n)
	Dose rate (Gy s^{-1})	Al$_2$O$_3$	20.97 (66%n)	2.94 (17%n)	0.79 (44%n)
		DP-951	21.22 (66%n)	2.86 (12%n)	0.70 (35%n)
		AlN	22.69 (69%n)	3.05 (21%n)	0.82 (46%n)
		MgO	23.15 (69%n)	3.04 (20%n)	0.85 (47%n)
		SiO$_2$	23.62 (70%n)	3.07 (21%n)	0.88 (49%n)
		Si$_3$N$_4$	28.43 (70%n)	3.20 (20%n)	0.87 (47%n)
without DSC	Neutron flux (n cm^{-2} s^{-1})		4.96×10^{12} (1.8%)	1.41×10^{12} (3.4%)	8.12×10^{11} (4.5%)
	Gamma Flux (γ cm^{-2} s^{-1})		8.15×10^{11} (3.9%)	1.98×10^{11} (8.4%)	7.68×10^{10} (13.4%)
	Nuclear heat load (mW cm^{-3})	Al$_2$O$_3$	76.97 (77%n)	7.43 (43%n)	3.41 (48%n)
		DP-951	60.47 (77%n)	5.24 (35%n)	2.30 (38%n)
		AlN	68.87 (80%n)	6.26 (45%n)	2.88 (50%n)
		MgO	77.99 (80%n)	7.28 (47%n)	3.31 (51%n)
		SiO$_2$	59.24 (80%n)	5.54 (49%n)	2.55 (53%n)
		Si$_3$N$_4$	86.10 (84%n)	6.57 (47%n)	2.98 (51%n)
	Dose rate (Gy s^{-1})	Al$_2$O$_3$	19.49 (77%n)	1.88 (43%n)	0.86 (48%n)
		DP-951	19.51 (77%n)	1.69 (35%n)	0.74 (38%n)
		AlN	21.13 (80%n)	1.92 (45%n)	0.88 (50%n)
		MgO	21.79 (80%n)	2.03 (47%n)	0.92 (51%n)
		SiO$_2$	22.37 (80%n)	2.09 (49%n)	0.96 (53%n)
		Si$_3$N$_4$	27.16 (84%n)	2.07 (47%n)	0.94 (51%n)

Comparing the two configurations, the heat loads and dose rates are very similar when the sensors are located between the blanket modules, as neutrons are the main contributors to the total heat loads and dose rates in that position and the neutron fluxes are similar between the two cases. Behind the DSC/BB (y = 10.5 cm) the heat loads and dose rates are 47% to 69% higher in the DSC configuration, depending on the material. In the position behind the middle of the BBs (y = 45 cm) the heat loads and dose rates are 5% to 9% higher in the case without DSC.

Comparing the three toroidal positions, the loads are much higher between blankets (y = 0) and much lower behind the middle of the blanket (y = 45 cm). Between the blankets the heat loads vary between 60 and 90 mW cm^{-3}, while behind the DSC/blanket (at y = 10.5 cm) they vary between 5 and 12 mW cm^{-3}. At y = 45 cm the heat loads vary from 2 to 4 mW cm^{-3}. Similarly, the dose rates vary between 20 and 30 Gy s^{-1} between blankets, 2 and 3 Gy s^{-1} behind the DSC/blanket (y = 10.5 cm) and 0.7 and 1 Gy s^{-1} at y = 45 cm.

As for the heat loads between different materials, DP-951 and SiO$_2$ are the ones with the lower values in general, although the differences are less pronounced behind the blankets. The maximum deviation between the values obtained with each material is ~30%, which is expected given the similar densities of the materials and the comparable atomic weights of their constituent elements. The contribution by neutrons to the loads and dose rates is dominant between the blankets (y = 0), while at y = 45 cm the contributions from neutrons and gammas are similar. At y = 10.5 cm, gammas play the most important role, especially in the case of the DSC, in which they contribute more than 80% of the total heat loads.

To finalize the comparison, the neutron flux spectra for the DSC case are presented in Figure 10. Although the statistical errors are large in most bins, and therefore unsuitable to be used as input in inventory or activation analyses, the ones related to the total flux are 1.8%, 3.6% and 4.7%, for the cases between blankets, behind the DSC and behind the blanket, respectively. The 14 MeV peak of uncollided neutrons is evident for the tally cell between the blankets (y = 0), vanishing completely in the remaining positions. For the case without the DSC the spectra are very similar and were not included in the plot.

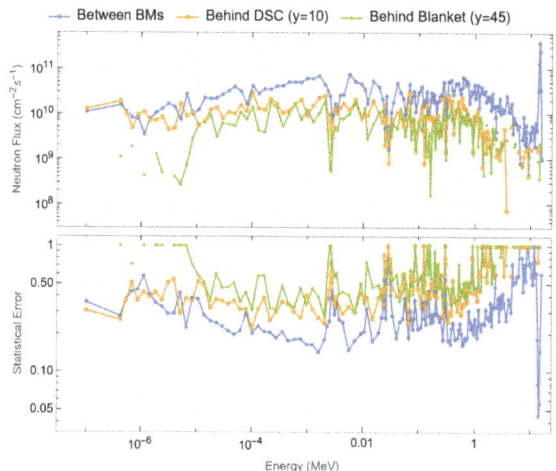

Figure 10. Neutron flux spectra (n cm^{-2} s^{-1}) and statistical error fraction behind the DSC and WCLL blanket.

3.1.3. Neutron and Gamma Fluxes in the Sensors

To assess the variation of the previous quantities with the poloidal location, FMESH tallies were defined for a thin slice between y = 10 cm and y = 11 cm. As shown in Figure 11, the mesh elements are very small (1.7 cm × 1 cm × 1.7 cm), to provide an accurate estimate of the fluxes and loads at each of the 60 positions described in Figure 9. Nevertheless, small deviations between the results presented in Table 1 and the ones obtained with the FMESH tallies are always to be expected, as the elements of the FMESH tallies have a different volume compared to the F4 tally cells. To keep the statistical errors as low as possible with such small elements, 5×10^{10} particles were run in each simulation.

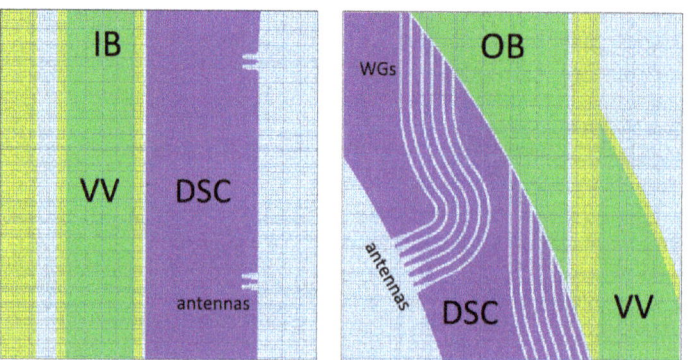

Figure 11. Detail of the meshes used in the simulations.

The neutron fluxes and statistical errors obtained for the configuration without the DSC are presented in Figure 12. As expected, the fluxes in the first wall are of the order of 3–4 × 10^{14} n cm^{-2} s^{-1}, while behind the blanket they vary between ~1 × 10^{11} and 1 × 10^{12} n cm^{-2} s^{-1}, depending on the position. In general, the statistical errors behind the DSC are less than 10%.

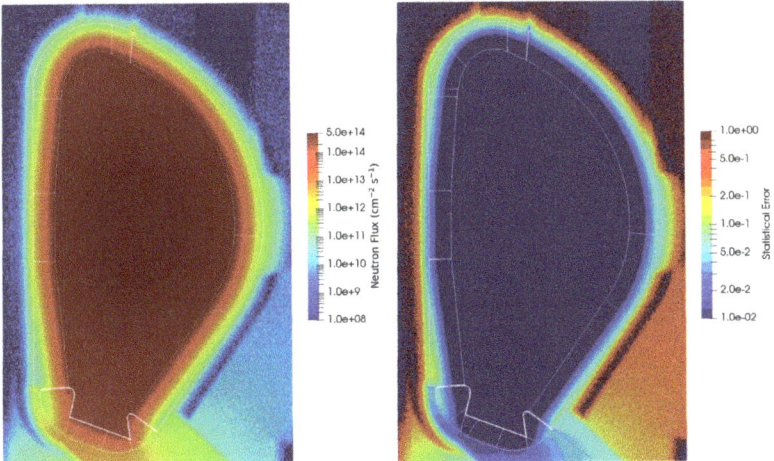

Figure 12. Neutron fluxes (n cm^{-2} s^{-1}) and statistical errors behind the WCLL blanket in plane y = 10.5 cm.

Figure 13 shows the values of the neutron fluxes at the 60 poloidal positions foreseen for magnetics sensors. Values vary from 1.1 × 10^{11} n cm^{-2} s^{-1} in position 6 (see Figure 9) to 5.2 × 10^{12} n cm^{-2} s^{-1} below the divertor (more than one order of magnitude of variation). In the equatorial plane, at position 14 (z = −13.5 cm), the neutron flux is 1.4 × 10^{12} n cm^{-2} s^{-1}, which agrees with the value provided in Table 1. The largest statistical error is 10.1%.

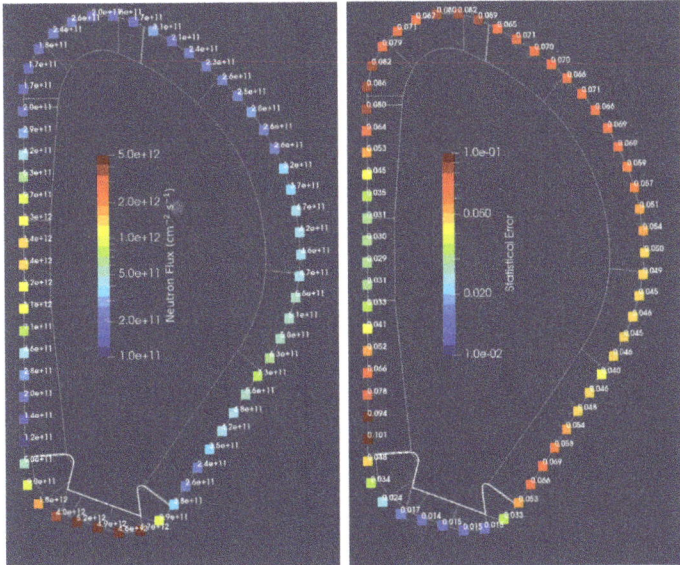

Figure 13. Neutron fluxes (n cm^{-2} s^{-1}) and statistical errors behind the WCLL blanket at 60 poloidal locations in plane y = 10.5 cm.

For comparison between the two cases, the ratios between the results obtained with the DSC and without the DSC are presented in Figure 14. The neutron fluxes with the DSC are slightly higher close to the first wall, but lower at almost all positions behind the DSC/blanket. The main conclusion is that the neutron fluxes are very close between configurations, with differences below 20% at most positions.

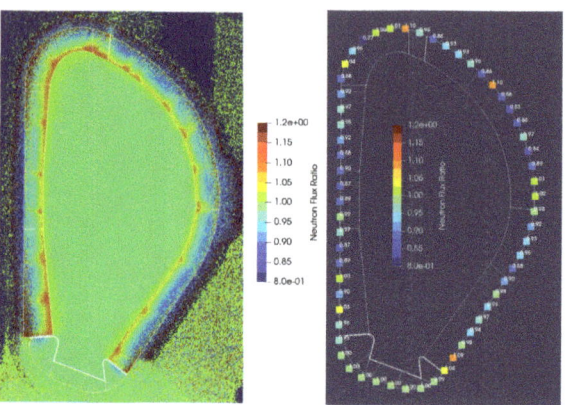

Figure 14. **Left**: Neutron flux ratios (with DSC/without DSC) in plane y = 10.5 cm. **Right**: Values at 60 poloidal locations.

The gamma fluxes behind the blanket are plotted Figure 15. They reach 2×10^{11} γ cm^{-2} s^{-1} at the equatorial plane (three orders of magnitude lower than at the first wall), while in the divertor region they reach 7×10^{12} γ cm^{-2} s^{-1}. These simulations were run with larger mesh elements (5 cm × 1 cm × 5 cm), to reduce the statistical errors of the simulations, which exceeded 10% in some positions with the initial mesh. Similar values were obtained between the two simulations (smaller and larger mesh elements), with statistical errors $\leq$ 10% for the larger elements.

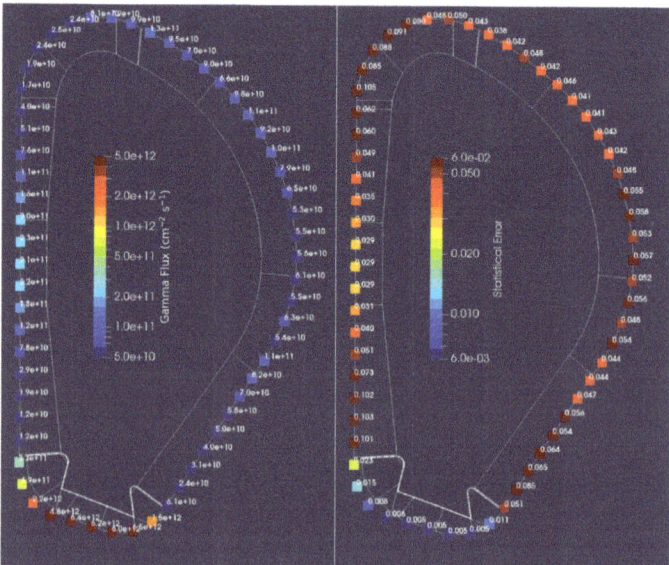

Figure 15. Gamma fluxes (γ cm^{-2} s^{-1}) and statistical errors behind the WCLL blanket at 60 poloidal locations.

The ratios between the gamma fluxes with and without the DSC are presented in Figure 16. Except for the divertor region—where the ratios are 1—the gamma fluxes increase with the DSC by a factor of 2–3. This is because more gammas are produced in the DSC than in the BB, due to the higher radiative capture cross-sections of iron and chromium (the main constituents of EUROFER) when compared to the radiative capture cross-section of lead (from the WCLL BB).

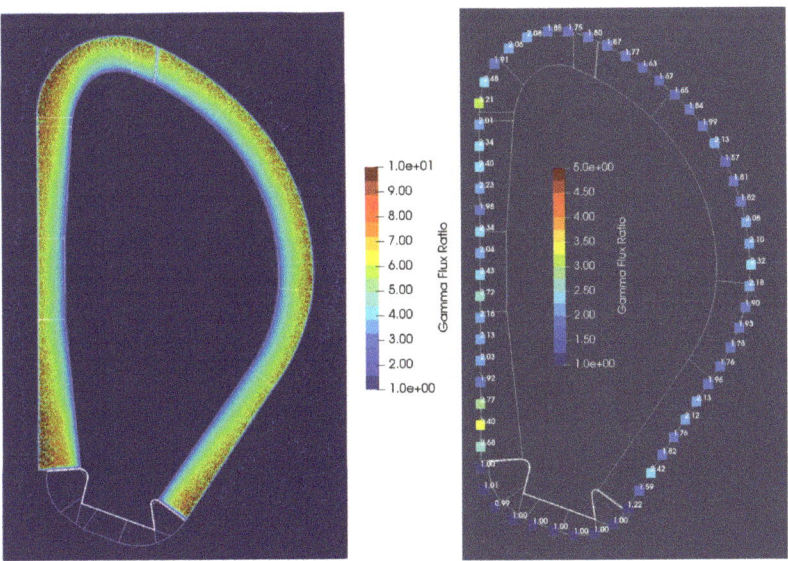

Figure 16. Left: Gamma flux ratios (with DSC/without DSC) in plane y = 10.5 cm. **Right**: Values at 60 poloidal locations.

Since the results of the next sections (nuclear heat loads, dose rates and dpa) were obtained as neutron and gamma fluxes multiplied by conversion factors, the statistical uncertainties presented up to now were considered acceptable and are omitted in the remaining results.

3.1.4. Displacements per Atom in the Vacuum Vessel

The dpa values in SS-316 (inner-vessel surface) were also calculated for the two configurations, as shown in Figure 17. The results are normalized per full power year (FPY), with DEMO scheduled to operate over 20 calendar years at an average availability of 30%, which results in a plant lifetime of 6 FPY (1.57 FPY in the first operation phase and 4.43 FPY in the second operation phase) [1]. The dpa values are very small behind either the blanket or the DSC, below 0.01 dpa/FPY in any position (except for the divertor region, where values are much larger). When both configurations are compared, the dpa values are smaller with the DSC, up to a factor of 2. This result indicates once again that the introduction of the DSC, with the current design, does not compromise the integrity of the VV.

3.1.5. Nuclear Heat Loads and Dose Rates

The final step of this analysis consisted in the estimation of the nuclear heat loads and dose rates in the six candidate materials foreseen for magnetics sensors. The nuclear heat loads for the two configurations are shown in Table 2.

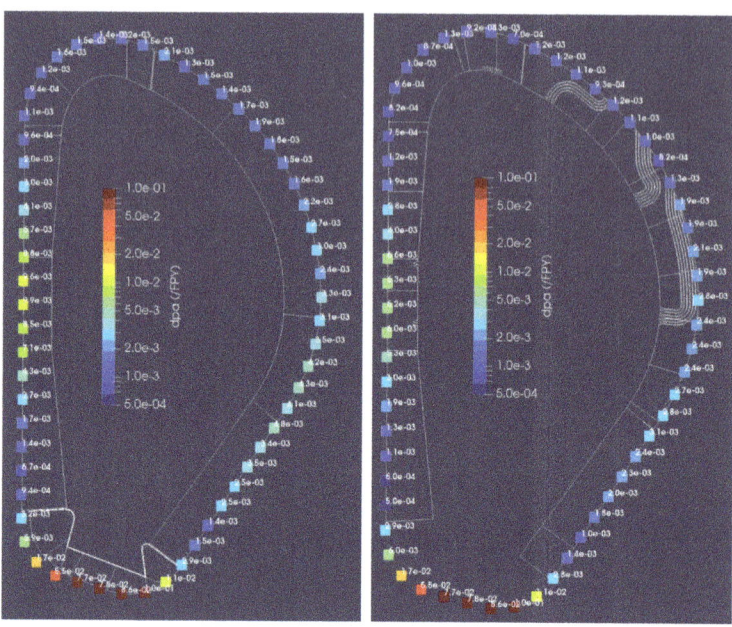

Figure 17. **Left**: dpa/FPY behind the WCLL blanket. **Right**: dpa/FPY behind the DSC.

Table 2. Total heat loads in six materials at 60 poloidal locations behind the WCLL blanket.

Pos	Total Heat Loads (mW cm^{-3})											
	with DSC						without DSC					
	Al$_2$O$_3$	DP-951	AlN	MgO	SiO$_2$	Si$_3$N$_4$	Al$_2$O$_3$	DP-951	AlN	MgO	SiO$_2$	Si$_3$N$_4$
1	165.4	131.0	143.6	157.1	116.6	152.6	165.8	131.3	143.9	157.5	116.9	152.9	
2	126.5	100.1	109.9	119.5	88.7	115.7	126.8	100.4	110.2	119.9	88.9	116.0	
3	49.12	38.73	42.43	45.89	34.02	43.55	48.80	38.45	42.16	45.61	33.81	43.27	Divertor
4	16.91	13.26	14.52	15.65	11.62	14.74	16.73	13.13	14.40	15.48	11.49	14.61	
5	9.03	7.09	7.67	8.40	6.23	7.78	8.73	6.83	7.42	8.15	6.04	7.53	
6	0.89	0.68	0.78	0.83	0.62	0.80	0.75	0.52	0.62	0.71	0.54	0.64	
7	1.27	0.98	1.10	1.18	0.88	1.11	0.57	0.43	0.49	0.54	0.41	0.50	
8	1.77	1.36	1.54	1.66	1.23	1.56	1.16	0.85	1.00	1.13	0.85	1.04	
9	1.84	1.38	1.61	1.74	1.31	1.65	1.65	1.19	1.36	1.60	1.21	1.43	
10	3.19	2.41	2.74	2.98	2.23	2.78	2.23	1.57	1.86	2.14	1.63	1.91	
11	5.13	3.87	4.36	4.80	3.60	4.42	3.37	2.34	2.81	3.28	2.50	2.89	
12	8.02	6.02	6.77	7.59	5.70	6.98	5.50	3.85	4.60	5.29	4.04	4.75	
13	10.66	8.14	9.08	10.05	7.50	9.32	5.95	4.10	4.95	5.77	4.44	5.21	
14	10.28	7.79	8.72	9.67	7.24	8.89	6.71	4.66	5.71	6.56	5.02	6.01	
15	9.49	7.17	8.21	8.96	6.71	8.40	7.09	4.97	6.09	6.92	5.27	6.32	Eq. plane IB
16	10.22	7.69	8.79	9.60	7.21	9.02	6.68	4.70	5.65	6.43	4.93	5.95	
17	6.52	4.95	5.57	6.10	4.57	5.68	4.96	3.42	4.13	4.77	3.66	4.24	
18	4.28	3.23	3.64	4.02	3.01	3.68	3.10	2.15	2.60	3.00	2.31	2.74	
19	3.80	2.88	3.21	3.55	2.65	3.24	2.35	1.64	1.94	2.33	1.78	2.08	
20	2.07	1.56	1.76	1.92	1.44	1.77	1.39	0.96	1.15	1.38	1.05	1.21	
21	1.28	0.96	1.07	1.20	0.90	1.08	0.71	0.50	0.63	0.69	0.53	0.63	
22	1.17	0.87	0.98	1.09	0.83	1.00	0.69	0.46	0.59	0.66	0.51	0.62	
23	1.37	1.01	1.11	1.30	0.98	1.15	0.75	0.52	0.63	0.71	0.55	0.64	
24	1.79	1.37	1.50	1.69	1.26	1.55	1.12	0.83	0.95	1.09	0.82	1.02	
25	1.59	1.21	1.36	1.49	1.11	1.35	1.21	0.85	1.03	1.19	0.90	1.05	
26	1.51	1.09	1.27	1.42	1.08	1.29	1.05	0.72	0.88	1.02	0.78	0.88	
27	2.90	2.28	2.52	2.68	1.98	2.55	1.99	1.51	1.74	1.87	1.40	1.78	
28	2.59	2.02	2.30	2.43	1.81	2.38	1.52	1.16	1.37	1.46	1.09	1.42	

Table 2. Cont.

Pos	Total Heat Loads (mW cm^{-3})											
	with DSC						without DSC					
	Al$_2$O$_3$	DP-951	AlN	MgO	SiO$_2$	Si$_3$N$_4$	Al$_2$O$_3$	DP-951	AlN	MgO	SiO$_2$	Si$_3$N$_4$
29	2.71	2.15	2.42	2.50	1.85	2.46	2.34	1.82	2.08	2.21	1.64	2.18
30	4.98	3.96	4.37	4.58	3.38	4.44	2.82	2.16	2.49	2.64	1.98	2.55
31	3.19	2.50	2.76	2.96	2.20	2.84	2.54	1.96	2.16	2.38	1.78	2.25
32	3.42	2.71	3.00	3.14	2.33	3.06	2.57	1.97	2.21	2.39	1.79	2.27
33	3.84	3.01	3.28	3.52	2.61	3.30	2.65	2.05	2.25	2.47	1.85	2.33
34	3.69	2.90	3.29	3.43	2.54	3.35	2.65	2.03	2.38	2.51	1.87	2.46
35	3.39	2.64	2.95	3.13	2.33	2.99	2.31	1.75	2.02	2.18	1.63	2.05
36	3.21	2.52	2.81	2.97	2.20	2.84	2.36	1.79	2.04	2.22	1.66	2.10
37	3.51	2.78	3.09	3.23	2.39	3.14	2.63	2.30	2.29	2.43	1.77	2.39
38	3.88	3.04	3.37	3.61	2.68	3.44	2.67	2.02	2.32	2.52	1.88	2.36
39	3.22	2.47	2.88	3.05	2.28	2.97	2.15	1.58	1.90	2.08	1.57	1.99
40	3.02	2.27	2.53	2.85	2.14	2.59	2.28	1.64	1.95	2.21	1.67	2.02
41	3.14	2.35	2.62	2.96	2.23	2.67	2.31	1.59	1.91	2.25	1.72	1.98	
42	2.85	2.15	2.42	2.70	2.03	2.51	1.90	1.31	1.56	1.86	1.41	1.62	
43	3.18	2.35	2.70	3.07	2.31	2.81	2.12	1.47	1.81	2.16	1.64	1.98	
44	2.95	2.16	2.50	2.76	2.09	2.53	2.09	1.40	1.72	2.04	1.57	1.76	Eq. plane OB
45	3.11	2.30	2.63	2.93	2.21	2.67	2.18	1.44	1.80	2.14	1.65	1.88	
46	3.69	2.80	3.17	3.48	2.61	3.27	3.23	2.28	2.74	3.14	2.41	2.94	
47	4.53	3.42	3.87	4.22	3.16	3.91	3.22	2.24	2.75	3.10	2.38	2.87	
48	5.14	3.91	4.36	4.77	3.57	4.44	3.84	2.76	3.25	3.62	2.76	3.33
49	5.50	4.19	4.65	5.13	3.84	4.75	3.52	2.49	3.05	3.40	2.60	3.23
50	4.63	3.53	3.96	4.30	3.21	4.01	2.95	2.14	2.48	2.86	2.17	2.59
51	3.35	2.50	2.89	3.11	2.34	2.92	2.57	1.83	2.13	2.47	1.89	2.26
52	3.15	2.38	2.70	2.97	2.22	2.75	2.04	1.45	1.73	1.97	1.50	1.79
53	3.06	2.33	2.67	2.85	2.14	2.73	1.75	1.21	1.51	1.68	1.29	1.53
54	1.72	1.29	1.45	1.59	1.20	1.45	1.06	0.75	0.89	1.03	0.78	0.91
55	2.17	1.64	1.82	2.02	1.52	1.84	1.39	1.01	1.16	1.32	1.00	1.17
56	3.70	2.77	3.19	3.49	2.63	3.28	3.20	2.39	2.70	3.07	2.32	2.87
57	34.54	27.33	29.48	32.30	23.94	30.51	34.19	27.04	29.14	31.96	23.70	30.16	
58	193.9	154.6	169.0	186.2	138.3	184.0	192.9	153.7	168.2	185.3	137.6	183.1	Divertor
59	170.4	135.2	148.7	163.1	121.0	159.6	170.1	134.9	148.4	162.7	120.8	159.2	
60	167.9	133.2	145.6	159.6	118.4	154.6	168.1	133.4	145.8	159.9	118.6	154.8	

As seen before, DP-951 and SiO$_2$ are the materials with lower values in general, although the variations between materials are mostly within 40%. As an example, the heat loads at the equatorial plane inboard in the configuration without the DSC vary from 5.3 mW cm^{-3} in SiO$_2$ to 7.1 mW cm^{-3} in Al$_2$O$_3$ (a variation of ~30%). Apart from the divertor region, the highest heat loads are at the equatorial plane inboard (a factor ~3 higher than at the outboard).

On average, excluding the points in the divertor region, the heat loads in the configuration with the DSC are 50% higher than the ones in the configuration without DSC. This is mostly due to the increased gamma fluxes coming from the DSC. The dose rates, obtained by dividing the nuclear heat loads (in mW cm^{-3}) by the material density, are very similar between materials, as shown in Table 3.

Table 3. Dose rates in six materials at 60 poloidal locations behind the WCLL blanket.

Pos	Dose Rate (Gy s^{-1})											
	with DSC						without DSC					
	Al$_2$O$_3$	DP-951	AlN	MgO	SiO$_2$	Si$_3$N$_4$	Al$_2$O$_3$	DP-951	AlN	MgO	SiO$_2$	Si$_3$N$_4$
1	41.87	42.26	44.04	43.88	44.01	48.13	41.97	42.36	44.14	43.98	44.11	48.23	
2	32.01	32.28	33.70	33.38	33.46	36.48	32.11	32.38	33.80	33.48	33.56	36.58	
3	12.43	12.49	13.01	12.82	12.84	13.74	12.35	12.40	12.93	12.74	12.76	13.65	Divertor
4	4.28	4.28	4.45	4.37	4.38	4.65	4.24	4.24	4.42	4.32	4.34	4.61	
5	2.29	2.29	2.35	2.35	2.35	2.45	2.21	2.20	2.28	2.28	2.28	2.37	

Table 3. Cont.

Pos	Dose Rate (Gy s^{-1})												
	with DSC						without DSC						
	Al$_2$O$_3$	DP-951	AlN	MgO	SiO$_2$	Si$_3$N$_4$	Al$_2$O$_3$	DP-951	AlN	MgO	SiO$_2$	Si$_3$N$_4$	
6	0.22	0.22	0.24	0.23	0.23	0.25	0.19	0.17	0.19	0.20	0.21	0.20	
7	0.32	0.32	0.34	0.33	0.33	0.35	0.14	0.14	0.15	0.15	0.15	0.16	
8	0.45	0.44	0.47	0.46	0.47	0.49	0.29	0.27	0.31	0.32	0.32	0.33	
9	0.47	0.44	0.49	0.49	0.49	0.52	0.42	0.38	0.42	0.45	0.46	0.45	
10	0.81	0.78	0.84	0.83	0.84	0.88	0.56	0.51	0.57	0.60	0.61	0.60	
11	1.30	1.25	1.34	1.34	1.36	1.39	0.85	0.76	0.86	0.92	0.94	0.91	
12	2.03	1.94	2.08	2.12	2.15	2.20	1.39	1.24	1.41	1.48	1.53	1.50	
13	2.70	2.63	2.79	2.81	2.83	2.94	1.51	1.32	1.52	1.61	1.67	1.64	
14	2.60	2.51	2.68	2.70	2.73	2.80	1.70	1.50	1.75	1.83	1.90	1.90	
15	2.40	2.31	2.52	2.50	2.53	2.65	1.79	1.60	1.87	1.93	1.99	1.99	Eq. plane
16	2.59	2.48	2.70	2.68	2.72	2.84	1.69	1.52	1.73	1.80	1.86	1.88	
17	1.65	1.60	1.71	1.70	1.73	1.79	1.25	1.10	1.27	1.33	1.38	1.34	
18	1.08	1.04	1.12	1.12	1.14	1.16	0.79	0.69	0.80	0.84	0.87	0.87	
19	0.96	0.93	0.99	0.99	1.00	1.02	0.60	0.53	0.59	0.65	0.67	0.66	
20	0.52	0.50	0.54	0.54	0.54	0.56	0.35	0.31	0.35	0.39	0.40	0.38	
21	0.32	0.31	0.33	0.34	0.34	0.34	0.18	0.16	0.19	0.19	0.20	0.20	
22	0.30	0.28	0.30	0.31	0.31	0.32	0.18	0.15	0.18	0.18	0.19	0.19	
23	0.35	0.33	0.34	0.36	0.37	0.36	0.19	0.17	0.19	0.20	0.21	0.20	
24	0.45	0.44	0.46	0.47	0.48	0.49	0.28	0.27	0.29	0.31	0.31	0.32	
25	0.40	0.39	0.42	0.42	0.42	0.43	0.31	0.27	0.32	0.33	0.34	0.33	
26	0.38	0.35	0.39	0.40	0.41	0.41	0.27	0.23	0.27	0.28	0.29	0.28	
27	0.73	0.73	0.77	0.75	0.75	0.80	0.50	0.49	0.54	0.52	0.53	0.56	
28	0.66	0.65	0.70	0.68	0.68	0.75	0.39	0.37	0.42	0.41	0.41	0.45	
29	0.69	0.69	0.74	0.70	0.70	0.77	0.59	0.59	0.64	0.62	0.62	0.69	
30	1.26	1.28	1.34	1.28	1.28	1.40	0.71	0.70	0.76	0.74	0.75	0.80	
31	0.81	0.81	0.85	0.83	0.83	0.90	0.64	0.63	0.66	0.66	0.67	0.71	
32	0.87	0.87	0.92	0.88	0.88	0.96	0.65	0.64	0.68	0.67	0.68	0.72	
33	0.97	0.97	1.01	0.98	0.98	1.04	0.67	0.66	0.69	0.69	0.70	0.73	
34	0.93	0.93	1.01	0.96	0.96	1.06	0.67	0.65	0.73	0.70	0.71	0.78	
35	0.86	0.85	0.90	0.87	0.88	0.94	0.59	0.56	0.62	0.61	0.61	0.65	
36	0.81	0.81	0.86	0.83	0.83	0.90	0.60	0.58	0.63	0.62	0.63	0.66	
37	0.89	0.90	0.95	0.90	0.90	0.99	0.67	0.74	0.70	0.68	0.67	0.75	
38	0.98	0.98	1.03	1.01	1.01	1.08	0.68	0.65	0.71	0.70	0.71	0.74	
39	0.82	0.80	0.88	0.85	0.86	0.94	0.54	0.51	0.58	0.58	0.59	0.63	
40	0.77	0.73	0.78	0.80	0.81	0.82	0.58	0.53	0.60	0.62	0.63	0.64	
41	0.79	0.76	0.80	0.83	0.84	0.84	0.58	0.51	0.59	0.63	0.65	0.62	
42	0.72	0.69	0.74	0.76	0.77	0.79	0.48	0.42	0.48	0.52	0.53	0.51	
43	0.81	0.76	0.83	0.86	0.87	0.89	0.54	0.48	0.56	0.60	0.62	0.63	
44	0.75	0.70	0.77	0.77	0.79	0.80	0.53	0.45	0.53	0.57	0.59	0.56	Eq. plane
45	0.79	0.74	0.81	0.82	0.83	0.84	0.55	0.47	0.55	0.60	0.62	0.59	
46	0.93	0.90	0.97	0.97	0.98	1.03	0.82	0.74	0.84	0.88	0.91	0.93	
47	1.15	1.10	1.19	1.18	1.19	1.23	0.81	0.72	0.84	0.86	0.90	0.91	
48	1.30	1.26	1.34	1.33	1.35	1.40	0.97	0.89	1.00	1.01	1.04	1.05	
49	1.39	1.35	1.43	1.43	1.45	1.50	0.89	0.80	0.94	0.95	0.98	1.02	
50	1.17	1.14	1.21	1.20	1.21	1.26	0.75	0.69	0.76	0.80	0.82	0.82	
51	0.85	0.81	0.89	0.87	0.88	0.92	0.65	0.59	0.65	0.69	0.71	0.71	
52	0.80	0.77	0.83	0.83	0.84	0.87	0.52	0.47	0.53	0.55	0.57	0.56	
53	0.77	0.75	0.82	0.80	0.81	0.86	0.44	0.39	0.46	0.47	0.49	0.48	
54	0.44	0.41	0.44	0.44	0.45	0.46	0.27	0.24	0.27	0.29	0.29	0.29	
55	0.55	0.53	0.56	0.57	0.57	0.58	0.35	0.32	0.36	0.37	0.38	0.37	
56	0.94	0.90	0.98	0.97	0.99	1.04	0.81	0.77	0.83	0.86	0.87	0.90	
57	8.74	8.82	9.04	9.02	9.03	9.62	8.66	8.72	8.94	8.93	8.94	9.51	
58	49.10	49.86	51.85	52.01	52.19	58.04	48.84	49.58	51.59	51.75	51.93	57.77	Divertor
59	43.13	43.60	45.60	45.55	45.67	50.33	43.05	43.51	45.52	45.45	45.57	50.23	
60	42.49	42.96	44.65	44.59	44.67	48.78	42.55	43.02	44.71	44.66	44.74	48.85	

The results presented in this section show that the DSC leads to an increase in the gamma fluxes and heat loads in the VV without compromising its integrity. The heat loads in the VV obtained with the DSC are well below the limit of 0.3 W/cm^3, while the dpa values, smaller than 0.01 dpa/FPY at any position, are lower than the ones obtained with the WCLL blanket.

For the magnetics sensors, the increase in the heat loads and dose rates by approximately 50% may have an impact on the integrity of the sensors. The absolute values of the loads and dose rates need to be evaluated by magnetics diagnostic development teams.

3.1.6. Neutron Fluence in Magnetics Sensors: Comparison with ITER

The neutron fluxes in the magnetics sensors presented in Figure 12 vary between 1.1×10^{11} n cm^{-2} s^{-1} and 1.4×10^{12} behind the blanket and are up to 5.2×10^{12} n cm^{-2} s^{-1} below the divertor. Table 4 presents these fluxes integrated over the DEMO operation phases (1.57 and 4.43 FPY), converted to n m^{-2} for comparison with ITER results. The neutron fluences in the ITER in-vessel magnetics sensors are expected to vary in the range 2.5×10^{24}–5×10^{24} n m^{-2} [48], with the fluences in the cable looms reaching up to 6.25×10^{24} n m^{-2} close to the upper port and 2.25×10^{24} n m^{-2} in the divertor [49]. The fluences behind the blankets presented in Table 4 for the whole DEMO lifetime are comparable to the ones expected for ITER, even though ITER will operate only for 0.54 FPY. This is due to the excellent shielding performance of the WCLL blanket. In the divertor region, the values obtained for DEMO are higher by a factor of 4, although the comparison here is not straightforward, since in the ITER simulations the sensors are installed inside the divertor cassette.

Table 4. Neutron fluxes and fluences in the magnetics sensors (behind the WCLL blanket).

	Neutron Flux (n m^{-2} s^{-1})	Neutron Fluence (n m^{-2})			
		1 FPY	1.57 FPY	4.43 FPY	6 FPY
Blanket (min)	1.10×10^{15}	3.47×10^{22}	5.45×10^{22}	1.54×10^{23}	2.08×10^{23}
Blanket (max)	1.40×10^{16}	4.42×10^{23}	6.94×10^{23}	1.96×10^{24}	2.65×10^{24}
Divertor	5.20×10^{16}	1.64×10^{24}	2.58×10^{24}	7.27×10^{24}	9.85×10^{24}

As mentioned before, with the HCPB blanket the fluences in the sensors would be increased up to a factor of 10 or more, which would increase the loads when compared to ITER. Furthermore, the WCLL blanket design changes every year, and a reduction of the blanket dimensions cannot be ruled out at this stage. It is also important to highlight that the present analysis is very preliminary, as it assumes homogeneous material compositions in the blanket and DSC. Therefore, it is important to repeat the analysis when the new models are available, if possible using a fully heterogeneous model of the WCLL blanket [50], with the required adaptations to provide space for the inclusion of magnetic sensors.

3.2. Ex-Vessel Diagnostics (Faraday Sensors)

As in ITER [51], the DEMO fiber optics current sensor (FOCS) diagnostic is planned to be installed on the outer surface of the VV, with the aim to provide information on the plasma current during long plasma discharges. To model this diagnostic, a thin layer was added close to the middle section of the VV in the reference model of Figure 3, as illustrated in Figure 18. This layer is a full poloidal segment, with 1 cm in the radial direction and 9.5 cm in the toroidal direction. It was used to calculate neutron and gamma fluxes at different poloidal positions, illustrated on the right side of Figure 18. The layer was first split at the equatorial level, and the remaining planes were obtained by rotating the preceding one by 5 degrees. In this way, 72 cells were added to the MCNP model, after conversion of this CAD model with SuperMC.

The only change made to the reference model was the addition of the 72 cells, which were filled with silica to calculate the heat loads in the optical fibers, with a density of 2.32 g/cm^3. The remaining modeling options were kept unchanged.

A preliminary simulation was run to determine the volumes and masses of each of the 72 cells, using a voided geometry. The statistical errors of the volumes were kept below 0.5%.

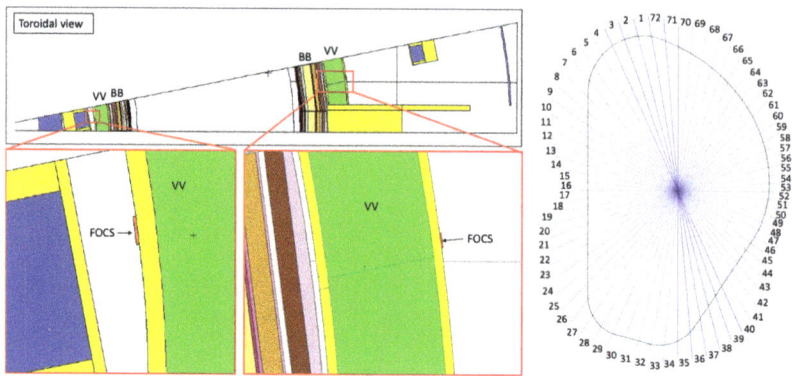

Figure 18. Left: Location of the FOCS in the MCNP model. **Right**: Poloidal positions used in the simulations to tally fluxes and energy deposition.

Fluxes, Heat Loads and Dose Rates in the Sensors

To obtain statistical errors below 10% in all cells, 4×10^{10} particles were simulated, using weight windows. The neutron and gamma fluxes in each cell, along with the nuclear heat loads and dose rates, are presented in Table 5. There are only two cases (gamma flux in position 54 and neutron flux in position 58) in which the statistical error was above 10%. The largest fluxes, heat loads and dose rates were obtained in the cells below the divertor, as expected from the previous simulations for magnetics sensors. Similar flux values were also obtained in a recent divertor study [52]. In position 32, the fluxes reach 1.31×10^{11} n/cm^2/s and 2.60×10^{10} γ/cm^2/s, while the nuclear heat load reaches 0.16 mW/cm^3 (29% by neutrons and 71% by gammas). The dose rate, calculated by dividing the nuclear heat load obtained with MCNP (1.63×10^{-4} W/cm^3) by the material density (2.32 g/cm^3) and multiplying it by 1000 (g/kg) $\times$ 3600 $\times$ 24 $\times$ 365.25 (s/FPY), reaches 2.2 MGy/FPY. Considering that the first DEMO operation phase corresponds to 1.57 FPY, and the second phase to 4.43 FPY, in the 6 FPY of DEMO lifetime that section of the FOCS would be exposed to 13.2 MGy (3.5 + 9.7). This value exceeds the 10 MGy considered as a conservative upper limit for the FOCS lifetime dose in ITER [53]. Nevertheless, this happens only in the divertor region, which is not modeled as accurately as the blanket in these simulations, and where the design is not well defined and the shielding has not been optimized. In the remaining regions, the dose rates are lower, by up to almost three orders of magnitude.

Table 5. Fluxes, heat loads and dose rates in the 72 positions of the FOCS (WCLL blanket).

Position	Neutron Flux (n/cm^2/s)	Stat. Error (%)	Gamma Flux (γ/cm^2/s)	Stat. Error (%)	Heat Load Total (W/cm^3)	n (%)	γ (%)	Dose Rate (Gy/FPY)
1	3.59×10^8	3.0	1.25×10^8	1.9	6.58×10^{-7}	18.8	81.2	8.94×10^3
2	2.92×10^8	3.2	9.93×10^7	1.9	5.27×10^{-7}	23.0	77.0	7.17×10^3
3	2.70×10^8	3.1	9.07×10^7	1.9	5.02×10^{-7}	23.2	76.8	6.82×10^3
4	2.32×10^8	3.1	8.31×10^7	2.0	4.57×10^{-7}	25.3	74.7	6.22×10^3
5	1.96×10^8	3.3	7.02×10^7	2.0	3.91×10^{-7}	24.6	75.4	5.32×10^3
6	1.79×10^8	3.3	7.51×10^7	1.9	3.96×10^{-7}	22.8	77.2	5.39×10^3
7	1.91×10^8	3.1	8.82×10^7	1.7	4.76×10^{-7}	22.4	77.6	6.48×10^3
8	2.74×10^8	2.7	1.38×10^8	1.4	7.05×10^{-7}	21.9	78.1	9.59×10^3
9	4.15×10^8	2.3	2.08×10^8	1.2	1.07×10^{-6}	22.0	78.0	1.46×10^4
10	5.65×10^8	2.1	2.82×10^8	1.1	1.53×10^{-6}	23.3	76.7	2.08×10^4
11	7.21×10^8	1.9	3.70×10^8	1.0	1.95×10^{-6}	22.1	77.9	2.65×10^4
12	8.69×10^8	1.8	4.52×10^8	0.9	2.40×10^{-6}	21.7	78.3	3.26×10^4

Table 5. Cont.

Position	Neutron Flux (n/cm²/s)	Stat. Error (%)	Gamma Flux (γ/cm²/s)	Stat. Error (%)	Heat Load Total (W/cm³)	n (%)	γ (%)	Dose Rate (Gy/FPY)
13	9.93×10^8	1.7	5.26×10^8	0.9	2.82×10^{-6}	22.1	77.9	3.83×10^4
14	1.08×10^9	1.7	5.54×10^8	0.9	2.94×10^{-6}	21.9	78.1	4.00×10^4
15	1.08×10^9	1.7	5.69×10^8	0.9	3.06×10^{-6}	22.4	77.6	4.16×10^4
16	1.09×10^9	1.7	5.80×10^8	0.9	3.11×10^{-6}	22.1	77.9	4.23×10^4
17	1.02×10^9	1.7	5.59×10^8	0.9	2.95×10^{-6}	21.2	78.8	4.01×10^4
18	9.74×10^8	1.8	5.36×10^8	0.9	2.93×10^{-6}	22.2	77.8	3.98×10^4
19	8.78×10^8	1.9	4.62×10^8	1.0	2.49×10^{-6}	21.4	78.6	3.38×10^4
20	7.23×10^8	2.0	3.70×10^8	1.1	1.97×10^{-6}	22.3	77.7	2.68×10^4
21	6.01×10^8	2.2	2.91×10^8	1.2	1.53×10^{-6}	22.8	77.2	2.08×10^4
22	4.56×10^8	2.4	2.21×10^8	1.3	1.14×10^{-6}	20.7	79.3	1.55×10^4
23	3.63×10^8	2.6	1.64×10^8	1.4	8.47×10^{-7}	21.9	78.1	1.15×10^4
24	3.40×10^8	2.7	1.33×10^8	1.6	6.97×10^{-7}	23.0	77.0	9.48×10^3
25	4.39×10^8	2.3	1.36×10^8	1.5	6.79×10^{-7}	18.2	81.8	9.23×10^3
26	1.10×10^9	1.4	3.12×10^8	0.9	1.63×10^{-6}	16.5	83.5	2.21×10^4
27	4.34×10^9	0.8	9.42×10^8	0.6	4.97×10^{-6}	17.7	82.3	6.76×10^4
28	1.70×10^{10}	0.4	3.34×10^9	0.4	1.84×10^{-5}	22.1	77.9	2.50×10^5
29	4.45×10^{10}	0.3	8.98×10^9	0.2	5.19×10^{-5}	26.6	73.4	7.07×10^5
30	8.16×10^{10}	0.2	1.81×10^{10}	0.2	1.08×10^{-4}	28.4	71.6	1.47×10^6
31	1.05×10^{11}	0.2	2.27×10^{10}	0.1	1.37×10^{-4}	28.5	71.5	1.86×10^6
32	1.31×10^{11}	0.2	2.60×10^{10}	0.1	1.63×10^{-4}	29.3	70.7	2.21×10^6
33	1.19×10^{11}	0.2	2.28×10^{10}	0.2	1.41×10^{-4}	27.6	72.4	1.92×10^6
34	9.15×10^{10}	0.2	1.55×10^{10}	0.2	9.85×10^{-5}	29.0	71.0	1.34×10^6
35	5.68×10^{10}	0.3	8.62×10^9	0.2	5.62×10^{-5}	29.9	70.1	7.65×10^5
36	3.35×10^{10}	0.4	4.67×10^9	0.4	3.13×10^{-5}	30.8	69.2	4.26×10^5
37	1.80×10^{10}	0.7	2.60×10^9	0.6	1.68×10^{-5}	28.6	71.4	2.28×10^5
38	1.24×10^{10}	1.1	1.82×10^9	1.0	1.15×10^{-5}	28.6	71.4	1.56×10^5
39	1.05×10^{10}	1.3	1.57×10^9	1.3	1.00×10^{-5}	29.2	70.8	1.36×10^5
40	8.93×10^9	1.7	1.41×10^9	1.4	8.78×10^{-6}	27.5	72.5	1.19×10^5
41	8.22×10^9	2.1	1.33×10^9	2.0	8.14×10^{-6}	26.5	73.5	1.11×10^5
42	7.31×10^9	2.4	1.24×10^9	2.6	7.44×10^{-6}	25.5	74.5	1.01×10^5
43	6.44×10^9	2.6	1.17×10^9	2.7	6.88×10^{-6}	23.6	76.4	9.35×10^4
44	6.10×10^9	3.4	1.07×10^9	2.9	6.08×10^{-6}	23.9	76.1	8.27×10^4
45	5.44×10^9	3.1	9.86×10^8	3.2	5.61×10^{-6}	24.4	75.6	7.64×10^4
46	5.21×10^9	3.1	9.60×10^8	3.2	5.56×10^{-6}	23.4	76.6	7.56×10^4
47	4.05×10^9	4.5	8.56×10^8	3.5	4.62×10^{-6}	19.6	80.4	6.29×10^4
48	3.69×10^9	4.2	7.69×10^8	4.0	4.01×10^{-6}	18.7	81.3	5.45×10^4
49	3.37×10^9	4.9	6.66×10^8	4.4	3.67×10^{-6}	15.1	84.9	4.99×10^4
50	3.12×10^9	5.7	6.06×10^8	5.2	3.13×10^{-6}	17.1	82.9	4.26×10^4
51	2.61×10^9	5.1	5.53×10^8	5.0	2.83×10^{-6}	19.8	80.2	3.85×10^4
52	2.32×10^9	5.2	4.57×10^8	4.2	2.26×10^{-6}	14.8	85.2	3.08×10^4
53	1.97×10^9	4.8	3.83×10^8	4.5	2.09×10^{-6}	16.3	83.7	2.85×10^4
54	1.97×10^9	5.6	4.18×10^8	17.9	2.16×10^{-6}	15.9	84.1	2.94×10^4
55	1.64×10^9	7.8	3.04×10^8	6.3	1.67×10^{-6}	18.0	82.0	2.27×10^4
56	1.45×10^9	6.9	2.67×10^8	6.0	1.41×10^{-6}	17.3	82.7	1.92×10^4
57	1.21×10^9	7.5	2.11×10^8	6.6	1.24×10^{-6}	18.3	81.7	1.69×10^4
58	1.09×10^9	10.4	2.23×10^8	6.7	1.06×10^{-6}	16.4	83.6	1.44×10^4
59	8.31×10^8	8.1	1.61×10^8	8.3	8.18×10^{-7}	14.5	85.5	1.11×10^4
60	5.59×10^8	9.2	1.46×10^8	6.0	7.17×10^{-7}	10.4	89.6	9.76×10^3
61	5.25×10^8	8.9	1.16×10^8	6.0	5.68×10^{-7}	15.3	84.7	7.73×10^3
62	5.01×10^8	7.0	1.34×10^8	5.2	6.83×10^{-7}	12.9	87.1	9.30×10^3
63	5.21×10^8	7.0	1.22×10^8	4.9	6.12×10^{-7}	10.7	89.3	8.32×10^3
64	3.60×10^8	6.3	1.11×10^8	5.6	5.57×10^{-7}	10.8	89.2	7.58×10^3
65	3.59×10^8	5.7	1.01×10^8	3.9	5.28×10^{-7}	10.6	89.4	7.18×10^3

Table 5. Cont.

Position	Neutron Flux (n/cm²/s)	Stat. Error (%)	Gamma Flux (γ/cm²/s)	Stat. Error (%)	Heat Load Total (W/cm³)	n (%)	γ (%)	Dose Rate (Gy/FPY)
66	3.73×10^8	7.9	1.01×10^8	3.5	5.03×10^{-7}	12.0	88.0	6.84×10^3
67	3.58×10^8	4.7	1.05×10^8	3.3	5.38×10^{-7}	13.1	86.9	7.32×10^3
68	4.25×10^8	3.8	1.30×10^8	2.4	6.38×10^{-7}	17.7	82.3	8.68×10^3
69	5.52×10^8	3.3	1.78×10^8	2.0	8.72×10^{-7}	20.1	79.9	1.19×10^4
70	9.79×10^8	2.1	3.58×10^8	1.2	1.78×10^{-6}	23.1	76.9	2.42×10^4
71	3.59×10^9	0.9	2.23×10^9	0.5	1.50×10^{-5}	30.5	69.5	2.04×10^5
72	1.53×10^9	1.5	6.09×10^8	0.9	3.50×10^{-6}	27.6	72.4	4.76×10^4

This is further illustrated in Figures 19 and 20, which show the neutron and gamma flux spectra in four positions, two at the equatorial port level (15 inboard and 53 outboard), one in the divertor region (31) and the remaining one above the plasma (70). In the divertor region the fluxes are clearly higher when compared to the other positions, for both neutrons and gammas. In position 53 (but also 15 and 31), the statistical errors in the bins are inevitably large, due to the very small binning and the blanket thickness in that area. Nevertheless, the statistical errors in the total neutron and gamma fluxes are only 4.8% and 4.5%, respectively.

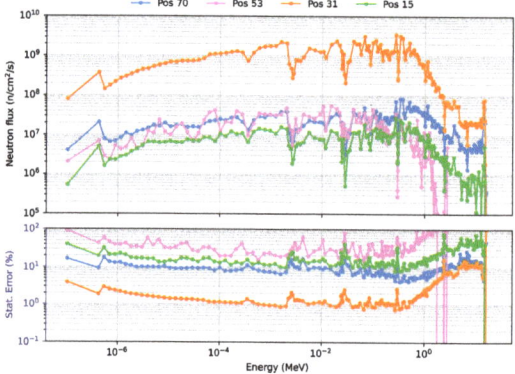

Figure 19. Neutron fluxes (n/cm²/s) and statistical errors (%) in 4 FOCS positions (WCLL blanket).

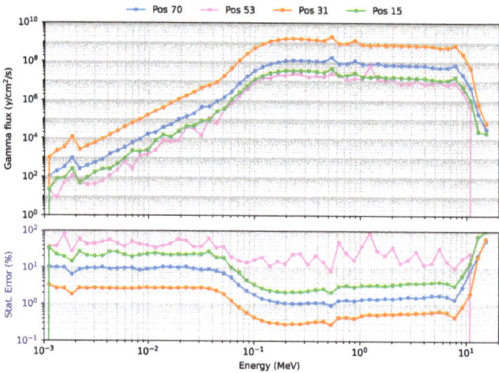

Figure 20. Gamma fluxes (γ/cm²/s) and statistical errors (%) in four FOCS positions (WCLL blanket).

The possibility of bringing the FOCS to the inner surface of the VV has also been discussed recently. However, the current results show that the sensors would not be able to withstand the radiation levels, as the neutron fluxes inside the vessel would be around three orders of magnitude higher than ex-vessel [1].

3.3. Equatorial Port Diagnostics

The EP configuration studied in this work was based on one of the port integration proposals presented in [54], for an EP housing the following three diagnostics:

- High-resolution core X-ray spectroscopy;
- Near-ultraviolet, visible and infrared divertor monitoring;
- Pellet monitoring.

In this integration proposal the six optical paths of the divertor monitoring and pellet monitoring systems are grouped together on the left side of the port (when looking towards the plasma) in two rows with three paths each. The X-ray spectroscope is placed on the right side, with the ducts angled slightly in the EP to increase the space for the port cell optical components of the other systems. This setup is presented in Figure 21.

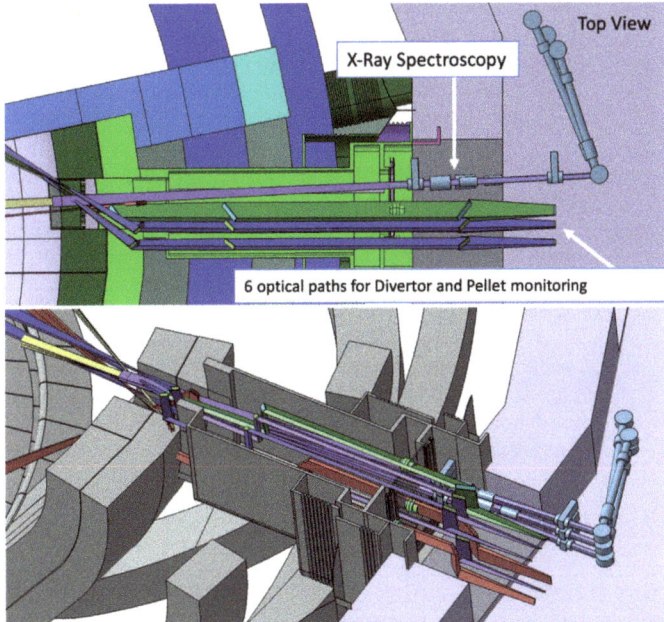

Figure 21. EP configuration with the X-ray spectroscopy (light blue), divertor monitoring (X-point and outer divertor tangential line-of-sight in purple, outer divertor surface views in red) and pellet monitoring (green) systems.

The objective of this study was to implement this EP configuration in MCNP and evaluate the neutron and gamma fluxes through these diagnostic ducts into the port cell, after the bioshield, testing possible shielding configurations based on the proposals of reference [54], illustrated in Figure 22. These proposals include the standard equatorial port plug shield block (reinforced if needed) and additional shielding in the mirror doglegs along the diagnostic ducts, in the middle of the port and in (or possibly before or after) the bioshield plug. EUROFER and stainless steel were considered for the EPP shield block, while boron carbide (B_4C) shielding trays similar to the ones foreseen for the EP diagnostics shielding modules (DSMs) of ITER [55] were considered for the middle of the port, due to their shielding efficiency and lower weight.

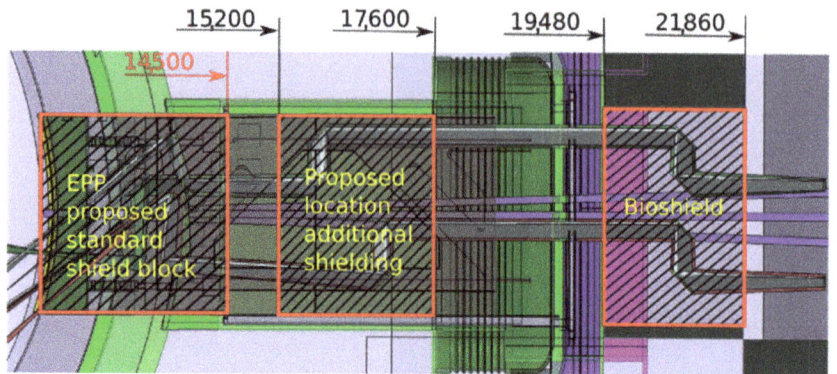

Figure 22. Possible radiation shielding locations proposed for the EP (units in mm).

The ducts from the diagnostics presented in Figure 21 were simplified in ANSYS SpaceClaim, through the removal of details from the vacuum windows and vacuum extensions (only the ducts and the mirrors were left in the model) and of all the spline surfaces present in the model. The design of the EP components was adapted from reference [56], which has the same shape and is compatible with the MCNP reference model.

All components of this model were filled, and the EP diagnostic ducts were carved inside. The result is shown in Figure 23. Some of the cells of Figure 23 were dimensioned to be filled with shielding in the MCNP model—a thickness of 2.4 m was reserved in the middle of the port for the B_4C shielding—while others were designed to be void cells. The complexity of the ducts in the neutronics CAD model is illustrated in Figure 24, where the cells are represented with transparency. One of the main challenges of this work was the generation of this CAD model, free from splines and small surfaces, ready to be converted to MCNP.

The diagnostic with the largest openings is the X-ray spectroscopy system, with a first wall opening of 23 cm × 10 cm (230 cm^2) that spreads into three ducts behind the first wall. As illustrated in Figure 25, these ducts are straight paths from the plasma to the port cell, with openings in the bioshield plug of 10.6 cm × 10 cm (106 cm^2). As there are no doglegs in this diagnostic, direct neutron streaming is expected through these ducts.

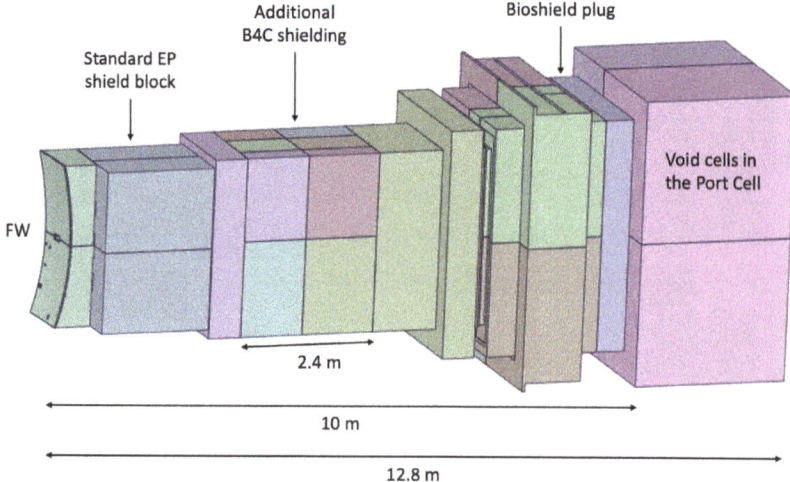

Figure 23. CAD model of the EP used in the simulations.

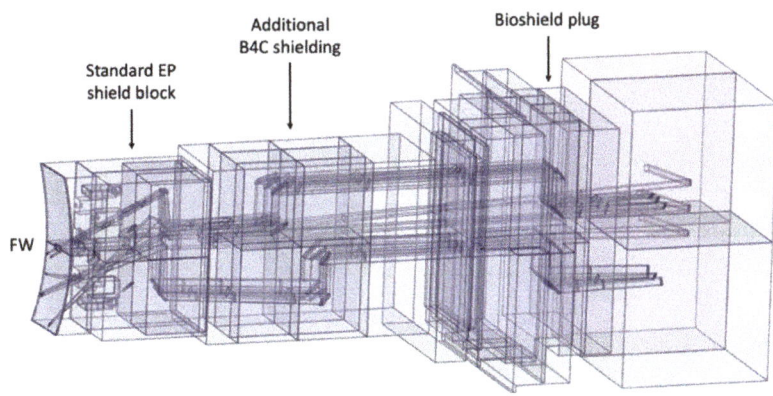

Figure 24. CAD model of the EP used in the simulations with transparent cells, showing the diagnostic ducts.

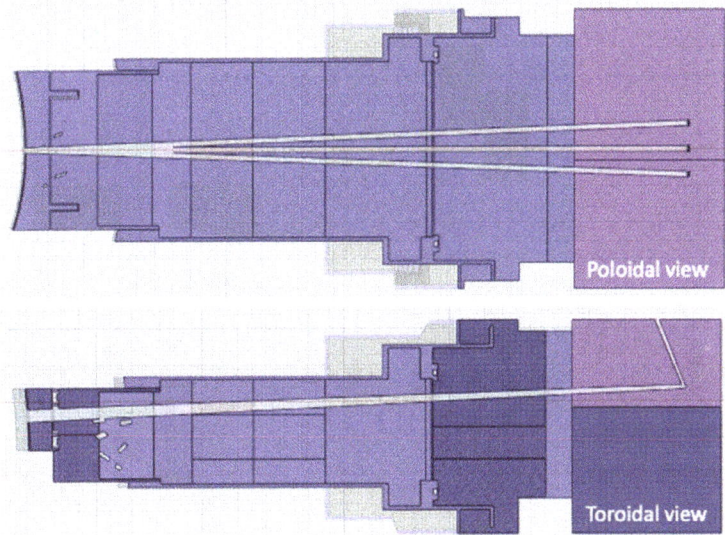

Figure 25. X-ray spectroscopy ducts along the EP.

The remaining diagnostics have much smaller openings in the first wall (all below 28 cm^2). Furthermore, they have doglegs, which will reduce streaming to the port cells, as shown before [25].

The converted model of the port was integrated in the reference model of Figure 4. The result is presented in Figure 26, for a plane in the middle of the X-ray spectroscopy ducts (left) and for plane y = −15 cm (with the near-ultraviolet and visible divertor spectroscopy ducts). It also illustrates the reasoning behind the shielding distribution inside the port: a first block was added to the standard shielding of the EPP (pink), which contains the first dogleg for all the diagnostics except the X-ray spectroscopy system, and a second block (of B_4C) was added in the middle of the port (yellow), to shield the second dogleg. Due to the low thickness of the bioshield plug, the third dogleg is not shielded in the studied configurations, although an additional shielding layer could be envisaged for this dogleg, placed in front of the bioshield plug.

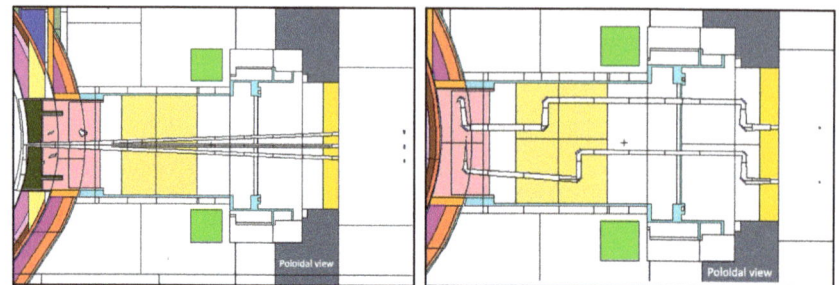

Figure 26. MCNP model of the EP. **Left**: X-ray spectroscopy ducts. **Right**: Near-ultraviolet (bottom) and visible (top) divertor spectroscopy ducts (poloidal view, plane y = −15 cm).

Most of the materials used in the equatorial port model are summarized in Figure 27. The first wall has 2 mm thick armor made of tungsten, with a second layer of 6.09% water and 93.91% EUROFER, taken directly from the definition of the WCLL BB which follows the material distribution set out in [57]. The definition for the first wall shield block behind it was adopted from the technical specification for the equatorial outboard limiter [58]: 60% EUROFER and 40% water. For the remaining shielding behind this block a mixture of 70% SS316L(N)-IG stainless steel and 30% water was assumed, while in the second shielding block (yellow in Figure 26, second dogleg) a homogenized mixture of 65% B_4C, 10% stainless steel and 25% void (to account for the spacings between the components) was defined, with an effective density of 2.28 g/cm^3 [55]. This mixture represents the B_4C shielding trays used in the ITER DSMs for the equatorial ports [55].

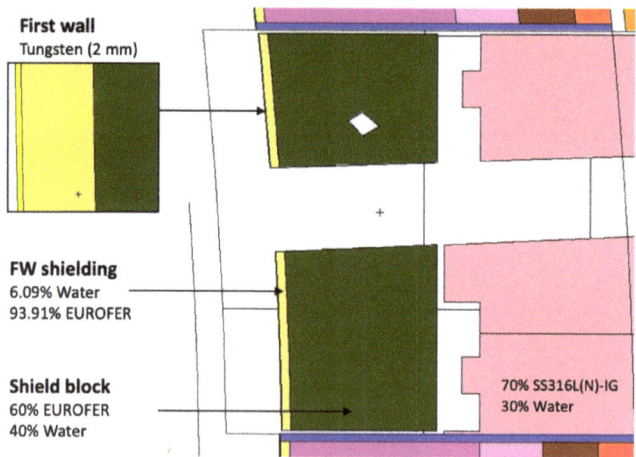

Figure 27. Materials used in the MCNP model of the equatorial port.

The mirrors were set to EUROFER, while the remaining components were kept with the same materials used in the equatorial port components of the MCNP reference model.

3.3.1. Neutron Fluxes, Gamma Fluxes and Dose Rates in the Port Cell

The neutron and gamma fluxes for this equatorial port configuration are presented in Figures 28 and 29 for several planes y and z, with the neutron flux statistical errors for the planes y = 80 cm and z = 0 presented in Figure 30. As expected, there is substantial neutron streaming through the X-ray spectroscopy ducts, reaching the port cell through the straight paths. This is visible mostly around planes y = 120 cm and z = 0. The neutron fluxes reaching the inner surface of the bioshield through these ducts (the one in z = 0 is used for this estimation) are of the order of 2×10^{10} n cm^{-2} s^{-1}, decreasing to 4×10^{9} n cm^{-2} s^{-1} in the mirror behind the bioshield. The gamma fluxes reach 3×10^{9} γ cm^{-2} s^{-1} at the inner surface of the bioshield and 2×10^{9} γ cm^{-2} s^{-1} and 2×10^{8} γ cm^{-2} s^{-1} in the mirror, while the dose rates in silicon obtained in these positions were 2E6 Gy/FPY and 5E5 Gy/FPY, respectively. The flux and dose rate values in the port cell are more than three orders of magnitude higher than those obtained with the reference model of the port without diagnostics (neutron fluxes below 1×10^{7} n cm^{-2} s^{-1} were obtained in the port cell with the reference model). For the remaining port diagnostics, the design of the first doglegs is effective to reduce the streaming, as shown in a previous study [25].

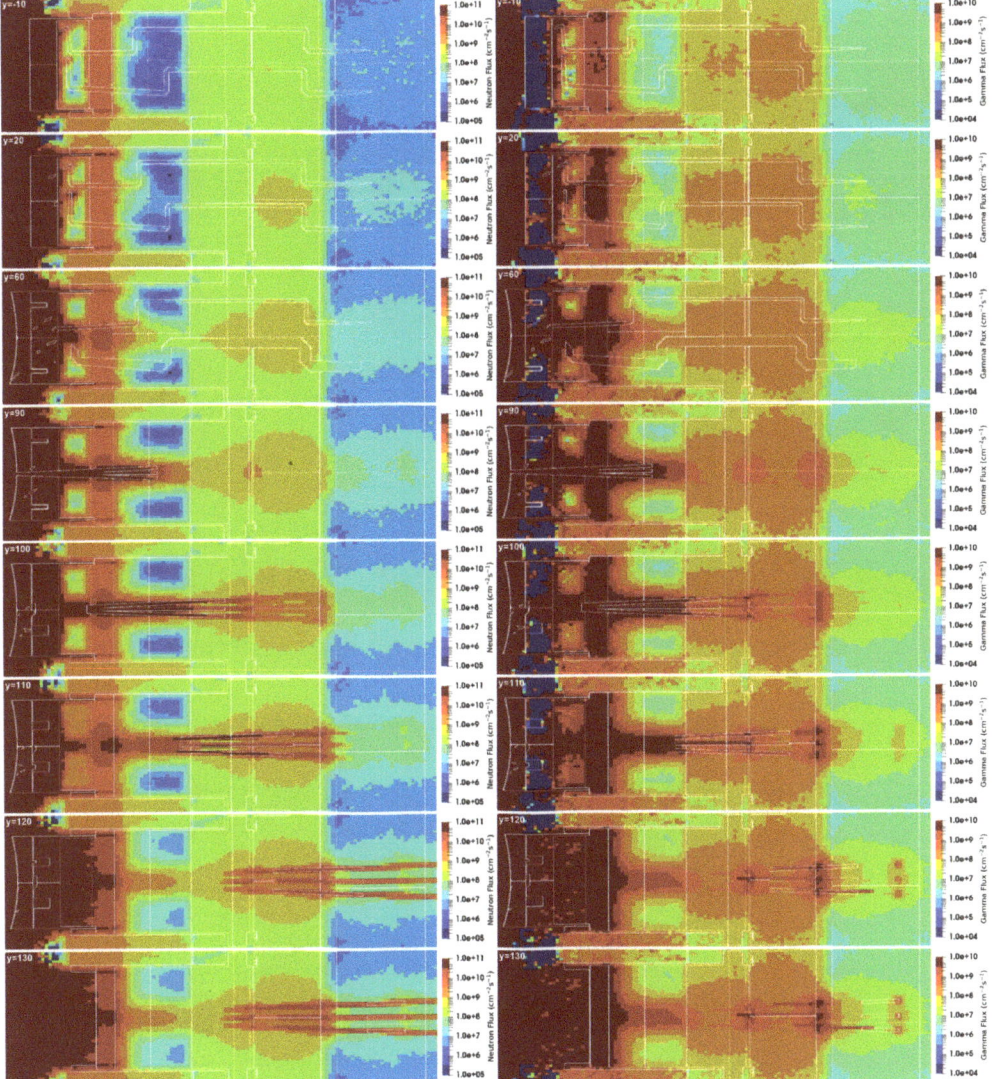

Figure 28. Neutron and gamma fluxes (cm^{-2} s^{-1}) in plane y for the configuration with diagnostics in the EP.

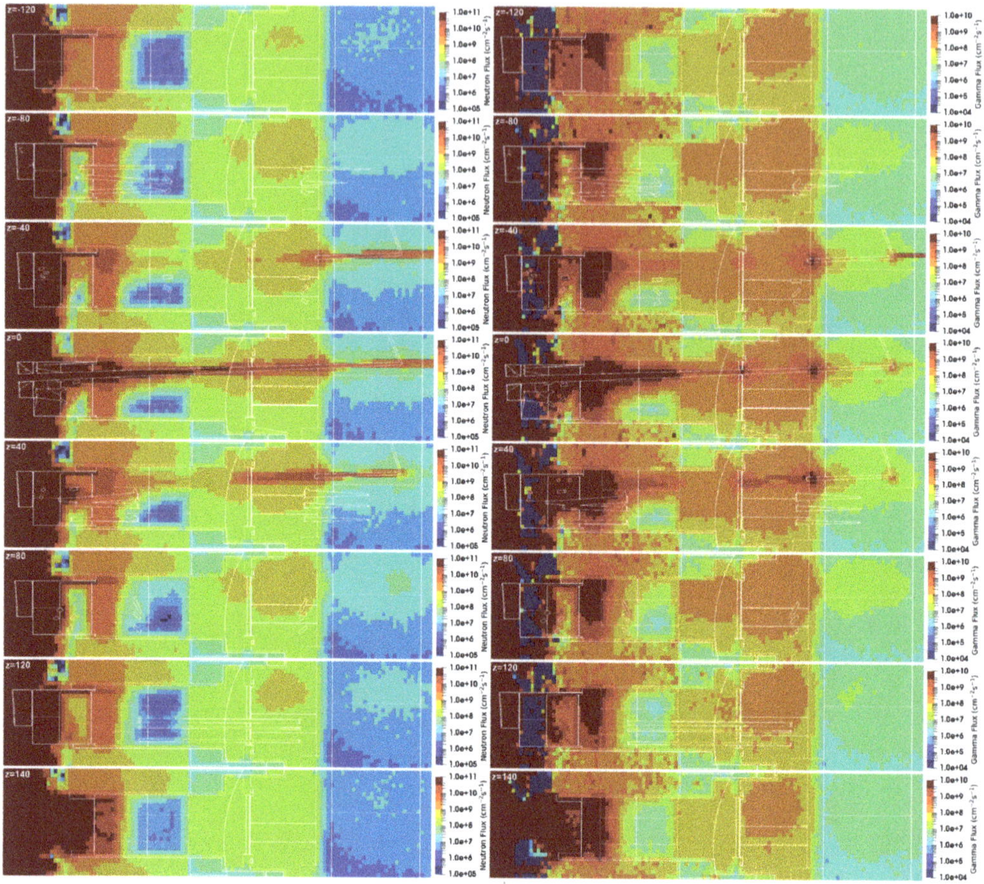

Figure 29. Neutron and gamma fluxes (cm^{-2} s^{-1}) in plane z for the configuration with diagnostics in the EP.

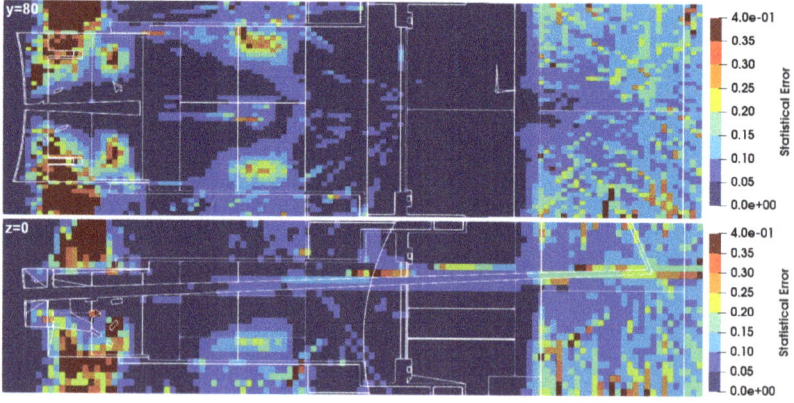

Figure 30. Statistical errors of the neutron fluxes, for the planes y = 80 cm and z = 0.

The statistical errors of the neutron fluxes, presented in Figure 30, are below 10% in most of the regions of interest. As expected, they increase along the ducts, even though the weight windows were fine-tuned to increase statistics in the port cell. Due to the distance of more than 12 m between the plasma and the port cell mirrors, it was not possible to have statistical errors below 10% in all the regions of the studied configurations. Nevertheless, F4 tallies were added at the main positions of interest (bioshield and port cell mirrors), and the flux values discussed in the previous paragraph were confirmed, with statistical errors between 3% and 10%.

The shutdown dose rates in the port cells were not calculated, for two reasons: (1) the lack of access, at this stage, to R2S/D1S codes for this kind of calculation in DEMO and (2) the fact that such a calculation would always be far from accurate, as the bioshield plugs and their penetrations, as well as the port cells, have not been designed yet (material activation would require accurate designs of the systems that will populate these rooms). On the other hand, there are no limits defined for the neutron and gamma fluxes in the port cells in the DEMO Nuclear Analysis Handbook [1], which defines a limit of 100 µSv/h in the port cell 12 days after shutdown. However, it can be anticipated, based on experience from ITER, that this limit will be greatly exceeded with the neutron streaming predicted for the X-ray spectroscopy ducts. Another open issue is the radiation limits that the vacuum windows and the optical fibers can withstand, as well as the locations where electronics are required, since in the present design the limits of 100 n cm^{-2} s^{-1} and 10 Gy of cumulative dose could only be enforced with large amounts of shielding in the port cell, or if the electronics are placed far from the streaming paths. In any case, before the other port diagnostics can be studied in more detail it is important to evaluate whether it is possible for the X-ray spectroscopy system to operate with smaller ducts or alternative configurations, to reduce streaming.

The priority was then to understand the effect of the duct cross-section on the neutron and gamma streaming through the port, to provide a guideline for diagnostic design. For this, a sensitivity analysis was carried out for straight ducts from the plasma to the port cell. The results are presented in the next section.

3.3.2. Sensitivity Analysis for Straight Ducts in the Equatorial Port

For this analysis, the equatorial port model presented above was used, with all the diagnostics removed and with only one duct (centered at z = 0 and y = 70 cm). Using that model, neutron and gamma flux spectra were calculated as a function of the duct size, for duct cross sections ranging from diameters of 3 cm up to the size of the X-ray ducts (23 cm toroidal × 10 cm poloidal). For these simulations, weight windows were used in conjunction with source biasing parameters produced by ADVANTG and F5 tallies, i.e., "next event estimators". Using F5 tallies, each time a source particle is created, or at any collision event, a deterministic estimation is made for the flux contribution at the detector point [59]. This makes F5 tallies ideal for the kind of simulation performed here, with straight ducts from the plasma to the port cell. They slow down the simulations considerably and do not allow the production of mesh tallies but yield accurate results with extremely low statistical errors. Due to the simulation time that F5 tallies require, only two were used in each simulation, for two points after the bioshield plug: at the outer surface of the plug (x = 2200 cm, y = 70 cm, z = 0) and 2 m away from that surface (x = 2400 cm, y = 70 cm, z = 0).

Figure 31 shows the neutron flux spectrum 2 m behind the bioshield plug for a circular duct with r = 1.5 cm. The total flux value is 3.7×10^8 n cm^{-2} s^{-1}, with a very small statistical error of 0.9%. The spectrum also shows that the statistical errors are smaller than 10% for some energy bins, and below 1% at 14 MeV. Also presented is the spectrum of uncollided neutrons—bins around 14 MeV—which corresponds to those neutrons that travel from the plasma to the port cell without any interactions. In these bins, the statistical errors are very small, in some cases below 1%. As will be shown later, the total flux value is compatible with the results obtained in the previous section.

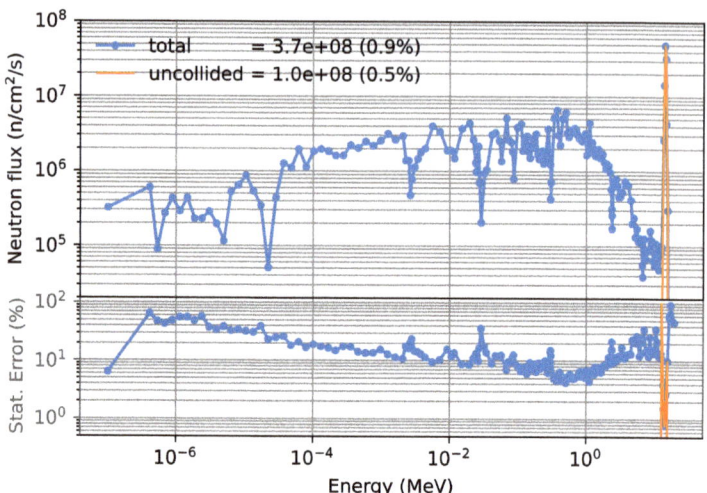

Figure 31. Neutron flux (n/cm^2/s) and statistical error 2 m behind the bioshield plug (x = 2400 cm, y = 70 cm, z = 0), for a circular duct with r = 1.5 cm.

The F5 tallies also allowed us to calculate the gamma spectra in the port cell, as shown in Figure 32 for the same case and position as before. The total gamma flux (4.5×10^7 γ cm^{-2} s^{-1}) is almost one order of magnitude lower than the neutron flux. The statistical errors are larger in this case, but still below 10%. The uncollided spectrum refers to gammas that were created somewhere in the geometry and traveled to the port cell without interactions. For both cases (neutron and gammas), reducing the statistical errors below 10% in all bins would be mandatory if these results were to be used as input in inventory or activation calculations; however, this would be prohibitive in terms of computational resources. Furthermore, the "collided" part of the spectrum depends heavily on the shielding configurations, which are very preliminary at this stage.

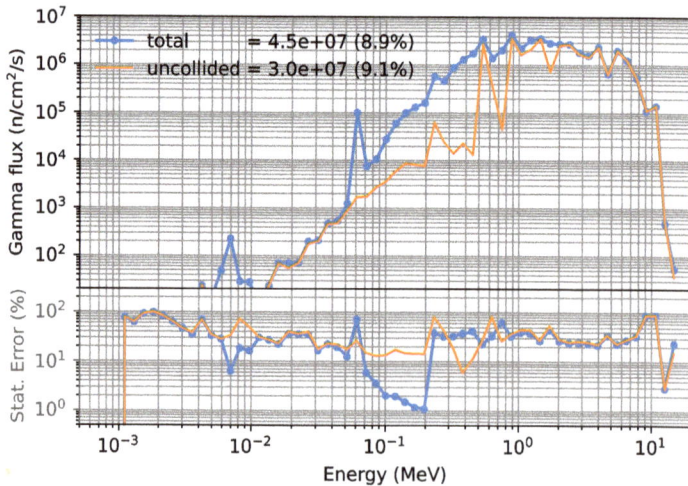

Figure 32. Gamma flux (γ/cm^2/s) and statistical error 2 m behind the bioshield plug (x = 2400 cm, y = 70 cm, z = 0), for a circular duct with r = 1.5 cm.

Similar spectra were calculated for several cases: circular ducts with radii between 1.5 cm and 4 cm (0.5 cm increments) and rectangular ducts of 10 cm in the poloidal direction and several toroidal lengths. The height of these ducts (10 cm) was chosen to be the one projected for the X-ray spectroscopy. The toroidal lengths were varied from 0.71 cm to 23 cm (the initial length foreseen for the X-ray ducts). The first five toroidal lengths, up to 5 cm, were selected to match the area of the circular ducts, to evaluate the effect of the duct shape on the fluxes. After 5 cm, four additional lengths were tested: 10 cm, 15 cm, 20 cm and 23 cm.

The results are presented in Table 6 (neutron fluxes) and Table 7 (gamma fluxes). Looking at the bioshield surface, the neutron fluxes vary from 5.38×10^8 n cm^{-2} s^{-1} for a circular opening with r = 1.5 cm to 3.91×10^9 n cm^{-2} s^{-1} for r = 4 cm. Comparing these values with the corresponding areas for rectangular ducts (toroidal lengths up to 5 cm), it becomes clear that the shape of the duct has no effect on the fluxes; very similar values were obtained for the same areas. When the toroidal length of the rectangular duct is increased to 23 cm, the fluxes increase to 1.9×10^{10} n cm^{-2} s^{-1} (or 1.2×10^{10} n cm^{-2} s^{-1} 2 m away from the bioshield). This is in excellent agreement with the results shown in the previous section, where fluxes of 1.2×10^{10} n cm^{-2} s^{-1} were estimated at the mirror location in the center duct of the X-ray diagnostic.

Table 6. Neutron fluxes as a function of the duct cross-sectional dimensions. BS refers to bioshield and M to a possible mirror location 2 m behind.

Neutron Flux (n cm^{-2} s^{-1})										
	Radius (cm)	Area (cm^2)	Total BS (n cm^{-2} s^{-1})	Error (%)	Uncollided BS (n cm^{-2} s^{-1})	Error (%)	Total M (n cm^{-2} s^{-1})	Error (%)	Uncollided M (n cm^{-2} s^{-1})	Error (%)
Circular Ducts	1.5	7.1	5.38×10^8	1.3	1.47×10^8	0.5	3.68×10^8	0.9	1.02×10^8	0.6
	2	12.6	9.51×10^8	0.7	2.63×10^8	0.3	6.55×10^8	0.7	1.82×10^8	0.3
	2.5	19.6	1.50×10^9	0.6	4.10×10^8	0.2	1.03×10^9	0.6	2.84×10^8	0.2
	3	28.3	2.17×10^9	0.5	5.90×10^8	0.2	1.48×10^9	0.5	4.09×10^8	0.2
	3.5	38.5	2.97×10^9	0.5	8.04×10^8	0.1	2.02×10^9	0.4	5.57×10^8	0.2
	4	50.3	3.91×10^9	0.5	1.05×10^9	0.1	2.64×10^9	0.4	7.26×10^8	0.1
	Dimensions (cm × cm)	Area (cm^2)	Total BS (n cm^{-2} s^{-1})	Error (%)	Uncollided BS (n cm^{-2} s^{-1})	Error (%)	Total M (n cm^{-2} s^{-1})	Error (%)	Uncollided M (n cm^{-2} s^{-1})	Error (%)
Rectangular Ducts	10 × 0.71	7.1	5.29×10^8	0.6	1.47×10^8	0.3	3.65×10^8	0.6	1.02×10^8	0.34
	10 × 1.26	12.6	9.71×10^8	1.2	2.62×10^8	0.2	6.60×10^8	0.6	1.81×10^8	0.26
	10 × 1.96	19.6	1.50×10^9	0.5	4.10×10^8	0.2	1.03×10^9	0.5	2.84×10^8	0.2
	10 × 2.83	28.3	2.18×10^9	0.7	5.90×10^8	0.2	1.49×10^9	0.4	4.09×10^8	0.17
	10 × 3.85	38.5	2.99×10^9	0.8	8.03×10^8	0.1	2.03×10^9	0.4	5.56×10^8	0.15
	10 × 5.03	50.3	3.86×10^9	0.5	1.05×10^9	0.1	2.63×10^9	0.4	7.27×10^8	0.13
	10 × 10	100	7.97×10^9	1.0	2.09×10^9	0.1	5.29×10^9	0.3	1.45×10^9	0.09
	10 × 15	150	1.20×10^{10}	0.4	3.13×10^9	0.1	7.91×10^9	0.3	2.17×10^9	0.07
	10 × 20	200	1.62×10^{10}	0.4	4.17×10^9	0.1	1.06×10^{10}	0.2	2.89×10^9	0.06
	10 × 23	230	1.90×10^{10}	0.9	4.80×10^9	0.1	1.22×10^{10}	0.2	3.32×10^9	0.06

Table 7. Gamma fluxes as a function of the duct cross-sectional dimensions. BS refers to bioshield and M to a possible mirror location 2 m behind.

Gamma Flux (γ cm^{-2} s^{-1})										
	Radius (cm)	Area (cm^2)	Total BS (γ cm^{-2} s^{-1})	Error (%)	Uncollided BS (γ cm^{-2} s^{-1})	Error (%)	Total M (γ cm^{-2} s^{-1})	Error (%)	Uncollided M (γ cm^{-2} s^{-1})	Error (%)
Circular Ducts	1.5	7.1	7.50×10^7	10.1	4.92×10^7	9.5	4.48×10^7	8.9	3.03×10^7	9.1
	2	12.6	1.49×10^8	9.3	8.09×10^7	6.0	9.89×10^7	10.6	5.47×10^7	6.5
	2.5	19.6	2.06×10^8	6.0	1.27×10^8	5.0	1.25×10^8	5.8	8.23×10^7	5.4
	3	28.3	3.91×10^8	9.8	2.08×10^8	5.9	2.55×10^8	10.8	1.37×10^8	6.7
	3.5	38.5	5.03×10^8	13.1	2.62×10^8	4.6	3.32×10^8	15.7	1.74×10^8	5.2
	4	50.3	6.14×10^8	4.4	3.55×10^8	3.0	3.57×10^8	5.0	2.24×10^8	3.3
	Dimensions (cm × cm)	Area (cm^2)	Total BS (γ cm^{-2} s^{-1})	Error (%)	Uncollided BS (γ cm^{-2} s^{-1})	Error (%)	Total M (γ cm^{-2} s^{-1})	Error (%)	Uncollided M (γ cm^{-2} s^{-1})	Error (%)
Rectangular Ducts	10 × 0.71	7.1	7.29×10^7	8.78	4.77×10^7	8.09	4.85×10^7	9.74	3.30×10^7	8.85
	10 × 1.26	12.6	1.61×10^8	13.1	9.49×10^7	6.85	1.01×10^8	15.7	6.33×10^7	7.44
	10 × 1.96	19.6	2.10×10^8	8.3	1.38×10^8	9.11	1.38×10^8	9.34	9.30×10^7	10.3
	10 × 2.83	28.3	3.41×10^8	6.76	2.00×10^8	4.32	2.23×10^8	7.6	1.32×10^8	4.2
	10 × 3.85	38.5	4.39×10^8	5.57	2.66×10^8	4.24	2.74×10^8	6.35	1.75×10^8	4.7
	10 × 5.03	50.3	6.14×10^8	5.23	3.44×10^8	3.95	3.80×10^8	6.15	2.22×10^8	4.58
	10 × 10	100	1.30×10^9	4.63	7.27×10^8	3.28	7.35×10^8	5.91	4.44×10^8	3.39
	10 × 15	150	2.17×10^9	4.25	1.16×10^9	2.46	1.20×10^9	5.57	7.00×10^8	2.91
	10 × 20	200	2.95×10^9	2.31	1.60×10^9	1.68	1.55×10^9	2.9	9.32×10^8	1.82
	10 × 23	230	3.47×10^9	2.52	1.83×10^9	1.8	1.83×10^9	3.04	1.06×10^9	1.8

When comparing the total flux with the uncollided flux, a ratio between 3.5 and 4 is found between the two for all cases. While the total flux will be affected by the EP shielding configuration, the uncollided flux will be similar regardless of the EP design.

Figure 33 shows the neutron fluxes plotted against the cross-sectional area of the ducts for the rectangular configuration (the results for the circular ducts are very similar and were omitted). The flux varies linearly with the duct area. The fits were obtained using Mathematica [36] for the simple expression $f = c\,A$, where f is the flux, A is the area of the duct and c is a constant, and they can be used as a first approximation to estimate the fluxes in ducts with different areas.

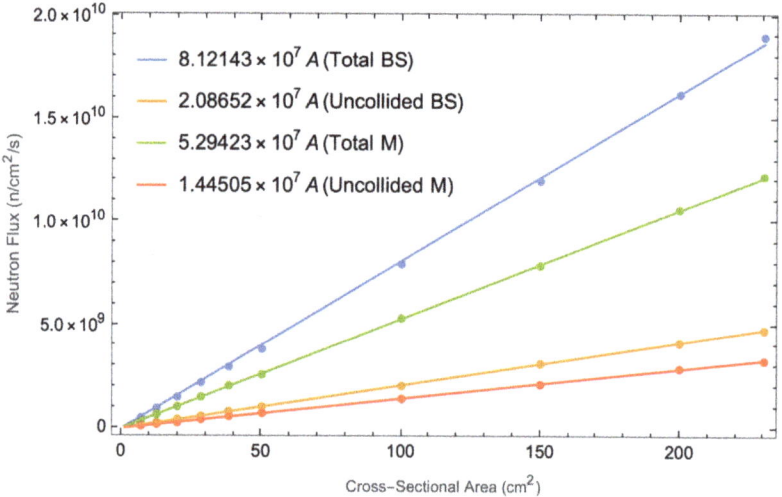

Figure 33. Neutron flux variation with the cross-sectional area, for rectangular ducts.

The gamma fluxes are a factor of 5–7 lower than the neutron fluxes. In the previous analysis, $2.3 \times 10^9\ \gamma\ \mathrm{cm}^{-2}\ \mathrm{s}^{-1}$ was obtained for the central duct of the X-ray system, with a statistical error of 15%, while here the flux is 50% higher: $3.47 \times 10^9\ \gamma\ \mathrm{cm}^{-2}\ \mathrm{s}^{-1}$ (2.5% statistical error). This variation can be explained by the 15% error in the previous simulations, which points to unreliable results.

3.3.3. Neutron/Gamma Cameras, Radiated Power and Soft X-ray Intensity

Two diagnostics are currently expected to use straight ducts with even smaller cross-sections than the ones simulated in the previous section (r = 1 cm): the neutron and gamma cameras [8,9] and the core radiated power and soft X-ray intensity system [10]. The aim of this section is to provide the neutron and gamma fluxes through the different ducts of these systems.

The CAD model of the neutron and gamma camera system is represented in Figure 34. It contains 13 ducts with a 1 cm cross-section radius as well as neutron and gamma detectors and a shielding/collimator block enclosing them. It is similar to the CAD model of the core radiated power and soft X-ray intensity system, presented in Figure 35, the main difference being that this updated design of the radiated power system contains 26 ducts instead of 13. The positions of the ducts are also not the same and intersect at different points. Despite these differences, these systems have relatively similar duct configurations, and since the simulations required to estimate the fluxes in the port cells are very CPU-intensive, the configuration presented in Figure 34 (neutron and gamma cameras) was adopted as representative for both systems in this analysis. The MCNP geometry used in the simulations, based on the EP design used in the previous section, is presented in Figure 36.

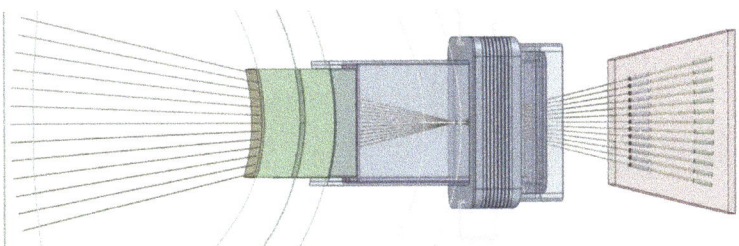

Figure 34. CAD model of the neutron and gamma cameras.

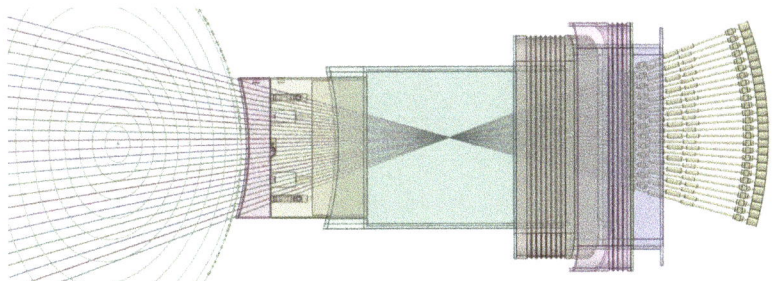

Figure 35. CAD model of the core radiated power and soft X-ray intensity system.

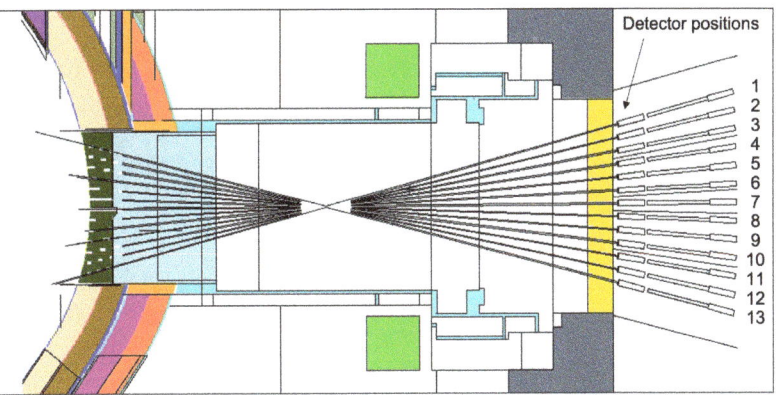

Figure 36. Neutronics model used in the simulations of the neutron/gamma cameras and core radiated power and soft X-ray intensity system (plane y = 50 cm).

All the cells within the ducts, including the detectors, were modeled as void, to prevent effects related to neutron or gamma scattering in the detector materials, which are different between the two systems. All materials of the EP were kept unchanged from the reference model.

As before, the simulations were run using the weight windows generated with the ADVANTG code. The source biasing parameters generated by ADVANTG were also added to the reference neutron source. As in the previous section, F5 tallies were used at the 13 detector positions to tally the neutron and gamma fluxes. The heat loads in beryllium, the material proposed for the vacuum window of the radiated power diagnostic, were also calculated with F5 tallies, using conversion factors.

The neutron and gamma fluxes in the 13 positions are summarized in Table 8. As expected, the uncollided neutron flux increases steadily from position 1 (top, 4.2×10^7 n cm^{-2} s^{-1}) to positions 6 and 7 (middle, 6.9×10^7 n cm^{-2} s^{-1}), decreasing afterwards until position 13 (bottom, 3.9×10^7 n cm^{-2} s^{-1}). These fluxes are mostly independent of the EP shielding configuration. The total fluxes are 2.3 to 2.7 times higher than the uncollided fluxes and have a similar trend (except at detector 3, where the flux is slightly higher than in detector 4). The total gamma fluxes are four to five times lower than the total neutron fluxes, except at detector 2, where the flux has an unexpectedly large statistical error. Even though the simulations were performed for 5–7 days with 720 processors in the MARCONI cluster, it was not possible to bring the statistical errors in the gamma fluxes below 10% for all positions. Nevertheless, a trend can be established from the results in the positions where the errors are smaller.

Table 8. Summary of the neutron and gamma fluxes at the 13 detector positions of Figure 36.

Detector	Neutrons				Gammas	
	Total		Uncollided		Total	
	Flux (n/cm^2/s)	Error (%)	Flux (n/cm^2/s)	Error (%)	Flux (n/cm^2/s)	Error (%)
1	1.07×10^8	1.2	4.15×10^7	0.3	1.94×10^7	8.6
2	1.26×10^8	1.1	5.07×10^7	0.3	5.05×10^7	46.4
3	1.41×10^8	0.7	5.83×10^7	0.2	2.78×10^7	8.3
4	1.38×10^8	0.7	5.89×10^7	0.2	3.01×10^7	13.5
5	1.54×10^8	0.7	6.66×10^7	0.2	3.18×10^7	7.8
6	1.61×10^8	0.6	6.94×10^7	0.2	3.66×10^7	9.2
7	1.60×10^8	0.6	6.92×10^7	0.2	3.65×10^7	14.4
8	1.57×10^8	0.7	6.70×10^7	0.2	3.62×10^7	9.6
9	1.52×10^8	0.7	6.43×10^7	0.2	3.08×10^7	7.9
10	1.42×10^8	0.8	5.95×10^7	0.2	3.48×10^7	13.5
11	1.33×10^8	0.8	5.46×10^7	0.3	3.16×10^7	14.0
12	1.21×10^8	1.0	4.66×10^7	0.3	2.67×10^7	18.3
13	1.04×10^8	1.3	3.85×10^7	0.4	2.23×10^7	14.7

The nuclear heat loads in beryllium are presented in Table 9. As these results were obtained with F5 tallies, the heat loads due to uncollided neutrons and gammas are also presented, along with the total heat loads obtained by summing the neutron and gamma contributions. Since beryllium is a neutron moderator and the neutron fluxes are higher than the gamma fluxes, neutrons have the highest contribution to the total heat loads, exceeding the gamma contribution by more than one order of magnitude. The total heat loads range from 3.0 µW/cm^3 at position 13 to 5.3 µW/cm^3 in the central detector positions (6 and 7). The statistical errors are below 1% for all positions.

Table 9. Heat loads in Be by neutrons and gammas at the 13 detector locations of Figure 36.

Detector	Neutron				Gamma				Total	
	Total		Uncollided		Total		Uncollided			
	Heat Load (W/cm^3)	Error (%)	Heat Load (W/cm^3)	Error (%)	Heat Load (W/cm^3)	Error (%)	Heat Load (W/cm^3)	Error (%)	Heat Load (W/cm^3)	Error (%)
1	3.02×10^{-6}	0.4	2.62×10^{-6}	0.4	1.94×10^{-7}	11.4	1.38×10^{-7}	9.4	3.21×10^{-6}	0.5
2	3.68×10^{-6}	0.3	3.22×10^{-6}	0.3	2.33×10^{-7}	9.4	1.73×10^{-7}	7.4	3.91×10^{-6}	0.4
3	4.18×10^{-6}	0.4	3.65×10^{-6}	0.3	2.34×10^{-7}	10.3	1.69×10^{-7}	6.8	4.41×10^{-6}	0.5
4	4.21×10^{-6}	0.3	3.70×10^{-6}	0.3	2.59×10^{-7}	7.5	2.02×10^{-7}	7.2	4.47×10^{-6}	0.3
5	4.77×10^{-6}	0.3	4.19×10^{-6}	0.2	2.59×10^{-7}	6.2	2.02×10^{-7}	5.7	5.03×10^{-6}	0.4
6	4.97×10^{-6}	0.2	4.38×10^{-6}	0.2	3.12×10^{-7}	8.1	2.16×10^{-7}	5.5	5.28×10^{-6}	0.3
7	4.96×10^{-6}	0.2	4.37×10^{-6}	0.2	2.91×10^{-7}	6.7	2.16×10^{-7}	5.5	5.25×10^{-6}	0.3
8	4.83×10^{-6}	0.3	4.25×10^{-6}	0.2	2.68×10^{-7}	7.0	2.16×10^{-7}	6.3	5.10×10^{-6}	0.4
9	4.67×10^{-6}	0.4	4.10×10^{-6}	0.3	2.91×10^{-7}	8.0	2.13×10^{-7}	5.7	4.96×10^{-6}	0.5
10	4.29×10^{-6}	0.4	3.75×10^{-6}	0.3	3.01×10^{-7}	12.1	2.03×10^{-7}	6.9	4.59×10^{-6}	0.5
11	3.94×10^{-6}	0.3	3.45×10^{-6}	0.3	2.36×10^{-7}	8.7	1.69×10^{-7}	7.5	4.17×10^{-6}	0.4
12	3.37×10^{-6}	0.4	2.93×10^{-6}	0.4	2.21×10^{-7}	9.7	1.72×10^{-7}	8.7	3.59×10^{-6}	0.5
13	2.84×10^{-6}	0.5	2.45×10^{-6}	0.4	1.55×10^{-7}	7.9	1.35×10^{-7}	8.6	2.99×10^{-6}	0.7

Due to the computational resources required to run these simulations, it can be anticipated that if more detailed results are required—nuclear heating in different detector volumes, for example—a different strategy should be followed, possibly involving the generation of a secondary source at the exit of the bioshield. The benchmark of such a source could be challenging; however, that work could be simplified by defining sources only at the bioshield openings. Such an approximation seems acceptable at this stage, considering that the contribution of uncollided neutrons (14 MeV) accounts for 86–88% of the neutron heat loads, or 82–83% of the total loads, as estimated with the F5 tallies in MCNP. This approach will be explored in future simulation work.

3.3.4. Alternative Configuration of the X-ray Spectroscopy Diagnostic

As stated in Section 3.3.1, the design proposed for the X-ray spectroscopy system leads to very high neutron and gamma streaming to the port cell. As shown in Figure 28, neutron fluxes up to 2×10^{10} n cm^{-2} s^{-1} were predicted to reach the port cell through the large straight ducts (23 cm × 10 cm) of the X-ray spectroscopy system, almost four orders of magnitude higher than in the default EP configuration without diagnostics. These fluxes, along with gamma fluxes one order of magnitude lower, would lead to high dose rates that would exceed the limits in the port cell and that could compromise the integrity of electronic devices in the port cell. The sensitivity study presented in Section 3.3.2 has allowed us to evaluate the effect of reducing the cross-section of the ducts on the neutron and gamma streaming to the port cell. In parallel, alternative diagnostic duct geometries have been investigated, based on flat highly oriented pyrolytic graphite (HOPG) pre-reflectors, as included in the design of a similar system for ITER [60]. Although the feasibility of these alternative configurations is still questionable—due to low reflectivity and the possibility of increased radiation streaming to the magnets [61]—it is worthwhile to evaluate if such configurations would address the radiation streaming issue. The neutronics simulations presented in this section aim to contribute to a better understanding of the different design options for the X-ray spectroscopy system.

The reference model used for the simulations is the same as presented in the previous sections. The neutronics CAD model of the system, including the HOPG mirrors to minimize streaming, is presented in Figure 37. The three ducts maintain the previous dimensions (23 cm × 10 cm), but not in straight paths from the plasma to the port cell, as before. Since the objective was to evaluate the streaming through the X-ray ducts, the other systems, which have very small contributions to the total fluxes in the port cell, were not included in the model.

The MCNP model is presented in Figure 38, for plane z = 1 cm. The crystal Bragg reflectors in the port cell were also included, and used to tally the neutron and gamma fluxes that cross the bioshield plug. All the simulations were run using weight windows generated with the ADVANTG code and further manipulated with the iWW-GVR tool.

The neutron and gamma fluxes obtained with the alternative duct configuration are presented in Figures 39 and 40, for planes y and z. The neutron fluxes in the three crystal Bragg reflectors of the port cell were 5.6×10^7, 1.2×10^8 and 7.4×10^7 n cm^{-2} s^{-1}, with statistical errors of 24%, 43% and 10%, respectively. Even though 2×10^{10} particles were simulated, for 5–7 days per simulation and with 720 processors per simulation, the statistical errors are very large in all but one of the mirrors. Nevertheless, the results indicate that with a configuration like this, and including some shielding optimization, it should be possible to reduce the neutron fluxes to below 1×10^8 n cm^{-2} s^{-1}, which means a reduction by more than two orders of magnitude when compared to the straight ducts.

Similar reductions were obtained in the gamma fluxes: 2.6×10^7, 6.2×10^7 and 4.3×10^7 γ cm^{-2} s^{-1} in the three mirrors, with statistical errors of 7%, 9% and 6%. Again, this is almost two orders of magnitude lower than the gamma fluxes obtained with the straight ducts.

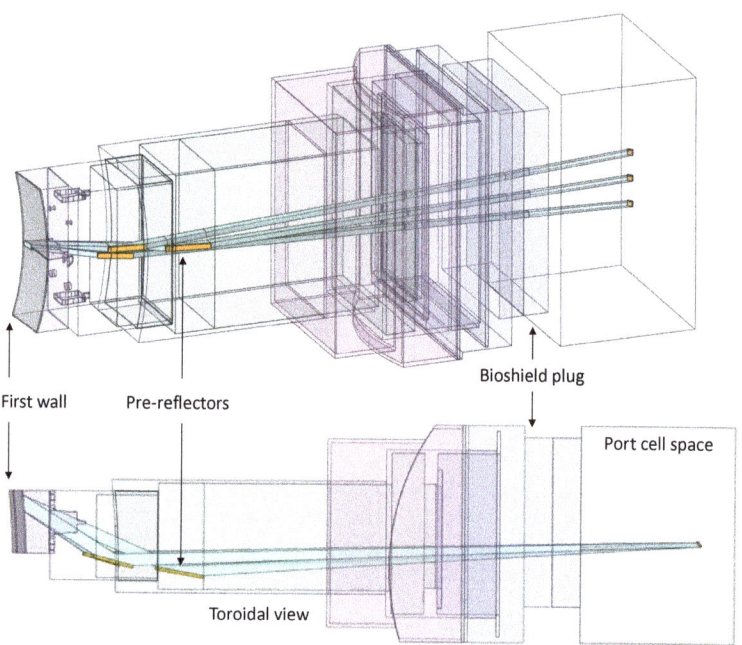

Figure 37. CAD model of the EP used in the simulations with transparent cells, showing the pre-reflectors and the diagnostic ducts of the X-ray spectroscopy system.

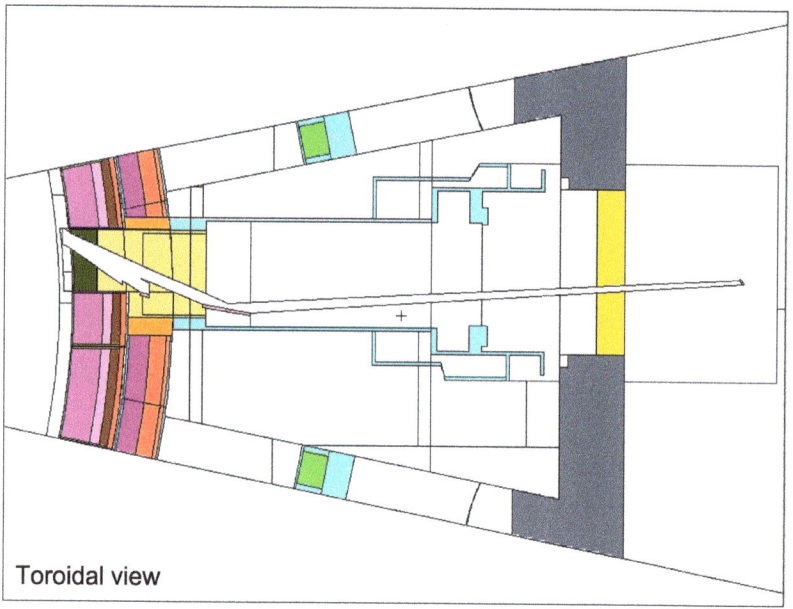

Figure 38. MCNP model of the EP, showing the X-ray spectroscopy ducts (plane z = 1 cm).

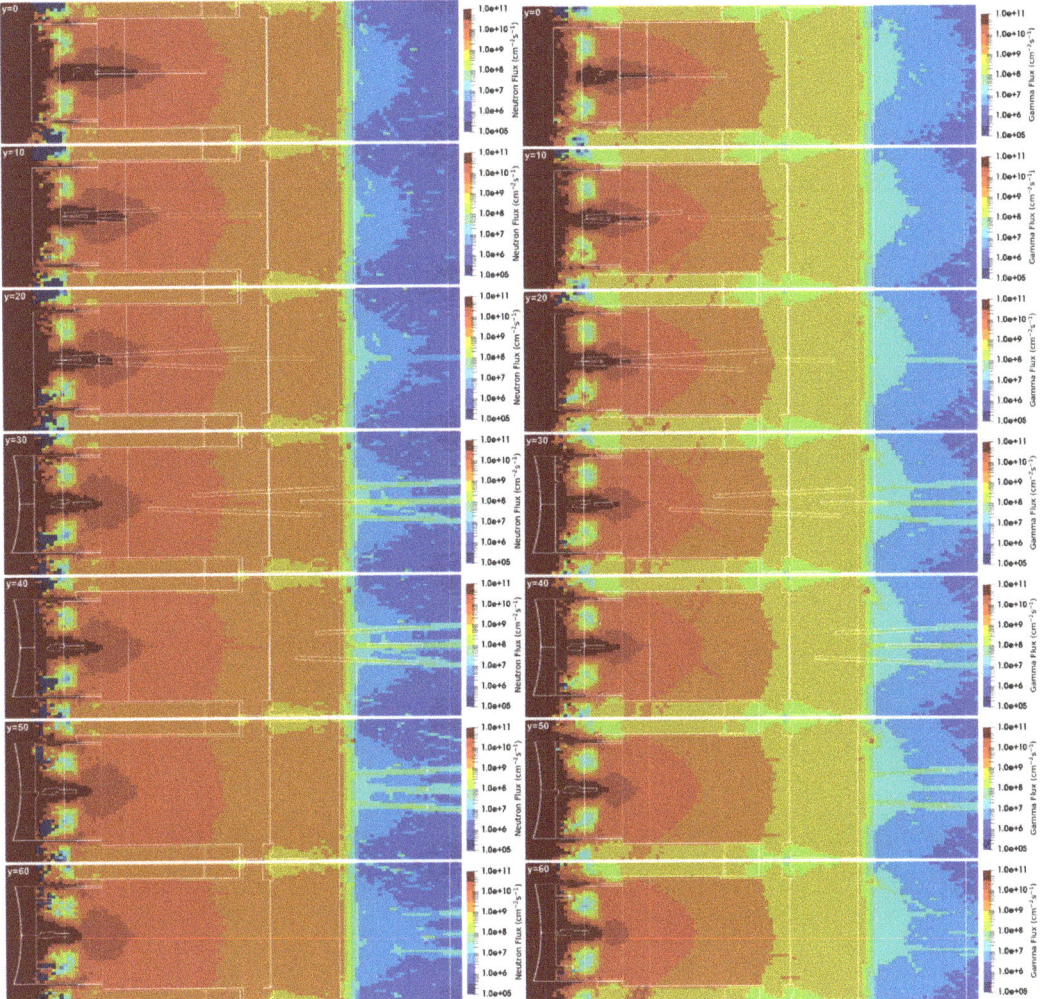

Figure 39. Neutron (n cm^{-2} s^{-1}) and gamma (γ cm^{-2} s^{-1}) fluxes in plane y with the alternative X-ray spectroscopy ducts.

As mentioned before, there are no limits defined for the neutron and gamma fluxes in the DEMO port cell, and no calculations of shutdown dose rates are provided here. However, it seems feasible, based on the simulation experience from ITER (although the dose rates will depend on the components present in the port cell and on the integrated fluxes) [62], to comply with the limit of 100 µSv/h in the port cell 12 days after shutdown with neutron fluxes below 1×10^8 n cm^{-2} s^{-1} reaching the port cell. Nevertheless, shutdown dose rate simulations with models of the port cell components are required to assess the compliance with the limit.

It should also be mentioned that without further shielding, the alternative duct configuration presented here is expected to increase the nuclear heat loads in the toroidal field coils. Further studies are therefore required to calculate these loads and to compare the results between configurations.

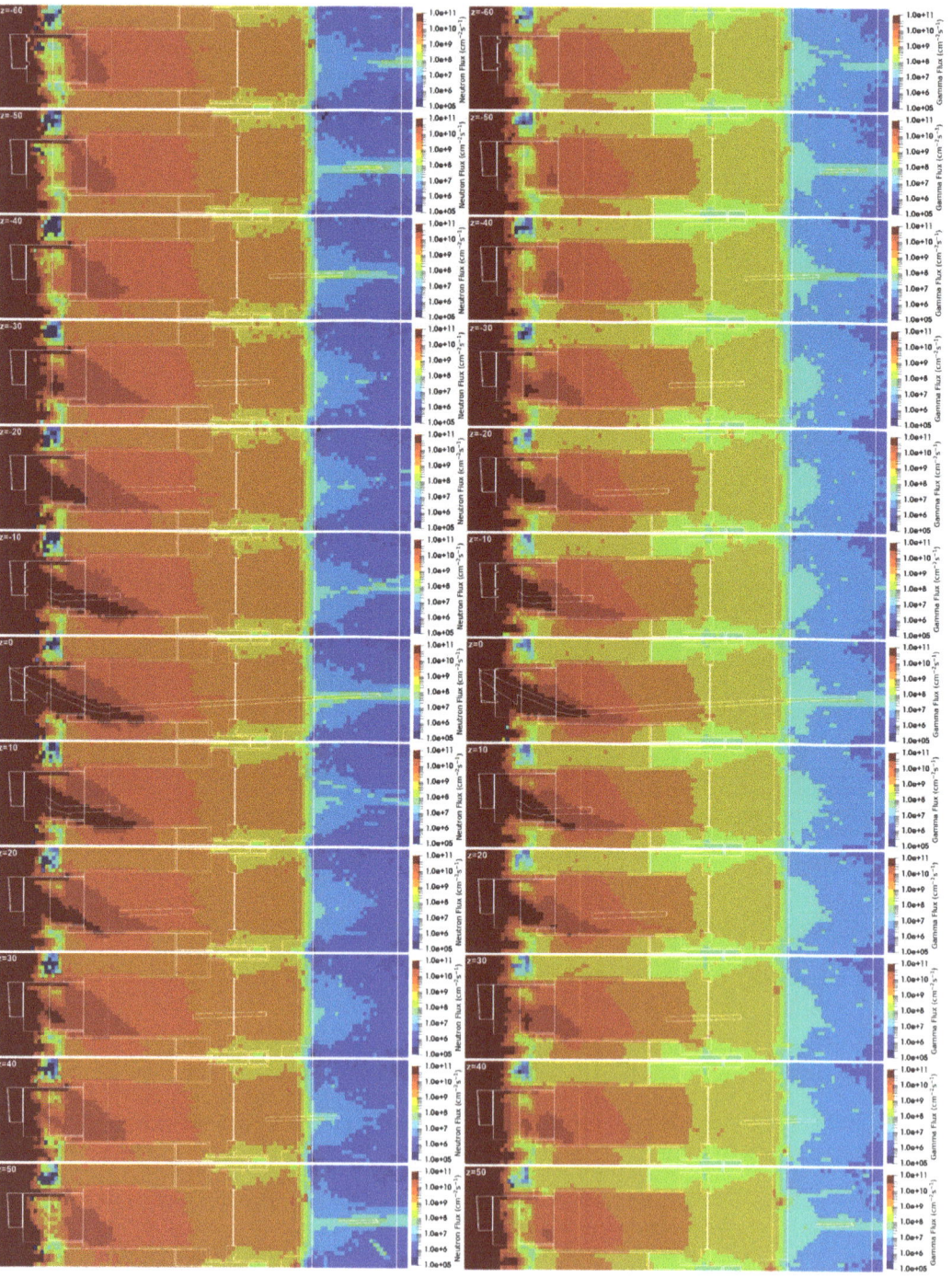

Figure 40. Neutron (n cm^{-2} s^{-1}) and gamma (γ cm^{-2} s^{-1}) fluxes in plane z with the alternative X-ray spectroscopy ducts.

4. Conclusions

This paper aimed to provide a broad view of the radiation environment that diagnostics in DEMO are expected to face, assuming as a reference the water-cooled lithium lead blanket (WCLL) configuration. Resorting to diagnostics representative of different integration approaches in DEMO—inner vessel, ex-vessel and equatorial ports—neutronics simulations were performed to estimate the fluxes, heat loads, dose rates and dpa in different sections of the tokamak, using pre-conceptual CAD models of the diagnostics.

The first simulations were related to inner-vessel diagnostics, distributed poloidally around the plasma: in-vessel magnetics sensors and the diagnostics slim cassette (DSC), projected for the integration of microwave reflectometry and the ECE. These simulations have shown that the introduction of the DSC designed for reflectometry will not compromise the mechanical integrity of the VV, as the fluxes and loads behind the DSC are comparable to the ones obtained behind the WCLL breeding blanket (BB) without the DSC. Another conclusion of this study is that the fluences in the magnetics sensors behind the blankets, integrated over the whole DEMO lifetime, are comparable to the ones expected for ITER, even though ITER will operate only for 0.54 FPY instead of the 6 FPY of DEMO. This is due to the excellent shielding performance of the current WCLL blanket design. It should be noticed, however, that the analysis presented here is not conservative: with the alternative helium cooled pebble bed (HCPB) blanket, the fluences in the sensors would increase up to a factor of 10, increasing the loads significantly in the in-vessel magnetics sensors in comparison with ITER. Additionally, the current WCLL blanket design is far from final, and a reduction in its shielding capability in the near future cannot be ruled out at this stage. Another important point is that the fluences below the divertor in DEMO are increased by at least a factor of 4 when compared to ITER. For all these reasons, R&D studies for magnetics sensors should still be based on the assumption that the loads in DEMO will exceed those expected for ITER.

Ex-vessel Faraday sensors were simulated next. As in ITER, this diagnostic is planned to be installed on the outer surface of the VV, with the aim to provide information on the plasma current during long plasma discharges. A maximum dose rate of 2.2 MGy/FPY was obtained in the simulations, which, integrated over 6 FPY, would exceed the 10 MGy considered as a conservative upper limit for the sensors' lifetime dose in ITER. Nevertheless, this happens only in the divertor region, which is not modeled as accurately as the blanket in these simulations (and where the shielding has not been optimized yet). In the remaining regions, the dose rates are up to three orders of magnitude lower.

Finally, an equatorial port containing three diagnostics—X-ray spectroscopy, divertor monitoring and pellet monitoring—was simulated in detail. Considerable neutron streaming to the port cell was predicted with the initial design of the X-ray spectroscopy diagnostic, which foresaw large (10 cm × 23 cm) straight ducts between the plasma and the port cell. With neutron fluxes up to 2×10^{10} n cm^{-2} s^{-1}, it can be anticipated that the current design would not comply with the dose rate limit of 100 µSv/h in the port cell 12 days after shutdown. A sensitivity analysis was then performed to evaluate the neutron streaming as a function of the duct cross-section, for diagnostics that require direct views of the plasma, without mirrors or doglegs. This study was extended to two such diagnostics, with 1 cm radius ducts: the neutron/gamma cameras and the radiated power and soft X-ray intensity diagnostic. Neutron fluxes of the order of $1-2 \times 10^8$ n cm^{-2} s^{-1} were obtained in the port cell for those diagnostics. Finally, an alternative design of the X-ray spectroscopy diagnostic, based on graphite pre-reflectors, was shown to reduce the neutron fluxes in the port cell to $\sim 1 \times 10^8$ n cm^{-2} s^{-1}. This configuration might, however, increase the nuclear heat loads in the toroidal field coils. Accurate shutdown dose rate calculations in the port cell should be carried out in future work, along with a detailed study of the effect of the diagnostic port configurations on the nuclear heat loads in the magnets.

Author Contributions: Conceptualization, R.L., Y.N., A.Q., A.M. and W.B.; methodology, R.L. and Y.N.; software, R.L., Y.N. and A.V.; validation, R.L., Y.N., F.C., E.P.C., M.C., B.B., A.V., J.B., A.S., A.G., A.Q., A.M. and W.B.; formal analysis, R.L. and Y.N.; investigation, R.L., Y.N., A.Q. and A.M.; resources, R.L., B.G. and A.V.; writing—original draft preparation, R.L. and Y.N.; writing—review and editing, all; visualization, R.L., Y.N., J.B., F.C. and B.B.; supervision, A.Q., A.G., A.S., E.P.C., M.C., A.M., B.G. and W.B.; project administration, B.G. and W.B.; funding acquisition, B.G. and W.B. All authors have read and agreed to the published version of the manuscript.

Funding: This work has been carried out within the framework of the EUROfusion Consortium, funded by the European Union via the Euratom Research and Training Programme (Grant Agreement No. 101052200—EUROfusion). Views and opinions expressed are, however, those of the author(s) only and do not necessarily reflect those of the European Union or the European Commission. Neither the European Union nor the European Commission can be held responsible for them. IPFN activities also received financial support from Fundação para a Ciência e Tecnologia (FCT) through projects UIDB/50010/2020 and UIDP/50010/2020 and the individual grant PD/BD/135230/2017 under the APPLAuSE Doctoral Program. This scientific paper has been published as part of the international project co-financed by the Polish Ministry of Science and Higher Education within the programme called "PMW" for 2022–2023.

Institutional Review Board Statement: Not applicable.

Informed Consent Statement: Not applicable.

Data Availability Statement: Not applicable.

Conflicts of Interest: The authors declare no conflict of interest.

References

1. Pereslavtsev, P.; Leichtle, D. DEMO Nuclear Analysis Handbook (NAH). 2020. Available online: http://idm.euro-fusion.org/?uid=2NXXCM (accessed on 2 May 2023).
2. Donné, T. The European roadmap towards fusion electricity. *Phil. Trans. R. Soc. A* **2019**, *377*, 20170432. [CrossRef]
3. Hirai, T.; Bao, L.; Barabash, V.; Chappuis, P.; Eaton, R.; Escourbiac, F.; Merola, M.; Mitteau, R.; Raffray, R.; Linke, J.; et al. High Heat Flux Performance Assessment of ITER Enhanced Heat Flux First Wall Technology after Neutron Irradiation. *Fusion Eng. Des.* **2023**, *186*, 113338. [CrossRef]
4. Richardson, M.; Gorley, M.; Wang, Y.; Aiello, G.; Pintsuk, G.; Gaganidze, E.; Richou, M.; Henry, J.; Vila, R.; Rieth, M. Technology Readiness Assessment of Materials for DEMO In-Vessel Applications. *J. Nucl. Mater.* **2021**, *550*, 152906. [CrossRef]
5. Biel, W.; Albanese, R.; Ambrosino, R.; Ariola, M.; Berkel, M.V.; Bolshakova, I.; Brunner, K.J.; Cavazzana, R.; Cecconello, M.; Conroy, S.; et al. Diagnostics for Plasma Control—From ITER to DEMO. *Fusion Eng. Des.* **2019**, *146*, 465–472. [CrossRef]
6. Biel, W.; Ariola, M.; Bolshakova, I.; Brunner, K.J.; Cecconello, M.; Duran, I.; Franke, T.; Giacomelli, L.; Giannone, L.; Janky, F.; et al. Development of a Concept and Basis for the DEMO Diagnostic and Control System. *Fusion Eng. Des.* **2022**, *179*, 113122. [CrossRef]
7. Gonzalez, W.; Biel, W.; Mertens, P.; Tokar, M.; Marchuk, O.; Mourão, F.; Linsmeier, C. Preliminary Study of a Visible, High Spatial Resolution Spectrometer for DEMO Divertor Survey. *J. Instrum.* **2020**, *15*, C01008. [CrossRef]
8. Cecconello, M.; Conroy, S.; Ericsson, G.; Hjalmarsson, H.; Franke, T.; Biel, W. Pre-Conceptual Study of the European DEMO Neutron Diagnostics. *J. Instrum.* **2019**, *14*, C09001. [CrossRef]
9. Giacomelli, L.; Nocente, M.; Perelli Cippo, E.; Rebai, M.; Rigamonti, D.; Tardocchi, M.; Cazzaniga, C.; Cecconello, M.; Conroy, S.; Hjalmarsson, A.; et al. Overview on the Progress of the Conceptual Studies of a Gamma Ray Spectrometer Instrument for DEMO. *J. Instrum.* **2022**, *17*, C08020. [CrossRef]
10. Chernyshova, M.; Dobrut, M.; Jabłoński, S.; Malinowski, K.; Fornal, T. Multi-Chamber GEM-Based Concept of Radiated Power/SXR Measurement System for Use in High Radiation Environment of DEMO. *J. Instrum.* **2022**, *17*, C05013. [CrossRef]
11. Brunner, K.J.; Marushchenko, N.; Turkin, Y.; Biel, W.; Knauer, J.; Hirsch, M.; Wolf, R.C. Design Considerations of the European DEMO's IR-Interferometer/Polarimeter Based on TRAVIS Simulations. *J. Instrum.* **2022**, *17*, C04001. [CrossRef]
12. Korsholm, S.B.; Chambon, A.; Gonçalves, B.; Infante, V.; Jensen, T.; Jessen, M.; Klinkby, E.B.; Larsen, A.W.; Luis, R.; Nietiadi, Y.; et al. ITER Collective Thomson Scattering—Preparing to Diagnose Fusion-Born Alpha Particles (Invited). *Rev. Sci. Instrum.* **2022**, *93*, 103539. [CrossRef]
13. Gonçalves, B.; Varela, P.; Silva, A.; Silva, F.; Santos, J.; Ricardo, E.; Vale, A.; Luís, R.; Nietiadi, Y.; Malaquias, A.; et al. Advances, Challenges, and Future Perspectives of Microwave Reflectometry for Plasma Position and Shape Control on Future Nuclear Fusion Devices. *Sensors* **2023**, *23*, 3926. [CrossRef]
14. Zerbini, M. Sailing on Far Infrared and Submillimeter Waves Plasma Diagnostics, towards THz-TDS and Beyond. In Proceedings of the 47th International Conference on Infrared, Millimeter and Terahertz Waves (IRMMW-THz), Delft, The Netherlands, 28 August 2022; pp. 1–4.

5. Malaquias, A.; Silva, A.; Moutinho, R.; Luis, R.; Lopes, A.; Quental, P.B.; Prior, L.; Velez, N.; Policarpo, H.; Vale, A.; et al. Integration Concept of the Reflectometry Diagnostic for the Main Plasma in DEMO. *IEEE Trans. Plasma Sci.* **2018**, *46*, 451–457. [CrossRef]
6. Belo, J.H.; Nietiadi, Y.; Luís, R.; Silva, A.; Vale, A.; Gonçalves, B.; Franke, T.; Krimmer, A.; Biel, W. Design and Integration Studies of a Diagnostics Slim Cassette Concept for DEMO. *Nucl. Fusion* **2021**, *61*, 116046. [CrossRef]
7. Luís, R.; Nietiadi, Y.; Belo, J.H.; Silva, A.; Vale, A.; Malaquias, A.; Gonçalves, B.; da Silva, F.; Santos, J.; Ricardo, E.; et al. A Diagnostics Slim Cassette for Reflectometry Measurements in DEMO: Design and Simulation Studies. *Fusion Eng. Des.* **2023**, *190*, 113512. [CrossRef]
8. Giannone, L.; El Shawish, S.; Herrmann, A.; Kallenbach, A.; Schuhbeck, K.H.; Vayakis, G.; Watts, C.; Zammuto, I. Shunt and Rogowski Coil Measurements on ASDEX Upgrade in Support of DEMO Detachment Control. *Fusion Eng. Des.* **2021**, *166*, 112276. [CrossRef]
9. Wuilpart, M.; Gusarov, A.; Leysen, W.; Batistoni, P.; Moreau, P.; Dandu, P.; Megret, P. Polarimetric Optical Fibre Sensing for Plasma Current Measurement in Thermonuclear Fusion Reactors. In Proceedings of the 22nd International Conference on Transparent Optical Networks (ICTON), Bari, Italy, 19–23 July 2020; pp. 1–4.
10. Quercia, A.; Pironti, A.; Bolshakova, I.; Holyaka, R.; Duran, I.; Murari, A.; Contributors, J. Long Term Operation of the Radiation-Hard Hall Probes System and the Path toward a High Performance Hybrid Magnetic Field Sensor. *Nucl. Fusion* **2022**, *62*, 106032. [CrossRef]
11. Quercia, A.; Albanese, R.; Fresa, R.; Minucci, S.; Arshad, S.; Vayakis, G. Performance Analysis of Rogowski Coils and the Measurement of the Total Toroidal Current in the ITER Machine. *Nucl. Fusion* **2017**, *57*, 126049. [CrossRef]
12. Vila, R.; Hodgson, E.R. A TIEMF Model and Some Implications for ITER Magnetic Diagnostics. *Fusion Eng. Des.* **2009**, *84*, 1937–1940. [CrossRef]
13. Luis, R.; Moutinho, R.; Prior, L.; Quental, P.B.; Lopes, A.; Policarpo, H.; Velez, N.; Vale, A.; Silva, A.; Malaquias, A. Nuclear and Thermal Analysis of a Reflectometry Diagnostics Concept for DEMO. *IEEE Trans. Plasma Sci.* **2018**, *46*, 1247–1253. [CrossRef]
14. Nietiadi, Y.; Luís, R.; Silva, A.; Ricardo, E.; Gonçalves, B.; Franke, T.; Biel, W. Nuclear and Thermal Analysis of a Multi-Reflectometer System for DEMO. *Fusion Eng. Des.* **2021**, *167*, 112349. [CrossRef]
15. Luís, R.; Nietiadi, Y.; Silva, A.; Gonçalves, B.; Franke, T.; Biel, W. Nuclear Analysis of the DEMO Divertor Survey Visible High-Resolution Spectrometer. *Fusion Eng. Des.* **2021**, *169*, 112460. [CrossRef]
16. Goorley, T.; James, M.; Booth, T.; Brown, F.; Bull, J.; Cox, L.J.; Durkee, J.; Elson, J.; Fensin, M.; Forster, R.A.; et al. Initial MCNP6 Release Overview. *Nucl. Technol.* **2012**, *180*, 298–315. [CrossRef]
17. Mckinney, G.W.; Brown, F.B., III; Hughes, G.H.; James, M.R.; Martz, R.L.; McMath, G.E.; Wilcox, T. MCNP 6.1.1—New Features Demonstrated. In Proceedings of the IEEE Nuclear Scientific Symposium, Seattle, WA, USA, 8–15 November 2014.
18. Plompen, A.J.M.; Cabellos, O.; De Saint Jean, C.; Fleming, M.; Algora, A.; Angelone, M.; Archier, P.; Bauge, E.; Bersillon, O.; Blokhin, A.; et al. The Joint Evaluated Fission and Fusion Nuclear Data Library, JEFF-3.3. *Eur. Phys. J. A* **2020**, *56*, 181. [CrossRef]
19. Fischer, U.; Kondo, K.; Angelone, M.; Batistoni, P.; Villari, R.; Bohm, T.; Sawan, M.; Walker, B.; Konno, C. Benchmarking of the FENDL-3 Neutron Cross-Section Data Library for Fusion Applications. *Nucl. Data Sheets* **2014**, *120*, 230–234. [CrossRef]
20. CATIA®V5, Dassault Systèmes, R2021 (R31) 2021. Available online: https://www.3ds.com/products-services/catia/ (accessed on 2 May 2023).
21. Ansys®Academic Research Mechanical, Release 2021R2. Available online: https://www.ansys.com/academic (accessed on 2 May 2023).
22. Li, Y.; Lu, L.; Ding, A.; Hu, H.; Zeng, Q.; Zheng, S.; Wu, Y. Benchmarking of MCAM 4.0 with the ITER 3D Model. *Fusion Eng. Des.* **2007**, *82*, 2861–2866. [CrossRef]
23. Wu, Y.; Song, J.; Zheng, H.; Sun, G.; Hao, L.; Long, P.; Hu, L. CAD-Based Monte Carlo Program for Integrated Simulation of Nuclear System SuperMC. *Ann. Nucl. Energy* **2015**, *82*, 161–168. [CrossRef]
24. Lu, L.; Qiu, Y.; Fischer, U. Improved Solid Decomposition Algorithms for the CAD-to-MC Conversion Tool McCad. *Fusion Eng. Des.* **2017**, *124*, 1269–1272. [CrossRef]
25. Iannone, F.; Bracco, G.; Cavazzoni, C.; Coelho, R.; Coster, D.; Hoenen, O.; Maslennikov, A.; Migliori, S.; Owsiak, M.; Quintiliani, A.; et al. MARCONI-FUSION: The New High Performance Computing Facility for European Nuclear Fusion Modelling. *Fusion Eng. Des.* **2018**, *129*, 354–358. [CrossRef]
26. Wolfram Research, Inc. Mathematica Edition: Version 13.0.0 2021. Available online: https://www.wolfram.com/mathematica/new-in-13/ (accessed on 2 May 2023).
27. Van Rossum, G.; Drake, F.L. *Python 3 Reference Manual*; CreateSpace: Scotts Valley, CA, USA, 2009; ISBN 1-4414-1269-7.
28. Ayachit, U. *The ParaView Guide: Updated for ParaView Version 4.3*; Avila, L., Ed.; Full Color Version; Kitware: Los Alamos, NM, USA, 2015; ISBN 978-1-930934-30-6.
29. Sartori, E. *VITAMIN-J, A 175 Group Neutron Cross Section Library Based on JEF-1 for Shielding Benchmark Calculations*; NEA: Washington, DC, USA, 1985.
30. You, J.H.; Mazzone, G.; Visca, E.; Greuner, H.; Fursdon, M.; Addab, Y.; Bachmann, C.; Barrett, T.; Bonavolontà, U.; Böswirth, B.; et al. Divertor of the European DEMO: Engineering and Technologies for Power Exhaust. *Fusion Eng. Des.* **2022**, *175*, 113010. [CrossRef]
31. Flammini, D. 2017 Generic DEMO MCNP Model at 22.5 Degree v1.3. 2019. Available online: http://idm.euro-fusion.org/?uid=2MCJ69 (accessed on 2 May 2023).

42. Moro, F. Analysis Model: 2018_DEMO_WCLL_layered_SM-Model. 2019. Available online: http://idm.euro-fusion.org/?uid=2LF65Z (accessed on 2 May 2023).
43. Moro, F.; Del Nevo, A.; Flammini, D.; Martelli, E.; Mozzillo, R.; Noce, S.; Villari, R. Neutronic Analyses in Support of the WCLL DEMO Design Development. *Fusion Eng. Des.* **2018**, *136*, 1260–1264. [CrossRef]
44. Cufar, A. 2017-DEMO_G_MCNP_HCPB+WCLL_Multi_PMI-3.3_V1.0_2020_port_neutronics.i. 2020. Available online: https://idm.euro-fusion.org/?uid=2NP5TU&version=v1.1 (accessed on 2 May 2023).
45. Cufar, A. MCNP DEMO 2020 Port Neutronics Model. 2020. Available online: https://idm.euro-fusion.org/?uid=2NQ7FD (accessed on 2 May 2023).
46. Fabbri, M.; Cubí, Á. IWW-GVR: A Tool to Manipulate MCNP Weight Window (WW) and to Generate Global Variance Reduction (GVR) Parameters. 2019. Available online: https://github.com/Radiation-Transport/iWW-GVR#iww-gvr-a-tool-to-manipulate-mcnp-weight-window-ww-and-to-generate-global-variance-reduction-gvr-parameters (accessed on 2 May 2023).
47. Mosher, S.W.; Bevill, A.M.; Johnson, S.R.; Ibrahim, A.M.; Daily, C.R.; Evans, T.M.; Wagner, J.C.; Johnson, J.O.; Grove, R.E. *ADVANTG—An Automated Variance Reduction Parameter Generator*; Oak Ridge National Laboratory: Oak Ridge, TN, USA, 2013.
48. Ma, Y.; Vayakis, G.; Begrambekov, L.B.; Cooper, J.-J.; Duran, I.; Hirsch, M.; Laqua, H.P.; Moreau, P.; Oosterbeek, J.W.; Spuig, P.; et al. Design and Development of ITER High-Frequency Magnetic Sensor. *Fusion Eng. Des.* **2016**, *112*, 594–612. [CrossRef]
49. Serikov, A.; Bertalot, L.; Clough, M.; Fischer, U.; Suarez, A. Neutronics Analysis for ITER Cable Looms. *Fusion Eng. Des.* **2015**, *96–97*, 943–947. [CrossRef]
50. Moro, F.; Arena, P.; Catanzaro, I.; Colangeli, A.; Del Nevo, A.; Flammini, D.; Fonnesu, N.; Forte, R.; Imbriani, V.; Mariano, G.; et al. Nuclear Performances of the Water-Cooled Lithium Lead DEMO Reactor: Neutronic Analysis on a Fully Heterogeneous Model. *Fusion Eng. Des.* **2021**, *168*, 112514. [CrossRef]
51. Gusarov, A.; Leysen, W.; Kim, S.M.; Dandu, P.; Wuilpart, M.; Danisi, A.; Barbero Soto, J.L.; Vayakis, G. Recent Achievements in R&D on Fibre Optics Current Sensor for ITER. *Fusion Eng. Des.* **2023**, *192*, 113626. [CrossRef]
52. Valentine, A.; Fonnesu, N.; Bieńkowska, B.; Łaszyńska, E.; Flammini, D.; Villari, R.; Mariano, G.; Eade, T.; Berry, T.; Packer, L. Neutronics Assessment of EU DEMO Alternative Divertor Configurations. *Fusion Eng. Des.* **2021**, *169*, 112663. [CrossRef]
53. Goussarov, A.; Balazs, L. Review of the Development Status of Relevant Faraday Sensor Measurement Systems. 2021. Available online: https://idm.euro-fusion.org/?uid=2NU853 (accessed on 2 May 2023).
54. Krimmer, A. DC-4-T050-D001 Final Report on Integration and Distribution of DEMO Port Based Plasma Diagnostic Systems. 2021. Available online: https://idm.euro-fusion.org/?uid=2NX42F (accessed on 2 May 2023).
55. Lopes, A.; Luís, R.; Klinkby, E.; Nietiadi, Y.; Chambon, A.; Nonbøl, E.; Gonçalves, B.; Jessen, M.; Korsholm, S.B.; Larsen, A.W.; et al. Shielding Analysis of the ITER Collective Thomson Scattering System. *Fusion Eng. Des.* **2020**, *161*, 111994. [CrossRef]
56. Späh, P.; Fanale, F.; Bruschi, A. Integration Model of the EC Equatorial Launcher, V1. 2020. Available online: http://idm.euro-fusion.org/?uid=2NT7FJ (accessed on 2 May 2023).
57. Del Nevo, A.; Oron-Carl, M. Internal Deliverable BB-3.2.1-T005-D001: WCLL Design Report 2018. 2019. Available online: http://idm.euro-fusion.org/?uid=2NUPDT (accessed on 2 May 2023).
58. Franke, T. Equatorial Outboard Limiter Technical Specification v1.3. 2019. Available online: http://idm.euro-fusion.org/?uid=2NSZDZ (accessed on 2 May 2023).
59. Shultis, J.K.; Faw, R.E. An MCNP Primer. 2011. Available online: https://www.mne.k-state.edu/~jks/MCNPprmr.pdf (accessed on 24 May 2023).
60. Krimmer, A. Design Description Document of the High Resolution Core X-Ray Spectroscopy System. 2021. Available online: https://idm.euro-fusion.org/?uid=2P9N92 (accessed on 2 May 2023).
61. Biel, W.; Krimmer, A.; Marchuk, O. DC-S.01.09-T001-D004 Final Report on Conceptual Studies for VUV and X-ray. 2022. Available online: https://idm.euro-fusion.org/?uid=2PSL8Y (accessed on 2 May 2023).
62. Chambon, A.; Luís, R.; Klinkby, E.; Nietiadi, Y.; Rechena, D.; Gonçalves, B.; Jessen, M.; Korsholm, S.B.; Larsen, A.W.; Lauritzen, B.; et al. Assessment of Shutdown Dose Rates in the ITER Collective Thomson Scattering System and in Equatorial Port Plug 12. *J. Instrum.* **2021**, *16*, C12001. [CrossRef]

Disclaimer/Publisher's Note: The statements, opinions and data contained in all publications are solely those of the individual author(s) and contributor(s) and not of MDPI and/or the editor(s). MDPI and/or the editor(s) disclaim responsibility for any injury to people or property resulting from any ideas, methods, instructions or products referred to in the content.

Article

Design and Development of a Diagnostic System for a Non-Intercepting Direct Measure of the SPIDER Ion Source Beamlet Current

Tommaso Patton [1,*], Alastair Shepherd [1,2], Basile Pouradier Duteil [1,3], Andrea Rigoni Garola [1], Matteo Brombin [1,4], Valeria Candeloro [1,5], Gabriele Manduchi [1,4], Mauro Pavei [1], Roberto Pasqualotto [1,4], Antonio Pimazzoni [1], Marco Siragusa [1], Gianluigi Serianni [1,4], Emanuele Sartori [1,6], Cesare Taliercio [1,4], Paolo Barbato [1,4], Vannino Cervaro [1], Raffaele Ghiraldelli [1,4], Bruno Laterza [1] and Federico Rossetto [1,4]

1. Consorzio RFX, 35127 Padua, Italy; alastair.shepherd@igi.cnr.it (A.S.); basile.duteil@igi.cnr.it (B.P.D.); andrea.rigoni@igi.cnr.it (A.R.G.)
2. Culham Centre for Fusion Energy, Culham Science Centre, United Kingdom Atomic Energy Authority, Abingdon, Oxfordshire OX14 3DB, UK
3. Swiss Plasma Center (SPC), Ecole Polytechnique Fédérale de Lausanne (EPFL), 1015 Lausanne, Switzerland
4. Institute for Plasma Science and Technologies of National Research Council (ISTP-CNR), 35127 Padua, Italy
5. Centro Ricerche Fusione (CRF), University of Padua, 35127 Padua, Italy
6. Department of Management and Engineering, University of Padua, 36100 Vicenza, Italy
* Correspondence: tommaso.patton@igi.cnr.it

Abstract: Stable and uniform beams with low divergence are required in particle accelerators; therefore, beyond the accelerated current, measuring the beam current spatial uniformity and stability over time is necessary to assess the beam performance, since these parameters affect the perveance and thus the beam optics. For high-power beams operating with long pulses, it is convenient to directly measure these current parameters with a non-intercepting system due to the heat management requirement. Such a system needs to be capable of operating in a vacuum in the presence of strong electromagnetic fields and overvoltages, due to electrical breakdowns in the accelerator. Finally, the measure of the beam current needs to be efficiently integrated into a pulse file with the other relevant plant parameters to allow the data analyses required for beam optimization. This paper describes the development, design and commissioning of such a non-intercepting system, the so-called beamlet current monitor (BCM), aimed to directly measure the electric current of a particle beam. In particular, the layout of the system was adapted to the SPIDER experiment, the ion source (IS) prototype of the heating neutral beam injectors (HNB) for the ITER fusion reactor. The diagnostic is suitable to provide the electric current of five beamlets from DC up to 10 MHz.

Keywords: neutral beam injector; negative ions accelerator; beam fluctuations; SPIDER; ITER

1. Introduction

SPIDER [1] (Source for the Production of Ions of Deuterium Extracted from a Radio frequency plasma) is the ion source (IS) full-scale prototype of the ITER heating neutral beam (HNB) and has been operating at the Padova Neutral Beam Test Facility (NBTF) since June 2018. The main purpose of the SPIDER experiment is to optimize the ITER HNB IS, and it is currently the largest radiofrequency-driven negative ion source operating in the world.

The main beam diagnostics are beam emission spectroscopy, visible imaging and the diagnostic calorimeter STRIKE [2], and none of those systems can provide a direct measurement of the beamlet current. In the past, the beamlet current has always been indirectly estimated by STRIKE; which consists of 16 unidirectional carbon fiber–carbon matrix (CFC) composite tiles, placed downstream of the set of grids that accelerate the beam. The tiles are exposed to the beam, and their temperatures are recorded using two infrared cameras. The beamlet current is then estimated from the thermal footprint of each beamlet via calorimetry, thanks to the moderate broadening of the temperature profile guaranteed by the anisotropy of CFC. However, this measurement also takes into account the thermal effect of neutral particles generated inside the accelerator (i.e., H^0 generated by stripping part of the negative ions) and passing through the grid apertures, and it has a low time resolution. For this reason, an electrical measurement was introduced to measure the current collected by each tile of STRIKE, which required a positive bias of the tiles to recollect the electrons emitted from the surface due to the impact of energetic particles from the beam. While this additional measurement provides useful information, it has some important drawbacks as only the average beamlet current of each tile can be extrapolated, losing the possibility of a spatial distribution of the beam on a scale smaller than the beamlet group. Moreover, since the tiles are positively biased, electrons generated between the last grid of the electrostatic accelerator and STRIKE, via the interaction of the beam with the background gas, are also collected from the electric measurement, leading to an overestimation of the beam current.

In this framework, the development of a new diagnostic system capable of precisely measuring the beamlet current with satisfactory time and spatial resolution is particularly advised and the purposes are several. Firstly, to directly measure the beamlet current to compare it to established SPIDER diagnostics. Secondly, for the possibility of assessing the beam uniformity, utilizing the beamlet separation. Finally, to investigate whether the RF field and a beating frequency in the plasma due to the differing oscillator frequencies affect the beamlet's current, and thus optics [3]. For these reasons, a new beam diagnostic called beamlet current monitor (BCM) was designed, developed, installed, commissioned and tested in SPIDER.

2. SPIDER

The ITER HNBs are required to supply 16.5 MW power each, with beam particles (hydrogen or deuterium) electrostatically accelerated up to 1 MeV and with a divergence lower than 7 mrad [4]. The plasma generation in neutral beam ion sources is typically conducted either by filament arc discharge (e.g., the negative ion source for neutral beam injection at JT-60 [5]) or via inductive coupling, as is performed at IPP-Garching's BUG [6] and ELISE [7] facilities. The plasma generation in SPIDER is based on the latter system, by upscaling the design of the RF-driven BUG and ELISE ion sources.

In SPIDER, the plasma is generated inside the IS by eight driver coils arranged in a 4×2 matrix; the driver coils of each row are connected in series and powered by a 200 kW self-excited push–pull RF generator. Each generator can oscillate within a 1 MHz $\pm$ 0.1 MHz range, and the self-oscillation frequency is imposed by the plasma parameters and by an adjustable variable capacitor C_v located at the output of each oscillator. For each oscillator, the value of C_v is adjusted to match the load impedance (transmission line, matching network, plasma equivalent impedance and stray parameters of the circuit), resulting in a variation of several kHz among the four generators since the load impedance is not identical for all the generators. Furthermore, the working frequency of each oscillator cannot assume any value (by adjusting C_v) within the oscillator frequency range due to the frequency flip phenomenon [8] which consists of an uncontrolled jump of the oscillator frequency when operating close to the resonant frequency of the load. This phenomenon further reduces the possibility to obtain good matching of the load with the same oscillation frequency for all generators.

The negative ions are generated inside the IS by volume or by surface production mechanisms, with the latter using evaporated cesium in the source to decrease the work function of the H^- production surfaces. With Cs, the extracted current density increases by around a factor of 5–10 [9]. To optimize the H^- extraction, a transverse magnetic filter is generated inside the ion source to confine fast electrons in the upstream region of the IS (driver region), avoiding the H^- destruction by high-energy electrons in the proximity of the plasma grid (PG), which closes the plasma chamber. This magnetic field is generated by DC flowing vertically inside the PG and returning via a busbar system, which can be regulated from 0 A up to 5 kA.

The negative ions are then extracted from the IS and accelerated by a three-grid electrostatic accelerator composed of the PG, extraction grid (EG) and grounded grid (GG) which can be biased up to −108 kV, −96 kV and 0 V, respectively, with respect to the ground potential. Each grid presents 1280 apertures divided into 16 beamlet groups, formed by 16 rows with five beamlets each, arranged in a 4 × 4 matrix. Embedded magnets are also present in the EG to suppress the co-extracted electrons [1] by bending their trajectory without substantially affecting the trajectories of the negative ions due to their much larger Larmor radius. The ion deflection by the suppression magnets is adjusted at the GG via electrostatic or magnetic compensation; the former via the displacement of the GG holes and the latter by embedded magnets inside the GG. Finally, a ferromagnetic sheet is placed in contact with the downstream side of the GG to confine the magnetic field inside the accelerator.

Since the beginning of the SPIDER operation, electric arcs involving the back of the IS started to arise when the vessel pressure was above 40 mPa (corresponding to 0.1 Pa inside the IS for H_2) under some experimental conditions due to large RF voltages [10]. Therefore, ahead of a major modification of the pumping system, a molybdenum mask of 0.25 mm thickness was fastened downstream of the PG (a set of pushers press the mask against the PG) to reduce the number of open apertures [11]. Initially, a mask with 80 open apertures was used, with a further reduction to 28 for Cs evaporation. This temporary solution provides a lower gas flow conductance between the IS and the vessel, allowing operation up to 0.45 Pa in the source with minimized RF discharge occurrence.

The presence of this newly available space between open grid apertures, a consequence of the use of the PG mask, was deemed a perfect opportunity to design and install a temporary diagnostic dedicated to the measurements of single beamlets.

3. BCM Concept Design

The design of the BCM is driven by the dual desire to measure both the DC and AC components of the accelerated beamlet current. The primary requirements, based on the experimental performance of SPIDER and the expected beam current with Cs evaporation, are summarized in Table 1.

Table 1. Identified main requirements for the BCM diagnostics.

Requirements	Value	Note
Current full-scale	≥40 mA	Max beamlet current I_b for the nominal accelerated D-current of 50 A
Current resolution	≤1 mA	Max allowed beam non-uniformity for I_b = 10 mA (early phase of surface production operation)
Sensitivity	≥5 mV/mA	Typical ADC sensitivity considering noise
Bandwidth	DC-10 MHz	Beatings in the kHz range and MHz harmonics due to RF
Clearance	≥20 mm	Beamlet cross-section at GG, divergence and deflection
External diameter	100 mm	Minimum distance between adjacent beamlets, divergence
Repeller voltage	≥100 V	STRIKE potential

Although the average current of a single beamlet was estimated to be about 3–5 mA via STRIKE calorimetry, the full scale of the BCM system was chosen by considering the SPIDER nominal beamlet current when operating with cesium close to the optimal parameters, since BCM is planned to operate in both the volume and surface production phases.

A resolution lower than 1 mA, i.e., the minimum current variation which can be resolved, was defined on the basis of the maximum non-uniformity allowed (10%) considering a beamlet current of 10 mA, corresponding to the expected accelerated current in the early phase of surface production with non-optimized source parameters. The sensitivity, i.e., the output voltage of the instrument over the input current, was chosen to be at least

5 mV/mA on the basis of the typical sensitivity of general-purpose oscilloscopes, also considering the installation of the diagnostic in a noisy environment such as SPIDER.

High-frequency fluctuations in the plasma, measured by a Langmuir triple probe [12], were identified at the fundamental frequency of the oscillators and their harmonics (Figure 1, highlighted blue). This observation is supported by the literature. High-frequency harmonics around the RF excitation field frequency may be present [13], and in particular, the ones up to three times the fundamental frequency seem to be the most relevant in terms of magnitude [14]. Peaks between 1 kHz and 100 kHz were also observed (Figure 1, highlighted green), with the lower frequency peaks compatible with the beatings of the power delivered to the plasma from each RF generator (blue cells of the table in Figure 1). For all the aforementioned reasons, a bandwidth from DC up to 10 MHz is a requirement of the BCM diagnostic.

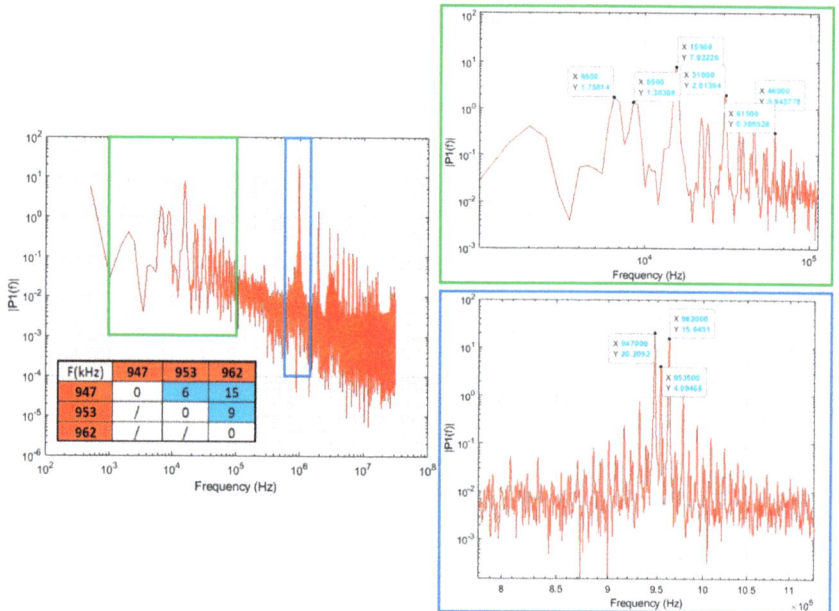

Figure 1. Langmuir triple probe signal, shot 8288, blip 3, volume operation with three out of four RF generators active. Generators working frequencies and related beating frequency are also shown in the red and light blue cells of the table (**left**). Zoom view around the beating frequencies (**top right**) and around the RF fundamental frequency (**bottom right**).

Beyond the electrical requirements of the current transducer, the measurement system shall also be equipped with a repeller which is a positively biased copper electrode crossed by the beam and placed downstream of the GG. The purpose of the repeller is to prevent positive ions, generated between the GG and STRIKE due to beam–gas interactions, from entering backwards to the current transducers of the BCM under the effect of the stray electric field penetrating through the GG apertures in the downstream region [15]. Furthermore, the STRIKE positive bias also helps positive ions move backward towards the GG. The requirement in terms of repeller voltage was chosen as $V_{rep} \geq 100$ V, which is a reasonable value for reducing the backstreaming ion current consistently since STRIKE is usually biased at 60 V.

Finally, the sensor size has also been assessed carefully, considering the available room, beam deflection, width and divergence downstream of the GG. This is a basic requirement since the interception of a steady-state high-energy beamlet (or part of it) would compromise not only the reliability of the BCM diagnostics but also the reliability of the whole SPIDER experiment. The sensor sizing was assessed in [2] by considering the worst conditions in terms of optics: 60 mrad divergence and 20 mrad horizontal deflection with a beamlet width at the exit of the GG equal to 16 mm, corresponding to the aperture diameter. In this conservative configuration, the minimum distance between the center of two adjacent apertures is 66 mm since the vertical pitch of the GG is 22 mm and a minimum of two apertures are masked between two beamlets, providing the maximum available space for the individual diagnostics (Figure 2). Therefore, the requirements concerning the sensor size were defined as clearance $\geq$ 20 mm and external diameter $\leq$ 100 mm depending on the distance from the GG (keeping below 200 mm).

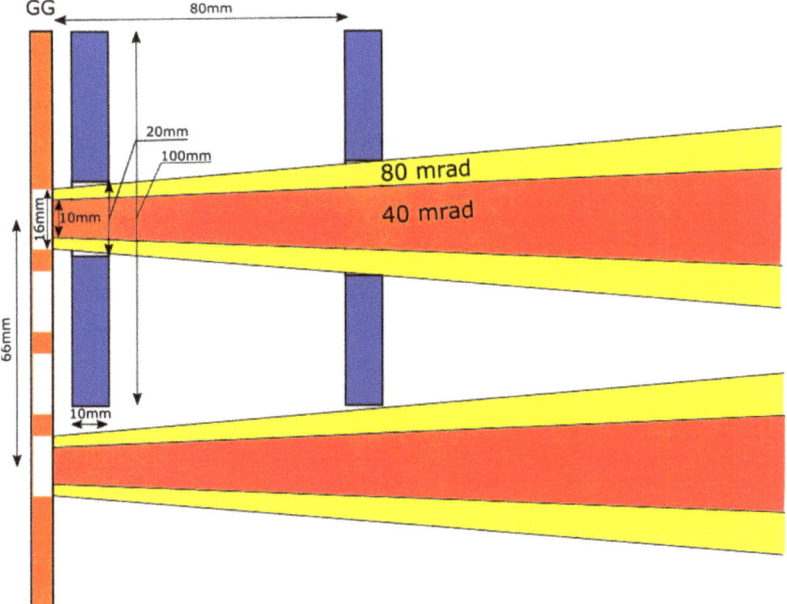

Figure 2. Vertical cross-section of 2 adjacent beamlets downstream of the GG (orange, apertures in white) with PG mask installed for the worst optics conditions (yellow cones) and for typical optics conditions (red cones); quotes are in mm. Example of minimum inner and outer diameters required for sensors (blue) depending on the distance from the GG.

4. Sensor R&D

4.1. AC Sensors

Two off-the-shelf current transformers (CTs) were chosen based on the high-frequency requirements, the BERGOZ ACCT-055 10 mA full scale (3 Hz–1 MHz) [16] and the Magnelab CT-F5.0_BNC (4.8 kHz–400 MHz) [17]. As an alternative to commercial CTs, a custom passive CT was designed and built in-house, based on the literature [18–20].

The current transformer (Figure 3a) consists of a magnetic core that surrounds the H$^-$ beamlet. The secondary winding is wound around the core for N turns and terminated by a parallel load resistance R_l. The secondary winding has a magnetizing inductance that, along with the core losses, can be expressed as a parallel inductance L_m and resistance R_m, with series resistance R_s, in the CT equivalent circuit (Figure 3b). To measure the transfer function, the sensor was connected to an HP 4194A Network Analyzer, with input impedance (R_0 parallel to C_0), via an RG58 coaxial cable with series inductance L_m and resistance R_m and parallel capacitance C_c, modeled as a π-junction.

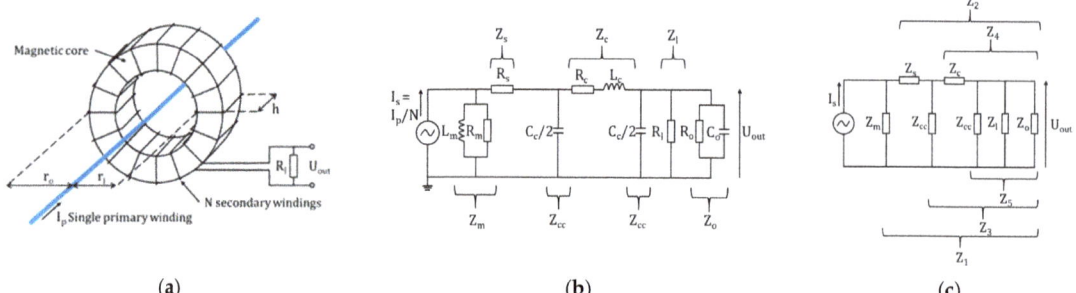

Figure 3. Current transformer: (**a**) diagram; (**b**) equivalent circuit; (**c**) equivalent circuit with grouped impedances for calculating the transfer function.

The current through the secondary winding of the CT I_s is the current of the primary winding I_p, in this case, the beamlet current, divided by the number of turns:

$$I_s = \frac{I_p}{N} \quad (1)$$

For a simplified ideal current transformer, the gain in the mid-frequency range is:

$$K = \frac{R_l}{N} \quad (2)$$

with the gain K dependent on the load resistance and the number of turns. A load resistance of 50 Ω and 10 turns were chosen to give the required 5 V/A gain. The low-frequency cutoff (−3 dB) f_{low} of a CT:

$$f_{low} = R_l / 2\pi L_m \quad (3)$$

is dependent on L_m, since R_l is chosen to set the gain. The impedance L_m and losses R_m of a magnetic core are dependent on the frequency, core permeability and geometry [21]. With the geometry limited by the available space around the beamlets, a high permeability material, nanocrystalline VITROPERM 500F [22,23], was chosen due to its high relative permeability $\mu_r = 50,000$.

The CT was modeled in both MATLAB and PSIM, with the impedances in the MATLAB model grouped as in Figure 3c. By treating the circuit as a voltage divider, the transfer function can be calculated (see Appendix A):

$$K = \frac{1}{N} \cdot \frac{Z_1 Z_3 Z_5}{Z_2 Z_4} \quad (4)$$

Using the MATLAB and PSIM models, a CT was designed to meet the gain, frequency and space requirements, with the parameters given in Table 2.

Table 2. CT parameters for model and measurement.

CT Parameter	Value
r_i [m]	0.03
r_o [m]	0.045
h [m]	0.02
N	10
μ_r	50,000
R_s [Ω]	0.22
R_c [Ω/m]	0.048

Table 2. *Cont.*

CT Parameter	Value
L_c [nH/m]	250
C_c [pF/m]	100
L_c [m]	15
R_l [Ω]	50
R_o [MΩ]	1
C_o [pF]	50
K [V/A]	4.9
f_{low} (−3 dB) [Hz]	1000

A CT prototype was constructed including a 15 m RG58 coaxial cable (the length required for the installation in SPIDER) connecting the CT to the 50 Ω load resistor. Figure 4 shows the measured response (purple crosses), which compares favorably with the MAT-LAB (yellow line) and PSIM (red points) models. At high frequencies, the gain and phase exhibit resonances, with the gain oscillations damped below 20 MHz. These resonances are due to the transmission line, which at 15 m was modeled as 50 junctions, giving a closer match to the measured frequency response.

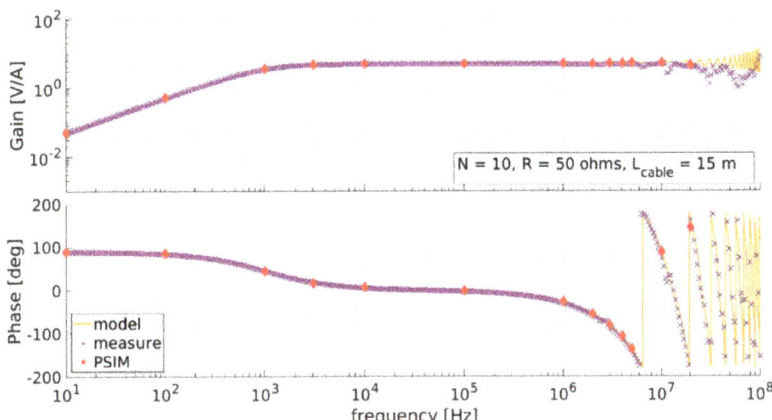

Figure 4. Measured CT response (purple) compared to MATLAB model (yellow) and PSIM model with 50 π junctions.

4.2. DC Sensors

The sensor chosen to fulfill the DC and low-frequency measurements is the LEM CTSR 0.3p [24], a closed-loop fluxgate [25] with a sensitivity of 4 mV/mA. While the sensor fulfills most of the design requirements, this sensitivity is insufficient since the noise picked up on the cables from the sensor to the acquisition system may easily exceed this value. Additionally, the sensor has a "natural" offset of 2.5 ± 0.005 V, which must be removed to utilize the full range of the measuring oscilloscope, since it is not convenient to read a small quantity (e.g., few mV) when the full scale of the acquisition system ADC is higher than three orders of magnitude (e.g., FS > 2500 mV). A custom conditioning circuit was designed both to cancel out the offset (Figure 5) and to amplify the output signal to obtain a sensitivity of the order of 250 mV/mA. The circuit's central component is the INA 114 instrumentation amplifier from Texas Instruments. Provided with an input signal V_{in}, a reference signal V_{ref} (2.5 V in our case, taken directly from the LEN internal reference) and an interchangeable resistor R_g, the amplifier produces a signal V_{out} given by:

$$V_{out} = (V_{in} - V_{ref})\left(1 + \frac{50k\Omega}{R_g}\right) \quad (5)$$

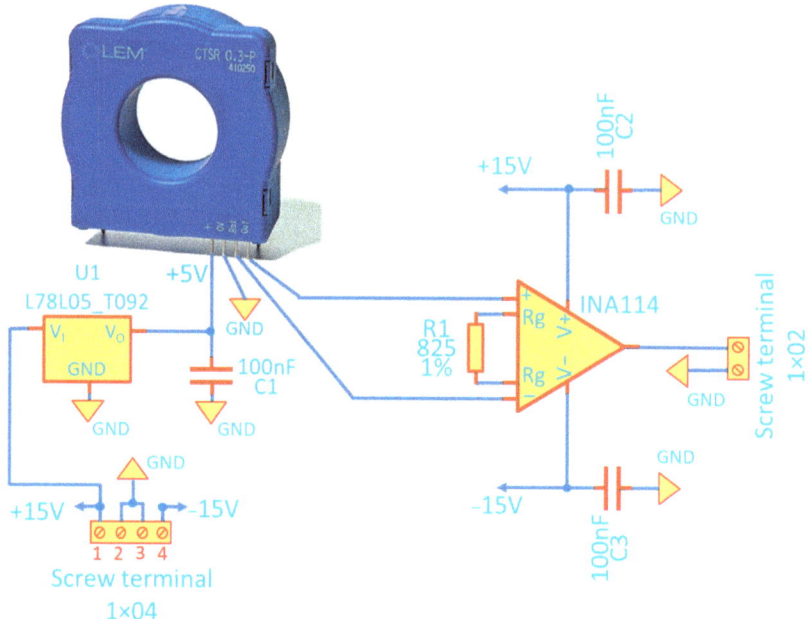

Figure 5. Schematic of the DC sensor with its signal conditioning circuit.

For example, a resistor of 813 Ω (chosen value) provides a multiplication factor of 62.5 and, taking into account the sensitivity of the sensor, a final gain of 250 mV/mA.

4.3. Mounting Structure

Each sensor group consists of a custom mounting structure containing the DC sensor, its conditioning circuit (protected from the plasma by a copper shield), the AC sensor and the repeller. The repeller is insulated from the grounded support structure by a PEEK spacer, and the structure is equipped with three PEEK locking dices and three centering screws to ensure a robust and adjustable ensemble (Figure 6). Each group is fixed to the PG mask pushers support structure downstream of the grounded grid, and an alignment tool was used to properly align the sensors with the beamlet aperture.

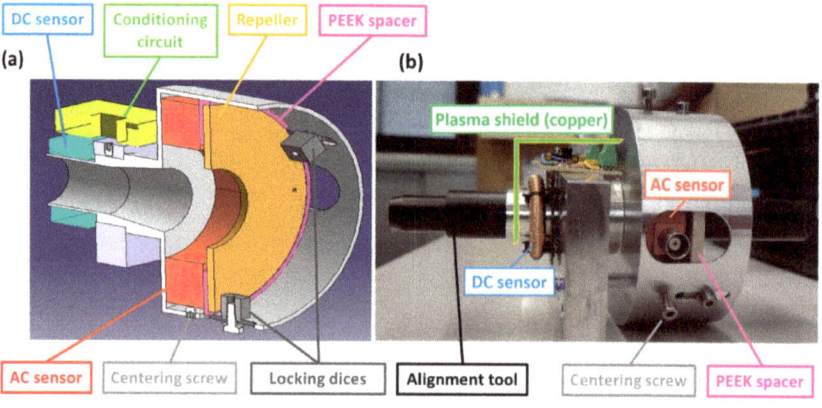

Figure 6. Mounting structure with AC and DC sensors: (**a**) model (**b**); assembled sensor.

4.4. Vacuum Testing

Preliminary tests to assess the compatibility of the instruments for the installation inside SPIDER were carried out with a twofold purpose. Firstly, to evaluate the outgassing as a function of the temperature and identify the nature of the contaminants (e.g., halogens must be absolutely avoided, since they react vigorously with cesium to produce salts). Secondly, to check the functionality of the DC sensor in HV, since active electronics are present and therefore local overheating, due to self-heating, may occur, leading to a change in the transfer function.

A dedicated test stand composed of a small stainless steel vacuum chamber (V ≈ 25.l, A = 0.75 m^2), a turbomolecular pump (Leybold Turbovac TMP 361, Sp_{nom}_N2 = 345.l/s), a full-range pressure gauge (Leybold ITR90) combining Bayard-Alpert and Pirani sensors and a mass spectrometer (Inficon Transpector 2) were used. The wall of the chamber was covered by a heating cable wrapped around the outer side of the bake-out system, and, finally, some thermocouples were placed both on the outer side of the wall and inside the chamber; a sketch of the experimental setup is shown in Figure 7.

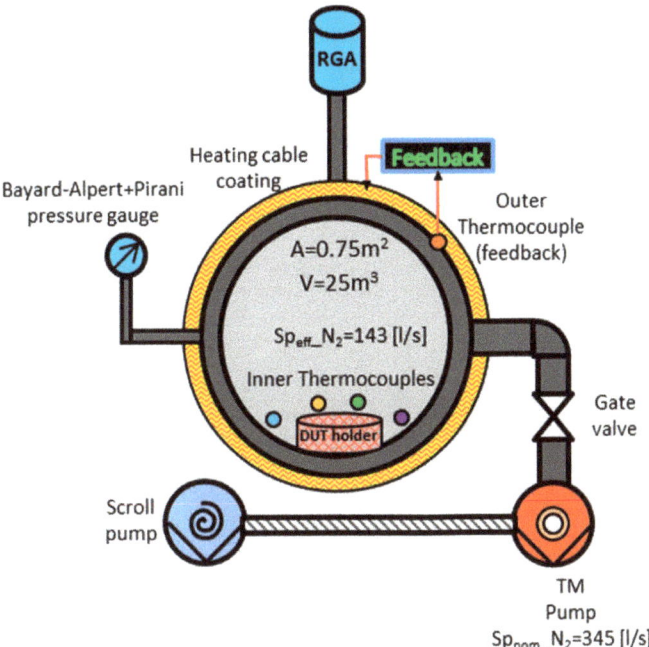

Figure 7. Scheme of the experimental setup adopted for the characterization of the instruments in a vacuum.

Before testing the instruments, the vacuum chamber was pumped down, and after 6 h, it was baked at 95 °C for 24 h, and then pumped at room temperature for up to 90 h until a base pressure of 2 × 10^{-7} mbar was reached. The bake-out temperature was chosen on the basis of the maximum rated temperature of the instruments to be tested so that the same cycle in principle could be applied to each sensor. The pump-down curve of the empty vacuum chamber is shown in Figure 8 (dashed line) where the typical t^{-1} dependency of the pressure due to the outgassing flux reduction over time is visible. The bake-out system, after an increase in pressure due to an increase in the desorbed gas flux stimulated by the temperature increase, allowed the chamber to approach its base pressure.

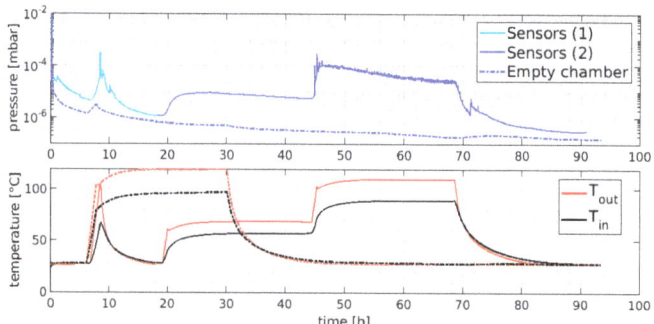

Figure 8. Pump-down curve of Magnelab CT-F5 (solid line) compared to that of the empty chamber (dashed line). Chamber temperature is shown in the bottom plot: red = outer thermocouple, black = inner thermocouple.

4.4.1. AC Sensor Outgassing Tests

The off-the-shelf AC sensors (Magnelab CT-F5), being completely passive, were tested to assess the outgassing rate as a function of the temperature and their capability to stay in a vacuum (e.g., without explosions due to trapped air at room pressure inside it). Additionally, this test enabled us to degas the sensors to reduce the contamination of the SPIDER vacuum as much as possible since no information about their vacuum behavior was provided by the manufacturer. The pump-down curve of the Magnelab CT-F5s compared to that of the empty chamber is shown in Figure 8. All three samples foreseen for the diagnostics implementation in SPIDER were inserted into the chamber after a cleaning phase using acetone; the inner thermocouples were placed between the sample holder and the instruments as shown in Figure 9a.

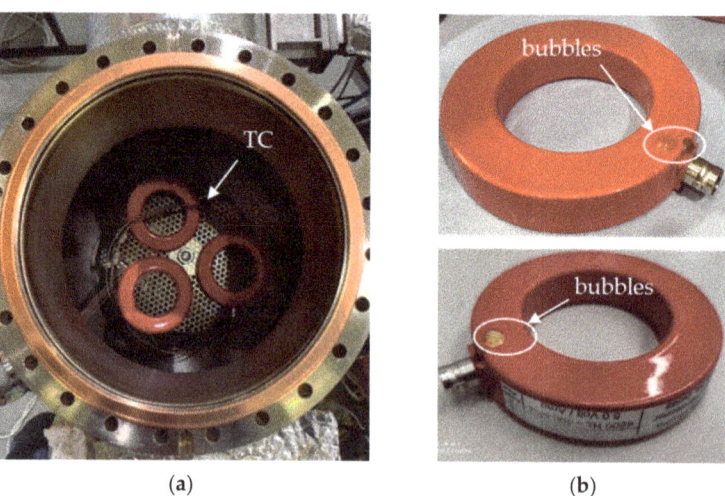

Figure 9. Magnelab CT: (**a**) in a vacuum chamber; (**b**) sample n° 3 (M3) after bake-out at 90 °C.

The experiment involved long periods with sustained external heating, and it was possible to deduce that the sensors would behave normally below 45 °C, as can be seen by the fact that the oscillation in the pressure signal (supposedly due to boiling) disappeared and the pressure approached the 10^{-7} mbar decade after a fast drop.

The residual gas analysis (RGA) during the pump-down curve is shown in Figure 10 where the main contaminants appeared to be water vapor (mass 17 and 18) and nitrogen (mass 28) both for the room temperature case and for the hot sensor case. Nevertheless, for this latter, other, heavier contaminants appeared, in particular, mass 44 (C0₂) and mass 92 (expected to be toluene, commonly used for fast-drying paints).

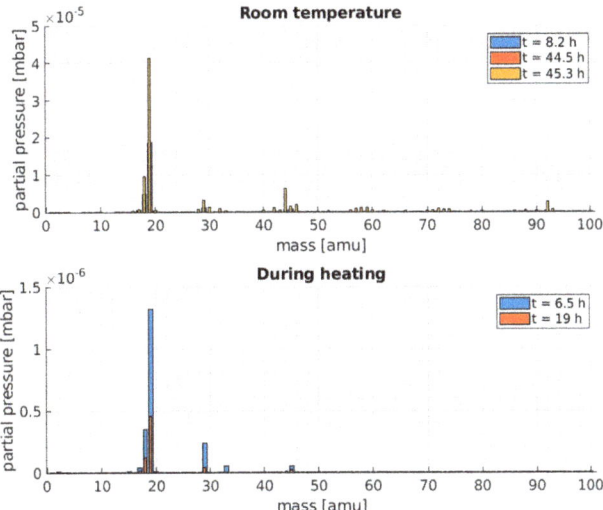

Figure 10. Residual gas analysis of Magnelab CT-F5: Partial pressure of all the species at different times during the bake-out experiment. The times at which the measurements were made are indicated on the labels.

On the basis of this test, an additional requirement in terms of the range of working temperatures in the vacuum inside SPIDER was defined as t < 45 °C; for this purpose, a thermal analysis was also carried out and is reported in Section 4.4.3.

After the test, the sensors were removed from the vacuum chamber and visually inspected: numerous microbubbles were observed on the painted surface, and a big bubble on sample n° 3 (M3) in the proximity of the BNC connector was also present, as shown in Figure 9b. The transfer functions of all three samples were measured with a spectrum analyzer (HP 4194A) to check if the nominal one was preserved after the test in UHV. This was confirmed for sensors M1 and M2, while the M3 sensor reported a slightly higher value for its low cutoff frequency.

The custom current transformer based on the VITROPERM 500F material was also tested in a high vacuum. The nanocrystal material is made of a very thin ferrite tape (14 μm to 20 μm) which is very brittle and is therefore provided already coated with epoxy resin or enclosed in a high-quality (UL94-V0, Class F) sealed plastic casing. The plastic casing type was chosen since epoxy resin (Class A) was expected to be less stable with the temperature; nevertheless, the sealed plastic case was not vacuum-tight, and, therefore, two small holes were carefully made in the casing to avoid trapped volumes (interspace between the core and the case) and therefore virtual leaks when exposed to a vacuum. The baseline pressure obtained during the pump-down test was 2×10^{-6} mbar, which was considered an acceptable value. The fact that this value is slightly higher than the off-the-shelf sensors might be due to the much greater overall surface of the tape exposed to the vacuum (>1 m²), considering the dimensions of the core and the tape thickness, as well as the fact that no bake-out was applied during this test. Unfortunately, the results when external heating was applied are not reliable since the pressure gauge was not working properly, and they have been excluded.

However, the absence of reliable outgassing data at high temperatures was not considered limiting for the installation, since from thermal simulations (already carried out on the basis of the results of Magnelab CT), the expected working temperature should not exceed 35 °C.

4.4.2. DC Sensor Vacuum Tests

After initial benchmarking tests, the vacuum and temperature responses of the LEM sensor (together with the custom signal conditioning circuit) were tested. After removing its plastic casing to avoid trapped volumes, one of the DC sensors and its custom circuit were fastened to the mounting structure used for the installation in SPIDER, which also acts as a heat sink. Thermocouples were fixed to the critical components of the sensor, as well as on the vacuum chamber walls both on the inner and outer sides as shown in Figure 11.

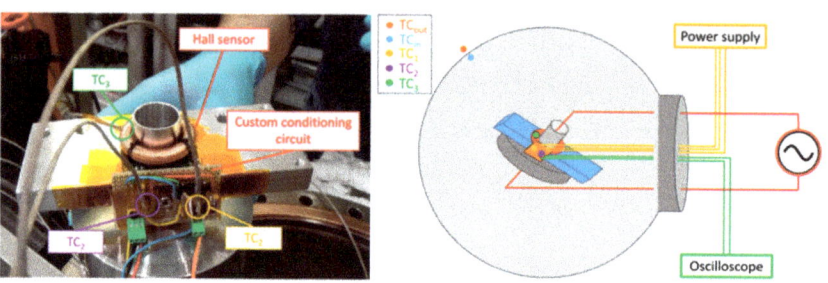

Figure 11. DC sensor experimental setup for the tests in a vacuum.

The pump-down curve is shown in Figure 12a, at room temperature (point A) to assess the outgassing rate and the thermal management when operating in a high vacuum, where the temperature increase in the sensor is only due to self-heating, as well as with external heating (point B) to check the sensor transfer function when subjected to external heating. The sensor outgassing rate was relatively low since the pressure reached values below 1×10^{-7} mbar. In addition, the identified solution to manage the heat dissipation of the sensor was found to be satisfactory also during the external heating (50 °C chamber temperature and self-heating of the instrument).

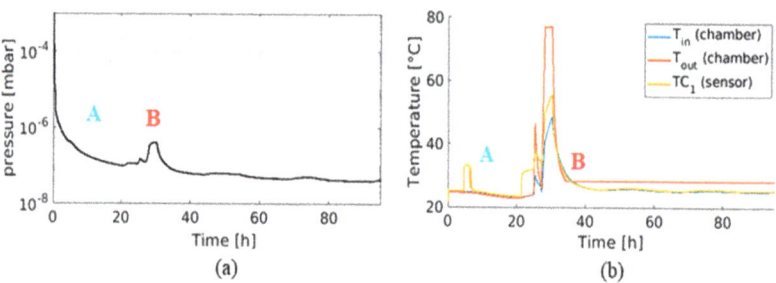

Figure 12. (a) Pump-down curves of DC sensor at room temperature and with external heating, (b) chamber temperature during the experiment (orange: outer thermocouple, blue: inner thermocouple, yellow: sensor thermocouple). Points A and B correspond to self-heating and external-heating phases, respectively.

The custom circuit designed to remove the sensor's "natural" offset before amplification is not perfect, as $V_{in}-V_{ref}$ with no external current is not precisely zero. An offset of approximately 10 mV remains, rising to several volts after amplification. In bench testing, the offset has proved to be unstable, changing each time the sensor is powered on/off but then remaining constant during operation, even with external heating up to 60 °C. Therefore, the behavior of the offset in vacuum was assessed both at room temperature and with external heating applied.

Figure 13a shows the evolution of the offset in the 80 min following the switching on of the sensor. In the first 10 min, the offset rises steadily until it reaches 3.5–4 V. It then remains at this level for the following 70 min. This behavior correlates to the rise in temperature of the sensor's components only due to the self-heating produced by the power dissipated by the electronics.

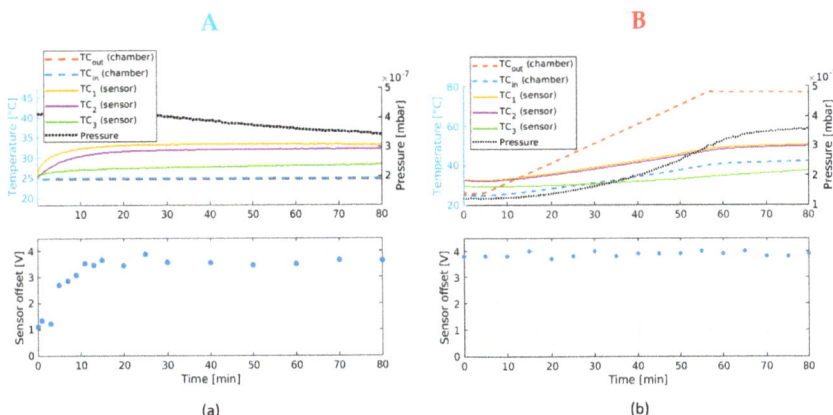

Figure 13. (a) DC sensor offset in vacuum at startup, with the vacuum chamber temperature kept constant at room temperature (point A of Figure 12), and (b) DC sensor offset in a vacuum while raising the ambient temperature (point B of Figure 12).

The sensor is expected to reach higher temperatures in SPIDER when exposed to the ion beam. The same test was therefore performed by heating the vacuum chamber up to temperatures around 70–80 °C (Figure 13b) after reaching the thermal equilibrium due to the self-heating process only. The sensor's temperature is sensitive to the raising temperature of its surroundings and reaches a maximum temperature of 50 °C, compared to 33–34 °C, when no external heating was applied to the chamber. However, this had no effect on the sensor's offset, which remained in the order of 3.8–4 V. Any variation in the offset of an individual sensor is thus attributed to the initial rise in temperature after being switched on, stabilizing once a steady temperature is reached. This behavior is mandatory to have a reliable measurement, as the external heating due to the beam presence should not affect the offset and the measurement.

Additional tests on the sensor's response to a current in a vacuum, at 25 °C and 80 °C, were carried out (Figure 14). Three cases were investigated, the DC response (left), the AC response for signals at different amplitudes but at a fixed frequency of 1 kHz (center) and the AC response for signals at different frequencies but with a fixed amplitude of 5 mA peak-to-peak (right). In all three cases, the sensor's performance seemed to change very little from one temperature to another. The sensor's gain in these experiments changed by a factor of 5% at most, which is comparable to the measurement error and not considered problematic; thus, good linearity is achieved when the sensor works in a vacuum in a range of temperatures compatible with that expected in SPIDER.

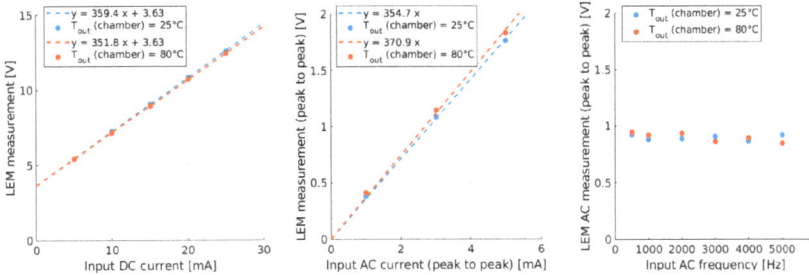

Figure 14. DC sensor response to flowing current at Tout = 25 °C and Tout = 80 °C.

4.4.3. Thermal Simulations

Thermal simulations were carried out using Ansys to assess the temperature of the sensors during SPIDER operation, in particular, driven by the temperature limitation of the Magnelab sensor when operating in a vacuum due to the large and uncontrolled outgassing rate when the temperature of the sensor exceeded 45 °C (see Section 4.4.1).

The potentially most critical issue is represented by the fact that the PG mask can reach a relatively high temperature due to the input power given by the plasma; therefore, the PG mask temperature could reach an equilibrium value up to 300 °C, where the supplied power is mainly dissipated by radiative thermal transfer [11]. Therefore, the radiated power might also reach the BCM sensors since many direct lines of sight between the mask and the sensors are allowed by the EG and GG apertures.

The thermal simulation was carried out (Figure 15) considering only radiative heat transfer (conservative hypothesis), assuming the following boundary conditions:

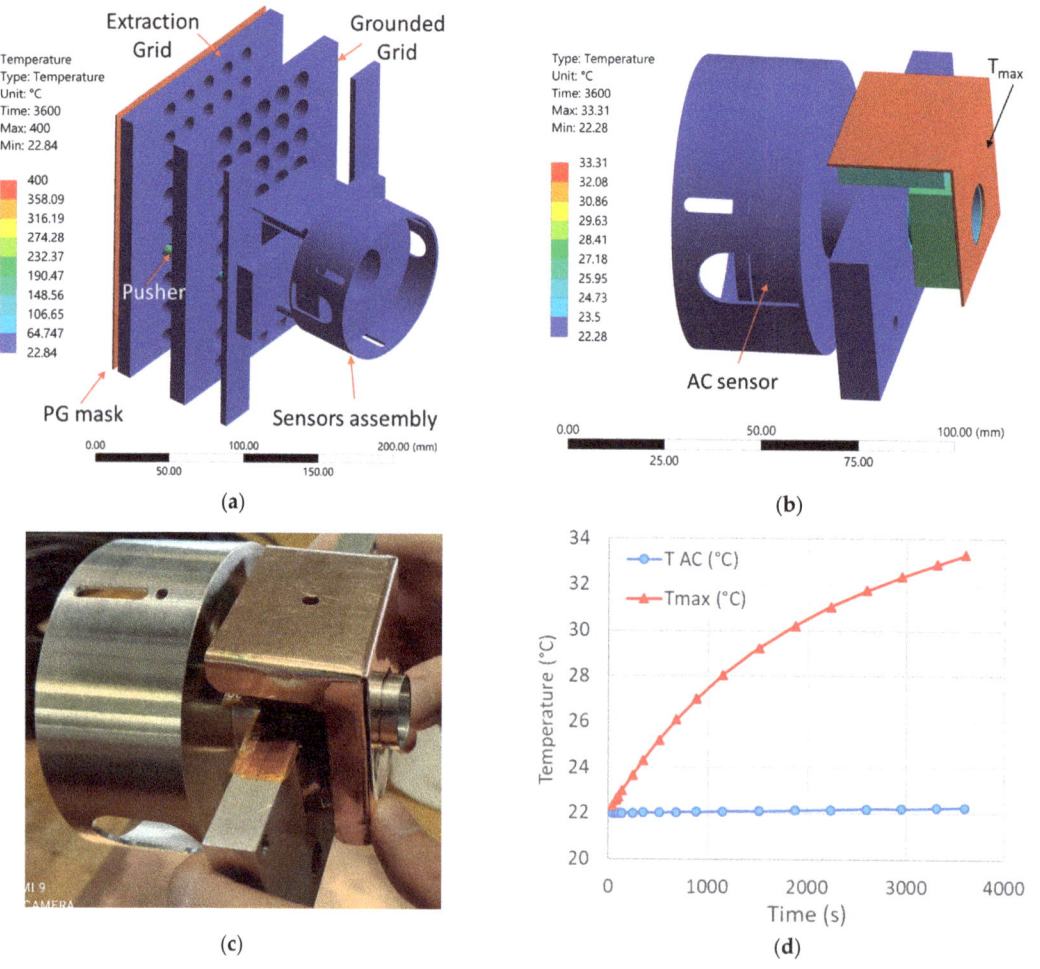

Figure 15. Transient thermal simulation of the BCM sensors assembly: (**a**) temperature at 3600 s in the whole simulation domain, (**b**) zoom view on the sensors assembly, (**c**) picture of the sensors assembly, (**d**) temperature on the hottest point (red curve) and on the AC sensor (blue curve).

- PG mask temperature set to 400 °C with emissivity equal to 1, representing the worst-case scenario (black body emitter, including a safety margin in the PG temperature);

- EG temperature set to 30 °C since it is actively cooled, and the emissivity of EG and GG set equal to 0.2 (slightly oxidized copper, the grids were exposed to air and humidity in the past [26]);
- Emissivity of pushers equal to 0.85 (pyrex glass [27]);
- Emissivity of the DC sensor and conditioning circuit equal to 0.75 (fiberglass of the electric circuit board [28]);
- Emissivity of plasma shield equal to 0.2 (slightly oxidized copper, the plate was exposed to air and humidity in the past [26]);
- Emissivity of sensors support structure equal to 0.1 (aluminum [27]);
- Emissivity of AC sensor equal to 0.03 (polished copper, the sensor surface was covered by copper tape to prevent plasma etching and to reduce the absorbed power by radiation [26]);
- Constant ambient temperature equal to 22 °C representing the vacuum vessel's inner surface which is at room temperature.

The simulation time was limited to 3600 s, which is the maximum pulse duration foreseen for SPIDER operation; however, from Figure 15d, it can be seen that values are well within the safety limits. The maximum temperature reached on the instruments due to the radiated power from the PG mask is about 33 °C, on the copper plasma shield upstream of the DC sensor (red line in Figure 15d), whereas the temperature on the AC sensor is almost stable around the room temperature due to the shielding effect of the sensor's mounting structure. In general, the radiated power reaching the sensors is relatively low due to the shadowing of the EG and GG grids; therefore, the installation of all the aforementioned sensors was considered reliable, considering the issues correlated to heat management.

5. Overview of the System

The BCM system is divided into the following main sub-systems: the sensors and the repeller disk (installed inside the vacuum vessel), cabling, the feedthroughs and surge arresters (installed on the vacuum/air feedthroughs), the data acquisition system, the power supply system and, finally, the CODAS database [29].

5.1. Sensor Installation

The position of the five beamlets chosen for the BCM measurements during SPIDER's first Cs campaign is shown in Figure 16. The beamlets are labeled according to the sensors making the group: the DC/low-frequency sensors (labeled H1 to H5) followed by the AC sensors (either Magnelab CT-F5s labeled M1 to M3, custom current transformer labeled F, or Bergoz ACCT 055 labeled Bz). North and south are indicated in the figure to better compare the positions of the sensor with the pictures taken inside the source during the installation (Figure 17).

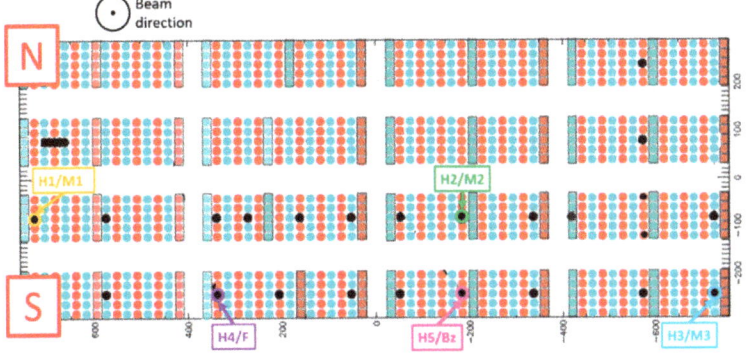

Figure 16. Positions of the five sets of sensors during the cesium campaign.

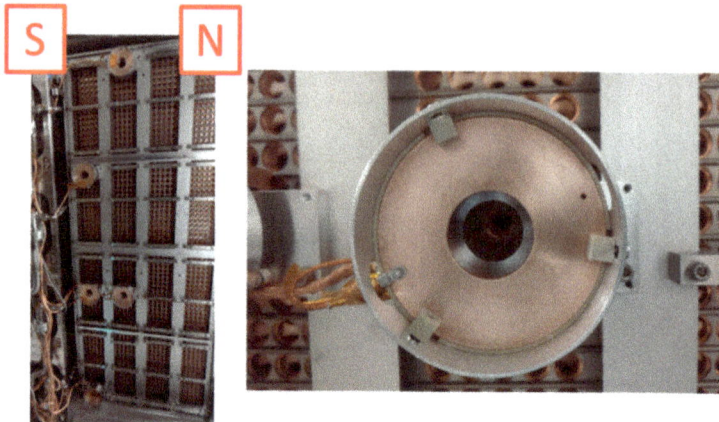

Figure 17. Installation in SPIDER (north and south are indicated to keep track of the orientation of the picture) showing the sensors fixed to the PG mask pusher structure.

The sensor's mounting structure is metallically connected to the GG, which, during the accelerator breakdowns, can oscillate between several kV values with respect to the ground potential due to the stray inductance of the related ground connections.

Therefore, a reliable design was implemented to mitigate the risk of failure of the BCM system, which could also compromise the reliability of the whole SPIDER experiment (surge propagation to other subsystems and vacuum leak due to the rupture of the vacuum feedthroughs).

To avoid any additional paths to ground for the breakdown current through the BCM system itself, the instruments were insulated from the mounting structure by Kapton pads, and the sensor's power supply as well as the data acquisition system were insulated from the ground. Finally, surge arresters were adopted to protect the feedthroughs (Figure 18).

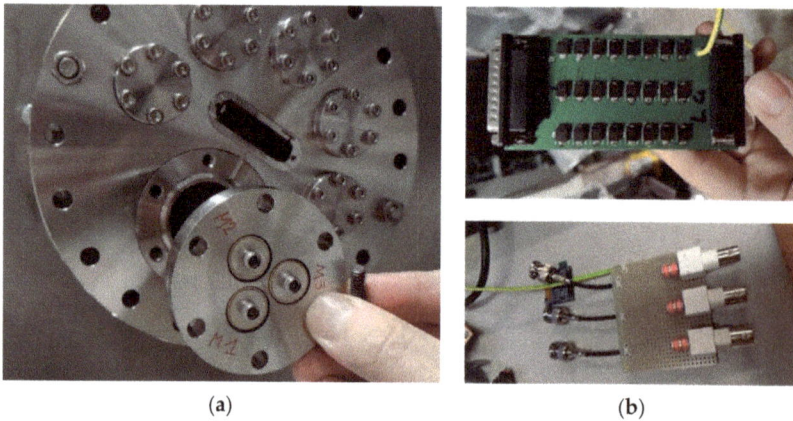

(**a**)   (**b**)

Figure 18. Pictures of: (**a**) the BCM feedthroughs; and (**b**) the related surge arresters: TVS PCB (**top figure**) and GDT board (**bottom figure**).

5.2. Feedthroughs and Surge Arresters

The DC sensors, the Bergoz, the Custom CT, and the repeller cables were connected to a D-Subminiature feedthrough (900 V DC voltage), which was protected against overvoltage by a dedicated PCB, hosting a set of transient voltage suppressors (TVS). The selected transient voltage suppressors (TVS) were connected between each pin and the vacuum vessel (Figure 18b top).

A different choice was made for the Magnelab AC sensors since a 50 Ω matched line is required to exploit the maximum bandwidth of the instruments. Therefore, they were connected to 50 Ω SMA coaxial feedthroughs floating shield type (1 kV DC voltage) (Figure 18a). For these feedthroughs, three electrode GDTs were chosen (instead of TVS) as voltage suppressors due to the much lower stray capacitance (1.5 pF with respect to 300 pF) to limit, as much as possible, the effect of the surge arrester in the high-frequency range (Figure 18b bottom). The DC spark-over voltage is 350 V, whereas the impulse spark-over voltage is <900 V for slew rates of 1 kV/μs. The main drawbacks of this type of surge arrester with respect to TVS are the strong dependence of the pulse breakdown voltage to the slew rate, the lower number of operations within the service life and after arc ignition, they remain in the conductive state until the applied voltage is enough to sustain the arc current (crowbar behavior) [30]. This type of GDT, like the TVS, becomes a "virtual short" at the end of its life, always providing protection to the devices connected in parallel; therefore, the protection of the system is always assured.

5.3. Data Acquisition and Power Supply System

The data acquisition system was designed to be as modular as possible and with insulated channels so that any damage to a part of the system would not compromise the whole data acquisition system.

A relatively cheap solution that also allows the components to be easily replaced was identified in the STEMlab 125-14 Red Pitaya boards (RPs). Those boards have two analog inputs (±1 V or ±20 V, selectable), an input bandwidth at −3 dB of 60 MHz, a maximum sampling rate of 125 Msps and a 14-bit ADC. These boards can be remotely controlled and can be interfaced with the network or via WiFi or 1 Gbit Ethernet protocol.

For the BCM system, six RPs were implemented to acquire the 10 sensors; the data communication for setting the parameters of the ADC and reading the acquired data is based on Ethernet (Figure 19).

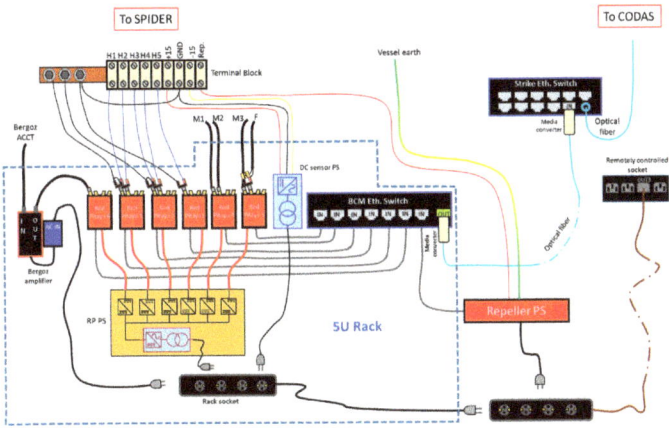

Figure 19. Overview of the BCM power supply and data acquisition system.

The six RPs were powered by a set of six DC/DC converters (24 V/5 V) capable of assuring 2 kV DC insulation both among the devices and to the ground. The primary of the DC/DC converters was fed by a 230 V/24 V DC power supply rated 4 kV AC to withstand voltage, also providing an additional insulation barrier to the ground. Finally, this power supply was connected to the electric grid available inside the SPIDER bunker via a remotely controlled socket, which can switch the whole BCM system on and off.

Additionally, the data transmission had to assure proper insulation for the RPs and to the ground; therefore, a network switch with insulated ports was installed; this type of Ethernet port includes isolation transformers with a minimum isolation rating of 1500 VRMS (2.1 kV peak) as required by the IEEE 802.3 standard for Ethernet interfaces.

The other power supplies are the DC sensor's power supply, consisting of an AC/DC 230 V/±15 V power supply developed in RFX, and the Bergoz ACCT power supply AC/DC 230 V/±15 V rated at 4 kV RMS I/O to withstand voltage feeding the ACCT amplifier. All of the aforementioned components were enclosed in a 5U rack and placed inside the SPIDER bunker about 4 m away from the VV. Finally, the repeller's power supply, a bipolar DC (-100 V $<$ V $<$ 100 V) power supply, was installed close to the BCM rack, and it was also connected to the Ethernet switch to allow it to be controlled remotely. One pole was connected to the VV and the other to the repeller disks so that only the plasma closed the circuit. The repeller power supply is shown in Figure 20a, along with the BCM rack, which houses the acquisition system in Figure 20b.

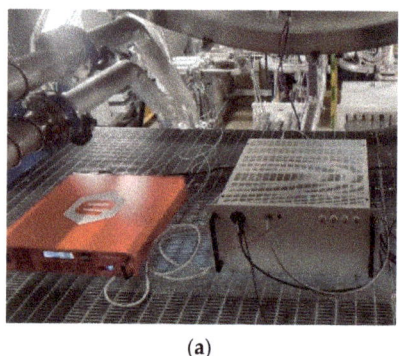

(a)   (b)

Figure 20. Picture of (**a**) the BCM power supply and (**b**) the data acquisition system.

5.4. CODAS

The internal organization of the RedPitaya acquisition system exploits the recent system of chip (SoC) solution proposed by Xilinx, which embeds, in the same chip, a programmable logic (PL) built on a state-of-the-art 28 nm high-k metal gate (HKMG) technology FPGA, and a dual-core ARM Cortex-A9 MPCore processing system (PS). In particular, the RedPitaya mounts the Zynq 7010 system that provides an Artix-7 middle-range device. The key factor that characterizes this family of chips is not actually the performance of the components themselves but the high bandwidth that the FPGA is capable of sustaining when communicating with the other internal SoC components, i.e., the CPU core and the DMA controller. In this way, we can think of a very responsive system that performs some small operations in a fast, strict, real-time environment, made possible with the internal programmable logic, and on the other hand, the high-level acquisition layer that is deployed within a complete operative system running on the processing system side.

The RedPitaya legacy software bundle (Red Pitaya OS 1.04-7) comes with several already-coded components such as a quite-fast oscilloscope, a frequency spectrum analyzer and some other useful tools. All of them are implemented using high-level software-defined components that, in turn, rely on one single specific FPGA firmware implementation that runs underneath. Although all these applications seem very promising, they effectively lack a comprehensive recording system fitting the actual performance that both the ADC and the Zynq could provide. The problem resides in the firmware implementation that is suited to record the input ADC data to an internal circular buffer that is eventually accessed by the oscilloscope application to plot the curve. To achieve a reliable complete recording of the overall input signal, the firmware had to be modified, enabling a well-fitted buffer handling between the internal logic part and the processing part. More specifically a complete recording pipeline of this kind that exploits a FIFO buffer and a DMA transaction to deploy acquired chunks of data directly into the system memory was implemented and is described in [31].

The complete handling of acquired data passes through the described internal high bandwidth communication that is possible thanks to the fast interconnection between the two subsystems (PL and PS) residing in the same chip. Besides this effective internal transaction of the acquired data from the ADC to the system memory, a specific Linux system kernel module handles all the custom parametrization of the acquisition system. Many different acquisition modes were implemented and can be then selected: a so-called "streaming" acquisition that is meant to acquire continuously from a slowly changing input signal, and a "triggered" acquisition where we adhere to the typical transient recorder feature of a classical fast DAQ device.

In both cases, the output complies with the MDSplus mdsip segmented protocol. Indeed, the overall SPIDER-CODAS acquisition system relies on the MDSplus framework to orchestrate the acquisition of the many diagnostics involved. In particular, the segmented acquisition system is a data stream that flows from the device to the experiment storage system scattered by chunks called segments. The idea of the segmented acquisition is that a segment over the data payload also has an idea of the acquisition time that those data relate to, so all the segments can be identified both from the sequential entry into the storage system but also from their time boundaries.

In the streaming implementation, all the segments are sent to the central CODAS at regular intervals to make the system able to refresh the data visualization plots with the newly acquired points. This makes it possible to see the experiment signals with a flow of plot updates and to track the recording at each instant.

On the other hand, there are situations where we want to record signals where all the information is contained in a very rapid time slot where the signal varies with a high-frequency spectrum, and then the information goes down all over long periods of time. In such a situation, the most compelling solution is the transient recording where either an internal or an external trigger is able to activate fast data acquisition that lasts in the internal memory, and that is then spooled to the central acquisition out of sync. If the internal triggering system is active, the recorder acts as a standard oscilloscope and the device records upon a particular change in the input signal; contrariwise, the external trigger can be set to fire the acquisition from both an electrical timing event (timing highway) or a network UDP multicast packet (MDSplus event).

So, depending on the nature of the input signal and the kind of sensor applied, the acquisition mode was also chosen. The DC sensor data were continuously acquired with a typical rate of 10 Ksps (CONTINUOUS acquisition mode), while the AC components acquired by the CTs were recorded either with fixed time frames during the pulse blips or with complete asynchronous MDSplus events that came from the synchronization of the experiment phases in CODAS (i.e., the start of the extraction grid power supply and others). In both cases, the acquisition modes are named SLOW or FAST, with typical sampling frequencies of 300 Ksps and 10–25 Msps, respectively. Alternatively, the AC sensors can be acquired at low frequency for the entire pulse using the CONTINUOUS mode if required.

5.5. Calibration

After the installation of the BCM in the vessel, a thorough calibration of each of the sensors was performed before closing the vacuum vessel. This was conducted by inserting a cable through the beam apertures of the sensor groups and applying various signals using a remote-controlled Red Pitaya function generator. A remote Red Pitaya oscilloscope monitored the sensors' responses, as well as the signals sent through the aperture, whose precise current values were determined using a shunt resistor. Figure 21 shows the calibration of the five DC sensors: although the offset values varied significantly from one sensor to another, the sensitivities were reasonably close. The gains calculated from the calibration ranged from 245 mV/mA to 266.6 mV/mA, keeping within 1% of the value of 250 mV/mA targeted by the circuit design.

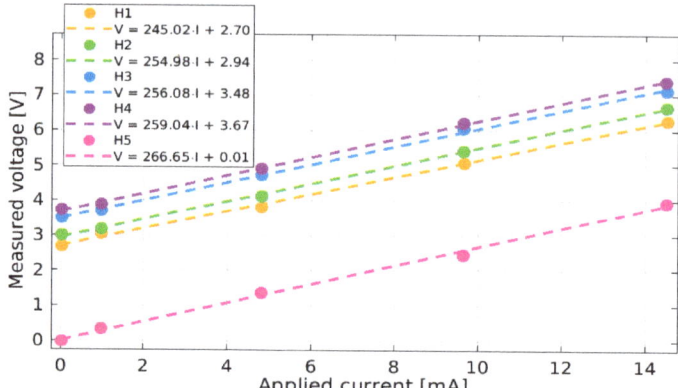

Figure 21. Calibration in SPIDER of the DC component of the DC sensors.

The frequency response of the AC sensors is shown in Figure 22. All of the sensors present a quasi-constant transfer function in the region 10 kHz–1 MHz, with sensitivities exceeding 5 mV/mA. The Bergoz offers remarkable sensitivity, 1000 mV/mA for a bandwidth of 10 Hz to 1 MHz (−3 dB). Finally, the custom transformer's frequency response highlights the importance of the choice of the parallel resistor: opting for 50 Ω instead of 100 Ω decreases the maximum sensitivity by a factor of two but helps to flatten the response in the central region.

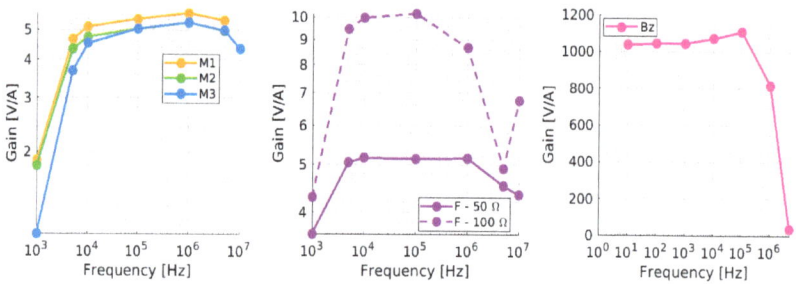

Figure 22. Calibration in SPIDER of the current transformers.

6. Experimental Results

The BCM diagnostic was partially operational during the SPIDER experimental campaigns S19 and S20 (December 2020 to March 2021—beam characterization without cesium) and fully operational during S21 (April 2021 to July 2021—cesium operation).

6.1. DC Results

Figure 23 shows an example of the acquired voltage signal from one of the DC sensors during a SPIDER beam extraction pulse with cesium in the source, and the subsequent analysis required to calculate the beamlet current. The raw BCM signal (black points) and a 500-point moving average (red line) are given in Figure 23a. The sensor has a non-zero voltage offset that varies from sensor to sensor as described in previous sections. Due to the magnetic nature of the DC sensor, the voltage also has a dependency on the plasma and filter stray field as well as the current of the H^- passing through it. Figure 23b shows the reference settings for the RF power, plasma grid current and extraction and acceleration voltages. As the RF power or plasma grid current changes (green and blue highlights in Figure 23a,b) the sensor voltage changes with it. To ensure a steady voltage baseline before and after beam extraction, at least two seconds of plasma with stable RF power and plasma grid current are required (time between the first and last pair of red dashed lines in Figure 23a). The increase in sensor voltage during the application of extraction and acceleration voltages is therefore solely attributed to the beam passing through the sensor. The beamlet current $I_{beamlet}$ (blue in Figure 23d) is given by (6):

$$I_{beamlet} = (V_{beamlet} - V_{baseline})/G_s \qquad (6)$$

where G_s is the individual sensor gain from the previous section. The voltage in the beam extraction $V_{beamlet}$ and the baseline voltage in the steady-state plasma $V_{baseline}$ are shown in red and black in Figure 23c, respectively.

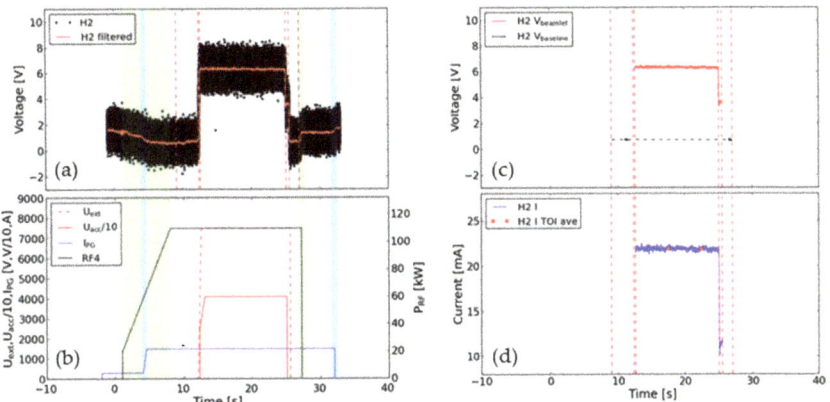

Figure 23. The first 40 s of shot 9145 for sensor H2 (with cesium in the SPIDER source): (**a**) sensor voltage signal (black) and 500 pt. average (red); (**b**) extraction, acceleration, filter field current and RF power signals (for the RF generator closest to the sensor); (**c**) averaged sensor voltage during beam extraction (red) and during plasma steady state for baseline (black); (**d**) beamlet current calculated from sensor voltage (blue) and ±1 s average at programmed times of interest (TOIs). The red dotted lines represent changes in power supply parameters shown in (**b**).

The noise on the raw BCM signal at ±2 V is quite large, which, with a gain of 250 V/A, corresponds to ±4 mA. In the case of surface H^- production with cesium, the beamlet currents of around 20 mA result in a clear increase in sensor voltage, as in Figure 23. In volume production, with lower beamlet currents, the change in the sensor voltage with beam extraction is less clear. However, with signal averaging, the beamlet current can still be measured successfully [32].

The diagnostic was used extensively during the SPIDER cesium campaign. Figure 24 shows the first day of cesium operation, with an approximately threefold increase in current density after the introduction of cesium, even with a lower total RF power. There is an observed inhomogeneity in the beamlet currents (Figure 24a). This is due to the differing availability of H^- at the point of extraction. There are several factors affecting the availability of H: magnetic drifts causing vertical asymmetry in the plasma density, inhomogeneous cesium deposition on the plasma grid and the position of the beamlet within the respective beamlet group (H2 and H4 in the beamlet group core and H1, H2 and H5 at the group edge) [33].

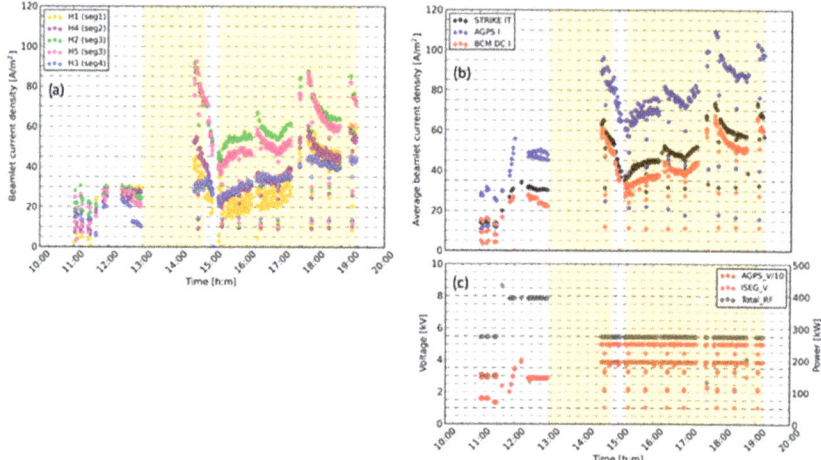

Figure 24. BCM DC results from the first day of SPIDER cesium operation: (**a**) beamlet current density for the five DC sensors; (**b**) average beamlet current density for the BCM, STRIKE electrical and AGPS electrical measurements; (**c**) total RF power and extraction and acceleration voltages. The times when the cesium ovens were open are highlighted in orange.

The average beamlet current measured by the BCM compares favorably with the STRIKE electrical measurement (Figure 24b). With the current PG mask STRIKE measures the electrical current, due to 28 beamlets, spread over 9 out of 16 tiles, while the BCM measures 5 individual beamlets. The STRIKE electrical measurement requires a positive bias of up to 100 V to collect electrons emitted from the surface due to beam particle impacts. This bias can also collect electrons created in the vessel due to beam–gas interactions, which increases the measured current above the beam current alone [34].

6.2. AC Results

For the AC sensors, the different acquisition modes (CONTINUOUS, SLOW and FAST) were all tested and provided a number of interesting results. When operating in the continuous acquisition configuration, the sensors (in particular, the Bz sensor, which has high sensitivity at low frequencies) are capable of measuring the sudden current rise at the beginning and end of extraction and acceleration. This can serve as an additional measurement of the DC at beam startup.

The continuous configuration is also useful as a tool to compare the measurements with and without beams. A clear difference can be observed when performing fast Fourier transforms (FFT) on the signals before or during extraction, the latter displaying a large number of peaks at amplitudes much higher than the former (Figure 25), therefore confirming that the frequencies revealed by the measurements are indeed present in the beam and not artifacts of the ambient noise. The maximum frequency shown on these particular FFTs is 5 kHz, i.e., half of the sampling frequency, as dictated by the Nyquist–Shannon sampling theorem.

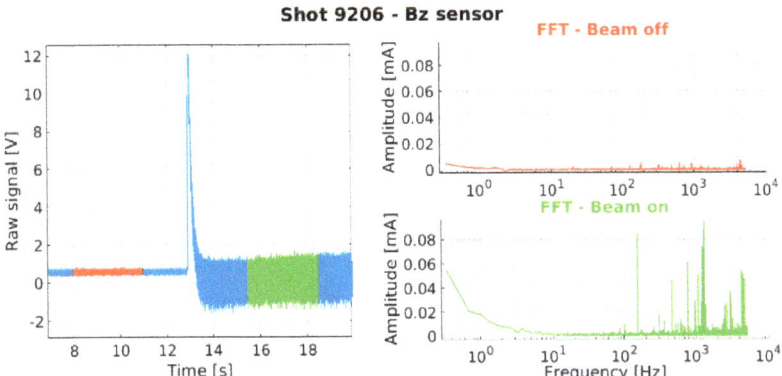

Figure 25. Bz measurements before extraction (orange) and during extraction (green). The FFT related to the same time intervals are reported with the same colors (right figure).

When operating in the SLOW or FAST acquisition modes, the sensors measure the data only during beam extraction. The sampling frequency is much larger than during continuous acquisition and allows for studies at much higher frequencies, provided that the measurement time window is reasonably short (Figures 26 and 27).

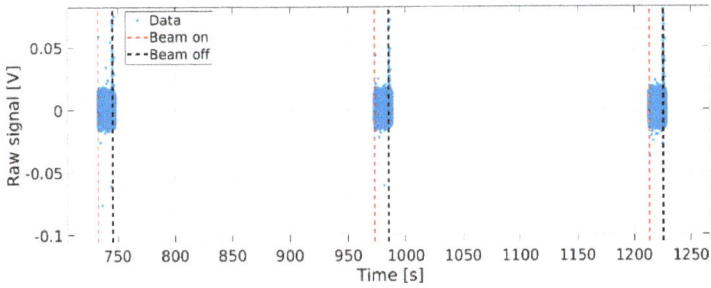

Figure 26. M1 measurement in SLOW acquisition mode—shot 9205.

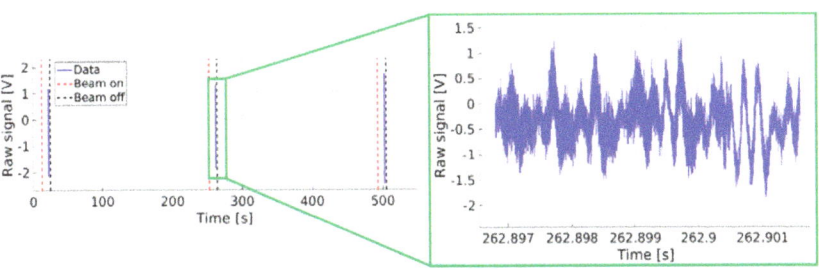

Figure 27. Bz measurement in FAST acquisition mode. Several frequencies can be discerned from the raw signal by the naked eye (e.g., at ∼4 kHz in this case)—shot 9342.

When examining the FFTs performed on the AC signals, monitoring the changes in the frequency and amplitude of the outstanding peaks from one blip to another is not a trivial task to perform with the naked eye. This exercise can be facilitated by producing spectrograms, where the FFTs are plotted side by side, using the y-axis to annotate the frequencies and color-coding to represent the amplitude of the peaks (Figure 28). The important peaks identified in the FFTs are easily recognized and can be correlated with the changes in the source parameters from one blip to another. Comparing the spectrograms of several sensors can also help to understand the similarities or differences between the individual beamlets. An example of a spectrogram of measurements using the SLOW acquisition mode is shown in Figure 28a, where peaks in the range of 50–100 kHz and others of the order of the kHz stand out. The RF generator frequencies (~1 MHz) are clearly visible on the spectrogram of measurements in the FAST acquisition mode (Figure 28b).

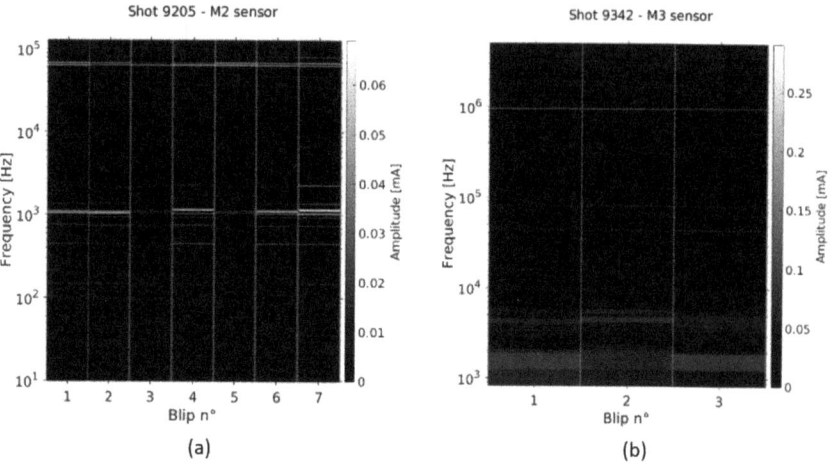

Figure 28. Examples of: (a) SLOW and; (b) FAST acquisition spectrograms.

The BCMs were used to perform an in-depth characterization of the AC component of the SPIDER beam during the first campaign with cesium evaporation. Recurring oscillations were found, their amplitudes were measured and compared to the DC values of the beamlet currents and the frequencies were correlated with those of the RF oscillators and other power supplies when possible [35].

7. Conclusions

A non-intercepting system aimed at measuring the current of individual beamlets of the SPIDER experiment across the extracting area was designed, installed and widely exploited during the SPIDER S20 (March 2021) and S21 (April–November 2021) experimental campaigns, referring to negative ion production by volume and surface production (cesium injection) mechanisms, respectively.

The design of this system was undertaken within a reduced time window due to the scheduled SPIDER shutdowns, representing the only possibility for the installation of this diagnostic inside a vacuum vessel since the operation with a reduced number of beamlets was foreseen only during these campaigns. For this reason, the available off-the-shelf instruments were considered and developed against the SPIDER operational requirements, and a custom wideband current transformer based on nanocrystal materials was developed in parallel with the procurement of the former. All the instruments were extensively tested before their installation in SPIDER to assess their behavior in a vacuum, considering heat management, outgassing, transfer function and overall behavior. After having identified a temperature threshold below which the instruments work properly, thermal simulations were also carried out to assess the maximum working temperature of the instruments inside SPIDER due to thermal radiation to verify if it was compatible with the proper operation of the sensor.

The BCM system was also designed considering the overvoltage of the grounded grid (where the sensors should have been fastened) expected during breakdowns; in particular, strategies based on sensor insulation, surge arresters, insulated power supplies (PSSs) and data acquisition systems (DASs) were adopted to assure adequate reliability. Furthermore, a cheap and modular DAS and PSS, whose components were easily procurable and replaceable, was designed.

The DAS design was based on six Red Pitaya boards, whose legacy software was completely rewritten from an in-house version capable of recording and storing the measurements on the SPIDER database in real time and synchronizing them with all of the other pulse parameters and diagnostics, with a trigger provided by the CODAS system. Furthermore, the implementation of several pre-defined features (sample rate, record length and delay time from trigger signal) provided a very flexible DAS.

This system provided measurements of the current of single beamlets for the first time within a bandwidth from DC up to more than 10 MHz. The first feature permitted the assessment of beam uniformity as a function of the plasma parameters. The AC bandwidth allowed this identification and thus confirmed the hypothesis that the beam current oscillates both around the RF fundamental frequency (1 MHz) as well as at the lower beating frequencies (kHz range) among the generators.

Dedicated data analyses exploiting the large amount of data collected by the BCM system during all the foreseen campaigns are in progress with the purpose of characterizing the beam features resolved in both time and space with respect to all the relevant pulse parameters.

Author Contributions: Conceptualization, A.R.G.; Methodology, T.P., E.S. and F.R.; Validation, B.P.D. and A.P.; Formal analysis, A.S., B.P.D. and M.S.; Investigation, A.S. and A.P.; Resources, A.R.G., M.B., V.C. (Valeria Candeloro), M.P., P.B., V.C. (Vannino Cervaro), R.G., B.L. and F.R.; Data curation, A.S. and B.P.D.; Writing—original draft, T.P.; Writing—review & editing, A.S. and B.P.D.; Supervision, T.P. and G.S.; Project administration, T.P. and A.R.G.; Software, G.M. and C.T.; Funding acquisition, R.P. All authors have read and agreed to the published version of the manuscript.

Funding: This work was carried out within the framework of the ITER-RFX Neutral Beam Testing Facility (NBTF) Agreement and received funding from the ITER Organization. The views and opinions expressed herein do not necessarily reflect those of the ITER Organization. This work has been carried out within the framework of the EUROfusion Consortium, partially funded by the European Union via the Euratom Research and Training Programme (Grant Agreement No 101052200—EUROfusion). The Swiss contribution to this work has been funded by the Swiss State Secretariat for Education, Research and Innovation (SERI). Views and opinions expressed are however those of the author(s) only and do not necessarily reflect those of the European Union, the European Commission or SERI. Neither the European Union nor the European Commission nor SERI can be held responsible for them.

Institutional Review Board Statement: Not applicable.

Informed Consent Statement: Not applicable.

Data Availability Statement: Not applicable.

Acknowledgments: This paper is dedicated to the memory of "golden-handed" mechanic Vannino Cervaro, one of the co-authors, who sadly passed away in February 2021.

Conflicts of Interest: The authors declare no conflict of interest.

Appendix A

The equation for the transfer function K of the current transformer circuit given in Figure 3b can be calculated by treating the circuit as a series of voltage dividers (Figure A1). Grouping the impedance of the entire circuit as Z_1 (Figure A1a), the voltage across the circuit U_1 is given by:

$$U_1 = I_s Z_1 \quad (A1)$$

U_1 is also the voltage drop across the magnetizing impedance Z_m and is equal to the sum of the voltages across Z_s and Z_3. Treating the circuit in Figure A1b as a voltage divider, the voltage across Z_3, U_2, can be expressed as:

$$U_3 = U_1 \frac{Z_3}{Z_3 + Z_s} \tag{A2}$$

In a similar manner the voltage across Z_5 (Figure A1c), which is the measured output voltage U_{out}, is:

$$U_{out} = U_3 \frac{Z_5}{Z_5 + Z_c} \tag{A3}$$

The grouped impedances Z_1, Z_2, Z_3, Z_4 and Z_5 used in the MATLAB model are calculated from Figure 3c using the series and parallel impedance combinations:

$$Z_5 = \left(\frac{1}{Z_{cc}} + \frac{1}{Z_1} + \frac{1}{Z_o}\right)^{-1} \tag{A4}$$

$$Z_4 = Z_c + Z_5 \tag{A5}$$

$$Z_3 = \left(\frac{1}{Z_{cc}} + \frac{1}{Z_4}\right)^{-1} \tag{A6}$$

$$Z_2 = Z_s + Z_3 \tag{A7}$$

Substituting (A1), (A2), (A4)–(A7) and (1) into (A3) gives the transfer function from (4):

$$K = \frac{1}{N} \cdot \frac{Z_1 Z_3 Z_5}{Z_2 Z_4} \tag{A8}$$

The individual impedances from Figure 3b used in the model are:

$$Z_o = \left(\frac{1}{R_o} + j\omega C_o\right)^{-1} \tag{A9}$$

$$Z_1 = R_1 \tag{A10}$$

$$Z_{cc} = \frac{2}{j\omega C_c} \tag{A11}$$

$$Z_c = R_c + j\omega L_C \tag{A12}$$

$$Z_s = R_s \tag{A13}$$

$$Z_m = \left(\frac{1}{R_m} + \frac{1}{j\omega L_m}\right)^{-1} \tag{A14}$$

The above formulas are valid when the transmission line is treated as a single π-junction. When n_c π-junctions are used in the model (Figure A1d), the transfer function becomes:

$$K = \frac{1}{N} \cdot \frac{Z_1' Z_3' Z_5'}{Z_2' Z_4'} \tag{A15}$$

where:

$$Z_5' = \left(\frac{1}{n_c Z_{cc}} + \frac{1}{Z_1} + \frac{1}{Z_o}\right)^{-1} \tag{A16}$$

Calculating Z_4' and Z_3' becomes an iterative process over n steps from 1 to n_c. For $n = 1$:

$$Z_{4n} = \frac{Z_c}{n_c} + Z_5' \tag{A17}$$

$$Z_{3n} = \left(\frac{2}{n_c Z_{cc}} + \frac{1}{Z_{4n}}\right)^{-1} \tag{A18}$$

For $n = 2 \to n_c - 1$:

$$Z_{4n} = \frac{Z_c}{n_c} + Z_{3n-1} \tag{A19}$$

$$Z_{3n} = \left(\frac{2}{n_c Z_{cc}} + \frac{1}{Z_{4n}}\right)^{-1} \tag{A20}$$

For $n = n_c$:

$$Z_{4n} = \frac{Z_c}{n_c} + Z_{3n-1} \tag{A21}$$

$$Z_{3n} = \left(\frac{1}{n_c Z_{cc}} + \frac{1}{Z_{4n}}\right)^{-1} \tag{A22}$$

Then:

$$Z_4' = \prod_1^{n_c} Z_{4n} \tag{A23}$$

$$Z_3' = \prod_1^{n_c} Z_{3n} \tag{A24}$$

$$Z_2' = Z_s + Z_3' \tag{A25}$$

$$Z_1' = \left(\frac{1}{Z_m} + \frac{1}{Z_2'}\right)^{-1} \tag{A26}$$

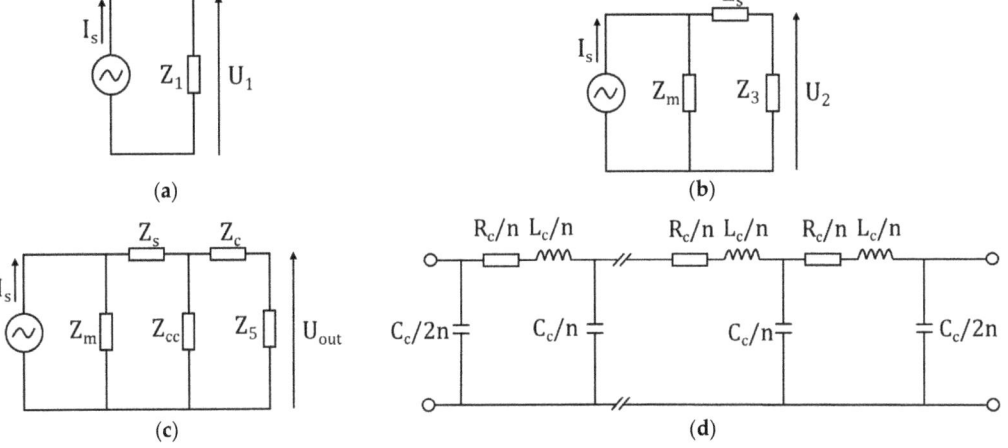

Figure A1. CT equivalent circuit using grouped impedances: (**a**) starting with the total impedance; (**b**,**c**) splitting the impedances by treating them as voltage dividers; (**d**) n π-junction transmission line.

References

1. Serianni, G.; Toigo, V.; Bigi, M.; Boldrin, M.; Chitarin, G.; Del Bello, S.; Grando, L.; Luchetta, A.; Marcuzzi, D.; Pasqualotto, R. SPIDER in the roadmap of the ITER neutral beams. *Fusion Eng. Des.* **2019**, *146*, 2539. [CrossRef]
2. Pimazzoni, A.; Brombin, M.; Canocchi, G.; Delogu, R.S.; Fasolo, D.; Franchin, L.; Laterza, B.; Pasqualotto, R.; Serianni, G.; Tollin, M. Assessment of the SPIDER beam features by diagnostic calorimetry and thermography. *Rev. Sci. Instrum.* **2020**, *91*, 033301. [CrossRef]
3. Brown, I.G. *The Physics and Technology of Ion Sources*; Wiley: New York, NY, USA, 2004.
4. Campbell, D.J. Preface to special topic: ITER. *Phys. Plasmas* **2015**, *22*, 021701. [CrossRef]
5. Kuriyama, M.; Akino, N.; Ebisawa, N.; Grisham, L.; Liquen, H.; Honda, A.; Itoh, T.; Kawai, M.; Kazawa, M.; Mogaki, K.; et al. Development of negative-ion based NBI system for JT-60. *J. Nucl. Sci. Technol.* **1998**, *35*, 739–749. [CrossRef]
6. Fantz, U.; Bonomo, F.; Fröschle, M.; Heinemann, B.; Hurlbatt, A.; Kraus, W.; Schiesko, L.; Nocentini, R.; Riedl, R.; Wimmer, C. Advanced NBI beam characterization capabilities at the recently improved test facility BATMAN Upgrade. *Fusion Eng. Des.* **2019**, *146*, 212–215. [CrossRef]
7. Heinemann, B.; Fantz, U.; Franzen, P.; Froeschle, M.; Kircher, M.; Kraus, W.; Martens, C.; Nocentini, R.; Riedl, R.; Ruf, B.; et al. Negative ion test facility ELISE—Status and first results. *Fusion Eng. Des.* **2013**, *88*, 512–516. [CrossRef]
8. Gasparini, F.; Recchia, M.; Bigi, M.; Patton, T.; Zamengo, A.; Gaio, E. Investigation on stable operational regions for SPIDER RF oscillators. *Fusion Eng. Des.* **2019**, *146*, 2172–2175. [CrossRef]
9. Belchenko, Y.; Dimov, G.; Dudnikov, V. Powerful injector of neutrals with a surface-plasma source of negative ions. *Nucl. Fusion* **1974**, *14*, 113. [CrossRef]
10. Serianni, G.; Toigo, V.; Bigi, M.; Boldrin, M.; Chitarin, G.; Bello, S.D.; Grando, L.; Luchetta, A.; Marcuzzi, D.; Tobari, H.; et al. First operation in SPIDER and the path to complete MITICA. *Rev. Sci. Instrum.* **2020**, *91*, 023510. [CrossRef] [PubMed]
11. Pavei, M.; Bello, S.D.; Gambetta, G.; Maistrello, A.; Marcuzzi, D.; Pimazzoni, A.; Sartori, E.; Serianni, F.D.A.G.; Franchin, L.; Tollin, M. SPIDER plasma grid masking for reducing gas conductance and pressure in the vacuum vessel. *Fusion Eng. Des.* **2020**, *161*, 112036. [CrossRef]
12. Candeloro, V.; Serianni, G.; Fadone, M.; Laterza, B.; Sartori, E. Development of a triple Langmuir probe for plasma characterization in SPIDER. *IEEE Trans. Plasma Sci.* **2022**, *50*, 3871–3876. [CrossRef]
13. Shinto, K.; Shibata, T.; Miura, A.; Miyao, T.; Wada, M. Observation of beam current fluctuation extracted from an RF-driven H-ion source. *AIP Conf. Proc.* **2018**, *2011*, 080016.
14. Wada, M.; Shinto, K.; Shibata, T.; Sasao, M. Measurement of a time dependent spatial beam profile of an RF-driven H− ion source. *Rev. Sci. Instrum.* **2020**, *91*, 013330. [CrossRef]
15. Veltri, P.; Cavenago, M.; Serianni, G. Spatial characterization of the space charge compensation of negative ion beams. *AIP Conf. Proc.* **2013**, *1515*, 541–548.
16. Bergoz Instrumentation Products Website. Available online: https://www.bergoz.com/products/acct/ (accessed on 22 June 2023).
17. MagneLab CT Current Transformer Webpage. Available online: https://gmw.com/product/ct/ (accessed on 22 June 2023).
18. Unser, K.B. Toroidal AC and DC current transformers for beam intensity measurements. *Atomkernenerg. Kerntech.* **1985**, *47*, 48–52.
19. Webber, R.C. Charged particle beam current monitoring tutorial. In Proceedings of the Beam Instrumentation Workshop, Vancouver, BC, Canada, 3–6 October 1994.
20. Zachariades, C.; Shuttleworth, R.; Giussani, R.; MacKinlay, R. Optimization of a high-frequency current transformer sensor for partial discharge detection using finite-element analysis. *IEEE Sens. J.* **2016**, *16*, 7526–7533. [CrossRef]
21. Fuzernova, J.; Fuzer, J.; Kollar, P.; Bures, R.; Faberova, M.J. Complex permeability and core loss of soft magnetic Fe-based nanocrystalline powder cores. *J. Magn. Magn. Mater.* **2013**, *345*, 77–81. [CrossRef]
22. Nanocrystalline VITROPERM ECM Products. Available online: https://www.mouser.com/pdfdocs/VACChokesandCoresDatasheet.pdf (accessed on 22 June 2023).
23. Vacuumshmelze Nanocrystalline VITROPERM. Available online: https://vacuumschmelze.com/products/soft-magnetic-materials-and-stamped-parts/nanocrystalline-material-vitroperm (accessed on 22 June 2023).
24. LEM CTSR 0.3-P Webpage. Available online: https://www.lem.com/en/ctsr-03p (accessed on 23 June 2023).
25. Tumanski, S. Modern Magnetic Field Sensors—A Review. *Organ* **2013**, *10*, 1–12.
26. Fluke Process Instruments—Metal Emissivities. Available online: https://www.flukeprocessinstruments.com/en-us/service-and-support/knowledge-center/infrared-technology/emissivity-metals (accessed on 22 June 2023).
27. The Engineering Toolbox–Surface Emissivity Coefficients. Available online: https://www.engineeringtoolbox.com/emissivity-coefficients-d_447.html (accessed on 26 June 2023).
28. Thermoworks–Infrared Emissivity Table. Available online: https://www.thermoworks.com/emissivity-table/ (accessed on 26 June 2023).
29. Luchetta, A.; Manduchi, G.; Taliercio, C. SPIDER CODAS Evolution toward the ITER compliant neutral beam injector CODAS. *IEEE Trans. Nucl. Sci.* **2017**, *64*, 2765–2769. [CrossRef]
30. Ardley, T. *First Principles of a Gas Discharge Tube (GDT) Primary Protector—Rev 2*; Bourns: Altadena, CA, USA, 2008.
31. Garola, A.R.; Manduchi, G.; Gottardo, M.; Cavazzana, R.; Recchia, M.; Taliercio, C.; Luchetta, A. A Zynq-Based Flexible ADC Architecture Combining Real-Time Data Streaming and Transient Recording. *IEEE Trans. Nucl. Sci.* **2020**, *68*, 245–249. [CrossRef]

32. Shepherd, A.; Duteil, B.P.; Patton, T.; Pimazzoni, A.; Garola, A.R.; Sartori, E.; Serianni, G. Initial results from the SPIDER beamlet current diagnostic. *IEEE Trans. Plasma Sci.* **2022**, *50*, 3906–3912. [CrossRef]
33. Shepherd, A.; Patton, T.; Duteil, B.P.; Pimazzoni, A.; Garola, A.R.; Sartori, E. Beam homogeneity of caesium seeded SPIDER using a direct beamlet current measurement. *Fusion Eng. Des.* **2023**, *192*, 113599. [CrossRef]
34. Shepherd, A.; Patton, T.; Duteil, B.P.A.P.; Garola, A.R.; Sartori, E.; Ugoletti, M.; Serianni, G. Direct current measurements of the SPIDER beam: A comparison to existing beam diagnostics. *arXiv* **2023**, arXiv:2305.18001.
35. Duteil, B.P.; Shepherd, A.; Patton, T.; Garola, A.R.; Casagrande, R. First characterization of the SPIDER beam AC component with the Beamlet Current Monitor. *Fusion Eng. Des.* **2023**, *190*, 113529. [CrossRef]

Disclaimer/Publisher's Note: The statements, opinions and data contained in all publications are solely those of the individual author(s) and contributor(s) and not of MDPI and/or the editor(s). MDPI and/or the editor(s) disclaim responsibility for any injury to people or property resulting from any ideas, methods, instructions or products referred to in the content.

Article

The Heavy-Ion Beam Diagnostic of the ISTTOK Tokamak—Highlights and Recent Developments

A. Malaquias [1,*], I. S. Nedzelskiy [1], R. Henriques [1] and R. Sharma [2]

[1] Instituto de Plasma e Fusão Nuclear, Instituto Superior Técnico, Universidade de Lisboa, Av. Rovisco Pais, 1049-001 Lisboa, Portugal; igorz@ipfn.tecnico.ulisboa.pt (I.S.N.); rhenriques@ipfn.tecnico.ulisboa.pt (R.H.)
[2] Culham Science Centre, Culham Centre for Fusion Energy, United Kingdom Atomic Energy Authority, Abingdon OX14 3DB, UK; ridhima.sharma@ukaea.uk
* Correspondence: artur.malaquias@ipfn.tecnico.ulisboa.pt

Abstract: The unique arrangement of the heavy-ion beam diagnostic in ISTTOK enables one to measure the evolution of temperature, density and pressure-like profiles in normal and AC discharges. The fast chopping beam technique provided the possibility to reduce the noise on the measurements of the plasma pressure-like profile and for the precise control of the plasma column position in real time. The consequent improvements in S/N levels allowed the observation of the effects of runaway beam magnetic energy conversion into plasma local heating. In addition, it made it possible to follow the evolution of the quiescent plasma maintained during AC transitions when the plasma current is null. The use of a new operation mode in the cylindrical energy analyzer provided an improved resolution up to five times in determining the fluctuations of the plasma potential as compared to the normal operation mode. Such analyzer is extremely compact (250 mm × 250 mm × 120 mm) and provides a unique geometry in order to cover the whole plasma diameter. The detector configuration choice gives the possibility for the simultaneous measurements of plasma poloidal magnetic field, plasma pressure-like and plasma potential profiles together with their fluctuations.

Keywords: nuclear fusion diagnostics; heavy-ion beam probe; heavy-ion beam diagnostic; plasma potential measurements; plasma poloidal field measurements

Citation: Malaquias, A.; Nedzelskiy, I.S.; Henriques, R.; Sharma, R. The Heavy-Ion Beam Diagnostic of the ISTTOK Tokamak—Highlights and Recent Developments. *Sensors* **2022**, 22, 4038. https://doi.org/10.3390/s22114038

Academic Editor: Hossam A. Gabbar

Received: 13 May 2022
Accepted: 25 May 2022
Published: 26 May 2022

Publisher's Note: MDPI stays neutral with regard to jurisdictional claims in published maps and institutional affiliations.

Copyright: © 2022 by the authors. Licensee MDPI, Basel, Switzerland. This article is an open access article distributed under the terms and conditions of the Creative Commons Attribution (CC BY) license (https://creativecommons.org/licenses/by/4.0/).

1. Introduction

The operation of magnetic confinement fusion devices requires the use of sophisticated diagnostic techniques in order to infer on the status of the plasma equilibrium and its evolution. The goal is to optimize the plasma pressure (the product of density and temperature) for maximum energy and particle confinement time in a quasi-stable equilibrium. There are many factors that can affect the performance of the fusion plasma such as impurity content, evolution of magnetohydrodynamics (MHD) modes, density and pressure limits, runaway electron (RE) formation, energetic ions loss, turbulence and transport and others. Diagnostic systems that are currently used are based on (i) magnetic and electric sensors, (ii) spectroscopic systems from visible to X-ray wavelengths, (iii) particle or laser-beam-aided diagnostics and (iv) neutron and gamma detection for fusion-rated plasmas. The ideal set of diagnostic data is the collection of plasma parameters from a full 2D poloidal plasma cross section with a few mm of spatial resolution and within tens of microsecond temporal resolution. In addition, magnetic flux surface reconstruction would benefit from 3D measurements based on a toroidal distribution of magnetic sensors (ideally toroidal distributed 2D sets of data). In practice, the experimental data coverage is limited by a combination of diagnostics that can provide measurements from one or a few parameters: (i) a reduced number of local sample volumes, (ii) an integrated tomographic measurement from a collection of lines of sight crossing the plasma into two or more directions (creating a mesh that covers the whole 2D plasma cross section) or (iii) a 1D radial profile or line integrated measurement from a beam- or laser-added diagnostic. This data collection is treated with the support of self-consistent equilibrium and transport codes in order to provide the status of the plasma and its evolution. Most useful in this evaluation process are diagnostics that can measure several plasma parameters from the same location at the same time instant. Moreover, if the measurements could cover the whole 2D plasma section, it would allow for a more precise calibration of MHD and transport codes' free parameters.

The heavy-ion beam diagnostic installed on the ISTTOK tokamak (Lisbon, Portugal) has been conceptualized in order to provide simultaneously the plasma radial profile evolution of the plasma temperature, electron density, plasma poloidal magnetic field and plasma potential. In fact, this diagnostic has the capability to scan the plasma in 2D, although in ISTTOK, it is limited in its coverage by geometrical factors. It can in practice provide a 1D full-diameter profile of each of the above parameters.

This paper describes the capabilities that have been developed in this diagnostic and includes a more detailed description of the physics basis for the measurement of the plasma poloidal magnetic field profile in ISTTOK.

2. The Heavy-Ion Beam Probe Operation Principle (Classic Configuration)

The original proposal for the heavy-ion beam probe (HIBP) was made in the US in 1970 [1]. A review of methods employed in this configuration of the diagnostic is given in [2]. At about the same time, the Russian Federation scientists started to employ the HIBP in several of their fusion devices. An overview of their work can be obtained in [3].

For the application in a tokamak or stellarator device, a beam of single ionized heavy ions I^+ (e.g., Cs^+, Rb^+, etc.) is injected into the plasma (this is called the primary beam) as depicted in Figure 1.

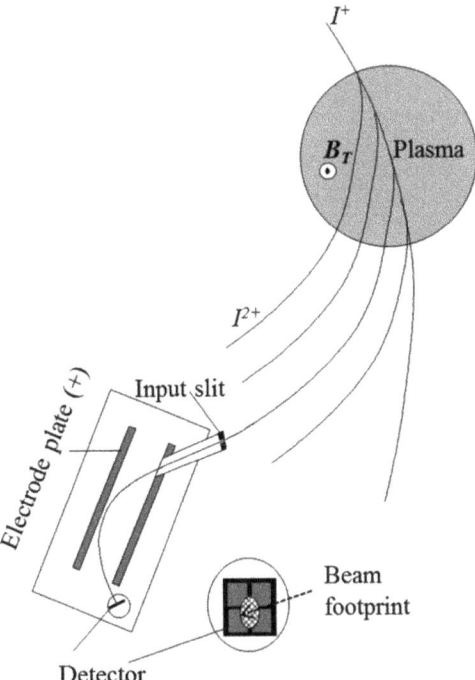

Figure 1. Illustration of the operation of a heavy-ion beam probe (HIBP). A heavy-ion beam I^+ is injected into the plasma perpendicularly to the toroidal magnetic field (B_T) producing a quasi-circular trajectory due to the Lorentz force. While colliding with the plasma electrons, the primary beam (I^+) is ionized and produces a fan of secondary beams along the plasma column (I^{2+}). The trajectory of each secondary beam is determined by half of the Larmor radius with respect to the primary beam.

The injection angle is chosen to be perpendicular to the toroidal magnetic field. While crossing the plasma, the primary beam will be further ionized due to collisions with the plasma electrons (and ions if the ion temperature exceeds 2 keV) producing a fan of secondary beams I^{2+} (e.g., Cs^{2+}, Rb^{2+}, etc.). The Lorentz force, due to the toroidal magnetic field, imposes a quasi-circular trajectory for the primary beam and for the emerging secondary beams, having half of the Larmor radius of the primary beam. Further down, the detection system collects a secondary beam that is produced over the primary beam elementary path length dl. This defines the sample volume $dV = dl \cdot A$ (here A is the beam cross-section area) at the position $r = \rho$ in the plasma. In order to scan the sample volume dV over the plasma cross section, the primary beam energy and angle are changed resulting in a grid of energy lines and angles [4]. An example of the scanning energy map is depicted in Figure 2 for the case of the HIBP installed in the TJ-II stellarator.

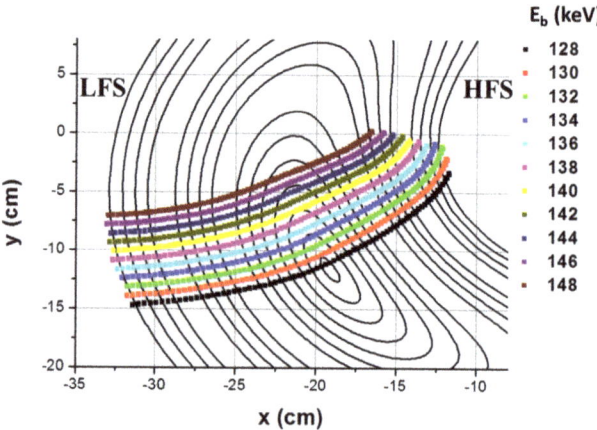

Figure 2. HIBP detector grid for the vertical cross section of TJ-II for Cs^+ probing ions. The detector grid consists of 11 detector lines (in different colors) of equal energy E_b. Each detector line scans the plasma cross section from high-field side to low-field side. (Reprinted from [4], with the permission of AIP Publishing.)

2.1. Plasma Potential Measurements

The most widely used detector in the HIBP configuration is the Proca-Green 30° parallel plate energy analyzer [5]. As illustrated in Figure 1, it consists of an input slit, a set of parallel conducting plates and an exit aperture followed by a split plate detector. The detector collects the beam intensity and allows one to evaluate its position based on the collected current in each individual cell. The secondary beam energy is obtained by the position of the beam in the detector as it results from the parabolic trajectory inside the electric field produced by the analyzing parallel plates.

The HIBP measures the local plasma potential at the sample volume location via the measurement of the secondary beam energy. These two quantities can be related by solving the energy conservation equation for the primary and secondary beams. The primary beam total energy before entering the plasma at source (s) equals the total energy at the ionization point (p). Further, the secondary beam total energy at ionization point (p) equals the total energy at detection point (d). Equating these two conditions, we obtain:

$$\begin{cases} K_{1s} + eV_s = K_{1p} + eV_p \\ K_{2p} + 2eV_p = K_{2d} + 2eV_d \end{cases} \quad (1)$$

Noting that at the ionization point, the kinetic energy of primary and secondary beams can be taken to be equal (as there is negligible momentum transfer from the electron to the primary ion), we can use $K_{1p} \approx K_{2p}$ and obtain, by subtraction of the above equations, the potential at the ionization point p,

$$V_p = (K_{2d} - K_{1s}) + (2V_d - V_s) \qquad (2)$$

Thus, V_p is only a function of the secondary beam energy as the injection energy of the primary beam is known and the potential at the exit of the source and detector is also known, usually kept at ground potential.

2.2. Plasma Density Fluctuation Measurements

The secondary beam carries also information about the plasma density on its current intensity. The collected current results from a source contribution and the integral attenuation of primary and secondary beams; it is therefore an integrated measurement. The current of each secondary beam detected is governed by

$$I_d^{2+} = 2 \cdot I_s^+ \cdot n(r) \cdot \sigma_{12}(T) \cdot dl \cdot A \cdot B, \qquad (3)$$

where the term on the right contains (i) the ratio of charge between primary and secondary beam ($I^{2+}/I^+ = 2$), (ii) the source factor of the secondary beam generation given by the product of the density at the ionization point and the effective cross section of the ionization process ($I^+ + e \rightarrow I^{2+} + 2e$), $n(r).\sigma_{12}(T)$, at the same location (this is usually called the 'n sigma' product), and (iii) the path length dl over which the ionization takes place (this is determined by the detector cell size projected over the primary beam trajectory). The terms A and B account respectively for the attenuation suffered by the primary beam from the plasma entry until the ionization volume location and the attenuation of the secondary beam from the ionization volume location until the plasma exit. These terms are explicitly given hereinafter. The effective cross section, $\sigma_{12}(T)$, is a function of plasma temperature and is calculated from the cross-section data (experimental and/or theoretical) for the reactions of interest. The calculation integrates in the velocity space the reactions' cross sections over the equilibrium Maxwellian velocity distribution of the plasma electrons. This integration is performed for each value of T covering the range of temperatures of interest.

The HIBP in this classical configuration has provided the measurements of local plasma potential and density fluctuations in many devices around the world. For example, the T-10 tokamak HIBP [6] can measure the cross phase of density oscillations, the poloidal phase velocity of turbulence rotation, the poloidal electric field and the radial turbulent particle flux. The use of spectral analysis techniques and cross correlations with magnetic probe measurements allows one to distinguish between various types of turbulence such as broadband, quasi-coherent, tearing and geodesic acoustic modes.

2.3. Plasma Poloidal Field Fluctuation Measurements

The trajectories of the primary and secondary beams are perturbed by the poloidal field produced by the plasma current. The Lorentz force imposes a deflection of the primary and secondary beams along the toroidal direction z (or φ, i.e., the toroidal magnetic field direction). This deflection can be measured by the position of the beam in the detector. In the case of the Proca-Green 30° parallel plate energy analyzer, the deflection is in the perpendicular direction to the deflection caused by the plasma potential measurements making these two measurements naturally independent. The final toroidal position of the secondary beam z_d^{2+} is governed by the following equation given in [3] (in cylindrical approximation)

$$z_d^{2+} = z_0^+ + \underbrace{\frac{e}{m}\int_{\rho_0}^{\rho_j} B_\rho(\rho)(t_j - \tau_1(\rho))d\rho}_{z^+} + \underbrace{\frac{2e}{m}\int_{\rho_j}^{\rho_d} B_\rho(\rho)(t_d - \tau_2(\rho))d\rho}_{z^{2+}} +$$
$$+ \underbrace{(t_d - t_j)\left(\frac{e}{m}\int_{\rho_0}^{\rho_j} B_\rho(\rho)d\rho + \dot{z}_0^+\right)}_{A} \quad (4)$$

The different terms on the right side contributing to this deflection are (i) the initial injection position of the primary beam, z_0^+, (ii) the deflection that the primary beam suffers from the source to the ionization point, z^+, (iii) the deflection that the secondary beam suffers from the ionization point to the detector, z^{2+}, and (iv) the term A corresponding to the contribution imposed by the increment in the secondary beam velocity at the ionization volume due to the local poloidal magnetic field times the travel time, plus the velocity in the z-direction of the primary beam at the ionization point.

In the particular case where local MHD activity is present at the ionization volume, having some characteristic fluctuation spectra, it is possible to link the detector beam position fluctuations with the oscillations of the poloidal field. This constitutes the basis for MHD fluctuation studies such as, for instance, the Alfvén eigenmode studies presented in [7]. In the most general case, the final position of the beam measured in the detector is integrated over many contributions and might not be uniquely related to the plasma poloidal field strength at the sample volume calling for a more in-depth evaluation of each term contribution.

3. The Heavy-Ion Beam Diagnostic Operation Principle

The path integral limitations mentioned above can be accounted for if a different configuration is used for the diagnostic. In order to distinguish it from the classical configuration, it is denominated the heavy-ion beam diagnostic (by replacing the word 'probe' with 'diagnostic'). It is based on a multiple-cell array detector (MCAD) that collects all the fan of secondary beams at the same time. This approach avoids the need to scan the primary beam in order to obtain 1D full-diameter profiles as they are provided quasi-instantaneously. It allows, in this way, preserving the phase information between the plasma fluctuations (up to the time limit that the beam takes to travel the plasma column: circa 0.5 microseconds in ISTTOK).

The operation of the HIB diagnostic is illustrated in Figure 3. The main difference from the HIB-probe arrangement is that all secondary and primary beams are 'instantaneously' collected in a multiple-cell array detector (MCAD). The beam finite width (in the toroidal direction) will be captured by the cells in the same row. Adding all the currents in one row provides the total current collected from the sample volume $dl \cdot A$, while the current distribution on the individual cells in one row allows the determination of the toroidal position of the beam.

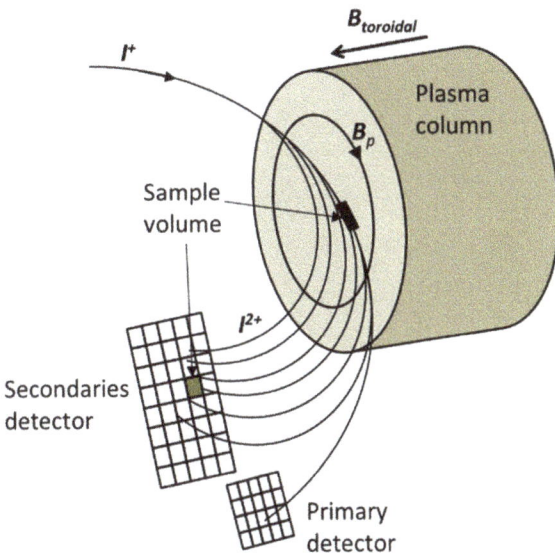

Figure 3. Schematic configuration of the heavy-ion beam diagnostic (HIBD). In this arrangement, all the secondary ions and the primary beam are detected in a multiple-cell array detector (MCAD).

In ISTTOK, this configuration was adopted. The ion gun injects a 3 micro-ampere, 20–25 keV energy beam of Xe^+ (or Cs^+, Hg^+) into the ISTTOK plasma. ISTTOK plasma has an 8.5 cm minor radius and 4–6 kA of plasma current confined in a 0.48 T toroidal magnetic field (Figure 4).

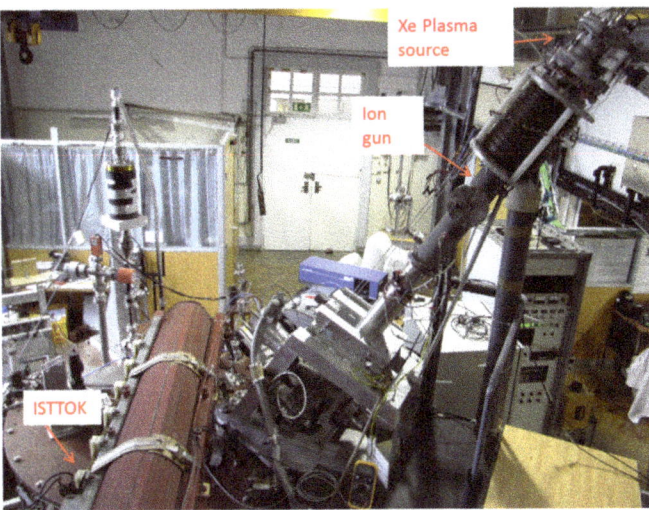

Figure 4. Set-up of the HIBD in ISTTOK tokamak.

In the following sections are presented the algorithms developed to account for the path integral effects and recover the local information along the primary beam path. Examples of experimental data retrieved by these methods are also presented.

3.1. The Determination of Local Values of the 'n sigma' Profile

The development of a retrieving method for the local values of 'n sigma' (here also referred to as 'pressure-like' due to the dependence of effective cross section σ_{12}, on the plasma temperature) has been detailed in [8]. The methodology is mainly based on the differentiation (starting from the first row of detector cells) of the currents collected on neighboring rows, being the difference related to the attenuation current of the primary beam between the two cells. In Figure 5, is depicted a simulation of the trajectories in ISTTOK (for a Xe+ beam, 22 keV) and the chosen detector line arrangement.

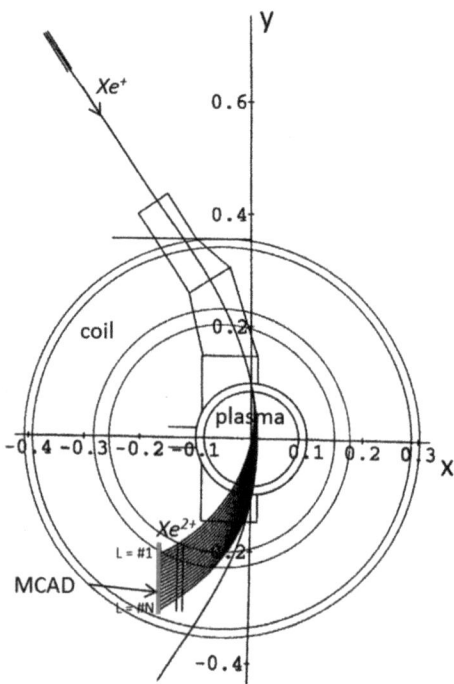

Figure 5. Trajectories (units in *m*) of a 22 keV primary Xe+ beam and corresponding cloud of secondary beams as calculated by the HIBD modeling code for the tokamak ISTTOK (B_T = 0.48 T). The detector, represented by the red line, has N rows in total, each indexed by the number L (#1, . . . , #N) (reprinted from [8]).

The attenuation processes that are relevant for ISTTOK comprise the electron impact ionization reactions of (i) $I^+ + e \rightarrow I^{2+} + 2e$, (ii) $I^+ + e \rightarrow I^{3+} + 3e$ and (iii) $I^{2+} + e \rightarrow I^{3+} + 2e$. The complete equation describing the contribution of these processes in the generation of the detected currents is (where the terms *A* and *B* of Equation (3) are here explicitly given)

$$I^{2+}_{j(det)} = 2I_0^+ \underbrace{\exp\left(-\int_{R_j}^{r_j} n_e(s_1)(\hat{\sigma}_{12}+\hat{\sigma}_{13})ds_1\right)}_{A} \times$$

$$n_e(r_j)\hat{\sigma}_{12}(r_j)dl_{(cell)} \underbrace{\exp\left(-\int_{r_j}^{R_p} n_e(s_2)\hat{\sigma}_{23}ds_2\right)}_{B} \quad (5)$$

As was demonstrated in [8], the determination of the $n\hat{\sigma}_{12}(r_j)$ relative profile can be performed by subtracting the primary beam current entering the ionization volume from the value of the detected secondary beam current obtained in the previous ionization volume (weight by the charge ratio between I^+ and I^{2+}). Therefore, the primary current available at each ionization volume ($j = 1, \ldots, N$) can be obtained from

$$I_j^+ = I_0^+ - \sum_{L=0}^{j-1} \frac{1}{2} I_{L(det)}^{2+} \tag{6}$$

with the definition that for $L = 0$, there is no current (no cell for $L = 0$). Making the ratio between the secondary current and the primary current corresponding to the same sampling volume, for each ionization volume j, we can write for the generation factor (ignoring, for now, the terms A and B in Equation (3))

$$n_e \hat{\sigma}_{12}(r_j) = \frac{I_{j(det)}^{2+}}{2 I_j^+ dl} \tag{7}$$

obtaining in this way its relative profile along the primary beam path (by scanning for all values of j). The absolute profile of 'n sigma' can be computed by using the relative 'n sigma' profile combined with the information about the total current lost by all secondary beams traveling in the plasma. This current is given by the difference between the primary beam injected current minus the primary beam collected current, measured after crossing the whole plasma (with the factor 2 accounting for charge conversion) and the total current of the secondaries collected at the detector

$$I_{total.lost}^{2+ \rightarrow 3+} = 2(I_0^+ - I_{det}^+) - I_{det.\,total}^{2+} \tag{8}$$

The current $I_{total.lost}^{2+ \rightarrow 3+}$ is distributed per each secondary beam weighted by the corresponding 'n sigma' integral. Each individual integral is taken along the secondary beam path from the ionization volume location until the plasma exit. The computation is performed using the following function

$$\aleph_j^{2 \rightarrow 3} = \frac{I_j^{2+ \rightarrow 3+}}{I_{total}^{2+ \rightarrow 3+}} = \frac{\left\{ I_j^{2+} \int_{r_j}^{R_p} n_e \hat{\sigma}_{23} dl \right\}}{\sum_{j=1}^{N} \left\{ I_j^{2+} \int_{r_j}^{R_p} n_e \hat{\sigma}_{23} dl \right\}} \tag{9}$$

The ratio in the above equation can be multiplied by the term $I_{total.lost}^{2+ \rightarrow 3+}$ allowing one to determine the current that was lost by each secondary beam (here we are using the approximation that the relative profile of $n\hat{\sigma}_{12}(r_j)$ and $n\hat{\sigma}_{23}(r_j)$ are identical, which is confirmed by numerical calculation). This information allows reconstructing the absolute values of $n\hat{\sigma}_{12}(r_j)$ at each ionization volume (to follow more details, including the contribution of the reaction $I^+ + e \rightarrow I^{3+} + 3e$, please consult [8]).

The importance of obtaining local measurements is not only beneficial for obtaining the absolute profiles but also manifests in improving the detection and characterization of fluctuations (in frequency and phase). To illustrate this, we modeled an equivalent perturbation in the plasma density and temperature induced by an MHD mode number $m = 1$ located at $r = 0.025$ m. The corresponding perturbation on the 'n sigma' value was $\pm 30\%$ of the local non-perturbed n sigma value. Elsewhere, the plasma presented a parabolic density profile and a peak temperature profile. The mode rotation frequency is 67 kHz with a sinusoidal imposed amplitude modulation rate of 6.7 kHz [8]. In Figure 6 left, is depicted the contour plot of 'n sigma' as inputted in the simulation code; on the right is presented the evolution of the inputted and retrieved $n\hat{\sigma}_{12}(r_j)$ values at a particular location of the central ionization volume ($r = 0.6$ cm) during one amplitude modulation cycle (period $T = 1/6700$ s). We can observe that the absolute value of 'n sigma' is retrieved within <2% absolute error. Moreover, the evolution in time of both 'n sigma' signals inputted and retrieved is in phase. On the same graph, we can see the evolution of the collected secondary current emerging from the ionization volume. The evolution of the collected secondary current preserves information about the frequency of the mode but shows a phase shift between the '*n sigma*' rotation signature of $\approx \pi$ (peak and valley relative position offset). This happens because the secondary current in this cell is influenced by the fluctuation of the primary beam current when it passes over the rotating $m = 1$ mode and propagates to the next secondary beam ionization volume.

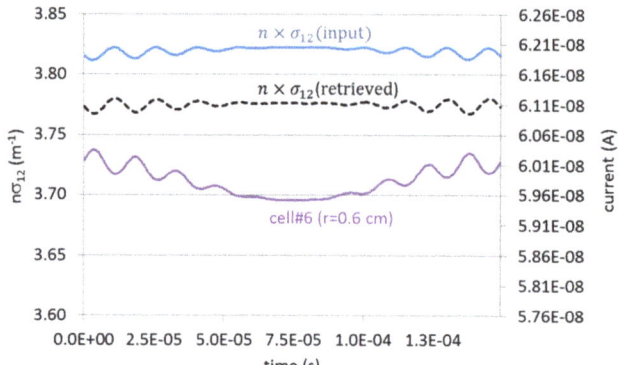

Figure 6. Left: illustration of the '*n sigma*' contour plot of the $m = 1$ mode (plasma radius = 8.5 cm); Right: the evolution of the inputted and retrieved '*n sigma*' (left scale) at the ionization volume in position $r = 0.6$ cm and the evolution of the detector current (right scale) for the secondary ions emerging from the ionization volume (reprinted from [8]).

Depending on the coexistence of other modes and size of their fluctuations, the contamination of the detector signals by the path integral effects of the primary beam can lead to 'ghost' coherence and spectral signatures if these integral effects are not removed in the signal analysis. It is important to point out that for the range of average densities in ISTTOK, below 5×10^{18} part/m^3, and due to the relatively small plasma cross section, these effects are not detectable. For higher density plasmas in larger devices, these effects are expected to show a detectable contribution.

3.1.1. Experimental Measurements of Pressure-like Profiles

Experiments have been conducted in ISTTOK [9] where the '*n sigma*' relative profile was measured during MHD activity. Combining the HIBD with the Mirnov coil data, it is possible to identify the mode number (Mirnov) and its location (HIBD).

In Figure 7, is depicted the evolution of an MHD instability measured by the HIBD.

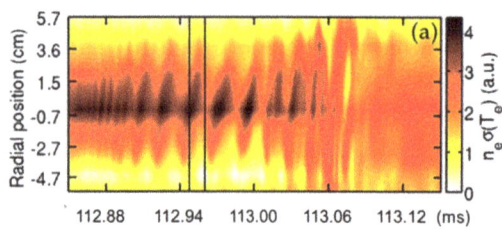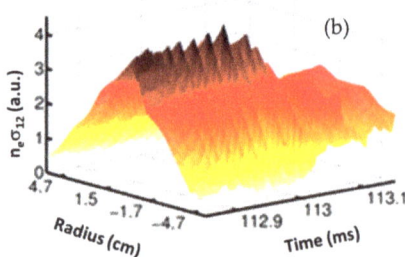

Figure 7. The contour plot of the '*n sigma*' (**a**) and the evolution of the relative profile (**b**) during MHD activity during MHD mode growth leading to the collapse of the plasma pressure profile after $t > 113.06$ ms (reprinted from [9], with the permission of AIP Publishing).

From the Mirnov coil signals, this has been identified as a $m = 2$ poloidal number mode. From the HIBD data, the evolution of the mode can be followed and the frequency of the mode rotation can be determined to be circa 63 kHz. This mode grows in size and ends with a disruption ($t > 113.06$ ms) leading to a flatness of the '$n\ sigma$' profile.

The relatively highly detailed information that can be retrieved with the HIBD can be used to visualize the radial structure of a tearing mode (TM). For instance, it is expected that in a TM, the inner flux and the outer flux surfaces would exhibit the characteristic oscillation frequency of the mode but with an opposite phase. In Figure 8, the HIBD signal is presented filtered with a central frequency of 65 kHz and a bandwidth of 20 kHz. Using the Mirnov coils, this mode is identified as $m = 4$, and, from the HIBD data, it is found to be located slightly off-center at around $r = 2$ cm. The bars indicate the location of the phase inversion as it is seen directly in the contour plot of the cell currents (peaks and valleys are offset by π rad).

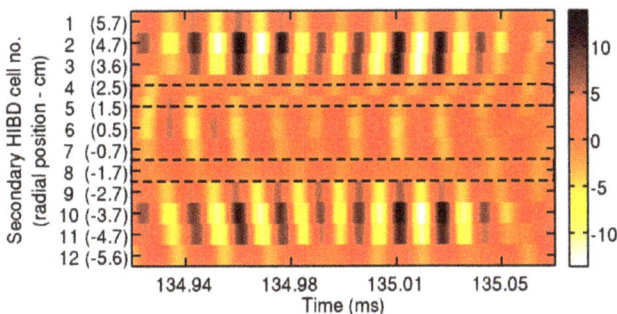

Figure 8. HIBD signals filtered with a 20 kHz bandwidth filter centered at 65 kHz. Dashed line rectangles indicate the location of the phase reversal. The scale is in a.u. (reprinted from [9], with the permission of AIP Publishing).

3.1.2. Control of Plasma Column Position

The '$n\ sigma$' profile can in principle be used to determine the plasma column position along the path of the primary beam by mapping the position in the plasma corresponding to the '$n\ sigma$' highest value. As the '$n\ sigma$' is a pressure-like measurement, it shall follow a function closely related to the plasma pressure (this is particularly true for the ISTTOK range of plasma density and temperature) and reflect the position of the plasma column center. In the case of monotonic peaked profiles, the peak value of '$n\ sigma$' could be used to follow the plasma column center movement (vertically, in ISTTOK). However, in the case of hollow pressure profiles and in the presence of relatively large pressure fluctuations, the first momentum of the '$n\ sigma$' distribution offers a more robust signal to follow. The determination of the first momentum of the '$n\ sigma$' function is simply made from the individual detector row positions weighted by the ratio of the local '$n\ sigma$' to the '$n\ sigma$' integral value (calculated from the contribution of all detector cells). This will give the centroid of '$n\ sigma$' that is then mapped from the detector into the plasma radius. This method was used to control the vertical (y) position of the plasma column of ISTTOK in real time [10].

In Figure 9 is depicted the evolution of the plasma column vertical position measured by the HIBD during AC operation. The red line corresponds to the pre-set point function given to the vertical position controller. The signals from the HIBD are processed in real time to provide the first momentum of the '$n\ sigma$' profile and feed it to the vertical controller. The controller compares the experimental results of plasma column position with the pre-set point function and acts in the plasma column (via vertical field coils) to correct the plasma position within a pre-defined interval of acceptance.

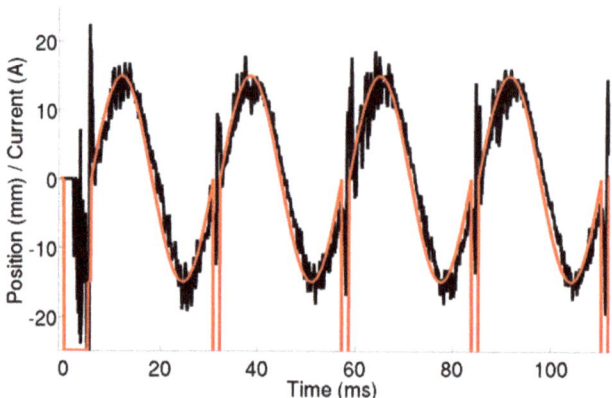

Figure 9. Control of the vertical plasma column position (black line) during an AC discharge by following closely the controller pre-set function (red line). The real-time feedback signal is obtained using the HIBD for retrieving the '$n\ sigma$' profile. The update frequency of the plasma column vertical position is f = 10 kHz (reprinted from [10]).

3.1.3. Pressure-like Profiles Measured during ISTTOK AC Operation

The ISTTOK tokamak is able to perform alternating discharges up to 1 s. duration with a full AC cycle lasting about 50 ms [11]. The AC transition is characterized by the instant when the total plasma current is null. It is observed that during this instant, the plasma density is not null and presents a local minimum which appears several hundreds of microseconds after the transition (Figure 10).

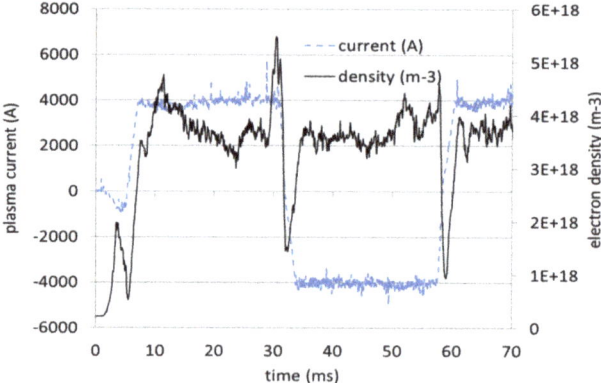

Figure 10. Evolution of plasma current (dashed line) and electron density during one AC cycle.

The HIBD has been used to study the plasma pressure-like profile evolution during the AC transition under edge electrode biasing [12]. In these experiments, the beam was chopped in the range of 180 kHz. The instants with the beam off were used to measure the noise in the detector cells and discount it during the beam-on phase (by interpolation between two off cycles). This technique allowed obtaining measurements from relatively low plasma pressures and revealing the evolution of the 'pressure-like' profiles during the AC transition. In Figure 11, is depicted the corresponding pressure-like profile evolution during the transition shown above, from positive to negative plasma current.

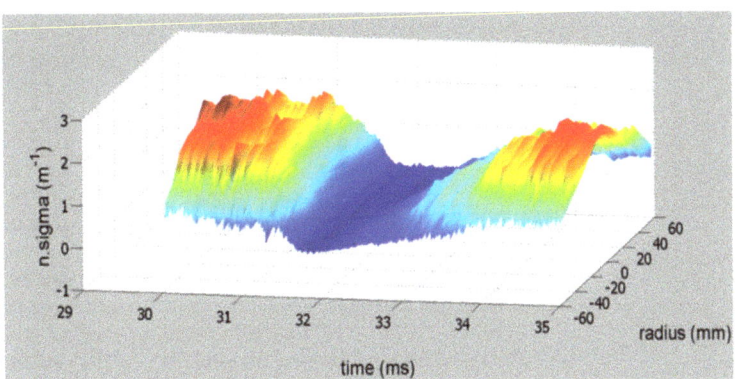

Figure 11. Evolution of the absolute value of '*n sigma*' profile during an AC transition from positive to negative plasma current.

During electrode-biasing-assisted AC operation, it was found that a noticeable amount of runaway electrons can be produced. These runaway electrons can form beams that collide with the limiter or the vessel. In this process, they can input current to the plasma, when converting the loss of magnetic energy associated with its own current. This effect could in principle be detected by the HIBD as a momentary increase in the plasma 'pressure-like' signal (due to the plasma temperature increase via Ohm heating). Figure 12 shows the clear increase in the '*n sigma*' (2D contour plot) value in the plasma core just after the X-ray burst ($t \approx 203.5$ ms). This observation is due to the conversion of the runaway beam magnetic energy into a plasma pressure increase. Further down in time ($t > 204$ ms) a second burst seems to have occurred as it displays the same kind of signature in the '*n sigma*' profile although has no X-ray signature associated, probably because the generated X-rays are out of the collection cone of the Si-based detector.

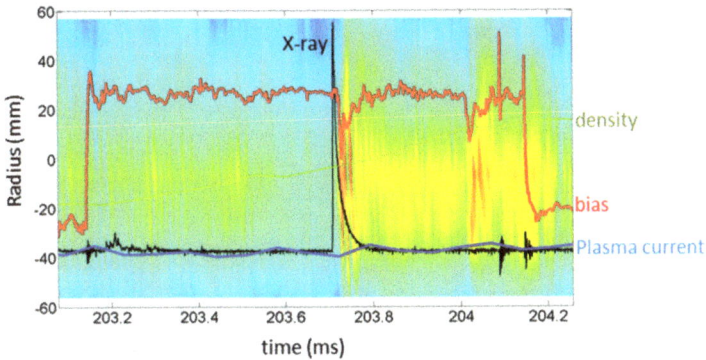

Figure 12. The colored background represents the contour plot of '*n sigma*' evolution (a.u.) covering the plasma radius from about ±50 mm. Superimposed signals show the plasma current, plasma density and electrode biasing voltage (a.u.). It is clearly seen that, following the X-ray signal, the '*n sigma*' profile increases in the plasma core region.

3.2. Determination of Plasma Density and Temperature Profiles

The retrieving of '*n sigma*' absolute values is most relevant for using the HIBD in temperature and density measurements, i.e., for obtaining the two profiles separately. One method used in ISTTOK [13] is based on the injection of two different beam species. Due to the type of ion source used, it is possible to produce individual or mixed beams of Cs^+, Xe^+ and Hg^+. If two different beam species are injected through the same path in the plasma and the corresponding 'n sigma' absolute profiles are retrieved, then their ratio is only a function of the corresponding ionization effective cross-section ratio. Since the density term is common for both species, we can write

$$\Re(T_e) = \frac{[n_e \times \hat{\sigma}_{1m}]_{I_1^+}}{[n_e \times \hat{\sigma}_{1m}]_{I_2^+}} = \frac{[\hat{\sigma}_{1m}(T_e)]_{I_1^+}}{[\hat{\sigma}_{1m}(T_e)]_{I_2^+}} \quad (10)$$

where m denotes the charge of the secondary beam. The use of this method calls for the largest possible ratio dependence between the effective cross sections involved such that the function $°(T)$ is strongly dependent on the plasma temperature. For the case of ISTTOK, the chosen reactions and species were the ones that produced the following ions: $Xe^+ \to Xe^{2+}$ and $Hg^+ \to Hg^{3+}$. The $°(T)$ function for these reactions is depicted in Figure 13 together with the ratio function for the processes $Xe^+ \to Xe^{2+}$ and $Hg^+ \to Hg^{2+}$, useful to estimate the current of the Hg^{2+} ions (which is not collected at the detector) from the collected Xe^{2+} ions.

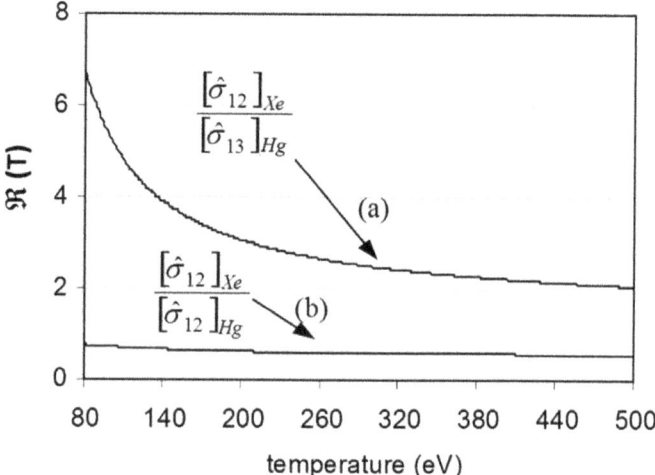

Figure 13. The effective cross-section ratio function for two ionizing reactions.

The trajectory of the ions of interest is evaluated by the simulation code. As depicted in Figure 14, the Hg^{3+} ions can be spatially discriminated from the Hg^{2+} ions (which are not detected). The Xe^{2+} ions are also spatially discriminated from the Xe^{3+} ions.

We may note that the reaction $Hg^+ \to Hg^{3+}$ only produces measurable signals (in ISTTOK) when the plasma temperature is above 80 eV. This condition constrains the determination of temperature to a typical plasma radius of $r \approx 3$ cm. By means of this mixed beam technique, the profiles of density and temperature were measured (Figure 15).

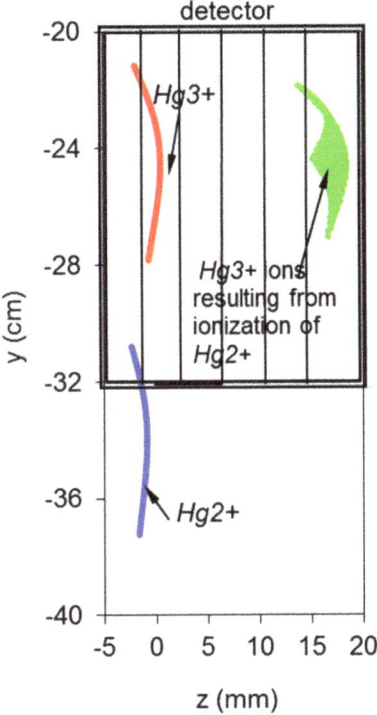

Figure 14. Footprint of Hg ions in the detector area. The boxed area corresponds to the MCAD real size.

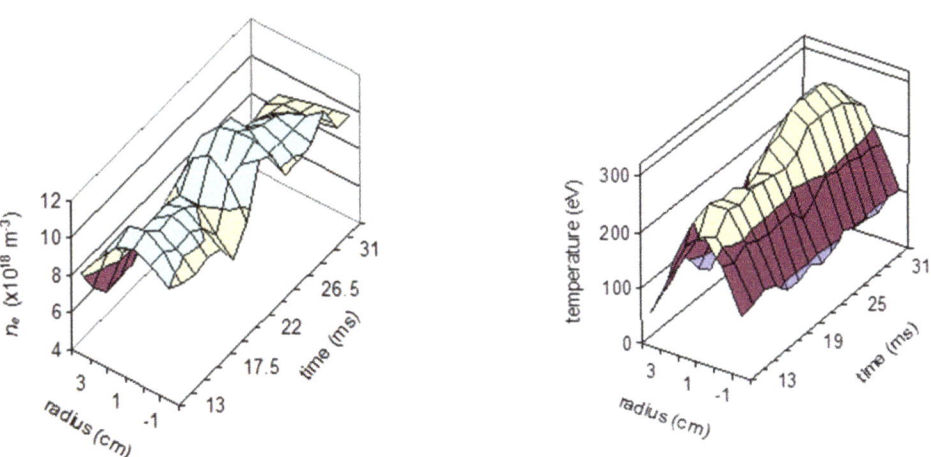

Figure 15. Density (**left**) and temperature (**right**) profiles obtained in ISTTOK with the HIBD using two species injection.

3.3. Poloidal Magnetic Field Retrieval and Measurements

The poloidal magnetic field B_p, produced by the plasma current, obeys the Maxwell Equation (Ampere's Law),

$$\mu_o \vec{j} = \vec{\nabla} \times \vec{B}_p \tag{11}$$

where j is the plasma current density. As mentioned above, some works using the HIB-probe configuration reported on the determination of local fluctuations of B_p. The measurement contains the integral effects of primary and secondary beam trajectories and cannot be considered absolutely local. However, in some special conditions, if the local magnetic field fluctuations at the sample volume exceed the path integral effects, then the frequency content of such fluctuations can be manifested in the beam position fluctuations. These are special conditions that require detailed analysis of the data in order to infer on its validity and on the non-contamination by other non-local sources.

On the other hand, the HIB-diagnostic configuration allows for the determination of the local poloidal magnetic field B_p, at each sample volume, by allowing the removal of the path integral effects. The total field at each sample volume will be the sum of the external coils' fields and the plasma current magnetic field. The external fields can be accounted by calculation or by specific measurements used for establishing the distribution and values of the machine vacuum fields. These known vacuum fields can be subtracted from the total measured field in order to obtain the plasma current poloidal field contribution.

Let us start by assuming cylindrical symmetry (which is a good approximation for ISTTOK) and represent the poloidal field flux lines as is depicted in Figure 16. We also assume that the beam trajectories are perpendicular to the flux surfaces (otherwise, the perpendicular component of the poloidal field to the beam trajectory is the one that is considered).

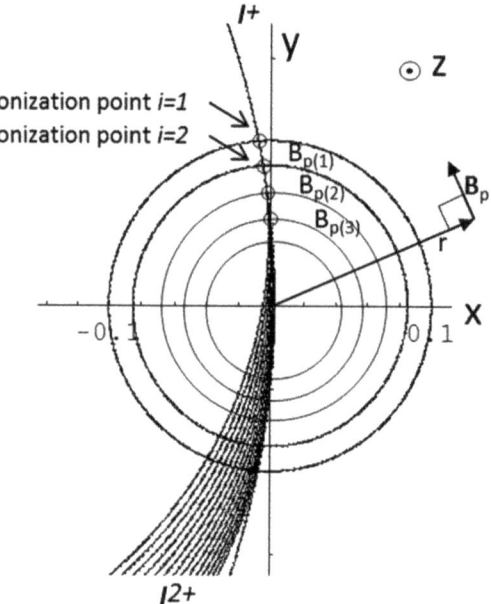

Figure 16. Ion trajectories assuming cylindrical geometry for the plasma current.

The integral form of Equation (11) can be given inside and outside the plasma column at point r as:

$$B_p = \frac{\mu_0 I(r)}{2\pi r} \quad (12)$$

where $I(r)$ represents the current enclosed in the cylindrical volume of radius r. The values of $I(r)$ inside the plasma are dependent on the plasma current density $j(r)$ profile by the relation (in cylindrical coordinates)

$$I(r) = \int_0^r j(r') 2\pi r dr' \quad (13)$$

Outside the plasma column, the value of $I(r)$ becomes constant and equal to the total plasma current I_{plasma}.

Once a current profile is defined, the profile of the poloidal magnetic field as a function of plasma radius r is given by the combination of Equations (12) and (13). The force exerted by the local poloidal field at each ionization volume imposes a local acceleration of the primary beam (and secondary) in the z-direction (the toroidal direction). If we manage to measure the primary beam local acceleration component that is perpendicular to the poloidal field (i.e., in the z or toroidal direction), then the corresponding local poloidal field can be obtained from:

$$|B_p| = \frac{m}{e}\left|\frac{dv}{dt}\right|\frac{1}{v_\perp} \qquad (14)$$

where dv/dt stands for the acceleration in the direction z which is perpendicular to both the field lines of $B_p(r)$ (see Figure 16) and the primary beam velocity $v_\perp$ (which is here defined as the velocity component lying in the plane XOY and perpendicular to $B_p(r)$ field lines). In ISTTOK, the geometry is such that for the assumed conditions, the primary beam velocity projection in the plane XOY is quasi-perpendicular to the poloidal magnetic field lines along all primary and secondary beam trajectories. This means that the acceleration z of the primary is induced by the total plasma current poloidal field at the ionization volume. In the case the angle between the primary beam velocity v_{XOY} and the field lines of $B_p(r)$ is not perpendicular, then the beam would feel a local z-acceleration weighted by the sine of the angle between these two vectors:

$$\left|\frac{dv}{dt}\right| = |\frac{e}{m}\vec{B}_p \times \vec{v}_{XOY}| = \frac{e}{m}|B_p|\cdot|v_{XOY}|\cdot \sin\left(\widehat{\vec{B}_p \vec{v}_{XOY}}\right) \qquad (15)$$

3.3.1. Retrieving of Local Values of Magnetic Poloidal Field

Considering the geometry in Figure 16, we may write for the secondary beam z-position in the detector, for ionization point $i = 1$,

$$z^{2+}_{d(1)} = z^+_1 + \bar{z}^{2+}_1 + v^+_1 T_{2(1)} + \underbrace{\frac{1}{2}\bar{a}^{2+}_{(1)}T^2_{p(1)}}_{S_1} \qquad (16)$$

where z^+_1 is the z-coordinate of the primary beam at the ionization point (it is known from the injection position of the source and the computation of the beam trajectory under the known magnetic fields, until the entrance in the plasma), and $\bar{z}^{2+}_1$ is the increment in the z-coordinate of the secondary ion from the plasma edge up to the detector (this term contribution can be accounted by using the total plasma current in Equation (12) as the poloidal field in this region is known). The term $v^+_1 T_{2(1)}$ accounts for the increment in the z-coordinate of the secondary ion due to the fact that it leaves the primary beam with finite z-velocity. At the ionization point, we can write $v^+_1 = v^{2+}_1$, and therefore its z-coordinate suffers an increment from the ionization point until the detector during its time of flight $T_{2(1)}$. The value of v^+_1 is obtained at the point the primary beam enters the plasma. It is computed by the simulation code and only depends on the total plasma current (as it is determined by the beam trajectory before it enters the plasma). The last term accounts for the z-increment due to the plasma integral poloidal field between the ionization point and exit of the plasma. It is written on the basis of an average acceleration on the secondary beam $\bar{a}^{2+}_{(1)}$, which accounts for the integral of the acceleration over the secondary beam trajectory from the ionization point until its exit of the plasma during the traveling time

$T_{p(1)}$. This average acceleration (which is unknown) can be related to the corresponding average z-velocity of the secondary beam $\bar{v}^{2+}_{(1)}$ using

$$\bar{a}^{2+}_{(1)} = \frac{\bar{v}^{2+}_{(1)}}{T_{p(1)}} \tag{17}$$

Combining Equations (16) and (17), we can obtain the average value of the secondary ion velocity counted from the ionization point 1 until the exit of the plasma,

$$\bar{v}^{2+}_{(1)} = 2\frac{z^{2+}_{d(1)} - z^{+}_{1} - \bar{z}^{2+}_{1} - v^{+}_{1}T_{2(1)}}{T_{p(1)}} \tag{18}$$

In fact, this first term can be calculated independently of the plasma current profile. It is given by the difference in the z-position of the first secondary beam at the ionization point (which can be taken equal to the primary beam z-position at the same point) and the z-position of the first secondary beam when it leaves the plasma, divided by the corresponding time of flight. This last position is obtained by taking the beam z-position on the detector after discounting the increment induced by the total plasma current from the exit point from the plasma until the detector. This is done using the total plasma current and the 3D simulation code. The flight times of the ions from the ionization point until the detector including the time spent between the plasma exit and the detector are given by the geometric installation of the HIBD in ISTTOK. They also can be obtained using the trajectory simulation code; in the present approximation, the flight times are considered not to be significantly affected by the poloidal field profile or by the total plasma current value.

Using this initialization result, let us consider now the motion equation of the secondary beam emerging from ionization point $i = 2$. We can write a similar relation for the z-position in the detector,

$$z^{2+}_{d(2)} = \bar{z}^{2+}_{(2)} + z^{+}_{(2)} + v^{+}_{(2)}T_{2(2)} + \underbrace{\frac{1}{2}\bar{a}^{2+}_{(2)}T^{2}_{p(2)}}_{s_2} \tag{19}$$

where $\bar{a}^{2+}_{(2)}$ represents the average z-acceleration of the secondary beam from ionization point 2 to the detector. In a similar way, the integral acceleration can be given by

$$\bar{a}^{2+}_{(2)} = \frac{\bar{v}^{2+}_{(2)}}{T_{p(2)}} \tag{20}$$

In Figure 17 is represented the above situation but for the general case of two secondary beams generated at two consecutive ionization points.

For simplicity, we assume that the detector line is just at the plasma exit, and therefore we can ignore in the above equations the z-position increment that each beam suffers from the plasma exit until the detector, $\bar{z}^{2+}_{i}$. As mentioned above, we can calculate these increments based on the total plasma current and build the equivalent geometry depicted in Figure 17. In this figure, the values of v are all representing velocity in the z-direction. If we subtract Equations (16) and (19) from each other and assume that (i) at the ionization point i, the z-velocity component of the primary beam and secondary beam are equal (it defines the secondary beam initial z-velocity) and (ii) that the path integral difference between the two consecutive secondary beams is negligible, i.e., the beams feel the same integral acceleration along their paths, $\bar{a}^{2+}_{i}$, then we can obtain the following algebraic relation:

$$z^{2+}_{d(i+1)} - z^{2+}_{d(i)} = v^{+}_{(i+1)}T_{2(i+1)} - v^{+}_{(i)}T_{2(i)} + z^{+}_{(i+1)} - z^{+}_{(i)} \tag{21}$$

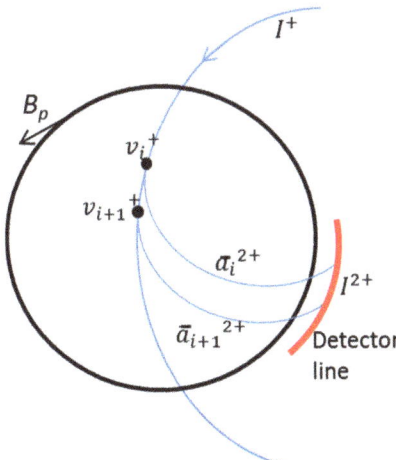

Figure 17. Trajectories of two consecutive secondary beams generated at two very close ionization points.

If we consider that the difference in the z-position of the two consecutive ionization points over the primary beam is much smaller than the difference in the z-position of the corresponding secondary ions on the detector, we can simplify the above equation to obtain

$$\Delta z_d^{2+} = \Delta v^+ \overline{T}_2 \qquad (22)$$

where $\overline{T}_2 = \frac{1}{2}\left(T_{2(i+1)} + T_{2(i)}\right)$ is the average flight time of the two secondary beams. Dividing both sides by dt, which is the elemental time increment that the primary beam takes to travel between the two ionization points, we finally obtain

$$\frac{\Delta z_d^{2+}}{\overline{T}_2 \cdot dt} = \frac{\Delta v^+}{dt} = a_i^+ \qquad (23)$$

This quantity is the primary beam z-acceleration taken at the location between the ionization points. Combining Equation (14) and the last equation above, we obtain the component of the magnetic poloidal field that is perpendicular to the primary beam velocity.

We tested this iterative process for several symmetric current density profiles with a total plasma current of 6 kA. The results are summarized in Figure 18. As the poloidal field changes sign from the top half of the plasma to the bottom half, we plot the absolute values of B_p. The proper sign can be recovered by considering the sense of the total plasma current. The method here presented can recover with great efficiency the absolute profile of the plasma magnetic poloidal field. An about 5–7% average of inaccuracy in absolute values can be observed for the radial range 2–2.5 cm of the plasma upper edge. This could be caused by the assumption that two secondary beams feel the same force while traveling over the plasma which may become less accurate when the travel times are longer.

In this evaluation of the method, we are not considering the detector spatial resolution. However, in practice, the finite size of the cells imposes an error of about 0.2–0.3 mm in the determination of the absolute z-position. This value already accounts for the beam profile distortion (that we can measure at the primary detector), the cell width (4 mm) and the noise on the detector cells.

For measuring fluctuations on the beams' positions, as the beam profile remains unchangeable, the spatial accuracy can become about 0.1 mm, limited by noise that can be minimized by either using fast beam chopping or by subtracting the signal from neighboring cells (with similar noise content but without beam current). For instance, the magnetic field fluctuations induced by the toroidal field power supplies (50 Hz and 300 Hz) were characterized using the primary beam multiple-cell detector toroidal fluctuations.

This method was first proposed for ISTTOK in 1994 [14]. The first application to experimental results in ISTTOK was published in 1997 [15]. The method was later presented with more mathematic detail and used to estimate the effect of the finite size of the detector cells and beam profile in the retrieval of the poloidal field [16].

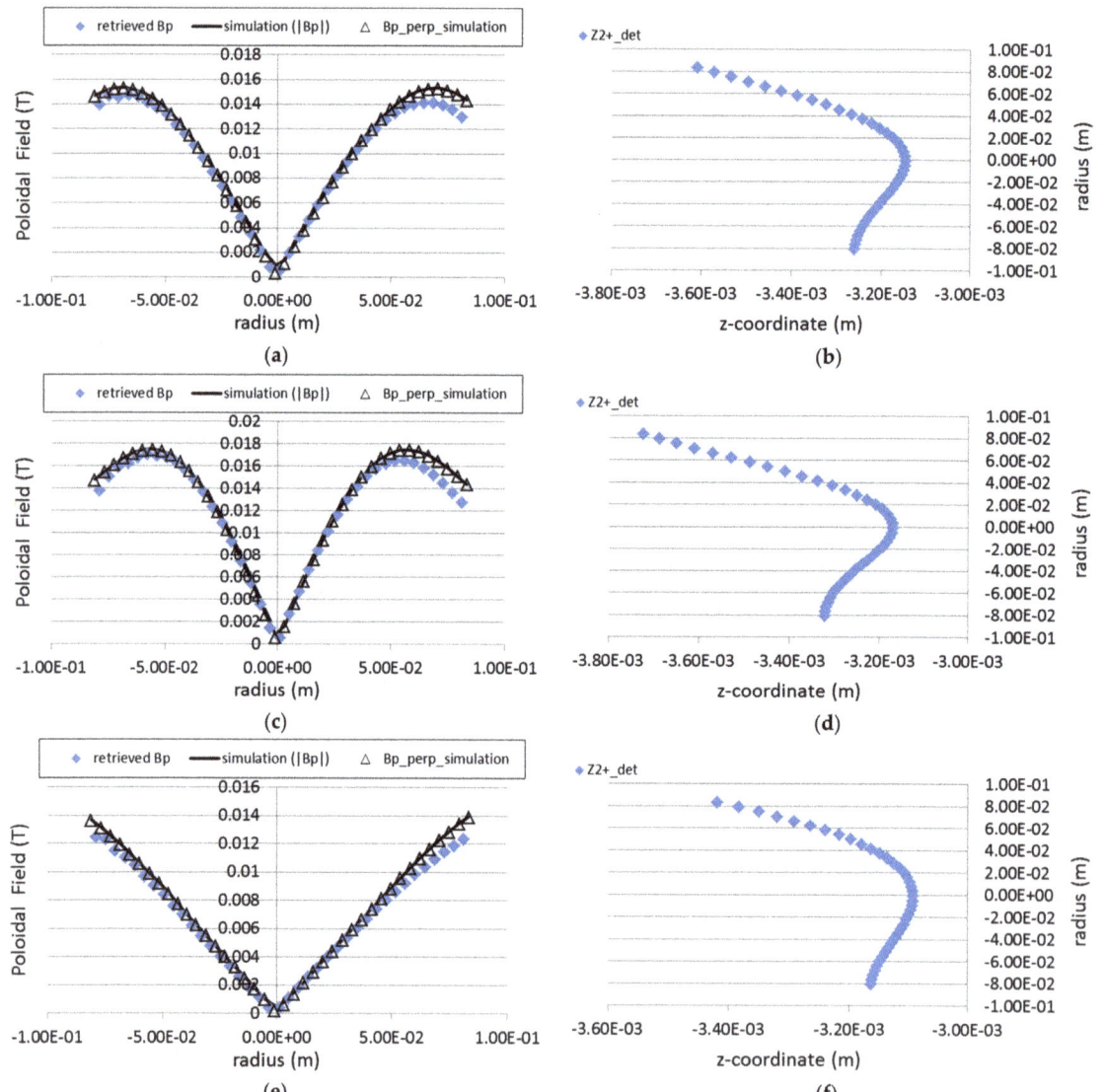

Figure 18. On the left side, figures show the retrieved values of the poloidal field profile (solid triangles) for three different current density profiles: (a) parabolic, (c) peaked (parabola square) and (e) flat. The values of the inputted profiles of the total poloidal filed and of the B_p z-component are also displayed in the same graph. The figures on the right (b–f) show the corresponding detector z-positions of the secondary beams as a function of the radial location of the ionization volume.

In spite of the initial assumptions for application to symmetric profiles, we tested the method for non-symmetric and for hollow current profiles. The results are depicted in Figure 19. They show that the method is still very effective in retrieving the corresponding poloidal field absolute values in such conditions.

In Figure 20 are depicted the retrieved poloidal field profiles for the same three plasma current density profiles as those used in Figure 18, considering the finite size of detector cells. The results show that even for the finite cell size, the retrieved values follow closely each of the poloidal field inputted profiles.

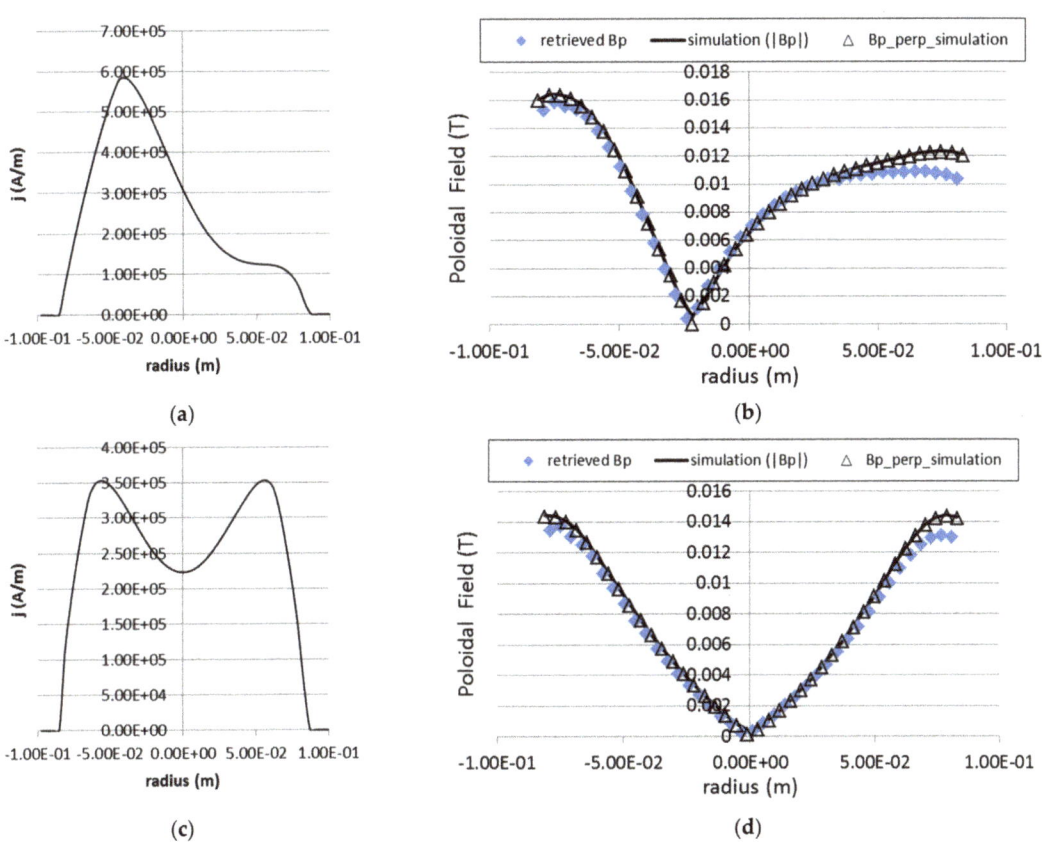

Figure 19. Retrieving of the poloidal field for an asymmetric and a hollow current density profiles. On the left graphs (**a**,**c**) are represented the inputted current density profiles and on the right (**b**,**d**) the retrieved values of the corresponding poloidal magnetic fields based only on the position of the secondary ions on the detector.

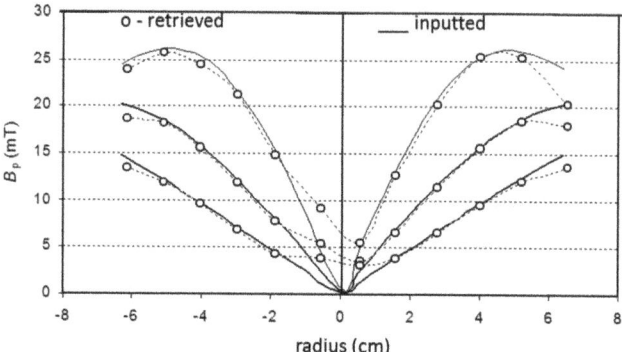

Figure 20. Retrieved poloidal field for three different poloidal field profiles (parabolic, parabolic squared and flat) taking into account beam profile distortion and finite cell size (12 × 5 MCAD with cell width of 4 mm).

3.3.2. Experimental Determination of the Evolution of the Plasma Current Magnetic Field Profile

The evolution of the plasma current magnetic field profile was measured during ISTTOK discharge ramp-up [15]. In these particular experiments, the measurements are confined to the core region of the plasma due to the low signal on the more peripheral cells. A detector of 9 × 6 cells was used with a cell width of 2 mm. The measurements were taken at four time instances as indicated in the reference discharge in Figure 21.

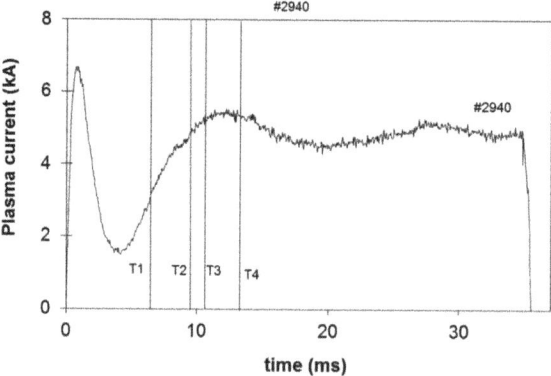

Figure 21. ISTTOK plasma current evolution and indication of the 4 instants when the poloidal field profile was determined (partially reprinted from [15]).

The measured detector positions are depicted in Figure 22a. The positions are interpolated by a continuous function from which is computed the poloidal field profile, shown in Figure 22b.

The main observable result is that the plasma current increase is well followed by the secondary beam positions showing that the plasma is not centered in the earlier stages of the discharge. Near the current plateau, the plasma occupies the central position of the chamber, and the slightly decreasing slope of the poloidal field profile indicates that the current channel is becoming less peaked as the discharge evolves.

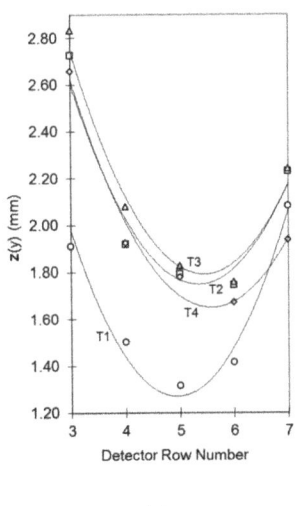 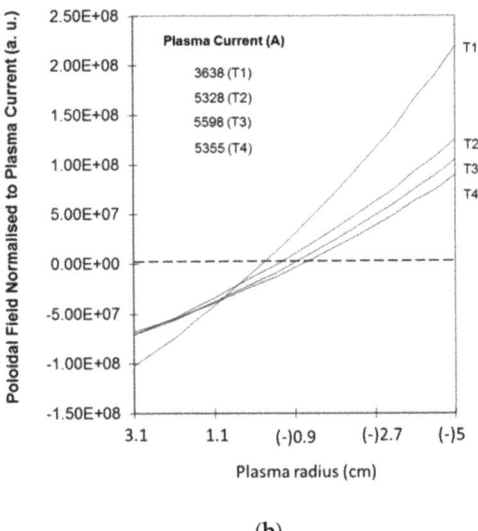

Figure 22. (**a**) The symbols indicate the measured z-positions of the secondary beams fitted with corresponding interpolation lines T1, T2, T3 and T4. (**b**) Estimated normalized poloidal field profiles from the detector characteristic lines at the time instantes T1, T2 T3 and T4 (reprinted from [15]).

3.4. Plasma Potential Measurements

As was demonstrated above in Equation (2), the plasma potential affects the energy of the secondary beams. The Proca-Green 30° parallel plate energy analyzer has been used for measuring the secondary ion beam energy in the HIB-probe configuration. However, in order to keep the possibility to collect the whole cloud of secondary beams in the ISTTOK HIB-diagnostic configuration, it is necessary to use a different detector arrangement. To that end, the direction of the analyzing field in the energy analyzer must be parallel to the tokamak toroidal field. The most efficient way to save space and perform the required analyzing capability is to use a cylindrical energy analyzer. There are several experiments that require measuring ion beam energies where this type of analyzer is employed. In particular, it has been studied for application on the large helical device (LHD) for plasma potential measurements [17] (not implemented). However, the possible application of this analyzer in the LHD was limited to measuring one secondary beam at a time.

In ISTTOK, we adopted a compact 90° energy analyzer with the aim of collecting all secondary beams at once. Simulations showed that the beams could be collected from the whole plasma diameter. However, the attained energy resolution was relatively poor (in the range of several tens of volts) due to the short path length inside the analyzing field. To overcome this difficulty, a new mode of operation was proposed and successfully tested. This consisted in retarding the beam while it travels inside the analyzer [18] by applying a retarding field between the entrance (grounded) and at the back detector plane (positively biased mesh). In Figure 23, is presented the equipotential map of the normal and deceleration modes of operation.

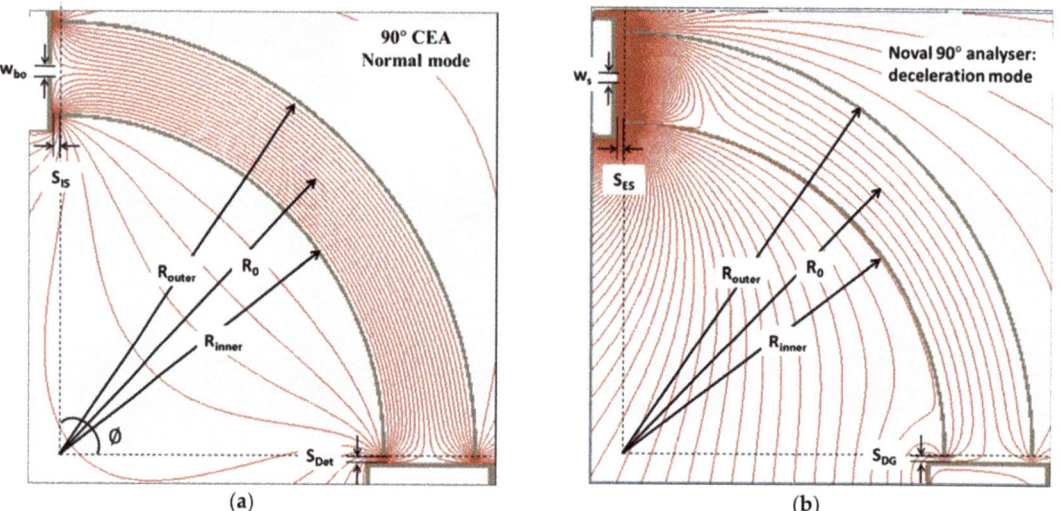

Figure 23. Representation of the equipotentials for the 90° cylindrical energy analyzer operating in normal mode (**a**) and the same analyzer operated in deceleration mode (**b**).

In the normal operation, the radial electric field $E(R)$ and the potential $\phi(R)$ inside the analyzer are described by

$$E_r(R) = E_0 \left(\frac{R_0}{R}\right)$$
$$\varphi(R) = E_0 R_0 \ln\left(\frac{R}{R_0}\right) + _0 \qquad (24)$$

where E_0 and ϕ_0 are the analyzer central values determined for the applied voltages to each electrode. For the normal mode operation, the required voltage applied between the electrodes in order to keep the beam in a specific trajectory is obtained by equalizing the electric and the centripetal forces

$$\Delta V = 2[(E_0 - q\varphi_0)/q(\frac{d}{R_0})] \qquad (25)$$

If the analyzer is operated with symmetric plate voltages, the value of ϕ_o is zero. In the deceleration mode, the value of the central potential ϕ_o is strongly increased by offsetting the plates' voltages. Consequently, the required electrode voltage difference to keep the beam in a particular trajectory is lowered in the deceleration mode as compared to normal mode operation. This is equivalent to obtaining a larger displacement on the beam orbit inside the analyzer in deceleration mode than in normal mode for a given change of beam energy if operating the analyzer at constant electrode voltages. In fact, the gain coefficient in beam displacement due to change in the beam energy k_E is given by (here we refer to the beam energy before it enters the analyzer)

$$k_E = \frac{E_{beam}}{E_{beam} - q\varphi_{0(dec.)}} \qquad (26)$$

For instance, for an input beam of 20 keV energy and a deceleration voltage of 8 kV, we obtain a coefficient gain of $k_E = 5$ ($q = 2$). This translates into an increased beam displacement in the detector of five times more than in normal mode. Other factors concur to limit this displacement gain such as beam profile stretching and the internal asymmetry of the electric field while the beam changes orbit. Taking all these factors into account, a magnification of M = 4 was obtained in the sensitivity to beam energy changes as compared to the normal case. A demonstration of this principle was performed using an electron beam, and the results matched the simulations [19].

In spite of it being possible to collect all secondary beams with this geometry, the practical placement of the analyzer was only possible away from the exit port of the tokamak. This implied the need to deflect the secondary beams from the port exit to the cylindrical energy analyzer entrance. The measurements are taken from four sample volumes in the plasma to fit in the available space in the deflection system. The full geometry is illustrated in Figure 24.

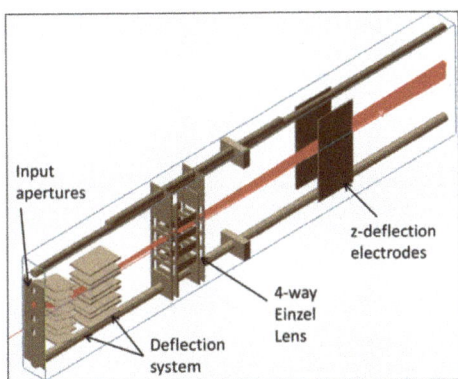
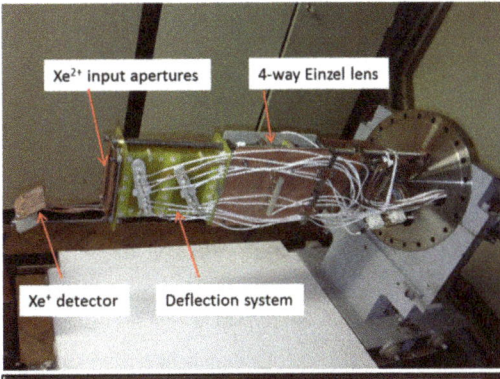
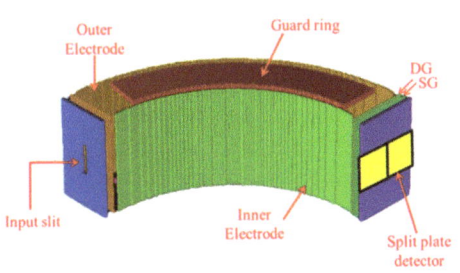
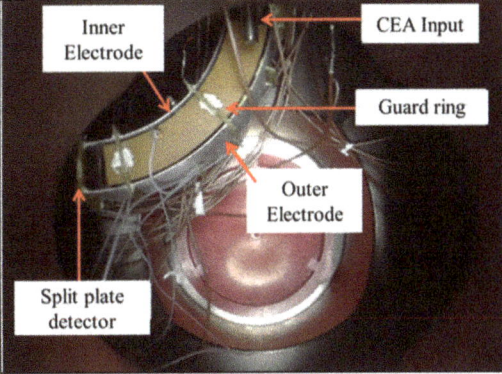

Figure 24. Components of the detection system: Xe^{2+} deflection system with Einzel lens (**top row**) and cylindrical energy analyzer (**bottom row**).

Experimental results (Figure 25) in the measurements of plasma potential in the core region are reported in [20]. In these test experiments, only the central channel was used due to limitations on the number of power supplies and amplifiers. The plasma was scanned by means of the vertical field coils, and a spatial range of values for the plasma potential was obtained. The plasma potential absolute values were calibrated with an indirect method implying up to ±70 V in absolute plasma potential error. As for the relative error, experiments demonstrate ±25 V or, equivalently, an upper limit of $\Delta E/E \sim 2 \times 10^{-3}$ for the CEA sensitivity to the plasma fluctuation measurements.

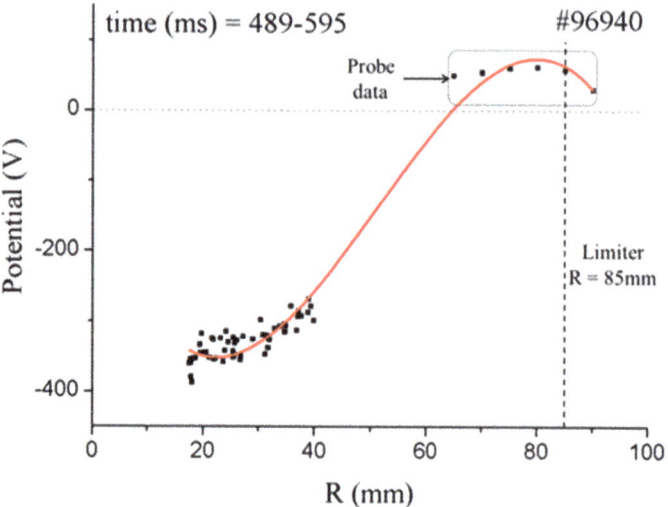

Figure 25. Plasma potential measurement by the HIBD in the plasma core of ISTTOK (reprinted from [20]).

3.5. Simultaneous Measurements of Pressure-like, B_p and V_p Fluctuations

The most recent developments on the ISTTOK HIBD were pursued aiming at achieving the possibility to measure simultaneously the profiles of plasma pressure-like (n sigma), plasma current magnetic field and plasma potential. This capability was tested by making the simultaneous measurements of the fluctuations of these three parameters via the fluctuations of the beam signals.

The detector for measuring the z-displacements of the secondary beams was made of 12 rows by 3 column cells, each with 4 mm width. This detector is placed at the exit port of the tokamak beyond the deflection electrode unit. Some of the central column cells were removed to let some secondary ions pass into the CEA (Figure 26). Signals were acquired in the CEA through the central opened channel number 3, counting from the top. The position of the secondary ions was measured by the balance of currents in the left and right cells in the front detector weighted by the total current collected at the front detector plus the current of the same beam collected at the back detector, behind the CEA exit. Due to the intrinsic deceleration operation mode properties, any changes in input angle in the z-direction at the CEA entrance are not transferred to a change in the z-position of the beam at the analyzer back detector [18] (angular first-order focus). These results allow us to separate the z-position displacements due to poloidal field changes at the front detector from those measured at the back of the CEA, which will be due to the changes in beam energy, i.e., plasma potential.

The raw results on the beam signals for three positive cycles of an AC discharge are presented in Figure 27. The data show the z-displacements of the secondary beam due to plasma potential variations (potential shift) and plasma poloidal field (toroidal shift) simultaneous with the pressure-like fluctuations (current fluctuations of the secondary beam). Cross-correlation analyses between the z-displacements measured at the front (plasma-current-driven) and back (plasma-potential-driven) detectors showed no correlation in the fluctuation of these signals, indicating that these two measurements can be performed independently.

Figure 26. Front MCAD for measuring the X^{2+} z-positions. Four apertures in the front MCAD allow for the passage of four secondary beams that are directed to the cylindrical energy analyzer, placed behind the circular flange shown in the picture.

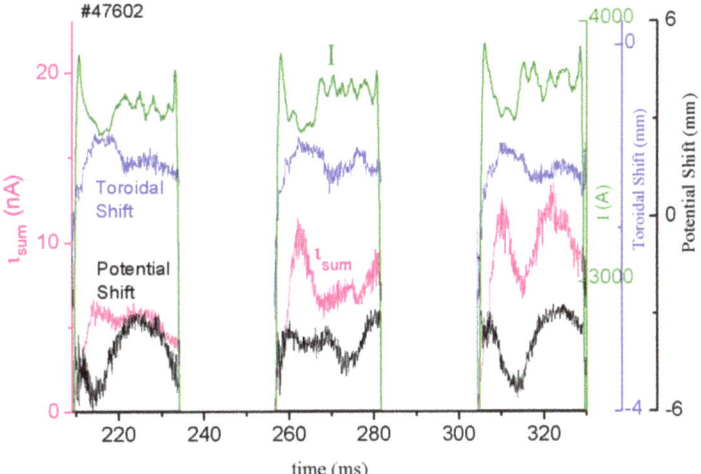

Figure 27. Raw signals of the fluctuations of the secondary beam position and intensity at the CEA front and back detectors.

4. Conclusions and Future Work

The ISTTOK HIBD is based on a unique configuration that allows the collection in a multiple-cell array detector of the probing secondary beams generated from the whole plasma diameter. The wealth of information obtained allows accounting for the path integral effects and retrieving the local values of the plasma parameters at the ionization volume in the plasma. The method allowed obtaining the profiles of plasma-like pressure with good spectral resolution. Exploitation of the plasma pressure-like measurements in AC discharges allowed for the identification and characterization of MHD activity and turbulent transport in edge polarization experiments. Further, the individual profiles of temperature and plasma density were obtained utilizing a Xe^+ and Hg^+ beam combination.

The real-time determination of the plasma column center using the distribution of the measured '*n sigma*' profile was successfully used to control the vertical position of the plasma column.

An innovative method to determine the plasma current poloidal magnetic field was here presented and tested. The accuracy of the method allows identifying different profiles of plasma current density. The method was successfully applied in the identification of the poloidal field profile evolution during the ramp-up phase of the plasma discharge.

The most recent developments addressed the measurements of the plasma potential using a cylindrical energy analyzer. The main difference from other experiments using this device is the introduction of the deceleration mode of operation. The predicted gains in the beam energy sensitivity can be tuned with the biasing voltage to reach nearly five times better resolutions. In principle, this analyzer configuration allows for measurements from several points along the primary beam path, inside the plasma. Experimental results for the core measurements were obtained with a relative precision of ± 25 V, whereas indirect calibration allowed an absolute with errors of ± 70 V. The preliminary results for the simultaneous measurements of the plasma potential, poloidal magnetic field and pressure-like fluctuations indicate that these measurements can be made without crosstalk. They pave the way for extending the coverage to more points in the plasma to allow for performing transport and turbulence analysis along the primary beam path with preservation of the phase.

The future work foresees the installation of additional amplifiers and power supplies in order to use the full capabilities of the CEA and cover more sample volumes in the plasma. When in full operation, it would be used to confront the experimental measurements of kinetic and magnetic pressure fluctuations with the theoretical and numerical predictions. By keeping the phase information, it adds a new ingredient which would be instrumental for providing time-correlated information on the interplay between turbulence, particle transport and MHD activity.

Author Contributions: Data curation, A.M., I.S.N., R.H. and R.S.; Formal analysis, A.M., I.S.N., R.H. and R.S.; Funding acquisition, A.M.; Investigation, A.M., I.S.N., R.H. and R.S.; Methodology, A.M., R.H. and R.S.; Project administration, A.M.; Software, A.M. and R.H.; Supervision, A.M.; Validation, A.M., I.S.N., R.H. and R.S.; Writing—original draft, A.M.; Writing—review & editing, I.S.N., R.H. and R.S. All authors have read and agreed to the published version of the manuscript.

Funding: This work was carried out within the framework of the EUROfusion Consortium and received funding from the Euratom research and training program 2014–2018 and 2019–2020 under grant agreement No 633053. IST activities also received financial support from 'Fundaçao para a Ciencia e Tecnologia' through projects UIDB/50010/2020 and UIDP/50010/2020. The views and opinions expressed herein do not necessarily reflect those of the European Commission.

Institutional Review Board Statement: Not applicable.

Informed Consent Statement: Not applicable.

Data Availability Statement: All data used is indicated in the references given or is explicitly included in the paper as the only main source. No other public sources of data exist.

Conflicts of Interest: The authors declare no conflict of interest.

References

1. Jobes, F.C.; Hickock, R.L. A direct measurement of plasma space potential. *Nucl. Fusion* **1970**, *10*, 195. [CrossRef]
2. Crowley, T.P. Rensselaer heavy ion beam probe diagnostic methods and techniques. *IEEE Trans. Plasma Sci.* **1994**, *22*, 291–309. [CrossRef]
3. Dnestrovskij, Y.N.; Melnikov, A.; Krupnik, L.; Nedzelskij, I. Development of heavy ion beam probe diagnostics. *IEEE Trans. Plasma Sci.* **1994**, *22*, 310–331. [CrossRef]
4. Sharma, R.; Khabanov, P.O.; Melnikov, A.V.; Hidalgo, C.; Cappa, A.; Chmyga, A.; Eliseev, L.G.; Estrada, T.; Kharchev, N.K.; Kozachek, A.S.; et al. Measurements of 2D poloidal plasma profiles and fluctuations in ECRH plasmas using the Heavy Ion Beam Probe system in the TJ-II stellarator. *Phys. Plasmas* **2020**, *27*, 062502. [CrossRef]

5. Solensten, L.; Connor, K.A. Heavy ion beam probe energy analyzer for measurements of plasma potential fluctuations. *Rev. Sci. Instrum.* **1987**, *58*, 516–519. [CrossRef]
6. Melnikov, A.V.; Drabinskiy, M.; Eliseev, L.; Khabanov, P.; Kharchev, N.; Krupnik, L.; De Pablos, J.; Kozachek, A.; Lysenko, S.; Molinero, A.; et al. Heavy ion beam probe design and operation on the T-10 tokamak. *Fusion Eng. Des.* **2019**, *146*, 850–853. [CrossRef]
7. Melnikov, A.V.; Eliseev, L.; Ascasibar, E.; Chmyga, A.; Hidalgo, C.; Ido, T.; Jiménez-Gomez, R.; Komarov, A.; Kozachek, A.; Krupnik, L.; et al. Alfvén eigenmode properties and dynamics in the TJ-II stellarator. *Nucl. Fusion* **2012**, *52*, 123004. [CrossRef]
8. Malaquias, A.; Henriques, R.; Nedzelsky, I. Inversion methods for the measurements of MHD-like density fluctuations by Heavy Ion Beam Diagnostic. *J. Instrum.* **2015**, *10*, P09024. [CrossRef]
9. Henriques, R.B.; Malaquias, A.; Nedzelskiy, I.S.; Silva, C.; Coelho, R.; Figueiredo, H.; Fernandes, H. Radial profile measurements of plasma pressure-like fluctuations with the heavy ion beam diagnostic on the tokamak ISTTOK. *Rev. Sci. Instrum.* **2014**, *85*, 11D848. [CrossRef] [PubMed]
10. Henriques, R.B.; Carvalho, B.B.; Duarte, A.S.; Carvalho, I.S.; Batista, A.J.N.; Coelho, R.; Silva, C.; Malaquias, A.; Figueiredo, H.; Alves, H.; et al. Real-time vertical plasma position control using the Heavy Ion Beam Diagnostic. *IEEE Trans. Nucl. Sci.* **2017**, *64*, 1431–1438. [CrossRef]
11. Fernandes, H.; Varandas, C.; Cabral, J.; Figueiredo, H.; Galvão, R. Engineering aspects of the ISTTOK operation in a multicycle alternating flat-top plasma current regime. *Fusion Eng. Des.* **1998**, *43*, 101–113. [CrossRef]
12. Malaquias, A.; Henriques, R.; Silva, C.; Figueiredo, H.; Nedzelskiy, I.; Fernandes, H.; Sharma, R.; Plyusnin, V.V. Investigation of the transition of multicycle AC operation in ISTTOK under edge electrode biasing. *Nucl. Fusion* **2017**, *57*, 116002. [CrossRef]
13. Malaquias, A.; Nedzelskii, I.S.; Varandas, C.A.F.; Cabral, J.A.C. Evolution of the tokamak ISTTOK plasma density and electron temperature radial profiles determined by heavy ion beam probing. *Rev. Sci. Instrum.* **1999**, *70*, 947. [CrossRef]
14. Cabral, J.A.C.; Malaquias, A.; Praxedes, A.; Van Toledo, W.; Varandas, C. The heavy ion beam diagnostic for the tokamak ISTTOK. *IEEE Trans. Plasma Sci.* **1994**, *22*, 350–358. [CrossRef]
15. Malaquias, A.; Cabral, J.; Varandas, C.; Canario, A. Evolution of the poloidal magnetic field profile of the ISTTOK plasma followed by heavy ion beam probing. *Fusion Eng. Des.* **1997**, *34–35*, 671–674. [CrossRef]
16. Malaquias, A. O Diagnóstico de Feixe de Iões Pesados do Tokamak ISTTOK Desenvolvimentos Tecnológicos e Algoritmos para Interpretação Dmedidas. Ph.D. Thesis, Instituto Superior Técnico, Lisbon, Portugal, January 2000. (In Portuguese).
17. Fujisawa, A.; Iguchi, H.; Taniike, A.; Sasao, M.; Hamada, Y. A 6 MeV Heavy Ion Beam Probe for the Large Helical Device. *IEEE Trans. Plasma Sci.* **1994**, *22*, 395–402. [CrossRef]
18. Nedzelskiy, I.S.; Malaquias, A.; Sharma, R.; Henriques, R. 90° cylindrical analyzer for the plasma potential fluctuations measurements by heavy ion beam diagnostic on the tokamak ISTTOK. *Fusion Eng. Des.* **2017**, *123*, 897–900. [CrossRef]
19. Sharma, R.; Nedzelskiy, I.; Malaquias, A.; Henriques, R. Characterization of modified 90° cylindrical energy analyser with electron beam. *J. Instrum.* **2020**, *15*, C01018. [CrossRef]
20. Sharma, R.; Nedzelskiy, I.; Malaquias, A.; Henriques, R. Plasma potential and fluctuations measurements by HIBD with 90° cylindrical energy analyzer on the ISTTOK tokamak. *Fusion Eng. Des.* **2020**, *160*, 112016. [CrossRef]

Article

Affect of Secondary Beam Non-Uniformity on Plasma Potential Measurements by HIBD with Split-Plate Detector

Igor Nedzelskiy [1,*], Artur Malaquias [1], Rafael Henriques [1] and Ridhima Sharma [2]

[1] Instituto de Plasma e Fusão Nuclear, Instituto Superior Técnico, Universidade de Lisboa, Av. Rovisco Pais, 1049-001 Lisboa, Portugal; artur.malaquias@ipfn.tecnico.ulisboa.pt (A.M.); rhenriques@ipfn.tecnico.ulisboa.pt (R.H.)
[2] Culham Science Centre, Abingdon OX14 3DB, UK; ridhima.sharma@ukaea.uk
* Correspondence: igorz@ipfn.ist.utl.pt

Citation: Nedzelskiy, I.; Malaquias, A.; Henriques, R.; Sharma, R. Affect of Secondary Beam Non-Uniformity on Plasma Potential Measurements by HIBD with Split-Plate Detector. *Sensors* **2022**, *22*, 5135. https://doi.org/10.3390/s22145135

Academic Editor: Hossam A. Gabbar

Received: 25 May 2022
Accepted: 5 July 2022
Published: 8 July 2022

Publisher's Note: MDPI stays neutral with regard to jurisdictional claims in published maps and institutional affiliations.

Copyright: © 2022 by the authors. Licensee MDPI, Basel, Switzerland. This article is an open access article distributed under the terms and conditions of the Creative Commons Attribution (CC BY) license (https:// creativecommons.org/licenses/by/ 4.0/).

Abstract: In a Heavy Ion Beam Diagnostic (HIBD), the plasma potential is obtained by measuring the energy of the secondary ions resulting from beam-plasma collisions by an electrostatic energy analyzer with split-plate detector (SPD), which relates the secondary ion beam energy variation to its position determined by the difference in currents between the split plates. Conventionally, the data from SPD are analyzed with the assumption that the secondary beam current is uniform. However, the secondary beam presents an effective projection of the primary beam, the current of which, as a rule, has a bell-like non-uniform profile. This paper presents: (i) the general features of the secondary beam profile formation, considered in the simplistic approximation of the circular primary beam and the secondary ions that emerge orthogonal to the primary beam axis, (ii) details of spit-plate detection and the influence of the secondary beam non-uniformity on plasma potential measurements, (iii) supported experimental data from the tokamak ISTTOK HIBD for primary and secondary beam profiles and the SPD transfer characteristic, obtained for the 90° cylindrical energy analyzer (90° CEA) and (iv) the implementation of a multiple cell array detector (MCAD) with dedicated resolution for the measurements of secondary beam profile and MCAD operation in multi-split-plate detection mode for direct measurements of the SPD transfer characteristic.

Keywords: nuclear fusion diagnostics; heavy ion beam probe; heavy ion beam diagnostic; plasma potential measurements; plasma poloidal magnetic field measurements

1. Introduction

The Heavy Ion Beam Diagnostic (HIBD) [1] is the only tool for direct measurements of plasma potential in magnetically confined fusion plasmas. The measurements of plasma potential fluctuations are of special interest and importance in investigations of turbulent transport [2]. Traditionally, the plasma potential is obtained from the measurements of secondary beam energy using an electrostatic energy analyzer with split-plate detector (SPD), which transforms the beam energy variation to the difference in currents between split plates [3]. Conventionally, with the assumption of relatively small electron density and temperature gradients inside the sample (primary beam ionization) volume, the data from SPD are analyzed with the assumption that the secondary beam current is uniform. However, because the secondary beam presents an effective projection of the primary beam inside the sample volume, it should be coupled with the profile of the primary beam, which, generally, is non-uniform.

This paper considers the possible influence of the secondary beam non-uniformity on plasma potential and its fluctuation measurements using SPD technique.

The paper is organized as follows. The coupling of primary and secondary beam profiles is examined in Section 2 for a simplistic model of circular primary beam and secondary ions that emerge perpendicular to the primary beam axis. Split-plate detection and SPD transfer characteristics for the non-uniform secondary beam are considered in Section 3. The influence of the secondary beam non-uniformity on plasma potential and its fluctuation measurements is analyzed in Section 4. The supported experimental data from the ISTTOK tokamak HIBD for primary and secondary beam profiles and SPD transfer characteristics obtained for the 90° cylindrical energy analyzer (90° CEA) are presented in Section 5. The multiple-cell and multi-split-plate detection for direct measurements of the secondary beam profile and SPD transfer characteristic is proposed in Section 6. Section 7 provides a summary.

2. Coupling of Primary and Secondary Beam Profiles

In HIBD, a beam of singly charged ions (primary beam, I^+) is directed through the plasma across the confining magnetic field with energy to provide a Larmor radius higher than the dimension of the plasma cross-section, as shown in Figure 1a. Colliding with plasma electrons, some primary ions are ionized to the double-charge state (secondary beam, I^{++}) along the primary beam path and are separated from the primary beam by the magnetic field. If to place a small aperture detector in the fan of the secondary ions that emerged from the plasma, it will only observe those secondaries that are created in a small sample volume along the primary beam, as schematically depicted in Figure 1c.

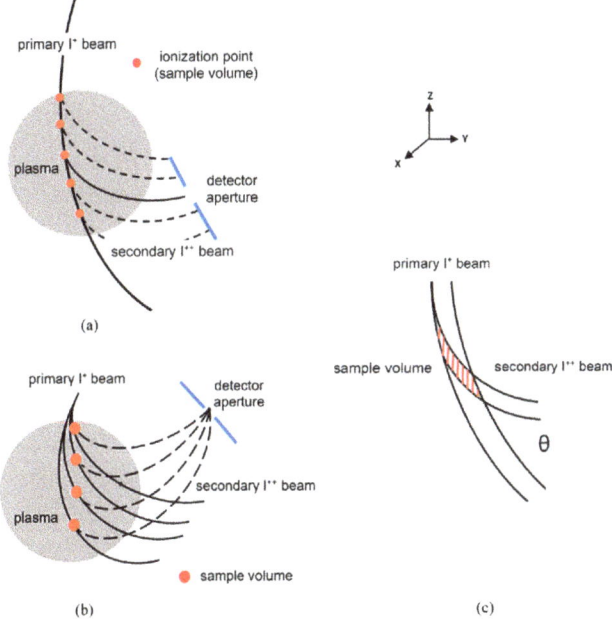

Figure 1. Schematics of HIBD primary and secondary beam trajectories (**a**,**b**) and sample volume formation (**c**).

In Figure 1a, the injection angle of the primary beam is fixed and its trajectory inside the plasma presents an effective detector line of the measurements. In traditional HIBD shown in Figure 1b, the primary beam injection angle is scanned during plasma shot, and a fan of secondary ions created inside the definite sample volume of every angle discriminated primary beam trajectory is focused in one spatially definite point by the toroidal magnetic field. The respective sample volumes in the angle fan of the primary beam trajectories inside the plasma form the detector line of the measurements. In HIBD configuration in Figure 1a, the plasma parameters along the detector line are measured simultaneously with multichannel detection of the secondary ions. This HIBD configuration is used on the ISTTOK [4]. On the other hand, in Figure 1b showing traditional HIBD configuration with detection of the focused secondary ions in one spatial point, the measurements of the plasma parameters in the sample volumes along the detector line are delayed in time. In both HIBD configurations, the formation of the sample volume is similar (as in Figure 1c) with the sample volume size determined by the dimensions of the primary beam and detector aperture.

Consider the primary beam of circular cross section and uniform current density. The sample volume then consists of a cylinder with parallel end planes that cut the cylinder at some angle θ with respect to the cylindrical (beam) axis. In the YZ plane projection in Figure 1c, the sample volume cross-section is roughly a parallelogram. Diagonal of the parallelogram is minimal at $\theta = 90°$, determining maximal spatial resolution of the measurements. For that hypothetical ideal situation, the 3D image of the sample volume of unit length (determined by detector slit width in the vertical Z direction) and circular primary beam of unit radius (in XY plane) is shown in Figure 2a with the emerged (in Y direction) secondary ions marked by arrows. As illustrated in Figure 2b the current distribution (profile) of the secondary beam is determined by the secondary ions created (integrated) along the chords in the XY cross-section of the primary beam. The respective 3D image of the secondary beam is shown in Figure 2c, demonstrating parabolic ($y = 1 - x^2$) and rectangular shapes of the beam in XY and XZ planes, determined, respectively, by chords lengths and the analyzer entrance slit width.

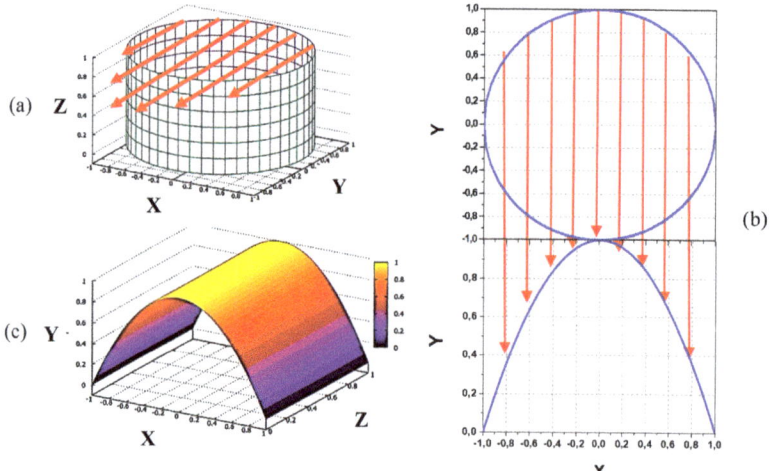

Figure 2. 3D image of sample volume of unit length and circular primary beam of unit radius (**a**), formation of the secondary beam profile in XY plane (**b**), and respective 3D image of the secondary beam (**c**).

These simplistic considerations demonstrate that even for the uniform primary beam, the secondary beam can attain non-uniformity as a result of the shape of the primary beam.

The 3D and 2D images of the above considered circular primary beam, but with parabolic, $y = 1 - x^2$, and bell-shape, $y = \exp(-5x^2)$, current density profiles are shown, respectively, in Figure 3a,b.

Figure 4a,b present the respective normalized 3D images and XY profiles of the secondary beam obtained in similar simplified considerations by integration in planes of primary beam chord slices, being:

$$\begin{array}{c} y = 1 - x^2, \\ y = 1 - 1.556x^2 + 0.558x^4 \\ y = \exp(-5x^2) \end{array} \quad (1)$$

for the, respectively, uniform ($y = 1$), parabolic ($y = 1 - x^2$) and bell-shape ($y = \exp(-5x^2)$) primary beam.

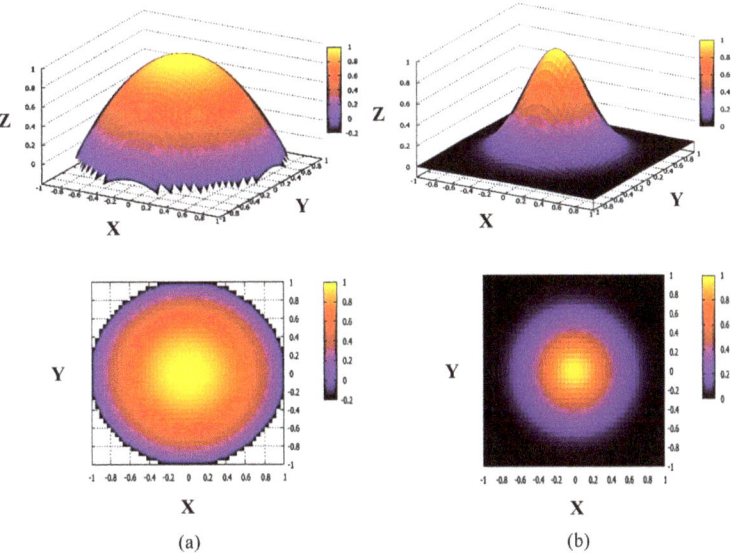

Figure 3. 3D and 2D images of circular primary beam with parabolic, $y = 1 - x^2$ (**a**) and bell-shape, $y = \exp(-5x^2)$ (**b**), current density profiles.

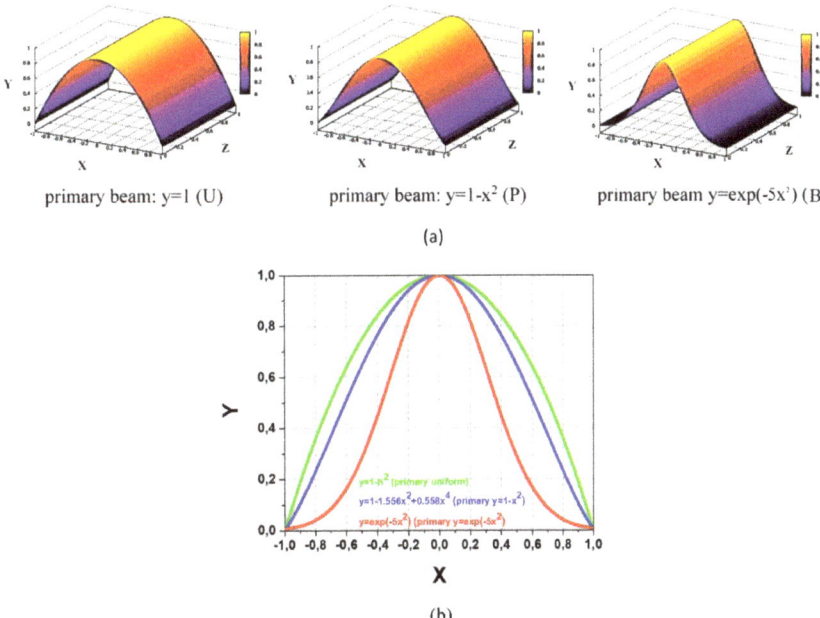

Figure 4. Normalized 3D images (**a**) and 2D profiles (**b**) of the secondary beam obtained by integration in planes of primary beam chord slices for uniform ($y = 1$), parabolic ($y = 1 - x^2$) and bell-shape ($y = \exp(-5x^2)$) primary beam.

The results in Figure 4 demonstrate the modifications of the secondary beam profile introduced by the non-uniformity of the primary beam. Additionally, notice that in real conditions of tangent propagation of secondary ions from the sample volume (as in Figure 1c), the respective XY cross-sections of the sample volume are rather ellipsoids than circles, with corresponding alterations in the secondary beam profile.

3. Split-Plate Detection and SPD Transfer Characteristics

In the HIBD standard split-plate detection, the secondary beam displacement, due to energy variation, is measured from the difference in the currents between the split plates. Conventionally, this can be explained according to the schematic of ideal case depicted in Figure 5a for the uniform rectangular beam of width w_b [3]. From simple geometrical considerations, the relationship between beam displacement Δz and split-plate current difference $\Delta i = i_1 - i_2$ between plates is:

$$\Delta z / w_b = h = \Delta i / i_\Sigma = (i_1 - i_2)/(i_1 + i_2) = \delta i, \qquad (2)$$

where $h = \Delta z / w_b$ is the normalized on the beam width displacement, $i_\Sigma = i_1 + i_2$ is the sum split-plates current and δi is the normalized beam current difference.

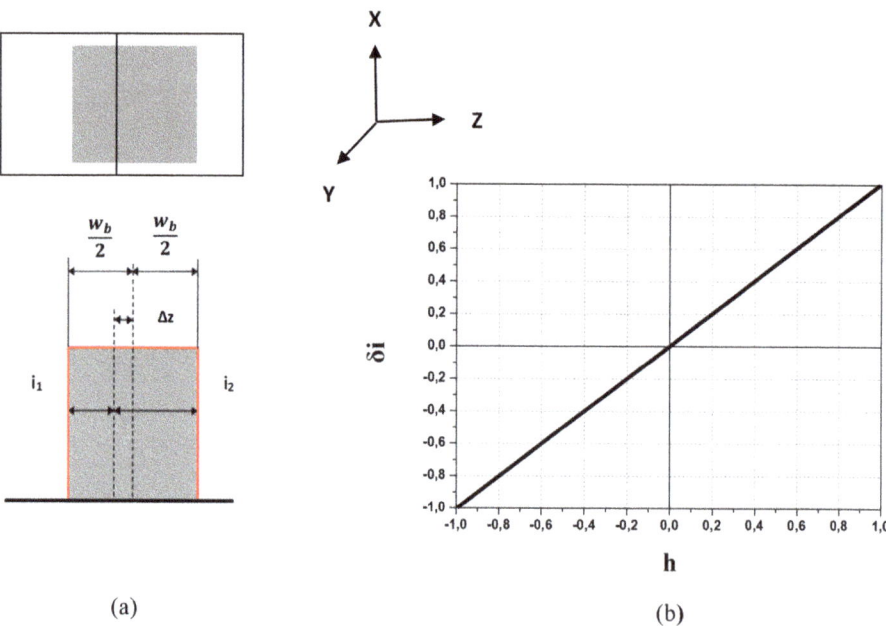

(a) (b)

Figure 5. Schematic of split-plate detection for ideal case of uniform rectangular beam of width w_b (**a**) and respective δi_I-h SPD transfer characteristic inside the dynamic range of $\delta i = \pm 1$ and $h = \pm 1$ (**b**).

The respective ideal secondary beam δi_I-h SPD transfer characteristic resulting from the effective scan of the beam across split-line is shown in Figure 5b and is linear inside the dynamic range of $\delta i = \pm 1$ and $h = \pm 1$.

Formally, the current on the split plates can be represented by triple integral $\int dx \int dy \int dz$ taken on the secondary beam current profile over the appropriate functions and bounds of x, y, and z. The beam current difference between split plates expressed in terms of current on one of the plates (for example, i_2) is:

$$\Delta i = i_1 - i_2 = i_\Sigma - 2i_2,$$
$$\delta i = 1 - 2i_2/i_\Sigma. \qquad (3)$$

Then, for the secondary beams in Figure 4 and beam displacement in the Z direction (split-line parallel to X axis as shown in Figure 6a), the current i_2 and δi_Z-h transfer characteristic are:

$$i_2 = \int_0^h dz \int_{-1}^1 dx \int_0^{f(x)} dy = Ch,$$
$$\delta i_Z = 1 - 2Ch/i_\Sigma \qquad (4)$$

where $y = f(x)$ is the secondary beam profile in the XY plane and $C = \int_{-1}^1 f(x)dx = $ Const.

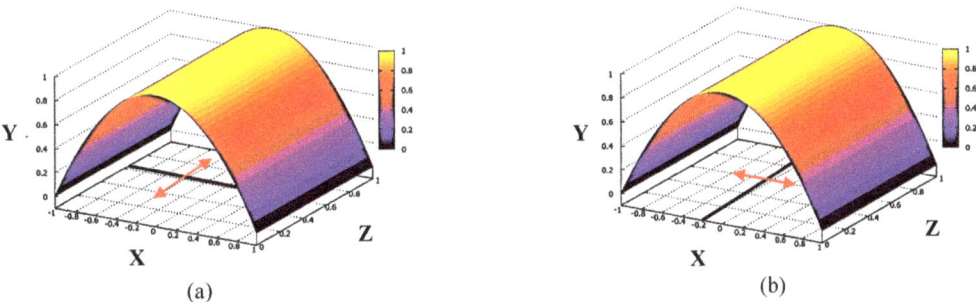

Figure 6. Orientation of split-line and secondary beam displacement: (**a**) beam displacement in Z direction with split-line parallel to X axis, and (**b**) beam displacement in X direction with split-line parallel to Z axis.

Contrary, for the beam displacement in the X direction (split-line parallel to Z axis as shown in Figure 6b), the respective current i_2 and δi_X-h transfer characteristic are:

$$i_2 = \int_{-1}^h dx \int_0^{f(x)} dy \int_0^1 dz = F(h),$$
$$\delta i_X = 1 - 2F(h)/i_\Sigma, \qquad (5)$$

where $F(h) = \int_{-1}^h f(x)dx$.

The results of Equations (4) and (5) demonstrate the orthotropy of the δi-h transfer characteristics for beam displacements in Z and X directions, or effective dependence on the orientation of the analyzer energy dispersion plane. As follows from Equation (4), for beam displacement in the Z direction and split-line parallel to the X axis (analyzer energy dispersion in YZ plane), the linearity of the δi_Z-h characteristic is persisted for any beam profile in the XY plane, determining the effective applicability of SPD data analysis based on the assumption of secondary ion beam's uniformity. This result applies for the HIBDs with the orientation of the analyzer energy dispersion plane coinciding with a plane of secondary beam propagation. This is typical for all traditional HIBDs operated with the injection of the angle-scanned primary beam (Figure 1b) and the mirror-type 30° Proca-Green electrostatic energy analyzer [3]. On ISTTOK, the HIBD is operated with the fixed injection angle of the primary beam (Figure 1a) and employs the 90° cylindrical energy analyzer (90° CEA) due to its multichannel operation ability along the analyzer non-dispersive direction [5]. Figure 7 shows the schematic of the 90° CEA arrangement on ISTTOK. Multichannel operation along non-dispersive Z direction is achieved by the orientation of the analyzer energy dispersion XY plane orthogonal to the YZ plane of secondary beam propagation. In that analyzer orientation, the secondary beam displacement due to energy variation is effectively converted to X direction (as shown in the inset in Figure 7) in Figure 6, and, in accordance with Equation (5), the SPD δi_X-h transfer characteristic should be sensitive to the secondary beam profile in the XY plane, violating the assumption of secondary beam uniformity in the analysis of the respective data from SPD.

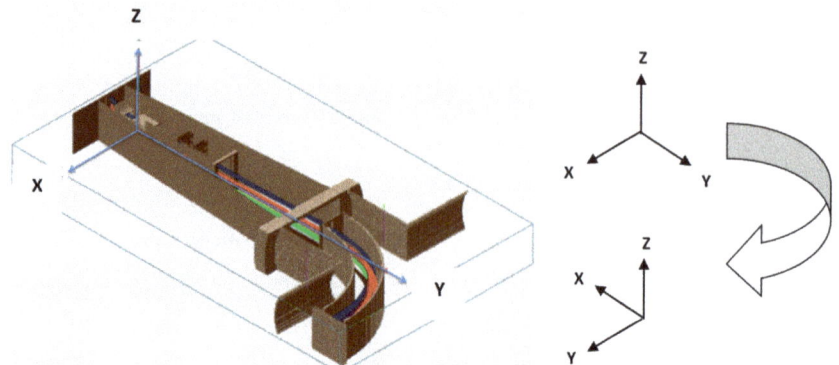

Figure 7. Schematic of the 90° CEA arrangement on ISTTOK.

Consider the secondary beams of the 3D images in Figure 4 obtained in the above simplified considerations. They are marked in relation to the primary beam profiles as U (uniform), P (parabolic) and B (bell-like).

In accordance with Equation (5) for beam displacement in the X direction and using Equation (1), the respective secondary beam δi_X-h transfer characteristics shown in Figure 8 are:

$$\delta i_{XU} = 2(h - (1/3)h^3)/1.333,$$
$$\delta i_{XP} = 2[h - (1.556/3)h^3 + (0.558/5)h^5]/1.169, \quad (6)$$
$$\delta i_{XB} = \mathrm{erf}(5^{1/2}h).$$

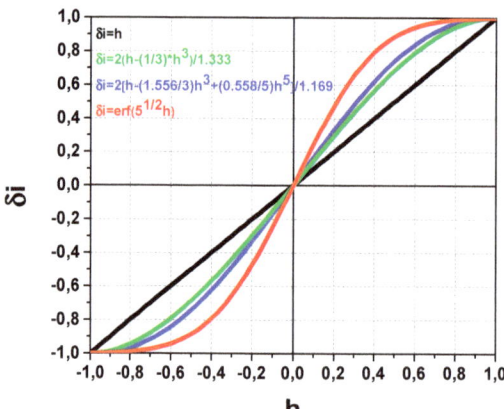

Figure 8. δi_X-h transfer characteristics for uniform (U), parabolic (P) and bell-like (B) profiles.

The results in Figure 8 demonstrate the deviation and non-linearity of the δi_{XU}-h. δi_{XP}-h and δi_{XB}-h transfer characteristics inside the dynamic range in comparison with ideal case of linear δi_I-h characteristics in Figure 5. Additionally, the following accompanying features should be pointed: (i) the effective narrowing of the dynamic range and (ii) magnification of the curves slope near $h = 0$. The last feature for the initially centered on split-line beam is equivalent to effectively increasing the sensitivity, which gives an advantage in the measurements of plasma potential fluctuations in such conditions.

Finally, it should be noticed here that investigations of the poloidal magnetic field and its fluctuations generated by the plasma current in standard HIBD applications with

30° Proca-Green energy analyzer are realized by the measurements of secondary beam displacement on SPD in a plane orthogonal to the analyzer energy dispersion plane [6,7]. These measurements are therefore sensitive to the secondary beam profile.

4. The Influence of the Secondary Beam Non-Uniformity on Plasma Potential and Its Fluctuations Measurements

For better a comparison, Figure 9 presents the respective differences in δi_U, δi_P, δi_B (subscript X is omitted) and δi_I: $\delta i_{U,P,B} - \delta i_I$ (i.e., the result for ideal case of uniform secondary beam in Figure 5 is subtracted from the results in Figure 8). The curves plotted in that way represent the deviations induced by beam profile in the measurements of plasma potential as a function of the secondary beam centerline position h. Note up to 40% deviation on δi_B in the case of the bell-shape beam profile (B), if the center of the secondary beam is positioned at $h_0 = 0.4$ inside the dynamic range window.

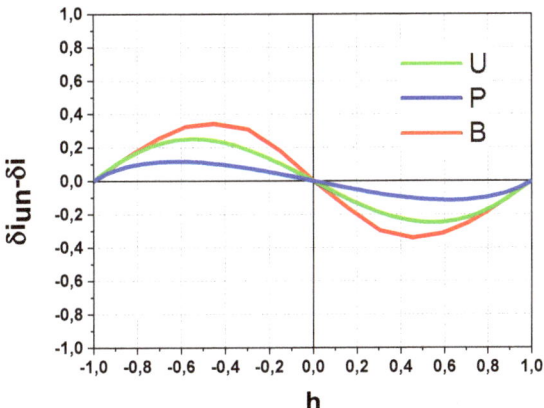

Figure 9. Deviations induced by the beam profile in measurements of plasma potential as a function of the secondary beam centerline position h for uniform (U), parabolic (P) and bell-shape (B) beam profile.

The plasma potential Φ_{pl} in measurements with 90° CEA is obtained from relation [5]:

$$(\Phi_{pl}/E_0) = [(w_b/2R_0)/C_E]\delta i, \qquad (7)$$

where w_b is the beam width, R_0 is the central radios of 90° CEA, C_E is the energy dispersion coefficient and E_0 is the beam energy.

For $w_b = 10$ mm, $R_0 = 105$ mm, $C_E = 1.56$ [8] and from the data in Figure 9 for the bell-shape profile of the secondary beam at $h_0 = 0.4$, the discrepancy in plasma potential value due to the non-linearity of δi-h characteristic is $\Delta(\Phi_{pl}/E_0) = 10^{-2}$, or 200 V in absolute value for the ISTTOK HIBD Xe$^+$ primary beam energy of $E_0 = 20$ keV.

The non-linear response of the SPD on the secondary beam non-uniformity also influences the measurements of plasma potential fluctuations. Figure 10 shows the δi response on harmonic variation of the secondary beam position on SPD, given by:

$$h = h_0 + a_h \sin(t), \qquad (8)$$

for bell-shape (δi_B, Figure 10a) and ideal uniform (δi_I, Figure 10b) beams, $a_h = 0.1$ and different h_0.

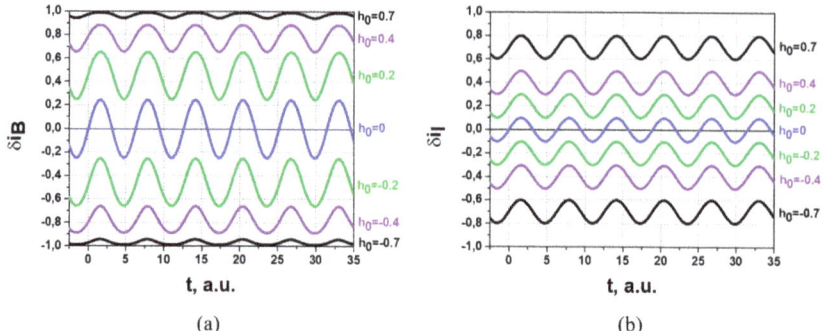

Figure 10. δi response on harmonic variation of beam position on SPD for different h_0 for bell-shape (**a**) and ideal uniform (**b**) beam.

Comparison of the results in Figure 10a for the bell-shape (δi_B) beam and in Figure 10b for the ideal (δi_I) uniform beam shows distortions (in amplitude and shape) in harmonic variation in the secondary beam position introduced by the non-linearity of the SPD response for different h_0.

It should be pointed out that in the cross-correlation analysis of the plasma fluctuations realized in simultaneous multichannel measurements, the difference in the background plasma potential in sample volumes determines the different relative positions of the analyzed secondary beams on SPD (different operation points h_0 on δi-h characteristic), attributing, therefore, a different response to the plasma potential fluctuations from different sample volumes.

5. Relevant Experimental Data from ISTTOK HIBD

5.1. Primary and Secondary Beam Profiles

The primary beam-line of the ISTTOK tokamak ($R_0 = 0.46$ m, $a = 0.085$ m, $B = 0.5$ T, $I_p = 7$ kA, $n_e = 5 \times 10^{18}$ m^{-3}, $T_e = 100$ eV) HIBD injector shown in Figure 11 [9] includes a set of electrostatic plates, wire monitor to control the primary beam profile in two directions (2 × 2 crossed in X and Y directions wires 0.3 mm of diameter) and defining slit (2 × 6 mm^2 of dimensions in the respective X and Y directions).

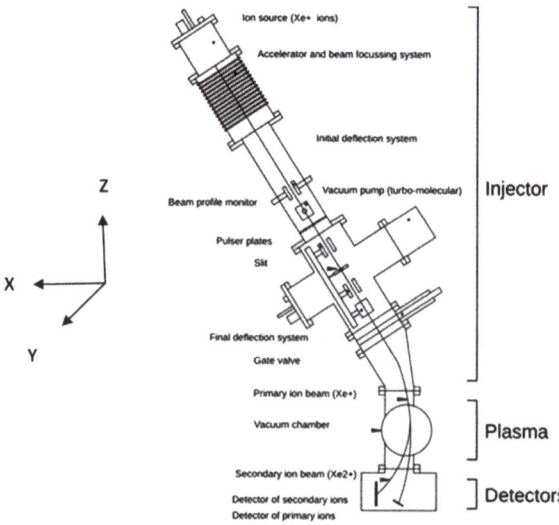

Figure 11. Primary beamline of ISTTOK HIBD injector.

The normalized profile of Xe⁺ primary ions obtained by scanning across the wires by electrostatic plates is shown in Figure 12 for two different background pressures (before and during plasma shot) in the primary beamline. It has a bell-like shape with full width at half maximum (FWHM) of FWHM = 1.5 mm at lower pressure, being widened at higher pressure due to the raising effect of beam scattering on the background gas particles.

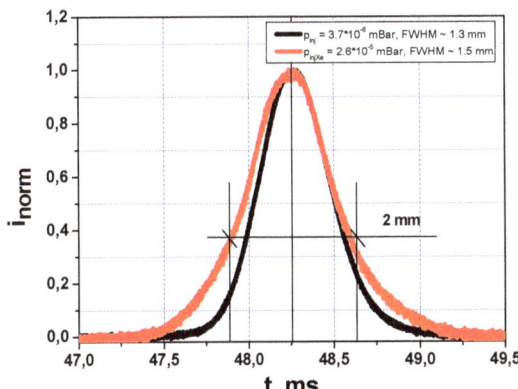

Figure 12. Normalized profiles of Xe⁺ primary ions across wire monitor.

Figure 13 presents the normalized profiles of the Xe⁺ beam observed in two (Y and X) directions on the primary beam multiple-cell detector (PBD) at the bottom of the tokamak chamber. The PBD presents a matrix of 15 cells (3 × 8 mm² of dimensions in respective X and Y directions) arranged into 5 rows (in X direction) and 3 columns (in Y direction), as shown in the insets in Figure 13. Despite the scrape-off by defining aperture, the primary beam profiles on PBD are close to the bell shape. The FWHM of the Gaussians fitting the experimental points are, respectively, FWHM = 6 mm in X and FWHM = 8 mm in Y directions, indicating approximately ~6 mrad (0.34°) of primary beam divergence (the difference in FWHMs is related to the difference in the respective dimensions of defining slit and PBD cells).

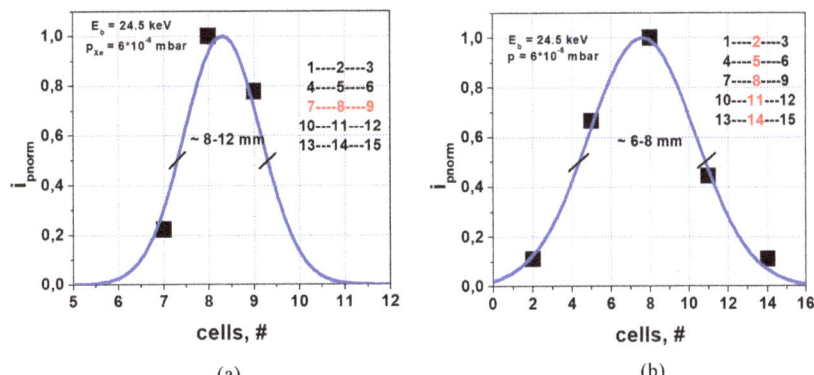

Figure 13. Normalized profiles of Xe⁺ primary ions on primary beam detector in Y (**a**) and X (**b**) directions.

Figure 14 shows the data obtained for the secondary Xe⁺⁺ beam profile in the XY plane in Figure 7 by electrostatic scanning across the cell of 3 mm width incorporated into the "stop" detector in ISTTOK HIBD experiments with time-of-flight (TOF) measurements of plasma potential [10]. The experimental points indicate a peaked non uniform secondary beam profile with rather wide wings. The FWHM of the fitting Gaussian is FWHM = 6 mm.

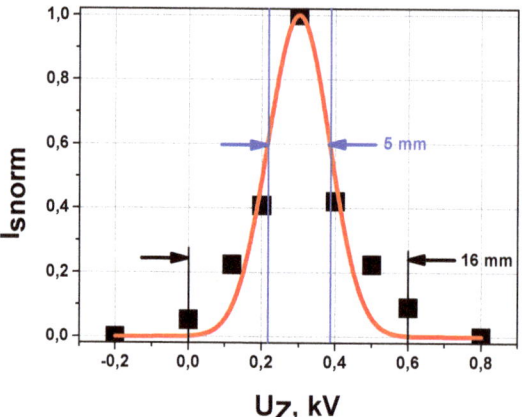

Figure 14. Secondary Xe^{++} beam profile on "stop" detector in ISTTOK HIBD experiments with time-of-flight measurements of plasma potential.

5.2. δi-h Characteristic

Figure 15 shows the δi-h SPD transfer characteristics of 90° CEA obtained on ISTTOK for the secondary Xe^{++} beam in several plasma shots by shot-to-shot variation in the voltage on the analyzer plates for the analyzer operation in normal, Figure 15a, and 2-times deceleration, Figure 15b, modes [8]. Despite the scattering of the data due to not absolute repeatability of the plasma parameters from shot-to-shot, the δi-h characteristics clearly differ, being almost linear in normal mode and non-linear in deceleration mode. For comparison, the graphics of the ideal δi_l-h and parabolic δi_P-h transfer characteristics in Figure 8 are also shown, respectively, in Figure 15a,b.

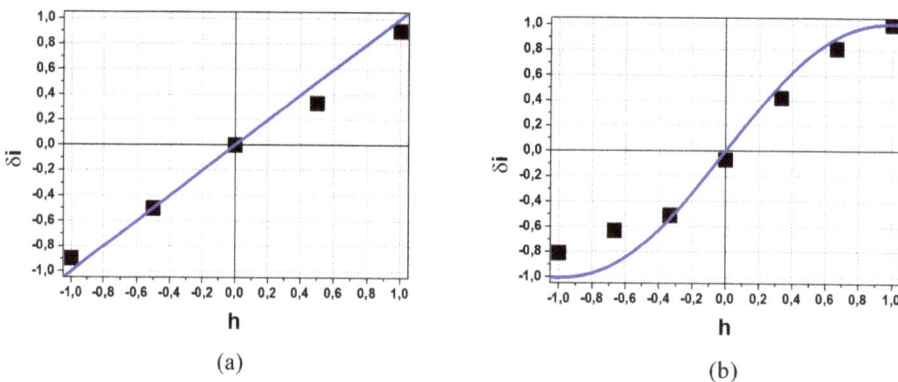

Figure 15. δi-h transfer characteristics of the secondary Xe^{++} beam for 90° CEA operation in normal (**a**) and 2-times deceleration (**b**) mode.

In ISTTOK HIBD, the positions and dimensions of sample volumes are determined by the positions and dimensions of the cells of the multiple cell array detector (MCAD) [4]. The existing physical and geometrical constraints dictate the layout of the secondary beam propagation from MCAD location to the 90° CEA provided (in similarity as in HIBD TOF experiments [9]) by the cylindrical electrostatic plates incorporated on the back of the MCAD dedicated apertures and followed by a multi-aperture einzel lens [11]. Obviously, in that way, the profile of the secondary beam at SPD can undergo transformations due to the ion optics aberrations and apertures-slits clippings. Matching the secondary beam profile in Figure 14 with the 90° CEA entrance slit width of 5 mm, the cutting of the profile wings (blue vertical lines in Figure 14) and transformation to an effectively more uniform profile may be expected, thus explaining the observed almost linearity of the δi-h characteristic in Figure 15a. On the other hand, as shown in [12], the operation of the 90° CEA in deceleration mode is characterized by modifications to the beam shape (elongation of beam image in energy dispersion direction proportional to the deceleration coefficient) and heterogeneity imposed by aberrations introduced by the analyzer. Such a property of the 90° CEA operation in deceleration mode can be a reason for the observed δi-h characteristic non-linearity in Figure 15b.

6. Multiple-Cell and Multi-Split-Plate Detection

The above considerations indicate the necessity of the secondary beam profile control. In real experimental conditions, it is also important in standard HIBD applications for plasma potential and its fluctuations measurements due to effects of misaligned and scraped-off secondary beam as a result of mechanical interferences in the secondary beam-line and analyzer entrance slits, aggravated by non-uniform components of the magnetic field along the beam propagation path [13,14]. The measurements of the secondary beam profile on SPD in one plasma shot are not trivial and to solve the issue, the implementation of MCAD with dedicated resolution may be suggested.

Figure 16 shows a number of MCAD cells of width l_{cell} distributed over the SPD dynamic range window. With that MCAD there is a possibility to determine the secondary beam profile in the energy dispersion plane in one plasma shot with a spatial resolution of the cell width. Additionally, there is a possibility to measure directly the δi-h characteristic by successive subtraction of the acquired signals from the cells, given by:

$$\delta i_n = \left({}_{j=1}\Sigma^n i_j - {}_{j=n+1}\Sigma^N i_j \right) / {}_{i=1}\Sigma^N i_j. \tag{9}$$

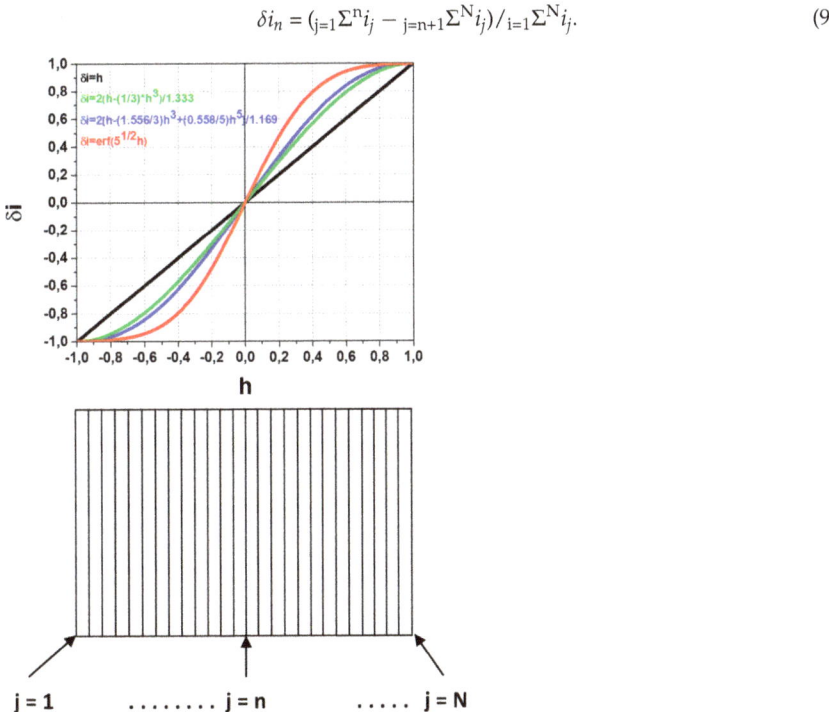

Figure 16. Schematic of multiple cell array detector for secondary beam profile and δi-h transfer characteristic measurements.

Such measurements are equivalent to effective scanning of δi-h characteristic, while MCAD operating in that mode may be considered as a multi-split-plate detector (MSPD).

The number and minimal width of the cells of MCAD in Figure 16 can be estimated from the analyzer dynamic range, $(\Delta E/E_0)_{DR}$, and the minimum detectable change of ion beam energy, $(\Delta E/E_0)_{min}$, determined by the signal-to-noise ratio (SNR) of the beam current measured on the cell. As an example, for the 90° CEA, the $(\Delta E/E_0)_{DR}$ and $(\Delta E/E_0)_{min}$ in terms of δi are [5]:

$$(\Delta E/E_0)_{DR} = [(w_b/R_0)/C_E], (\delta i = \pm 1) \qquad (10)$$

$$(\Delta E/E_0)_{min} = [(w_b/2R_0)/C_E](SNR)^{-1}, (\delta i = (SNR)^{-1}). \qquad (11)$$

Estimation of cell number (and therefore the number of reliable points N_{est}, or resolution of the δi-h characteristic measurements) is given by the ratio of $(\Delta E/E_0)_{DR}/(\Delta E/E_0)_{min}$, being from Equations (10) and (11) $N_{est} = 2(SNR)$. Then, estimation for the corresponding cell width is $l_{cell} = w_b/N_{est}$, and, particularly for SNR = 10 and w_b = 10 mm (for 90° CEA operated in 2-times deceleration mode [12]); this results in the minimal cell width of l_{cell} = 0.5 mm.

7. Summary

The coupling of the primary and secondary beam profiles inside the sample volume is examined in the simplistic model of circular primary beam and secondary ions that emerge orthogonal to the primary beam axis. These considerations demonstrate that even for the uniform primary beam, the secondary beam can attain non-uniformity determined by the shape of the primary beam with additional modifications, if the primary beam is non-uniform. In the frame of that simplistic model, the δi-h transfer characteristics of the non-uniform secondary beam are orthotropic in directions of beam displacement depending on the orientation of the analyzer energy dispersion plane. In the case of energy dispersion and beam propagation planes coincidence, the linearity of δi-h characteristic is persisted for any secondary beam profile in a plane orthogonal to the plane of beam propagation. It is correct for practically all traditional HIBDs operated with angle-scanned primary beam and the mirror-type 30° Proca-Green electrostatic energy analyzer. In the case of energy dispersion and the beam propagation planes orthogonality, the SPD δi-h transfer characteristic is non-linear, determined by the respective secondary beam profile. This situation takes place for ISTTOK HIBD operated with fixed injection angle primary beam and 90° CEA. The non-linearity of δi-h transfer characteristic was observed experimentally for 90° CEA operated in 2-times deceleration mode. In that case, the assumption of secondary beam uniformity in the analysis of the respective data from SPD is violated and requires knowledge of the secondary beam profile. The corresponding measurements can be realized with MCAD of dedicated resolution. In addition, for that MCAD operated in multiple-split-plate configuration, the direct measurements of δi-h transfer characteristic are available. Additionally, for the non-uniform profiles of the secondary beam, the MCAD data can be used as a supplement for plasma potential measurements by evaluation the shift in center of mass of the secondary beam current distribution.

Finally, it should be noted that knowledge of the secondary beam profile and its influence on the measurements is also important in traditional HIBD applications in investigations of the poloidal magnetic field and its fluctuations generated by plasma current, and if the secondary beam misalignment and scrape-off are suspected.

Author Contributions: I.N.: conceptualization, methodology, investigation, writing-original draft preparation. A.M.: validation, formal analysis, writing-rview and editing, supervision. R.H.: software, formal analysis. R.S.: software, visualization. All authors have read and agreed to the published version of the manuscript.

Funding: This work has been carried out within the framework of the EUROfusion Consortium and has received funding from the Euratom research and training program 2014–2018 and 2019–2020 under grant agreement No. 633053. IST activities also received financial support from "Fundaçao para a Ciencia e Tecnologia" through projects UIDB/50010/2020 and UIDP/50010/2020. The views and opinions expressed herein do not necessarily reflect those of the European Commission.

Conflicts of Interest: The authors declare no conflict of interest.

References

1. Jobes, F.C.; Hickok, R.L. A direct measurement of plasma space potential. *Nucl. Fusion* **1970**, *10*, 195. [CrossRef]
2. Melnikov, A.V. *Electric Potential in Toroidal Plasmas*; Springer Series in Plasma Science and Technology; Springer International Publishing: Cham, Switzerland, 2019.
3. Solensten, L.; Connor, K.A. Heavy ion beam probe energy analyzer for measurements of plasma potential fluctuations. *Rev. Sci. Instrum.* **1986**, *58*, 516. [CrossRef]
4. Cabral, J.A.C.; Malaquias, A.; Praxedes, A.; van Toledo, W.; Varandas, C.A.F.; Dias, J.M. The heavy ion beam diagnostic for the tokamak ISTTOK. *IEEE Trans. Plasma Sci.* **1994**, *22*, 350. [CrossRef]
5. Nedzelskiy, I.S.; Malaquias, A.; Sharma, R.; Henriques, R. 90° cylindrical analyzer for the plasma potential fluctuations measurements by heavy ion beam diagnostic on the tokamak ISTTOK. *Fusion Eng. Des.* **2016**, *123*, 123. [CrossRef]
6. Malaquias, A.; Cabral, J.A.C.; Varandas, C.A.F.; Canario, A.R. Evolution of the poloidal magnetic field profile of the ISTTOK plasma followed by heavy ion beam probing. *Fusion Eng. Des.* **1997**, *34–35*, 671. [CrossRef]
7. Shimizu, A.; Fujisawa, A.; Nakano, H. Consideration of magnetic field fluctuation measurements in torus plasma with a heavy ion beam probe. *Rev. Sci. Instrum.* **2005**, *76*, 043504. [CrossRef]
8. Sharma, R.; Nedzelskiy, I.S.; Malaquias, A.; Henriques, R. Plasma potential and fluctuations measurements by HIBD with 90° cylindrical energy analyzer on the ISTTOK tokamak. *Fusion Eng. Des.* **2020**, *160*, 112016. [CrossRef]
9. Cabral, J.A.C.; Nedzelskiy, I.S.; Malaquias, A.J.; Gonçalves, B.; Varandas, C.A.F. Improved 20 keV injection system for the heavy-ion-beam diagnostic of the tokamak ISTTOK. *Rev. Sci. Instrum.* **2003**, *74*, 1853. [CrossRef]
10. Nedzelskiy, I.S.; Malaquias, A.J.; Tashchev, Y.I.; Silva, C.; Figueiredo, H.; Fernandes, H.; Varandas, C.A.F. Multichannel time-of-flight technique for plasma potential profile measurements by heavy ion beam diagnostic on the tokamak ISTTOK. *Rev. Sci. Instrum.* **2006**, *77*, 033505. [CrossRef]
11. Sharma, R.; Nedzelskiy, I.S.; Malaquias, A.; Henriques, R.B. Design and optimization of the electrostatic input module for the ISTTOK Tokamak HIBD cylindrical energy analyzer. *JINST* **2017**, *12*, C11018. [CrossRef]
12. Nedzelskiy, I.S.; Sharma, R.; Malaquias, A.; Henriques, R. Properties of the 90° cylindrical energy analyzer with internal deceleration of the beam in real conditions of operation. *Fusion Eng. Des.* **2021**, *170*, 112709. [CrossRef]
13. Zhang, X.; Lei, J.; Connor, K.A.; Demers, D.R.; Schoch, P.M.; Shah, U. Analysis of heavy ion beam probe potential measurement errors in the Madison Symmetric Torus. *Rev. Sci. Instrum.* **2004**, *75*, 3502. [CrossRef]
14. Keating, L.M.; Schoch, P.M.; Crowley, T.P.; Russell, W.G.; Schatz, J.G., Jr. High sensitivity detectors for measurement of space potential fluctuations. *IEEE Trans. Plasma Sci.* **1994**, *22*, 424–429. [CrossRef]

Article

First Results of the Implementation of the Doppler Backscattering Diagnostic for the Investigation of the Transition to H-Mode in the Spherical Tokamak Globus-M2

Anna Ponomarenko [1,*], Alexander Yashin [1,2], Gleb Kurskiev [2], Vladimir Minaev [2], Alexander Petrov [1], Yuri Petrov [2], Nikolay Sakharov [2] and Nikita Zhiltsov [2]

[1] Plasma Physics Department, Peter the Great St. Petersburg Polytechnic University, 195251 St. Petersburg, Russia
[2] Plasma Research Laboratory, Ioffe Institute, 195251 St. Petersburg, Russia
* Correspondence: annap2000dreeonn@gmail.com or ponomar_am@spbstu.ru

Abstract: This paper presents the first results of a study of the LH transition on the new spherical Globus-M2 tokamak using the Doppler backscattering (DBS) diagnostic. New data characterizing the H-mode of discharges with higher values of the plasma parameters, such as magnetic field B_t up to 0.9 T and plasma current I_p up to 450 kA, were collected and analyzed. An upgraded neutral beam injection (NBI) system was used to initiate the LH transition. DBS allows the measurement of the poloidal rotation velocity and the turbulence amplitude of the plasma. The multi-frequency DBS system installed on Globus-M2 can simultaneously collect data in different areas spanning from the separatrix to the plasma core. This allowed for the radial profiles of the rotation velocity and electric field to be calculated before and after the LH transition. In addition, the values and temporal evolution of the velocity shear were obtained. The associated turbulence suppression after the transition to the H-mode was investigated using DBS.

Keywords: tokamak; plasma diagnostics; H-mode; Doppler backscattering; plasma turbulence

Citation: Ponomarenko, A.; Yashin, A.; Kurskiev, G.; Minaev, V.; Petrov, A.; Petrov, Y.; Sakharov, N.; Zhiltsov, N. First Results of the Implementation of the Doppler Backscattering Diagnostic for the Investigation of the Transition to H-Mode in the Spherical Tokamak Globus-M2. Sensors 2023, 23, 830. https://doi.org/10.3390/s23020830

Academic Editor: Hossam A. Gabbar

Received: 9 November 2022
Revised: 24 December 2022
Accepted: 9 January 2023
Published: 11 January 2023

Copyright: © 2023 by the authors. Licensee MDPI, Basel, Switzerland. This article is an open access article distributed under the terms and conditions of the Creative Commons Attribution (CC BY) license (https://creativecommons.org/licenses/by/4.0/).

1. Introduction

Developing an accurate and reliable diagnostic system for the successful study of the various phenomena taking place in a hot plasma of a magnetic confinement device, such as a tokamak, is a task of utmost importance. For this reason, a wide range of methods and techniques have been proposed and tested on different machines over the years [1]. The effective implementation of any diagnostic is a priority for any experimentalist, as their results will have a great influence on the capabilities of any future fusion device.

An example of the successful implementation of the diagnostics of Doppler backscattering (DBS) for the investigation of plasma processes during the transition to H-mode is presented in this paper. The H-mode is an intriguing and illustrative phenomenon to discuss. The results obtained and analyzed can highlight the usefulness and showcase the range of data that is possible to obtain using such a diagnostic system. The transition to the H-mode has been the topic of a multitude of studies throughout the years on various devices all over the world because of its significant importance and interest to the plasma physics community. Ever since its discovery [2], a lot of effort has been made to reproduce it in all major toroidal confinement devices and discover its properties along with the main causes of the transition to this high confinement regime. A lot of progress has been made in understanding the H-mode, and it is now proposed that it will be the operational mode for future fusion devices, such as the International Thermonuclear Experimental Reactor (ITER) [3]. The H-mode is an enhanced confinement regime characterized by an increase in energy confinement time. A large variety of methods are used to study the phenomena that accompany the LH transition [4]. Much of the experimental work conducted indicates that the H-mode seems to be defined by the anomalous transport in plasma. It was shown that the suppression of such turbulence perturbations leads to the transition to improved confinement and it was also recognized that the E × B shear plays a key role in this process [5,6]. However, the mechanisms responsible for the formation of this sheared rotation vary drastically and remain somewhat unclear. This is why detailed theoretical work and various simulations alongside experimental research are aimed at gaining insight into the mechanisms responsible for this operational mode.

It is important to point out, as it was also noticed, there is a distinct difference between the improved confinement in spherical tokamaks (ST; characterized by a smaller aspect ratio) and that witnessed in other types of devices. This makes the topic a subject worth investigating. One key observed difference between the devices Mega Ampere Spherical Tokamak (MAST), National Spherical Torus Experiment (NSTX), and Globus-M/M2 is the fact that the energy confinement time has a strong dependence on the values of the toroidal magnetic field, which has not previously been observed [7–9]. It is worthy of note that all of the aforementioned tokamaks differ in parameters and yet still show a similar dependence on the magnetic field, which contradicts the conventional $\tau_{E,98y,2}$ scaling [8]. It was also determined that there seems to be a strong correlation between the increase in energy confinement time and the decrease in collisionality, which is seen as an indication of favorable plasma parameters for the future [10]. All this leads to the question of what other properties of the H-mode or the LH transition deviate in spherical tokamaks in comparison to more conventional devices.

The results of experimental research on the transition to the H-mode on the spherical Globus-M2 tokamak using the DBS method are presented in this paper, which is structured as follows. The next section is devoted to the description of the DBS method and its basic principles. This includes a description of the multi-frequency DBS system installed on the Globus-M2 and used in the experiments. After that, the LH transition on Globus-M2 is discussed. Along with that, the DBS measurements of the velocity and turbulence behavior are analyzed, as the LH transition is thoroughly investigated. Finally, a summary of all the results is presented.

2. Doppler Backscattering

Doppler backscattering (DBS, also referred to as Doppler reflectometry) is a diagnostic method that is based on the probing of the plasma using microwave beams at oblique incidence [11,12]. In the case of the existence of a cut-off layer for the probing beam (i.e., the point that limits the trajectory of the probing beam), one can detect backscattering of microwave radiation, which takes place predominantly near the cut-off (see the schematic representation in Figure 1). This means that this diagnostic allows for local measurements of the plasma parameters.

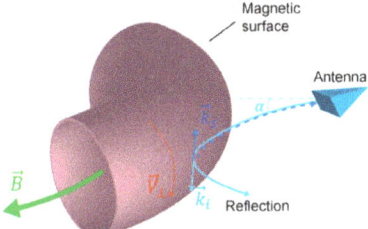

Figure 1. Schematic of a microwave beam incident on the reflection layer.

Analysis of the reflected probing beam signals leads to the observation that the Doppler shift in the backscattered frequency spectra is directly proportional to the perpendicular rotation velocity of the density turbulence moving with the plasma:

$$\omega_{Doppler} = \vec{V} \cdot \vec{k} = V_\perp k_\perp \qquad (1)$$

This velocity component $V_\perp$ corresponds to a rotation in the direction of the diamagnetic or the E × B drift (see Figure 1). Due to this, the radial electric field can also be calculated using DBS measurements. To construct a velocity or electric field profile, separate measurements of the cut-off layer positions of different probing beam frequencies are required. This would allow following the inhomogeneity of the velocity. This means the velocity shear can be calculated using DBS, which is one of the key parameters that influence the transition to the H-mode. $k_\perp$ is the wave vector of scattering fluctuations and its size can be determined by Bragg's law:

$$\vec{k}_\perp = \vec{k}_i - \vec{k}_s \quad (2)$$

where $\vec{k}_i$ is the wave vector of the incidence wave, and $\vec{k}_s$ is the wave vector of the scattered wave of electromagnetic radiation. These vectors are shown in Figure 1.

In order to obtain the Doppler frequency shift, the Doppler backscattering method records two signals, the in-phase (I) and quadrature (Q) signals, which are then formed into a complex IQ signal. Then, the method calculates the derivative of the phase of the complex signal (time dependence of the frequency shift of the scattered signal) or the time dependence of the shift of the spectra of short parts of the IQ signal [12]. Both approaches exhibit approximately the same temporal behavior and are thus deemed equally appropriate for use.

Apart from that, the amplitude of the complex IQ signal directly corresponds to the plasma turbulence behavior. It determines the backscattered power in a given wave number region near the cut-off. This means using DBS it is possible to investigate the properties of turbulence during the LH transition and in the H-mode.

This method has been used on a variety of tokamaks and has garnered results regarding the rotation velocity and radial electric field of the plasma [12–19]. On Globus-M, this diagnostic has also been successfully utilized but only for the study of plasma oscillatory processes, not to investigate the transition to H-mode [20]. Moreover, a multitude of other phenomena was discovered and fruitfully studied, such as geodesic acoustic modes (GAMs), limit-cycle oscillations (LCOs), Alfven eigenmodes (AEs), filaments, and others [21–24]. As a result, it was deemed appropriate to continue research on the LH transition and the H-mode using the DBS method on the new and upgraded Globus-M2 tokamak.

2.1. Doppler Backscattering to Study the LH Transition and the H-Mode

On the Globus-M2 tokamak, the installed DBS system that uses dual-frequency probing investigated the LH transition [25]. Two pairs of fixed frequencies were chosen: 20, 29 GHz and 39, 48 GHz. Each frequency channel includes a microwave circuit with dual homodyne detection, which allows for the quadrature detection of backscattered radiation. Two steerable antennas are used to probe the plasma with O-mode microwaves. One of the antennas is used for frequencies of 20 and 29 GHz, while the second one is for 39 and 48 GHz frequencies. The antennas can be rotated in both the poloidal and toroidal directions. They are also connected to stationary DBS equipment using flexible waveguides. In initial experiments, the DBS system was positioned in the same way as on the Globus-M. The system was located in the equatorial plane of the tokamak. An example of ray tracing performed for the four DBS channels, calculated using a special ray tracing code, written in the Wentzel–Kramers–Brillouin (WKB) approximation and for the Globus-M2 geometry, is shown in Figure 2 [26]. The code requires EFIT data about the magnetic field configuration and density measurements using Thomson scattering diagnostics to allow for an accurate calculation of the probing beam trajectory. Figure 2 depicts the poloidal cross-section of the Globus-M2 tokamak with a series of blue lines indicating the trajectory of the probing beams that correspond to the four probing frequencies available: 20, 29, 39, and 48 GHz. The detection region covered a considerable interval of normalized small radii $\rho = 0.7–1.1$ for typical discharges, discussed in the following section.

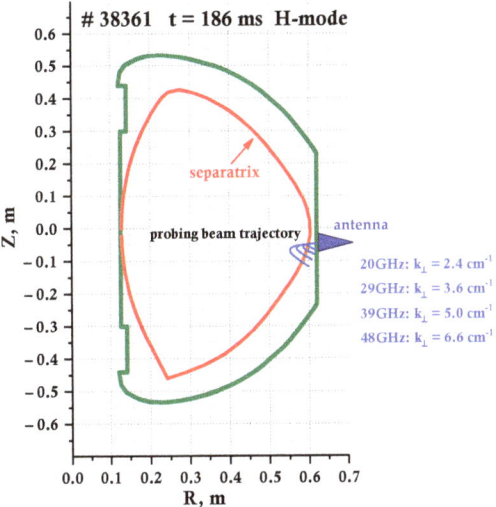

Figure 2. Ray tracing for the four-frequency DBS system used to study the LH transition.

3. LH Transition and H-Mode on Globus-M2

In experiments on the Globus-M2 tokamak (major radius R ≈ 0.35 m, minor radius a ≈ 0.22 m), a two-fold increase in the toroidal magnetic field (compared to the Globus-M) and well-pronounced LH transitions with all the characteristic features of the transition were observed. The key indicators of the transition to H-mode include the reduction in D_α, a positive break in the slope of plasma density, and periphery pressure gradients.

An example of an observed LH transition for discharge #39174 is presented in Figure 3. In this case, it was achieved with the following main parameters of the deuterium plasma: toroidal magnetic field B_t = 0.7 T; plasma current I_p = 280 kA; averaged electron density $\langle n_e \rangle$ = (3–6) × 10^{19} m^{-3}; electron temperature in the plasma core T_e = 0.6–1.1 keV; elongation $k \approx 1.9$; triangularity $\delta \leq 0.5$. The direction of the magnetic field was chosen in such a manner that the toroidal ion drift was directed toward the X-point (lower null magnetic configuration).

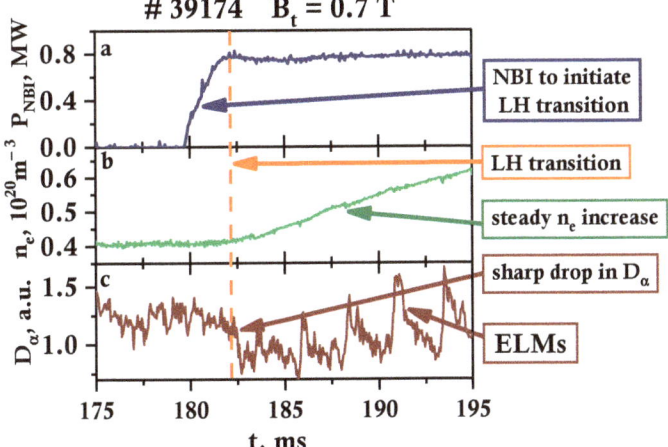

Figure 3. Temporal evolution of plasma parameters on Globus-M2 for discharge #39174: (**a**) NBI power P_{NBI}, (**b**) electron density n_e, and (**c**) D_α signal.

For discharge #39174, the LH transition was initiated by the deuterium neutral beam injection (NBI) with particle energy 28 keV and heating power 0.8 MW (Figure 3a). The beam impact parameter is equal to 32 cm and the average pitch angle of the deposited fast ions is approximately 40°. The LH transition started at 182 ms, 2 ms after the neutral beam injection, which is indicated by the vertical orange line in Figure 3. On Globus-M2 deuterium, NBI was not the only method to achieve LH transition, as the transition to H-mode was also observed when using hydrogen NBI, as well as two NBI sources, were introduced into the plasma at different times. This method used does not dramatically affect the LH transition itself. The NBI allowed to effectively heat the plasma, as the electron and ion temperatures in the plasma core exceeded 1 keV.

During the LH transition, one may observe the expected increase in electron density n_e in Figure 3b and a drop in D_α in Figure 3c. The transition to the H-mode also leads to the appearance of edge localized modes (ELMs), which are a sign of the formation of a strong pressure gradient at the plasma edge [27]. They can be seen in the form of periodical D_α bursts (Figure 3c).

After the transition to H-mode, an increase in the plasma density was followed by the rise of the total stored energy W_{DIA}, measured by the diamagnetic loop. Compared to the experiments on Globus-M, carried out with a magnetic field of 0.4 T, the total stored energy increased by a factor of four [10]. The h-factor also changes after the LH transition and increases from 0.5 to 1.2–1.4, which is the result of the energy confinement time τ_E in H-mode doubling and reaching 7 ms. While the NBI pulse is active, a quasi-steady phase state of the H-mode was established, characterized by a $\frac{dn}{dt} \approx 0$ and $\frac{dW_{DIA}}{dt} \approx 0$ for 10–15 ms, which is 1.5–2 times more than the τ_E for the discussed discharge.

3.1. Velocity Behavior during the LH Transition

The in-phase and quadrature (IQ) signals obtained using DBS were used for calculating the Doppler frequency shift, which corresponds to the poloidal rotation velocity of the plasma. The temporal evolution of the plasma rotation velocity during the LH transition was also investigated.

An example of the rotation velocity calculated for discharge #38361 by the "phase derivative" method smoothed over a time interval of 128 μs is presented in Figure 4. Two velocity evolutions are depicted: Figure 4c shows the behavior inside the separatrix at $\rho = 0.86$, while Figure 4d shows outside at $\rho = 1.04$. The vertical dashed line at 182.5 ms indicates the LH transition. One may observe that the poloidal rotation velocity inside the separatrix was around 3 km/s in the direction of the electron diamagnetic drift during the L-mode and increased rather dramatically during the transition to the H-mode, reaching the average value of 9 km/s. The behavior of the temporal evolution for plasma outside the separatrix differs. The velocity value does not exceed 2 km/s and is predominantly directed toward the ion diamagnetic drift pre-LH transition. Afterward, its value decreases closer to 0 km/s and its direction becomes harder to judge, which is highlighted by the error bars in Figure 4d.

It is also worthy of note that in H-mode a periodic change in velocity values was observed. The comparison of the D_α signal (Figure 4b) and the velocity leads to the observation that there is a correlation between ELMs and the bursts of increasing velocity seen. The appearance of each ELM leads to a rapid increase in values of up to 11 km/s in Figure 4c, while during the inter-ELM periods, the velocity decreases once again; nonetheless, it remains around 7 km/s. In the case of Figure 4d, the direction of the rotation is also affected, with it changing during an ELM burst. Modeling of the reaction of the DBS signal to filaments was performed in [28], and the results highlighted that filaments that can develop during an ELM burst can impact the velocity values due to the backscattering being of a non-linear nature.

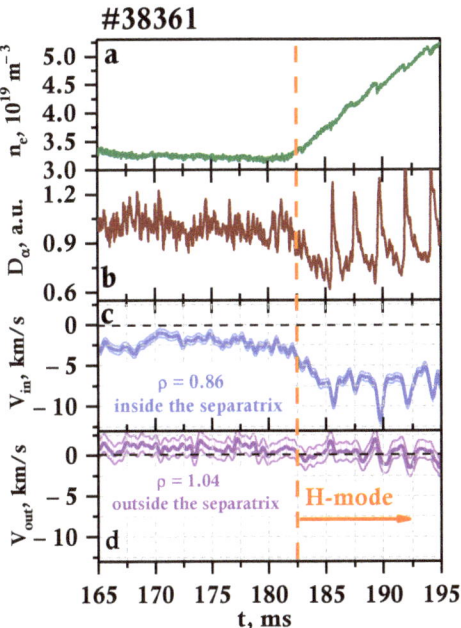

Figure 4. Temporal evolution of different plasma parameters on Globus-M2 during the LH transition: (**a**) electron density n_e, (**b**) D_α signal, (**c**) plasma rotation velocity at $\rho = 0.86$, and (**d**) plasma rotation velocity at $\rho = 1.04$.

The radial profiles of the plasma rotation velocity are also worth investigating before and after the LH transition. It is possible to obtain these measurements using the DBS system described earlier. Due to the existence of multiple probing frequencies, the values are gathered simultaneously from different radii in the tokamak plasma. As has already been mentioned, the probing frequencies for these experiments were 48, 39, 29, and 20 GHz, with 48 GHz corresponding to the deeper part of the plasma and 20 GHz to the periphery of the plasma.

The radii are obtained by means of a ray tracing code, described in Section 2.1, as it is designed not only to calculate the probing beam trajectory but to also present information regarding the cut-off, such as the wave vector values and radii of the cut-off [26]. Two different moments in time were chosen to calculate the velocity profiles before and after the LH transition. The first profile was obtained for $t_1 = 180$ ms during the L-mode, while the second one for $t_2 = 190$ ms during H-mode. Table 1 contains the data of the calculated radii and corresponding wave vector values used for the profiles presented. One can observe a change in the cut-off radii associated with a change in electron density after the transition. The position of the cut-offs shifts toward the inner plasma region by approximately 1 cm, with the 48 GHz channel being the exception. Its radius value increases, meaning it moves slightly closer to the periphery.

Table 1. Radii and wave vector values for discharge #38361 calculated by a ray tracing code.

#38361	180 ms (before the LH Transition)		190 ms (after the LH Transition)	
f, GHz	R, cm	$k_\perp$, cm^{-1}	R, cm	$k_\perp$, cm^{-1}
20	61.2	2.2	60.4	2.4
29	60.1	3.4	59.1	3.6
39	58.2	4.8	57.6	5.0
48	55.3	6.4	56.2	6.6

The resulting profiles are presented in Figure 5. The figure shows the radial profile of both the rotation velocity and electric field, which were calculated using DBS measurements. The electric field can easily be calculated assuming that the velocity measured is indeed the $E \times B$ drift. Two different scales are used to indicate the obtained values: the one on the left for the rotation velocity, and the one on the right for the electric field. Additionally, different colors are used to indicate the different moments in time for which the profiles are calculated. The black line corresponds to 180 ms before the LH transition and the pink line to 190 ms after the LH transition. A clear peak is present on both profiles at around r = 58 cm. The absolute values of the velocity and electric field decrease closer to the periphery and the core. One can see a significant increase in the values for all radii after the LH transition, for example, in the case of the rotation velocity from 2.5 km/s to 9.2 km/s at its maximum. It is also worthy of note that for the 180 ms profile, there is a change in the sign of the velocity values at some radius, as at r = 61.2 cm the value is positive, and negative at r = 60.1 cm. This change in the direction of the rotation is not observable in the 190 ms case in the data available.

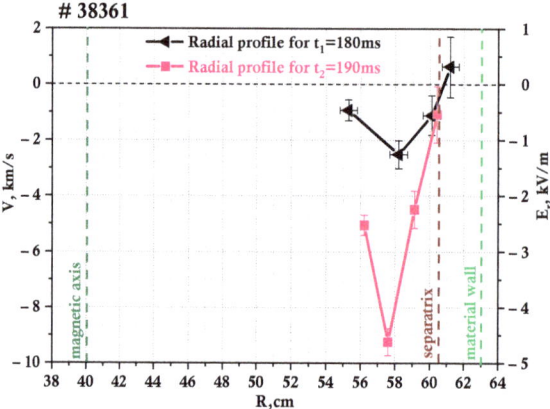

Figure 5. Plasma rotation velocity and electric field profile, measured by DBS.

The multi-frequency DBS system allows the following of the inhomogeneity of the poloidal rotation velocity, which is necessary for the determination of the rotation velocity shear. The analysis of the change in its behavior associated with the LH transition is of importance. Figure 6a demonstrates the temporal evolution of the rotation velocity shear for discharge #38361, calculated at the separatrix at $\rho = 1.0$. During L-mode, the value was around 3×10^{-5}, which is believed to be the threshold value that initiates the transition to H-mode in Globus-M/M2. This effect was previously predicted in modeling works and a similar shear value was observed in other devices [29,30]. After the LH transition, there is a nearly two-fold increase in velocity shear. This is accompanied by a simultaneous decrease in the amplitude of the backscatter signal in Figure 6b, which is proportional to the amplitude of turbulence with wave number values in Table 1. This can also be observed in the form of a decrease in the intensity of all frequency components in the spectrogram of DBS amplitude in Figure 6c. This phenomenon is due to the fact that the shear plays a significant role in the suppression of anomalous transport and contributes to the plasma's stable H-mode.

3.2. Turbulence Behavior during the LH Transition

As was made evident in Figure 6, suppression of plasma turbulence is observed, so it is important to investigate the properties of turbulence for discharges with a transition to H-mode. For the purpose of investigating the core plasma during the LH transition, a 55 GHz frequency channel was also added to the DBS system, shown in Figure 2 [31]. This was conducted for the #39171 discharge, which is presented in Figure 7. The spectrograms of the complex DBS signals were calculated for the available frequencies of 20, 29, 39, and 55 GHz, and can be seen in Figure 7a. The intensity of each of these spectrograms is proportional to the turbulence amplitude, and the position of the center of gravity of the spectrum at each moment in time is proportional to the plasma rotation velocity [12]. This is why the temporal evolution of the spectral center of gravity was investigated as well, presented in Figure 7b. Different signs of the shift of the center of gravity correspond to the rotation of the plasma in opposite directions.

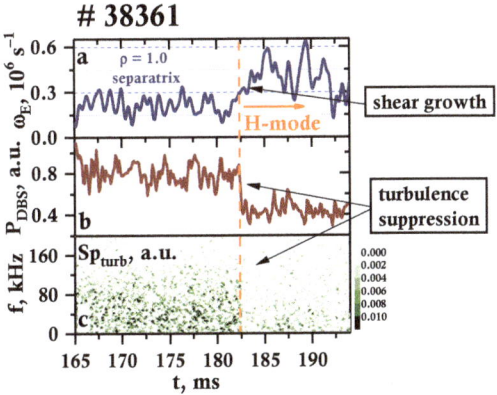

Figure 6. Temporal evolution of: (a) velocity shear, (b) DBS amplitude, and (c) DBS amplitude spectrogram.

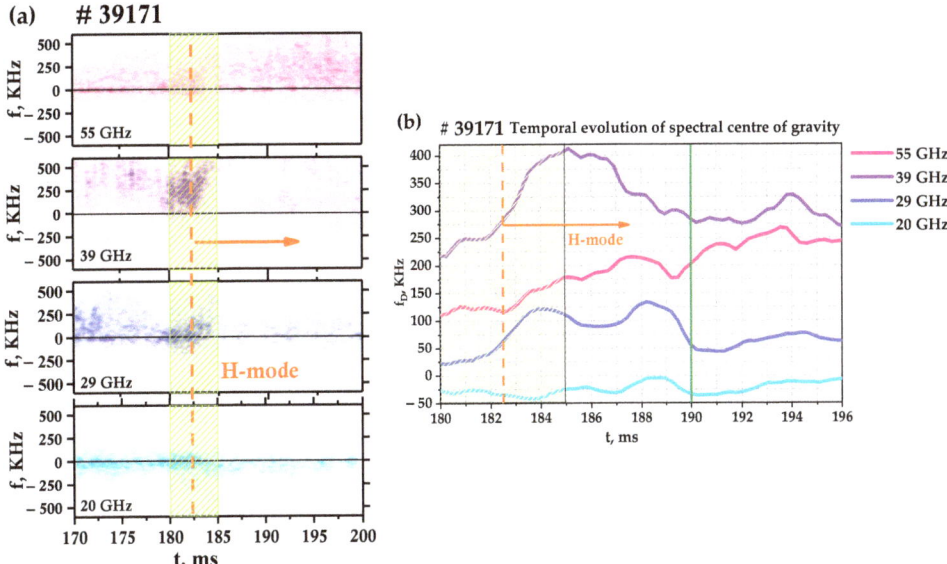

Figure 7. (a) Spectrograms of DBS complex signals for discharge #39171 and (b) temporal evolution of spectral center of gravity.

The characteristics of discharge #39171 were similar to #38361 (see Figures 4–6), however, the NBI power was reduced. This led to a slight change in the LH transition. The main differences were the plasma density being lower during the LH transition, and the D_α emission decreasing and the plasma density increasing slower during the transition to H-mode. However, the data collected regarding turbulence behavior remained consistent despite this.

The LH transition is marked by the orange vertical line at 182.5 ms in Figure 7a. Right after the transition to H-mode (see the segment in the green box in Figure 7), there is a noticeable shift of the spectra to a higher-frequency region for the 29, 39, and 55 GHz channels (inner plasma region), while this is not observable for the 20 GHz channel (close to the separatrix). For example, in Figure 7b, it can be seen how frequency shift values increase dramatically from 250 to 410 kHz in a matter of 3 ms after the LH transition. Such a steady growth of the frequency shift values is also true for the 29 and 55 GHz probing frequencies during this time. However, after that, the frequency shift values decrease in the 29 and 39 GHz frequencies case. This can be explained by the fact that during this time, in the H-mode, the electron density continues to increase, leading to the movement of the cut-off position closer to the separatrix. This behavior of the center of gravity of the spectrograms corresponds to the rotation velocity behavior, which also grows after the LH transition, as shown in Figure 4c,d. It is also worthy of note that the spectral power decreases after a several-milliseconds delay (when a quasi-stationary H-mode is achieved after 185 ms), primarily on the peripheral channels. This can be seen as a decrease in color intensity in the spectrograms. The turbulence suppression is most pronounced for the 39, 29, and 20 GHz channels or in the 0.55–0.58 m radii interval. One can see that the spectral power does not change significantly for the 55 GHz channel.

These observations are confirmed by the obtained temporal evolution of the spectral power of the fluctuations that is presented in Figure 8. The different DBS probing frequencies are depicted using different colors as indicated in the figure. The LH transition is also marked using an orange vertical line at 182.5 ms. As the absolute power levels of the backscattered radiation were not measured and the sensitivity of the detectors differs for different probing frequencies, the results were obtained in the form of normalized power levels at t = 190 ms. The temporal evolution of each channel is presented in the corresponding calculated relative units. The change in fluctuation level is virtually identical for 39, 29, and 20 GHz, as it drops by almost half in the span of 3 ms right after the transition to H-mode. The 55 GHz channel exhibits dramatically different behavior with a slight increase in spectral power observed after the LH transition, meaning no turbulence suppression takes place in this area of the plasma.

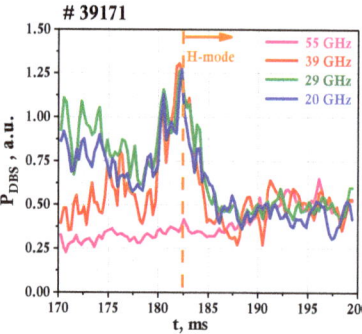

Figure 8. Temporal evolution of DBS power for discharge #39171.

Additionally, Figure 9 depicts the spectra of the DBS complex signals at two different moments in time. They were calculated at 180 ms before the transition to the H-mode, shown in Figure 9a, and at 190 ms after the transition when a quasi-stationary H-mode is achieved, shown in Figure 9b. Additionally, vertical lines that correspond to the center of gravity of the spectrum are introduced (see Figure 7b for comparison). The observable frequency shift of the spectra can be attributed to the rotation of the plasma. The positive frequency shift corresponds to the movement of the fluctuations in the electron diamagnetic drift direction. The largest shift is observed for the 39 GHz probing frequency and it even increases in H-mode. In addition, it can be easily determined that the Doppler shift at the 20 GHz frequency is not significant before and after the transition to the H-mode. It is also observable that there is a decrease in spectral power at 190 ms for the 39, 29, and 20 GHz channels.

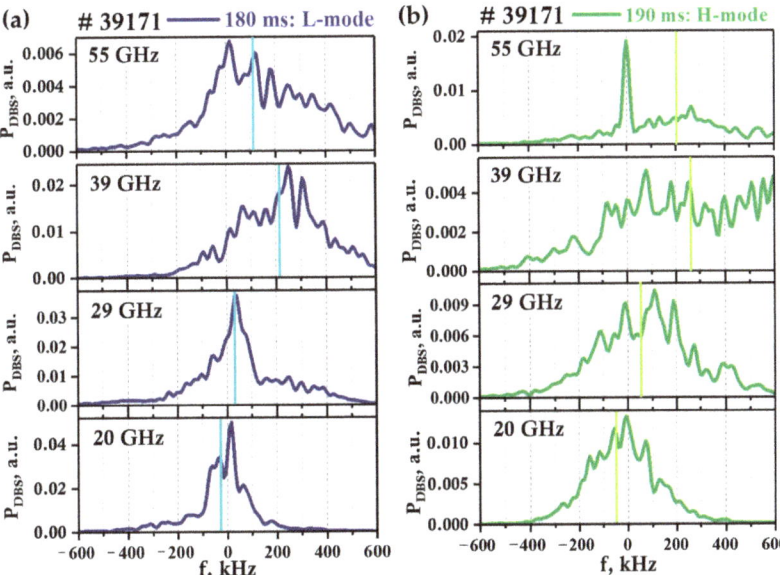

Figure 9. Spectra of DBS complex signals for discharge #39171 at: (**a**) 180 ms in L-mode; (**b**) 190 ms in H-mode.

To highlight the repetitive nature of the observed turbulence suppression after the LH transition, the temporal evolution of the spectral power for several discharges was compared. The results are presented in Figure 10 for the probing frequency of 29 GHz (plasma periphery) for a series of discharges (#39171 shown in yellow, #39173 shown in red, and #39177 shown in green) with a transition to H-mode. The moment of the LH transition at around 182.5 ms is indicated by the orange vertical line. Here, a drop in the turbulence level is observed in all discharges and it is almost identical in behavior and values. It can be noted that the relative level decreases by about a factor of 2.

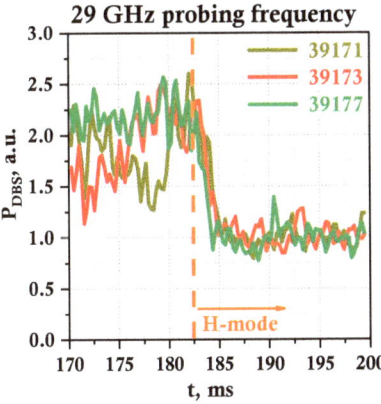

Figure 10. Temporal evolution of DBS power for 29 GHz probing frequency for discharges #39171, #39173, and #39177.

4. Summary

The Doppler backscattering diagnostic was installed on the upgraded spherical tokamak Globus-M2 and has been successfully used since the beginning of the experimental studies on the device. Its description, as well as an example of its implementation are presented and discussed. The use of DBS for the study of the LH transition is demonstrated. DBS measurements of the poloidal plasma velocity and small plasma turbulence allowed for it to be investigated by observing the behavior of these parameters.

First, measurement results of the poloidal rotation velocity profile using a multi-frequency DBS system on Globus-M2 are presented in the work. In experiments, for $\rho = 0.7$–1.0, the plasma was observed to rotate in the direction of electron diamagnetic drift, while outside the separatrix in the direction of the ion diamagnetic drift. In the pedestal region, the absolute value of the velocity was seen to increase significantly during the LH transition from 3 km/s in L-mode to 9 km/s in H-mode. The velocity at the separatrix did not change much after the transition and remained close to zero. The multi-frequency probing approach of the DBS method allowed investigation of the spatial inhomogeneity of the rotation velocity and, thus, the calculation of the velocity shear, which is an important characteristic associated with the LH transition. The velocity shear value was found to be close to 3×10^{-5}, right before the transition to H-mode. This value is believed to be the threshold value for the LH transition on the Globus-M2 tokamak.

Additionally, the plasma density fluctuations with wave vector values of 2–7 cm^{-1} have been studied on Globus-M2 using DBS. During the LH transition, the suppression of such small-scale plasma turbulences was observed in the temporal evolution of the spectral power of the DBS signals. The information obtained about the turbulence amplitude in different discharge areas indicates that the turbulence suppression is of a peripheral nature. These results were compared for a series of discharges, and the repetitive nature of the turbulence suppression in the periphery after the LH transition was also noted.

The value of the measured velocity shear being close to the threshold value, together with the rapid decrease in the level of peripheral turbulence, fits well into the concept of the transition to H-mode by shear turbulence suppression.

Author Contributions: Conceptualization, A.P. (Anna Ponomarenko) and A.Y.; methodology, A.P. (Alexander Petrov) and G.K.; software, A.P. (Anna Ponomarenko) and N.Z.; validation, V.M., Y.P. and N.S.; formal analysis, A.Y.; investigation, A.P. (Anna Ponomarenko) and N.Z.; resources, V.M., Y.P., and N.S.; data curation, A.Y.; writing—original draft preparation, A.Y. and A.P. (Anna Ponomarenko); writing—review and editing, A.Y., G.K., Y.P. and N.S.; visualization, A.P. (Anna Ponomarenko); supervision, A.Y.; project administration, A.Y.; funding acquisition, A.Y. All authors have read and agreed to the published version of the manuscript.

Funding: This research was funded by the Ministry of Science and Higher Education of the Russian Federation: 0784-2020-0020.

Institutional Review Board Statement: Not applicable.

Informed Consent Statement: Not applicable.

Data Availability Statement: Not applicable.

Acknowledgments: This work was done using the Federal Joint Research Center "Material science and characterization in advanced technology", including the unique scientific facility "Spherical tokamak Globus-M".

Conflicts of Interest: The authors declare no conflict of interest.

References

1. Donné, A.J.H.; Costley, A.E.; Barnsley, R.; Bindslev, H.; Boivin, R.; Conway, G.; Fisher, R.; Giannella, R.; Hartfuss, H.; Hellermann, M.G.; et al. Chapter 7: Diagnostics. *Nucl. Fusion* **2007**, *47*, S337. [CrossRef]
2. Wagner, F.; Becker, G.; Behringer, K.; Campbell, D.; Eberhagen, A.; Engelhardt, W.; Fussmann, G.; Gehre, O.; Gernhardt, J.; Gierke, G.V.; et al. Regime of Improved Confinement and High Beta in Neutral-Beam-Heated Divertor Discharges of the ASDEX Tokamak. *Phys. Rev. Lett.* **1982**, *49*, 1408. Available online: https://journals.aps.org/prl/abstract/10.1103/PhysRevLett.49.1408. (accessed on 10 July 2022). [CrossRef]
3. Doyle, E.J.; Houlberg, W.A.; Kamada, Y.; Mukhovatov, V.; Osborne, T.H.; Polevoi, A.; Bateman, G.; Connor, J.W.; Cordey, J.G.; Fujita, T.; et al. Chapter 2: Plasma confinement and transport. *Nucl. Fusion* **2007**, *47*, S18. Available online: https://iopscience.iop.org/article/10.1088/0029-5515/47/6/S02. (accessed on 10 July 2022).
4. Wagner, F. A quarter-century of H-mode studies. *Plasma Phys. Control. Fusion* **2007**, *49*, B1. Available online: https://iopscience.iop.org/article/10.1088/0741-3335/49/12B/S01/meta. (accessed on 10 July 2022). [CrossRef]
5. Burrell, K.H. Tests of causality: Experimental evidence that sheared $E \times B$ flow alters turbulence and transport in tokamaks. *Phys. Plasmas* **1999**, *6*, 4418. [CrossRef]
6. Schmitz, L. The role of turbulence—Flow interactions in L- to H-mode transition dynamics: Recent progress. *Nucl. Fusion* **2017**, *57*, 025003. [CrossRef]
7. Valovič, M.; Akers, R.; Cunningham, G.; Garzotti, L.; Lloyd, B.; Muir, D.; Patel, A.; Taylor, D.; Turnyanskiy, M.; Walsh, M.; et al. Scaling of H-mode energy confinement with I_p and B_T in the MAST spherical tokamak. *Nucl. Fusion* **2009**, *49*, 075016. [CrossRef]
8. Kaye, S.M.; Connor, J.W.; Roach, C.M. Thermal confinement and transport in spherical tokamaks: A review. *Plasma Phys. Control. Fusion* **2021**, *63*, 123001. [CrossRef]
9. Kurskiev, G.S.; Bakharev, N.N.; Bulanin, V.V.; Chernyshev, F.V.; Gusev, V.K.; Khromov, N.A.; Kiselev, E.O.; Minaev, V.B.; Miroshnikov, I.V.; Mukhin, E.E.; et al. Thermal energy confinement at the Globus-M spherical tokamak. *Nucl. Fusion* **2019**, *59*, 066032. [CrossRef]
10. Kurskiev, G.S.; Gusev, V.K.; Sakharov, N.V.; Petrov, Y.U.V.; Bakharev, N.N.; Balachenkov, I.M.; Bazhenov, A.N.; Chernyshev, F.V.; Khromov, N.A.; Kiselev, E.O.; et al. Energy confinement in the spherical tokamak Globus-M2 with a toroidal magnetic field reaching 0.8 T. *Nucl. Fusion* **2022**, *62*, 016011. [CrossRef]
11. Hirsch, M.; Holzhauer, E.; Baldzuhn, J.; Kurzan, B.; Scott, B. Doppler reflectometry for the investigation of propagating density perturbations. *Plasma Phys. Control. Fusion* **2001**, *43*, 1641. [CrossRef]
12. Conway, G.D.; Schirmer, J.; Klenge, S.; Suttrop, W.; Holzhauer, E.; The ASDEX Upgrade Team. Plasma rotation profile measurements using Doppler reflectometry. *Plasma Phys. Control. Fusion* **2004**, *46*, 951. [CrossRef]
13. Bulanin, V.V.; Lebedev, S.V.; Levin, L.S.; Roytershteyn, V.S. Study of plasma fluctuations in the Tuman-3m tokamak using microwave reflectometry with an obliquely incident probing beam. *Plasma Phys. Rep.* **2000**, *26*, 813–819. [CrossRef]
14. Hennequin, P.; Honoré, C.; Truc, A.; Quéméneur, A.; Lemoine, N.; Chareau, J.M.; Sabot, R. Doppler backscattering system for measuring fluctuations and their perpendicular velocity on Tore Supra. *Rev. Sci. Instrum.* **2004**, *75*, 3881–3883. [CrossRef]
15. Peebles, W.A.; Rhodes, T.L.; Hillesheim, J.C.; Zeng, L.; Wannberg, C. A novel, multichannel, comb-frequency Doppler backscatter system. *Rev. Sci. Instrum.* **2010**, *81*, 10D902. [CrossRef]
16. Oyama, N.; Takenaga, H.; Suzuki, T.; Sakamoto, Y.; Isayama, A. Density Fluctuation Measurements Using a Frequency Hopping Reflectometer in JT-60U. *Plasma Fusion Res.* **2011**, *6*, 1402014. [CrossRef]
17. Hillesheim, J.; Crocker, N.; Peebles, W.; Meyer, H.; Meakins, A.; Field, A.; Dunai, D.; Carr, M.; Hawkes, N. Doppler backscattering for spherical tokamaks and measurement of high-k density fluctuation wavenumber spectrum in MAST. *Nucl. Fusion* **2015**, *55*, 73024. [CrossRef]
18. Hu, J.Q.; Zhou, C.; Liu, A.D.; Wang, M.Y.; Doyle, E.J.; Peebles, W.A.; Wang, G.; Zhang, X.H.; Zhang, J.; Feng, X.; et al. An eight-channel Doppler backscattering system in the experimental advanced superconducting tokamak. *Rev. Sci. Instrum.* **2017**, *88*, 073504. [CrossRef]
19. Vermare, L.; Hennequin, P.; Honoré, C.; Peret, M.; Dif-Pradalier, G.; Garbet, X.; Gunn, J.; Bourdelle, C.; Clairet, F.; Morales, J.; et al. Formation of the radial electric field profile in the WEST tokamak. *Nucl. Fusion* **2022**, *62*, 026002. [CrossRef]
20. Yashin, A.; Bulanin, V.; Petrov, A.; Ponomarenko, A. Review of Advanced Implementation of Doppler Backscattering Method in Globus-M. *Appl. Sci.* **2021**, *11*, 8975. [CrossRef]
21. Yashin, A.Y.; Bulanin, V.V.; Gusev, V.K.; Khromov, N.A.; Kurskiev, G.S.; Minaev, V.B.; Patrov, M.I.; Petrov, A.V.; Petrov, Y.U.V.; Prisyazhnyuk, D.V.; et al. Geodesic acoustic mode observations in the Globus-M spherical tokamak. *Nucl. Fusion* **2014**, *54*, 114015. [CrossRef]
22. Yashin, A.Y.U.; Bulanin, V.V.; Gusev, V.K.; Kurskiev, G.S.; Patrov, M.I.; Petrov, A.V.; Petrov, Y.V.; Tolstyakov, S.Y. Phenomena of limit-cycle oscillations in the Globus-M spherical tokamak. *Nucl. Fusion* **2018**, *58*, 112009. [CrossRef]
23. Bulanin, V.V.; Gusev, V.K.; Kurskiev, G.S.; Minaev, V.B.; Patrov, M.I.; Petrov, A.V.; Petrov, Y.V.; Yashin, A.Y. Application of the Multifrequency Doppler Backscattering Method for Studying Alfvén Modes at a Tokamak. *Tech. Phys. Lett.* **2019**, *45*, 1107–1110. [CrossRef]
24. Bulanin, V.V.; Gusev, V.K.; Khromov, N.A.; Kurskiev, G.S.; Minaev, V.B.; Patrov, M.I.; Petrov, A.V.; Petrov, M.A.; Petrov, Y.V.; Prisiazhniuk, D.; et al. The study of filaments by the Doppler backscattering method in the 'Globus-M' tokamak. *Nucl. Fusion* **2019**, *59*, 096026. [CrossRef]

25. Bulanin, V.V.; Yashin, A.Y.; Petrov, A.V.; Gusev, V.K.; Minaev, V.B.; Patrov, M.I.; Petrov, Y.V.; Prisiazhniuk, D.V.; Varfolomeev, V.I. Doppler backscattering diagnostic with dual homodyne detection on the Globus-M compact spherical tokamak. *Rev. Sci. Instrum.* **2021**, *92*, 033539. [CrossRef]
26. Yashin, A.Y.; Bulanin, V.V.; Petrov, A.V.; Petrov, M.A.; Gusev, V.K.; Khromov, N.A.; Kurskiev, G.S.; Patrov, M.I.; Petrov, Y.V.; Tolstyakov, S.Y.; et al. Multi-diagnostic approach to geodesic acoustic mode study. *J. Instrum.* **2015**, *10*, P10023. [CrossRef]
27. Leonard, A.W. Edge-localized-modes in tokamaks. *Phys. Plasmas* **2014**, *21*, 090501. [CrossRef]
28. Yashin, A.; Teplova, N.; Zadvitskiy, G.; Ponomarenko, A. Modelling of Backscattering off Filaments Using the Code IPF-FD3D for the Interpretation of Doppler Backscattering Data. *Sensors* **2022**, *22*, 9441. [CrossRef]
29. Cavedon, M.; Birkenmeier, G.; Pütterich, T.; Ryter, F.; Viezzer, E.; Wolfrum, E.; Dux, R.; Happel, T.; Hennequin, P.; Plank, U.; et al. Connecting the global H-mode power threshold to the local radial electric field at ASDEX Upgrade. *Nucl. Fusion* **2020**, *60*, 066026. [CrossRef]
30. Liang, A.S.; Zou, X.L.; Zhong, W.L.; Ekedahl, A.; Duan, X.R.; Shi, Z.B.; Yu, D.L.; Yang, Z.C.; Wen, J.; Xiao, G.L.; et al. Critical velocity shear flow for triggering L-H transition and its parametric dependence in the HL-2A tokamak. *Nucl. Fusion* **2020**, *60*, 092002. [CrossRef]
31. Yashin, A.Y.; Bulanin, V.V.; Gusev, V.K.; Minaev, V.B.; Petrov, A.V.; Petrov, Y.V.; Ponomarenko, A.M.; Varfolomeev, V.I. Doppler backscattering systems on the Globus-M2 tokamak. *J. Instrum.* **2022**, *17*, C01023. [CrossRef]

Disclaimer/Publisher's Note: The statements, opinions and data contained in all publications are solely those of the individual author(s) and contributor(s) and not of MDPI and/or the editor(s). MDPI and/or the editor(s) disclaim responsibility for any injury to people or property resulting from any ideas, methods, instructions or products referred to in the content.

Article

Modelling of Backscattering off Filaments Using the Code IPF-FD3D for the Interpretation of Doppler Backscattering Data

Alexander Yashin [1,2,*], Natalia Teplova [1,2], Georgiy Zadvitskiy [3] and Anna Ponomarenko [1]

1 Plasma Physics Department, Peter the Great Saint Petersburg Polytechnic University, 195251 Saint Petersburg, Russia
2 Plasma Research Laboratory, Ioffe Institute, 195251 Saint Petersburg, Russia
3 Max-Planck-Institut für Plasmaphysik, 85748 Garching, Germany
* Correspondence: alex_yashin@list.ru

Abstract: Filaments or blobs are well known to strongly contribute to particle and energy losses both in L- and H-mode, making them an important plasma characteristic to investigate. They are plasma structures narrowly localized across a magnetic field and stretched along magnetic field lines. In toroidal devices, their development is observed to take place in the peripheral plasma. Filament characteristics have been studied extensively over the years using various diagnostic techniques. One such diagnostic is the Doppler backscattering (DBS) method employed at the spherical tokamak Globus-M/M2. It has been observed that the DBS signal reacts to the backscattering from filaments. However, the DBS data have proven difficult to analyze, which is why modelling was undertaken using the code IPF-FD3D to understand what kind of information can be extrapolated from the signals. A circular filament was thoroughly investigated in slab geometry with a variety of characteristics studied. Apart from that, the motion of the filaments in the poloidal and radial directions was analyzed. Additionally, other shapes of filaments were presented in this work. Modelling for the real geometry of the Globus-M/M2 tokamak was performed.

Keywords: filaments; tokamak; plasma diagnostics; Doppler backscattering; simulations

1. Introduction

Fluxes of power and particles to plasma-facing components are a concerning phenomenon for magnetic confinement devices (tokamaks and stellarators), particularly for future fusion devices, such as ITER [1], as these fluxes can result in damage to the machine that can significantly affect the core plasma performance [2]. A variety of factors play a role in energy and particle transport with it resulting from a competition between sources of parallel and perpendicular losses. Filaments or blobs (filament-like plasma perturbations) are well known to strongly contribute to particle and energy losses both in L- and H-mode regimes [3], making them an important plasma characteristic to investigate. They are plasma structures that are narrowly localized across a magnetic field while being stretched out for up to several meters along magnetic field lines. In toroidal devices, their development is observed to take place in the peripheral plasma, close to the last closed flux surface (LCFS) or separatrix [4–6].

Filament characteristics have been studied extensively over the years using various diagnostic techniques [7]. One such diagnostic technique (unconventional though it may be for this purpose) is the Doppler backscattering (DBS) method [8], which was initially employed at the spherical tokamak Globus-M [9–13] and later the ASDEX Upgrade tokamak [14] for the purpose of studying filaments. It has been observed that the DBS signal reacts to backscattering from filaments. The reaction manifests itself as a burst of coherent fluctuations (CF) of the measured IQ (In-phase and Quadrature) signals. However, the measurements collected using DBS have proven difficult to analyze, which is why modelling needs to be undertaken to understand what kind of information can indeed be extrapolated from the signals. The first results of such modelling were obtained in the work in [15] describing the reaction of the DBS signal to a filament with set parameters such as its shape, direction of its motion and trajectory; it contains data regarding the possible reaction of the DBS diagnostics to changes in these parameters. While the results allowed for some understanding of the filaments in Globus-M, it should be noted that the modeling presented in the work in [15] possessed a number of erroneous conclusions, which are discussed in this paper.

This work built on the recent observations of filaments that have illuminated that there is much work still to be done in understanding the types of filaments observed in Globus-M2 [16]. For this reason, modelling was undertaken using the code IPF-FD3D [17] in slab geometry. One aspect that needs to be more thoroughly investigated is the circular filament interpretation. While some simulation data were analyzed so as to understand the experimentally detected filaments (this, however, highlighted that this proposed model cannot be applicable for all types of filaments observed), there are other characteristics that need to be modelled when it comes to circular filaments. Apart from that, the motion of filaments has been experimentally investigated in detail [18], which provided an insight into the fact that the filaments can move in different directions (even having rather complex trajectories). This makes it an important aspect to be modelled for DBS diagnostics, as the data collected can provide characteristics such as the direction of the moving filaments and their velocity. Additionally, experimental measurements [19] have raised several questions that include the possibility of the development of filament structures of different shapes, which are presented in this work.

The paper is structured as follows. The next section contains the conditions of the modelling that was undertaken. After that, the circular filament was studied in detail, with the impact of its various parameters as well as its different motion on the DBS signals presented. Then, the strip filament model was investigated. After that, data for different types of stretched filaments are presented. All these findings are then compared to data obtained for a different density profile. Finally, the DBS signals for the real Globus-M2 tokamak geometry are discussed.

2. Materials and Methods

Two-dimensional full-wave simulations were conducted with the finite-difference time-domain (FDTD) code IPF-FD3D [17] in slab geometry. The FDTD scheme is a time-steeping iterative scheme for solving Maxwell's equations on a regular grid. The FDTD method is used in the IPF-FD3D code to simulate electromagnetic wave propagation in cold plasma, which is described by the following system of equations:

$$\frac{\partial J}{\partial t} = \varepsilon_0 \omega_{pe}^2 E - \omega_{ce} J \times \widehat{B_0} \tag{1}$$

$$\frac{\partial E}{\partial t} = \frac{1}{\varepsilon_0} \nabla \times H - \frac{1}{\varepsilon_0} J \tag{2}$$

$$\frac{\partial H}{\partial t} = -\frac{1}{\mu_0} \nabla \times E \tag{3}$$

where J is the plasma current, the plasma density is described by plasma frequency ω_{pe}^2, the strength of the background magnetic field is given by frequency ω_{ce}, and its direction is given by $\widehat{B_0}$. The electric E and magnetic H fields of the electromagnetic wave are described by Maxwell's Equations (2) and (3). In the code, these partial differential equations are translated into finite difference equations, and then the fields E and H and the current J are calculated consecutively in a leap-frogging time scheme. The IPF-FD3D code implements a one-, two-, and three-dimensional solver of Equations (1)–(3) on a cartesian grid. In this paper, the 2D solver was used. In addition to the equation-solving engine, the code incorporates such elements as numeric antennas, in phase/quadrature (I/Q) detectors for optical mixing and phase measurement, and ideally absorbing boundary conditions at the edge of the computation grid. The code is mainly applied to Doppler reflectometry and the investigation of microwave components; however, it can be used for other purposes as well. The main weaknesses of the FDTD method are, first, the inflexible grid geometry that does not allow for adaptation to the geometry of the modeled structures, and second, the need to calculate many time steps until a stationary state is reached. While the former is not relevant to the modeling of smoothly varying plasma densities, the latter, in combination with the need for gathering statistics on turbulent processes, leads to substantial use of computational resources, as a high number of realizations is computed. Therefore, the supercomputer of the Supercomputer Center of the Peter the Great St. Petersburg Polytechnic University was used for the calculations.

The simulation setup is organized as follows. For a given series of conditions set for the filament, the simulations obtain the I and Q signals of the quadrature detector, which can be compared to the experimental measurements obtained using DBS. For the purpose of understanding a variety of data collected, real experimental scenarios were examined. The electron density profile (referred to as the high-density profile) used in the simulations was taken from Thomson measurements and is presented on the left in Figure 1. The experimental plasma density profile used for the computations was close to linear, with values reaching $n_e = 5.8 \times 10^{-19}$ m^{-3} at y = 0.15 m (shot no. 36569 at Globus-M).

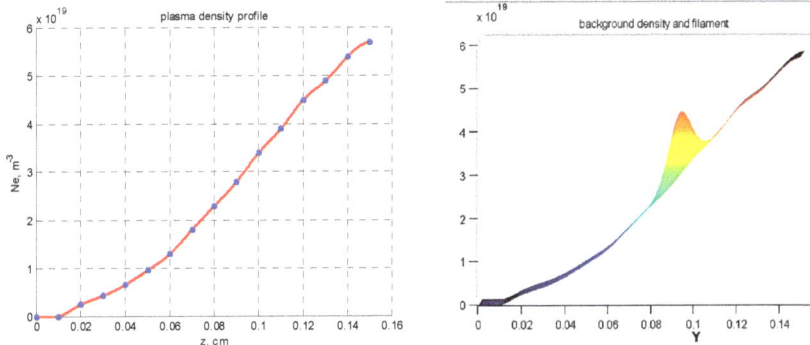

Figure 1. (**left**) Experimental electron density profile for Globus-M2, (**right**) experimental electron density profile for Globus-M2 with introduced filament distribution.

The approach to modeling the filaments involves introducing a filament density distribution on top of the chosen density profile. This is demonstrated on the right in Figure 1, which contains the resulting density profile of the combined background and filament density. The model of the filament was relatively simple with, a Gaussian cross-section that allowed freedom in choosing the shape, size, amplitude, and position of the filament.

The parameters that corresponded to the conditions of the performed modelling are presented in Table 1. These include the size of the box, step size, coordinates of the equatorial plane and plasma edge, position of the antenna, its tilt angle, and others.

Figure 2 schematically represents the position and size of the box where the calculations were performed in regard to the Globus-M/M2 poloidal cross-section. The antenna is demonstrated with its position in the equatorial plane by the grey triangle, and the series of black lines are the calculated probing beam trajectories using a ray-tracing code for the Globus-M/M2 geometry. The box was placed to entirely include the DBS system, i.e., the antenna and the probing beam trajectories. The parameters of the DBS antenna of the installed system were as follows: antenna tilt angle $\alpha = 13°$, antenna horn mouth 5.5 cm, and a Gaussian beam with a flat wave front in the antenna mouth. The coordinate system of the box is also demonstrated, with the X axis corresponding to the poloidal direction (parallel to the magnetic field lines), and the Y axis to the radial direction in the tokamak.

Table 1. Parameters of the performed modelling using the code IPF-FD3D.

Box size	40 × 24 cm = 1538 × 923 pts
Step size	dx = 1 pt = 0.026 cm
Equatorial plane	X = 10 cm = 384 pts
PML	Y = 0.625 cm = 24 pts
Plasma edge	Y = 2.08 cm
Antenna coordinates	X = 4 cm = 154, Y = 0
Antenna tilt angle	6 deg
Horn mouth	5.5 cm
Wave front	flat, Gaussian beam, w0 = 2.5 cm, z0 = 0

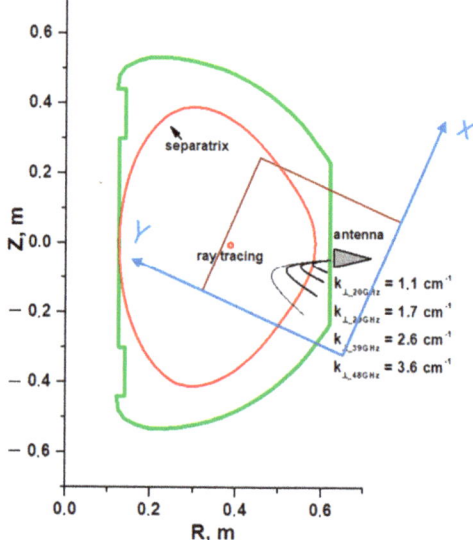

Figure 2. Poloidal cross-section of the Globus-M2 tokamak with performed ray tracing.

Computations were performed for a range of probing frequencies, i.e., 20, 29, 39, 48, 50, 55, 60, 65, 70, and 75 GHz, in O-mode, which corresponded to the one in the installed systems in the Globus-M2 tokamak. Their schematics are presented in Figure 3. The system on the left is the low-frequency system, which uses homodyne detection, and the one on the right is the high-frequency system that employs heterodyne detection. The systems are described in detail in the works in [20,21].

One of the goals of this work was to investigate different circular filaments, which meant that a level of flexibility in their parameters was necessary. Different values of their size, amplitude (percentage of density at the cutoff of the probing wave), position of the filament, and a range of probing frequencies (a variety of cut-off radii) were necessary to increase the database of circular filaments in comparison to the results in paper [15]. All the parameters of the circular filaments are presented in Table 2, where all the values in italics represent the additional data obtained. All these various parameters were analyzed to observe the transition from linear to nonlinear scattering, as the purely linear model was unable to explain all experimental data.

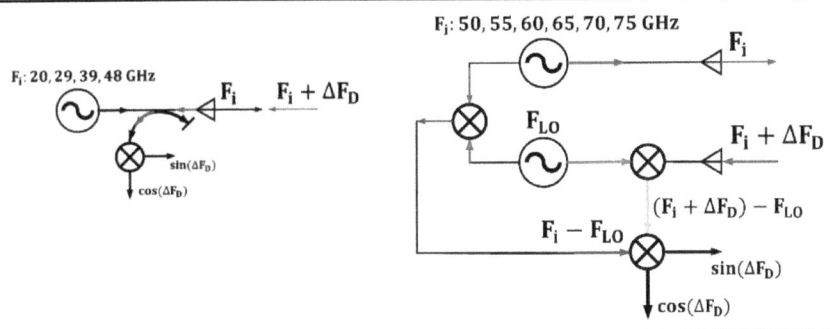

Figure 3. Schematics of DBS systems in Globus-M2.

Table 2. Parameters of the performed modelling for circular filaments.

Frequencies	20, 29, 39, 48, 50, 55, 60, 65, 70, 75 GHz
Antenna tilt angle	13 degrees
Horn mouth	5.5
Wave front	flat
Diameter of the filament	0.5 cm, 1.5 cm, 3 cm
Amplitude of the filament	0.1%, 1%, 10%, 25%, 50%, 80%, 100%
Position of the filament	In the vicinity of the cut off 48 GHZ

Apart from changes to the filament itself, its motion was investigated. As recent research suggests, depending on the location where the filament is developed in, it can move in different directions, with rather complex trajectories also being possible [17]. Thus, it was important to include this aspect into the modelling, with the poloidal and radial directions being added to the calculations (Table 3). The code also allowed for complex trajectories (meaning not in just one direction) to be implemented. In the case of purely poloidal or purely radial motion, the position of the filament was changed with a change in the probing frequency. The filament motion was simulated by independent snapshots with a spatial step of 1 mm. If a time interval was assigned equal to this step, the velocity of the filament could be determined. For each snapshot, the values of IQ signals (or amplitude and phase) were calculated to obtain a time dependency.

Table 3. Parameters of the performed modelling for the motion of circular filaments.

Frequencies	48 GHz
Antenna tilt angle	13 degrees
Horn mouth	5.5
Wave front	flat
Diameter of the filament	0.5 cm, 1 cm, 1.5 cm, 5 cm
Amplitude of the filament	0.1%, 1%, 5%, 10%, 50%, 75%, 100%
Position of the filament	4 cm, 6 cm, 8 cm, 10 cm, 12 cm

In addition, to provide a more complete analysis, the shape of the filament was changed in the performed simulations. This was conducted based on the experimental evidence that suggests that a simple circular filament model may not be accurate for all conditions [18]. The forms included strip filaments and filaments stretched in both the poloidal and radial directions. A range of their parameters were investigated and are presented in Tables 4 and 5.

Table 4. Parameters of the performed modelling for the strip filament.

Frequencies	30, 35, 40, 45, 50, 55, 60, 65, 70, 75 GHz
Antenna tilt angle	13 deg
Horn mouth	5.5 cm
Wave front	flat
Diameter of the filament	0.5 cm, 1 cm, 1.5 cm, 3 cm, 5 cm, 6 cm
Amplitude of the filament	0.1%, 1%, 5%, 10%, 20%, 30%, 40%, 50%, 60%, 70%, 80%, 90%, 100%, 110%, 150%
Position of the filament	In the vicinity of the cut off

Table 5. Parameters of the performed modelling for the stretched filament.

Frequencies	20, 29, 39, 48, 50, 55, 60, 65, 70, 75 GHz
Antenna tilt angle	13 deg
Horn mouth	5.5 cm
Wave front	flat
Diameter of the filament	0.5 cm, 1.5 cm, 3 cm
Amplitude of the filament	0.1%, 1%, 10%, 25%, 50%, 80%, 100%
Position of the filament	In the vicinity of the cut off

3. Circular Filaments

Modelling in the work in [15] was undertaken to answer the question how DBS signals of a given probing frequency in Globus-M would react to the presence of a circular filament in the proximity of its cut-off, which in that case was 48 GHz. However, this was not enough to explain all the data collected using the Globus-M/M2 DBS system, which lead to keen interest in how other channels (more specifically of higher frequencies with cut-offs in deeper regions, but the SOL region was also investigated) could be influenced by a circular filament in the periphery where they develop [4–6,17]. In this work, the calculations of DBS signals for various probing frequencies (Table 1) were performed. The filament was positioned near the separatrix at the cut-off radius of the 48 GHz probing frequency, while the signals at other radii were analyzed.

Density perturbations with a circular filament near the cut-off of the 48 GHz probing frequency introduced on top of the bulk plasma density are presented in Figure 4. The radial position of the filament was not changed, but the filament moved with a constant velocity of 10 km/s in the poloidal direction (in the direction of the x axis).

For the conditions previously described, a variety of DBS signals were calculated, with several examples presented in Figure 5. They included two pairs of signals for the probing frequency of 48 GHz obtained for circular filaments of different diameters (left—0.5; right—3 cm) and amplitudes (black line—1%; red line—100%). After a certain critical size of the circular filament was reached, a significant delay in the formation of the filament between the low and high amplitude case was seen (right in Figure 5). This was not seen for the 0.5 cm filament and was not observed for any other intermediate diameter values.

Another example of the influence of the size of the circular filament on the DBS signals is presented in Figure 6. It contains the IQ signals obtained for a filament with a 5 cm diameter. The comparison of the low (1%) and high (50%) amplitude cases led to the observation that the signal frequency significantly increased after transitioning to a non-linear regime.

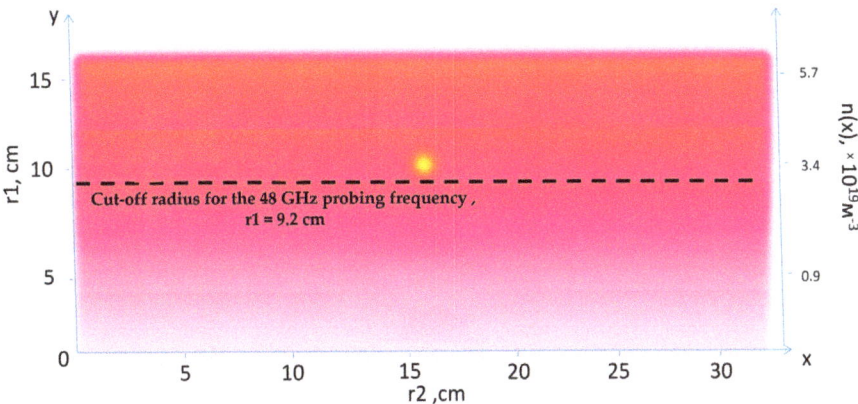

Figure 4. Density perturbations with an introduced circular filament (yellow circle).

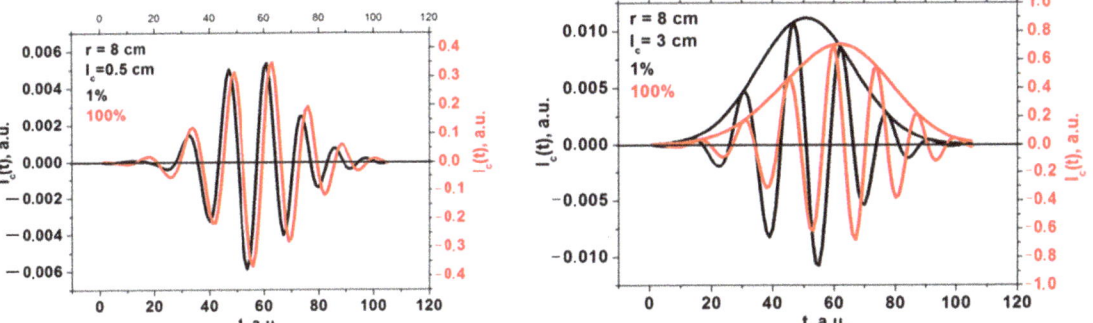

Figure 5. DBS signals for the 48 GHz probing frequency in the case of a circular filament of (**left**) 0.5 cm diameter and (**right**) 3 cm diameter. The black signals correspond to filaments with an amplitude of 1% of density at cutoff of the probing wave, and the red signals correspond to an amplitude of 100%.

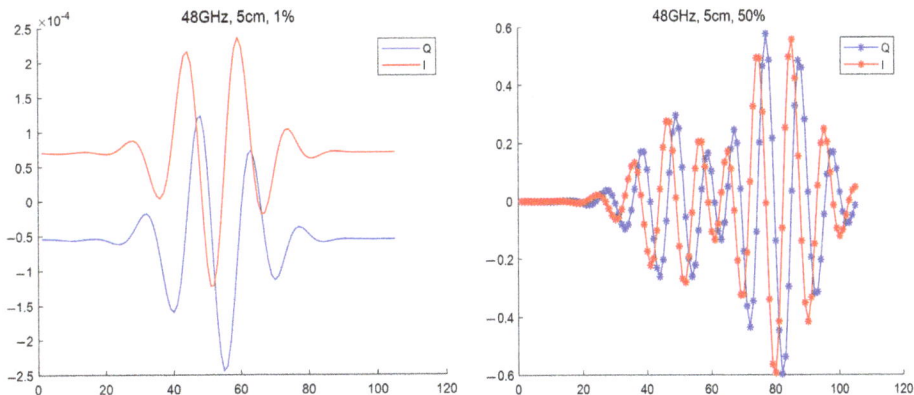

Figure 6. DBS IQ signals for the 48 GHz probing frequency for a circular filament with a 5 cm diameter: (**left**) amplitude of 1% of density at cutoff of the probing wave, (**right**) amplitude of 100% of density at cutoff of the probing wave.

A more systematic approach was then taken to determine how signals of different probing frequencies react to the presence of circular filaments of different sizes. The linear case was investigated with calculations undertaken for the 0.1% amplitude filament. The results of these calculations are presented in Figure 7. The dependency of the maximum of the signal amplitude, as well as its main frequency component, is demonstrated on the left with the bold lines representing the change in signal amplitude depending on the probing frequency (scale on the left), while the other three lines with pentagon shapes represent the signal frequency behavior (scale on the right). The colors correspond to different circular filament diameters, with navy-blue describing the 3 cm diameter filament, wine-red describing that of 1.5 cm, and olive-green describing that of 0.5 cm. The vertical lines indicate the probing frequencies installed on the DBS system in the Globus-M2 tokamak (i.e., 20, 29, 39, 48, 50, 55, 60, 65, 70, and 75 GHz). Additionally, the orange dashed vertical line highlights the position of the filament at the cut-off radius of the 48 GHz probing frequency. In the linear case (0.1% amplitude), we observed a steady decrease in the signal frequency, while the signal amplitude experienced a steady increase until a certain peak was reached, after which a gradual decrease followed. The maximum of the signal amplitude took place around the 48 GHz frequency, where the filament was located in the model. The values of the signal amplitude also suggested that the DBS diagnostic would only be able to detect the filaments using channels with a range of probing frequencies of 40–55 GHz.

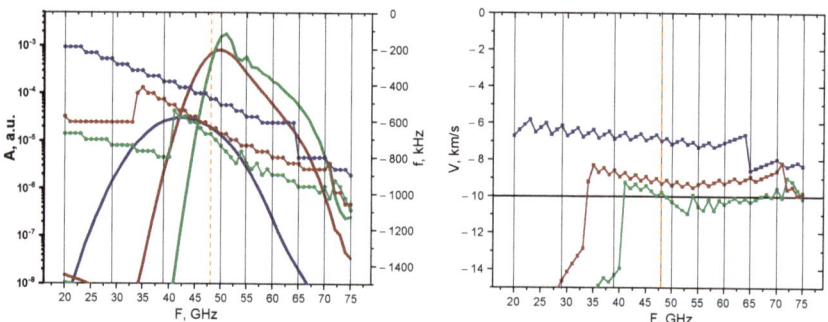

Figure 7. For the circular filament with an amplitude of 0.1%: (**left**) dependency of the DBS signal amplitude (bold lines with scale on the left) and frequency (lines with pentagon shapes with scale on the right) on the probing frequency; (**right**) dependency of the calculated velocity on the probing frequency. The navy-blue lines are the circular filament with a 3 cm diameter, the wine-red are that of 1.5 cm, and the olive-green are that of 0.5 cm. The black vertical lines indicate the probing frequencies available in Globus-M2, and the vertical orange dashed line indicates the position of the filament at cutoff of the 48 GHz probing wave. The horizontal line marks the set 10 km/s filament velocity.

The Doppler frequency shift (signal frequency) was then used to calculate the filament velocity by the formula $V = \Delta\omega/k$, where $k = 2k_0 \sin\alpha$ is the wave vector detected by the DBS method. The motion of the filament in the poloidal direction was set at 10 km/s (horizontal line on the right), which allowed us to meaningfully analyze the estimated velocities. In the case of the 0.5 cm diameter filament, the results were accurate for the 35–75 GHz probing frequencies, while the other two cases (blue line—3 cm and red line—1.5 cm) provided velocity measurements of higher values.

An analysis was carried out for the non-linear case (50% filament amplitude), and the results of the calculations are presented in Figure 8. The behavior of the signal amplitudes and frequency remained similar to the linear case; however, the values of the signal amplitude implied that the filaments would be observable using a wider range of probing frequencies of 45–65 GHz. Additionally, filaments of all sizes exhibited very similar behaviors and values, but in the linear case, a greater discrepancy was observed between the 3 cm filament and the others. The velocity for all the filaments was slightly above the set 10 km/s value after 47 GHz, but before that the values were lower and differed.

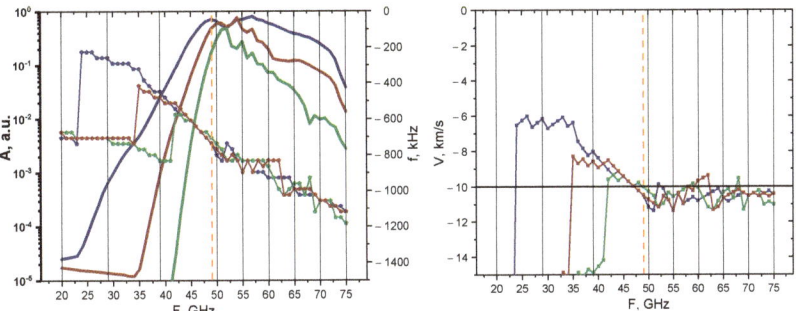

Figure 8. For the circular filament with an amplitude of 50%: (**left**) dependency of the DBS signal amplitude (bold lines with scale on the left) and frequency (lines with pentagon shapes with scale on the right) on the probing frequency; (**right**) dependency of the calculated velocity on the probing frequency. The navy-blue lines are the circular filament with a 3 cm diameter, the wine-red are that of 1.5 cm, and the olive-green are that of 0.5 cm. The black vertical lines indicate the probing frequencies available in Globus-M2, and the vertical orange dashed line indicates the position of the filament at cutoff of the 48 GHz probing wave. The horizontal line marks the set 10 km/s filament velocity.

3.1. Motion of the Filament in the Poloidal Direction from Different Radial Positions

Apart from a fixed radial filament position, calculations were also performed for filaments positioned at different radii (Table 2). Figure 9 presents the density perturbations with a series of circular filaments positioned at different radii and introduced on top of the bulk plasma density. The circular filament moved in the poloidal direction (in the direction of the x axis) at a 10 km/s velocity.

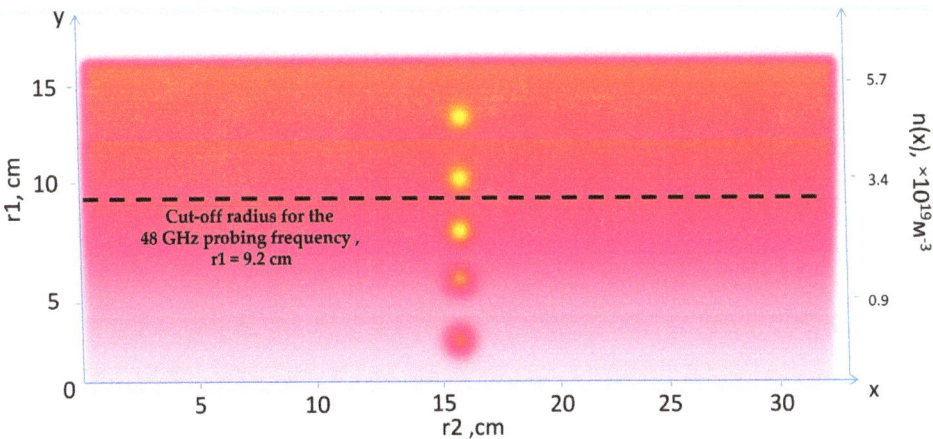

Figure 9. Density perturbations with an introduced series of circular filaments (yellow circles) positioned at different radii.

To investigate the influence of the position of the filament, the DBS signals for a given probing frequency with the circular filament at different radii were calculated, and the signal amplitude was analyzed. The results for the 48 GHz probing frequency are presented in Figure 10. Filaments of different sizes (left—0.5 cm, right—3 cm) and for different amplitudes (red line—1%; black line—100%) were also compared. One may note that a good resolution was only observed in the linear case for the small filament. With increasing the filament size and amplitude, the spatial resolution dropped.

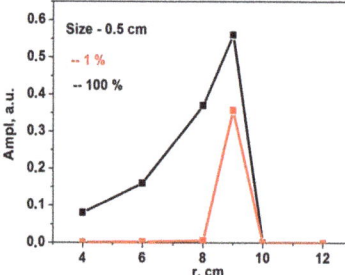 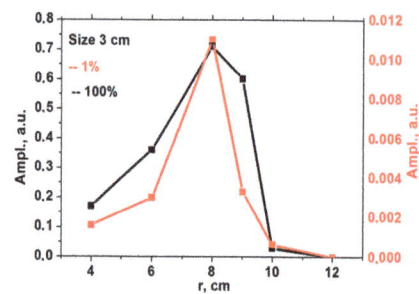

Figure 10. The dependency of the signal amplitude on the radial position of the circular filament of (**left**) 0.5 cm diameter and (**right**) 3 cm diameter. The red line corresponds to the filament with an amplitude of 1% of density at cutoff of the probing wave 48 GHz where the filament was positioned, and the black line corresponds to an amplitude of 100%.

The signal frequency was investigated for the 0.5 cm filament. It is presented in Figure 11 and is depicted by the red line. In both the linear and non-linear cases, the signal frequency coincided with the frequency predicted by the formula for the Doppler shift in the Born approximation (red horizontal line) only when the filament was positioned in the cutoff of the given frequency. For all other positions, the frequency values were lower.

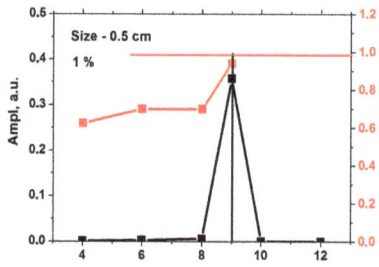 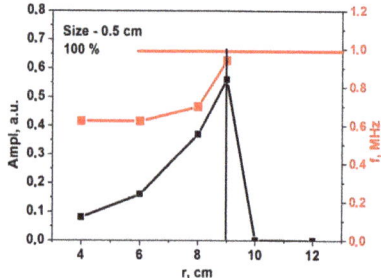

Figure 11. For the circular filament, the dependency of the signal amplitude (black line, scale on the left) and frequency (red line, scale on the right) on the radial position of the circular filament (**left**) with an amplitude of 1% of density at cutoff of the probing wave 48 GHz where the filament was positioned, and (**right**) a filament with an amplitude of 100%. The horizontal red line is the frequency predicted by the formula for the Doppler shift in the Born approximation.

3.2. Two-Dimensional Motion of the Filament

Experiments on various tokamaks demonstrated that filaments can travel both in the poloidal and the radial directions [17]. Complex trajectories were also observed, where the filaments had both velocity components. These observations made this an aspect of interest to model. In this work, we additionally investigated the possible effect that the radial velocity component has on DBS signals. The circular filament positioned at a radius 0.54 m of the 48 GHz probing frequency was investigated. Two-dimensional maps of the velocity signals were obtained, and the results are presented in Figure 12. The red lines indicate the analyzed trajectory. DBS signals were extracted at a chosen radius, which corresponded to a specific probing frequency.

The 48 GHz signal for poloidal motion (left in Figure 12) was investigated and presented in Figure 13. It was similar to the ones presented in previous sections and was used as reference for the signals calculated with added radial components. Its spectrum was calculated and is presented on the left. The main frequency component was 590 kHz.

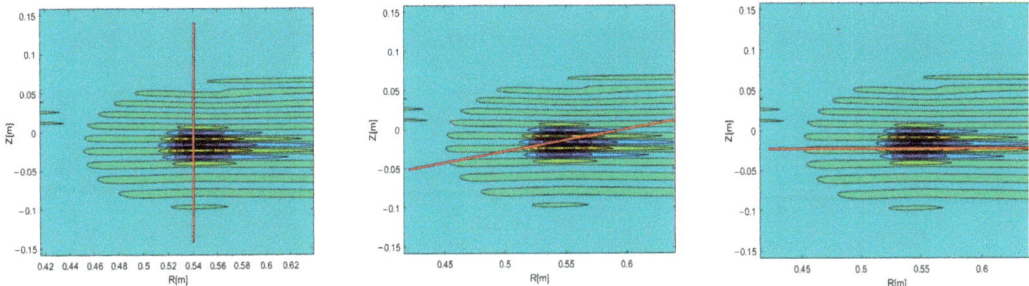

Figure 12. Two-dimensional DBS signal map for the circular filament positioned at the 48 GHz probing frequency. The red line indicates the trajectory: (**left**) poloidal motion, (**middle**) motion in both the poloidal and radial directions, (**right**) radial motion.

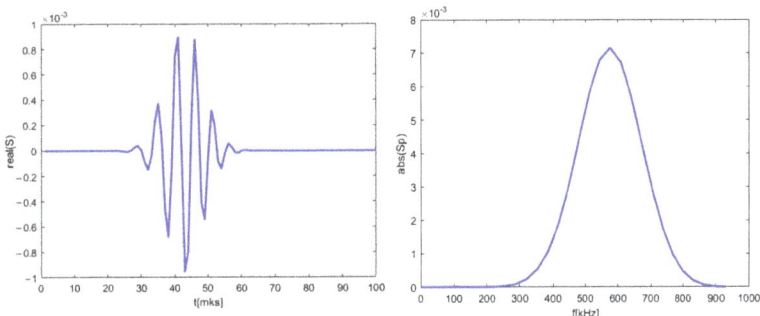

Figure 13. (**left**) DBS signal for the 48 GHz probing frequency; (**right**) calculated spectrum of the signal on the left.

A radial velocity component was introduced, and calculations were carried out for the new trajectory. As an example, the 2D map is presented in the middle in Figure 12. The new DBS signal is shown in the form of a red line in Figure 14. Along with that, the reference signal from Figure 13 is presented for comparison as a blue line. Some differences were observable for the new trajectory. The length of the new signal decreased significantly. In addition, while the largest peaks remained the same, the smaller ones decreased in amplitude and shifted closer toward the main ones. This was also highlighted using the calculated spectrum on the right. The main signal frequency component stayed at 590 kHz, but the spectrum widened in comparison to the poloidally moving filament.

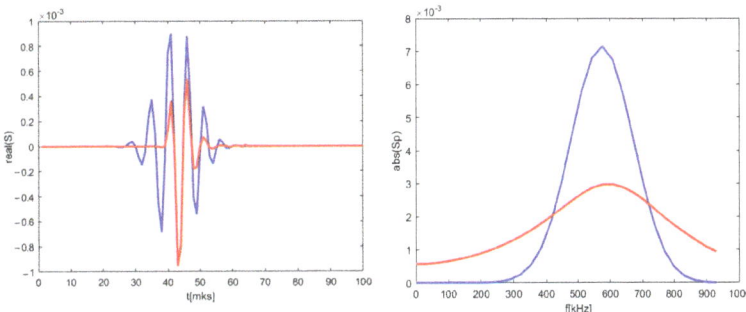

Figure 14. (**left**) DBS signals for the 48 GHz probing frequency; (**right**) calculated spectra of the signal on the left. The blue lines correspond to the reference signal from Figure 13. The red lines correspond to the calculations for trajectory in the middle in Figure 12.

The strictly radial motion presented on the right in Figure 12 was investigated. The obtained DBS signal is depicted in Figure 15 as a red line alongside the reference signal for comparison. For the case of the purely radial motion, the signal was always a single peak spanning over the whole signal length. The signal frequency was much lower than in the case of the circular filament moving poloidally, as the calculated spectrum on the right has its maximum at 1 kHz.

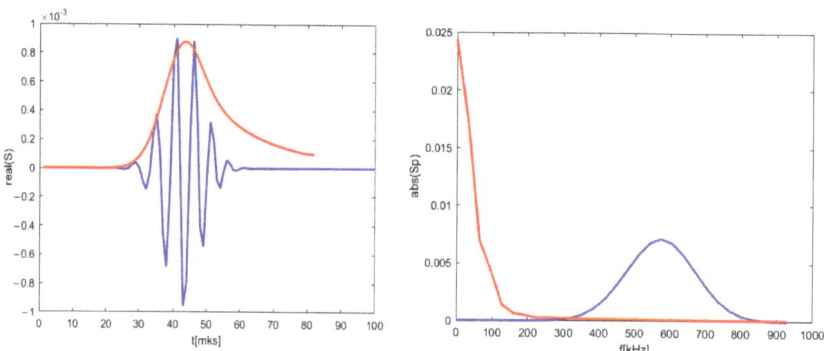

Figure 15. (**left**) DBS signals for the 48 GHz probing frequency; (**right**) calculated spectra of the signal on the left. The blue lines correspond to the reference signal from Figure 13. The red lines correspond to the calculations for trajectory in Figure 12.

Signals of other frequencies were investigated while the filament was still positioned at the radius 0.54 m of the 48 GHz probing frequency so as to observe the influence of the radial component on the DBS signals. The strictly poloidal motion at a velocity of 10 km/s was studied. The 2D maps of the signals for the 39 GHz (left in Figure 16) and 55 GHz (right in Figure 16) were calculated.

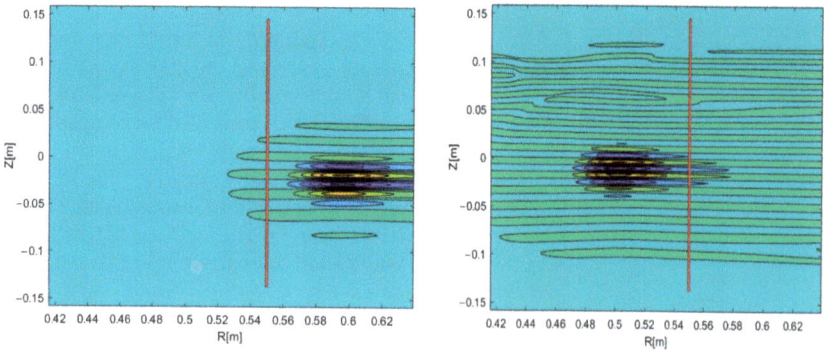

Figure 16. Two-dimensional DBS signal map for the circular filament positioned at the 48 GHz probing frequency: (**left**) 39 GHz, (**right**) 55 GHz probing frequency. The red line indicates the trajectory.

The differences in the signals at various probing frequencies (or radii) were investigated and are presented in Figure 17. The blue line corresponds to the 39 GHz probing frequency, while the red one corresponds to 55 GHz. One may observe that the signals differed in amplitude and frequency in the spectra on the right. There was a shift of 3 ms between the two signals, with the 39 GHz signal forming earlier than the 55 GHz one. This shift cannot be explained by the motion in the radial direction, however there is data to explain this phenomenon. It is believed to be associated with the stretching and tilt of turbulence eddies [22].

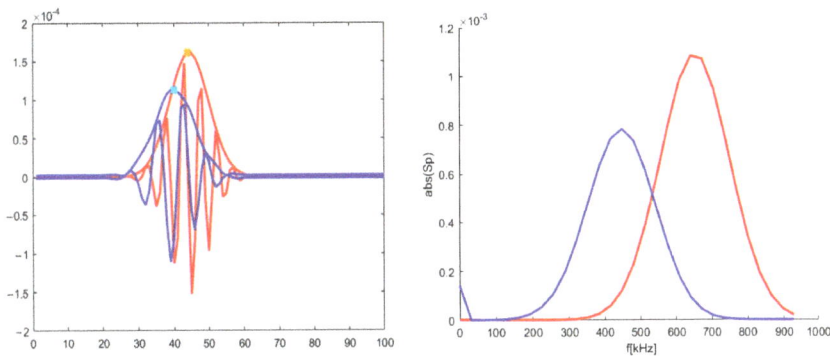

Figure 17. (**left**) DBS signals; (**right**) calculated spectra for the signal on the left. The blue lines correspond to the 39 GHz probing frequency, and the red lines correspond to 55 GHz.

These parameters were also calculated for the case of an added radial velocity of 10 km/s with the circular filament moving in the direction of the increasing radius. The 2D maps of the DBS signals for the 39 GHz (left) and 55 GHz (right) probing frequencies are presented in Figure 18.

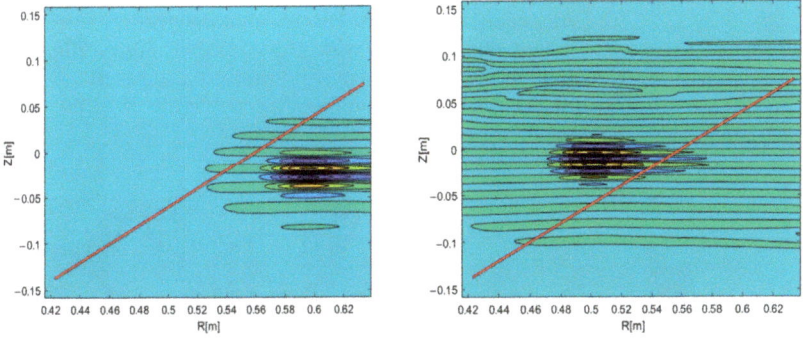

Figure 18. Two-dimensional DBS signal map for the circular filament positioned at the 48 GHz probing frequency: (**left**) 39 GHz; (**right**) 55 GHz probing frequency. The red line indicates the trajectory.

The signals for the 39 and 55 GHz frequencies were extracted from the 2D maps in Figure 18. They are presented in Figure 19. Just as in the previous scenario, the amplitude of the 39 GHz signal was smaller than the 55 GHz one. In the spectrum on the right, one can see that the value of the 39 GHz signal frequency is now around 510 kHz rather than 450 kHz as was the case in the purely poloidal motion scenario. The shift between the two signals was also different with the radial motion introduced. The 55 GHz frequency signal now forms earlier than the 39 GHz one with a 7 ms delay. This is explained by the motion of the filament from the inner to the outer plasma regions. The 7 ms value was larger than the 3 ms delay between the signals in the case of the poloidal motion, meaning that the turbulence stretching and tilt cannot be playing the only key role in the formation of the DBS signal.

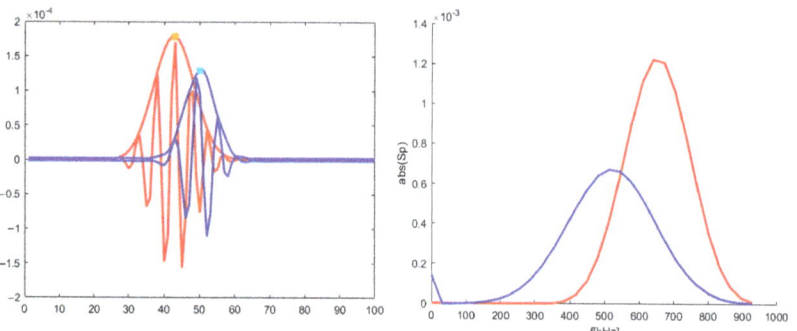

Figure 19. (**left**) DBS signals; (**right**) calculated spectra for the signal on the left. The blue lines correspond to the 39 GHz probing frequency, and the red line corresponds to 55 GHz.

4. Strip Filaments

The circular model for filaments was unable to explain several types of detected filaments. This raised questions as to what other models could be applied and investigated. For instance, there were observations of filaments radially localized in a large area with no delay between the detected filaments between DBS channels. This led to the idea of introducing filaments in the form of a strip stretching across the whole box where the modelling takes place. This ensured that the influence of the filament could be, to some extent, observed in a wide range of radii. Figure 20 presents the density perturbations with the strip model filament introduced on top of the bulk plasma density spanning the radial direction at a set poloidal coordinate. Additionally, the electric field distribution is presented.

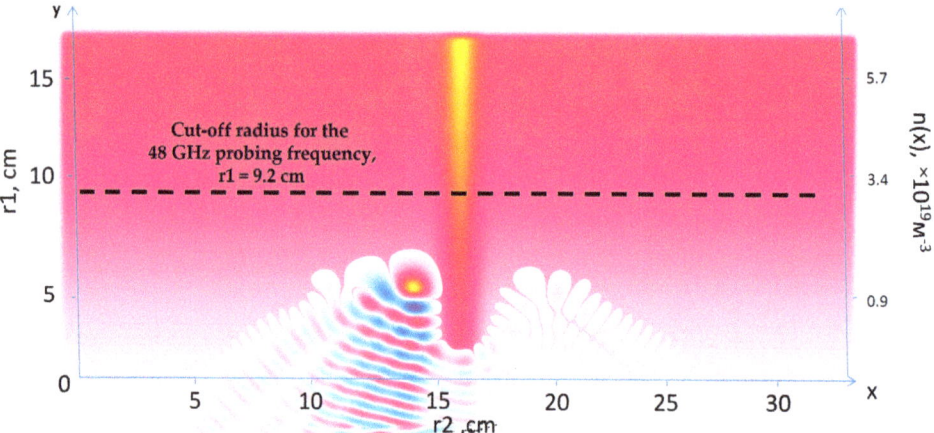

Figure 20. Density perturbations with an introduced strip filament (yellow vertical line). The electric field distribution is also presented.

DBS signals were calculated for the conditions, and the parameters are presented in Table 4. An example of the obtained signals is presented in Figure 21. Two cases were analyzed with the red line representing the linear one (1% filament amplitude), and the blue line—the non-linear case (100% filament amplitude). One may observe that there was a delay between the development of the filament in the non-linear regime in comparison to the linear one.

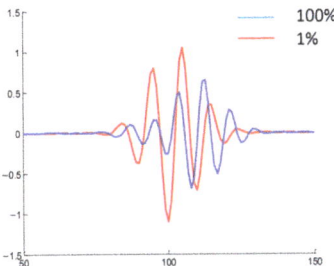

Figure 21. The DBS signal for the strip filament. Red line corresponds to the filament with an amplitude of 1% of density at cutoff of the probing wave 48 GHz where the filament was positioned, and the blue line corresponds to an amplitude of 100%.

This model was also compared to the filaments observed in Globus-M, and the results are presented in Figure 22. The calculated signal for the filament in the form of a strip with a width of 1.5 cm and amplitude of 1% is shown in red, while the black lines are the experimental data collected using DBS. The main issue with this model was the fact that the modelled signal was shorter than the signal in the experiment. Additionally, in the experiment, the frequency of the signal changed for different probing frequencies, while that did not take place in the calculations.

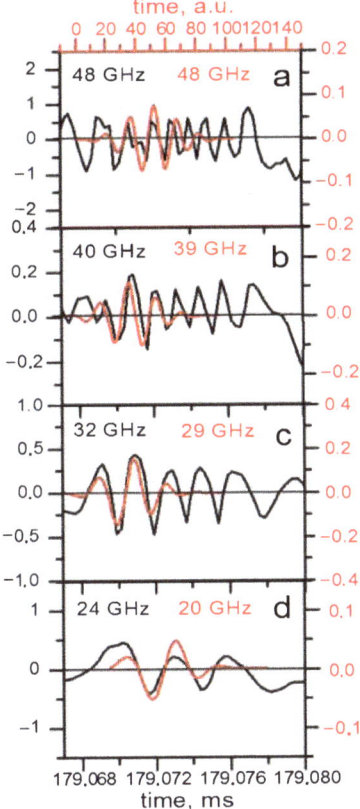

Figure 22. Experimental DBS signals (black lines) and calculated DBS signals for a strip filament model (red lines) for different probing frequencies.

5. Stretched Filaments

Experiments on various tokamaks have highlighted that filaments of other shapes can form in plasma [18]. The gathered data showed that filaments can stretch in both the radial and poloidal directions. This has caused interest, as some filaments observed in Globus-M2 could not be explained by the circular and strip filament models. The filament density used for the performed simulations is presented in Figure 23. For comparison, the circular filament is shown in the middle, and the stretched filaments are shown on either side of it. For a given size of the filament, it was stretched with proportions of 1:2 or 1:4.

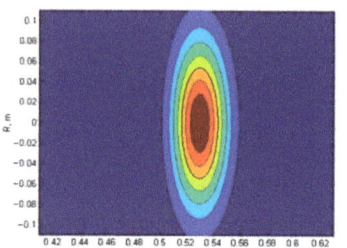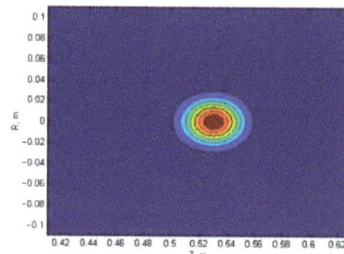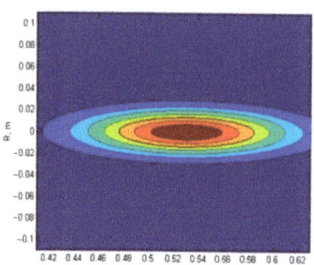

Figure 23. Filament density distribution: (**left**) radially stretched, (**middle**) circular, and (**right**) poloidally stretched filament.

5.1. Radially Stretched Filaments

The radially stretched filament with proportions of 1:2 was compared to the circular one. Differently sized filaments and amplitudes (linear and non-linear regimes) were investigated. The filaments were positioned at the cut-off radius of the 48 GHz probing frequency, and the signals for the 48 GHz frequency were obtained. The filament moved poloidally at a set velocity of 10 km/s. Figure 24 depicts the signals and amplitude for various conditions that were compared. The signal amplitude for the stretched filament was always greater. In the linear case (1% amplitude) there was no delay between the signals of the circular and stretched filament, while in the non-linear case (50% amplitude) a time offset was clearly visible, where the stretched filament signal developed later. The blue dashed vertical line can be used to observe this phenomenon.

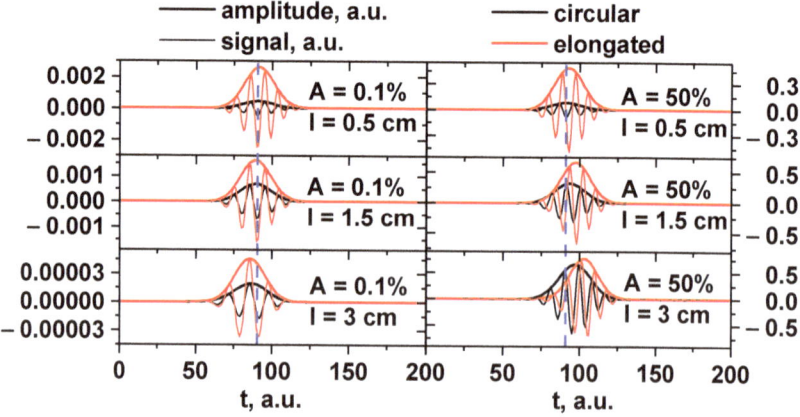

Figure 24. DBS signals (bold lines) and amplitude (regular lines) for the circular filament (black lines) and radially stretched filament (red lines): (**left**) filament amplitude of 0.1% of density at cutoff of the probing wave 48 GHz where the filament was positioned; (**right**) filament amplitude of 50%. Different filament diameters: (**top**) 0.5 cm, (**middle**) 1.5 cm, and (**bottom**) 3 cm.

The change in signal frequency was also of interest, so the spectra for the signals presented in Figure 24 were calculated. In Figure 25, the spectra of the filament with a 0.5 cm diameter for the linear (left) and non-linear (right) cases are presented. The circular filament frequency (black line) did not change with the transition from the linear to the non-linear regime, but the stretched filament frequency was always slightly larger and grew even more for the 50% amplitude.

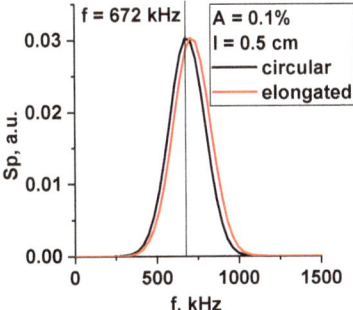 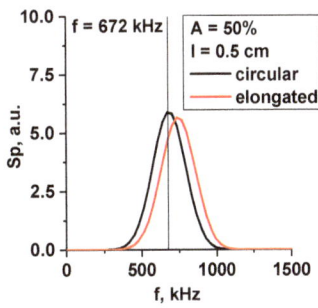

Figure 25. Spectrum of DBS signals for the 0.5 cm filament: (**left**) filament amplitude of 0.1% of density at cutoff of the probing wave 48 GHz where the filament was positioned; (**right**) amplitude of 50%. The black lines correspond to the circular filament, and the red lines correspond to the radially stretched filament.

The largest filament with a 3 cm diameter is presented in Figure 26. The frequency in the linear case (0.1% amplitude) was smaller, with it being the same for the circular and stretched filaments. For the 50% amplitude case, the frequency was greater, with a value of 704 kHz, and its values were closer for the circular and stretched filaments. This was similar to the 1.5 cm filament.

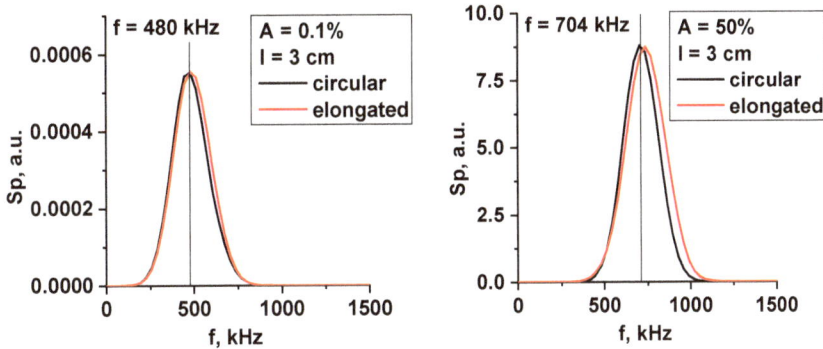

Figure 26. Spectrum of the DBS signals for the 3 cm filament: (**left**) filament amplitude of 0.1% of density at cutoff of the probing wave 48 GHz where the filament was positioned; (**right**) filament amplitude of 50%. The black lines correspond to the circular filament, and the red lines correspond to the radially stretched filament.

Signals for several probing frequencies were obtained and are presented in Figure 27. The filament was positioned at the cut-off radius for 48 GHz, and the signals for frequencies 48, 39, 29, and 20 GHz (DBS system in Globus-M/M2) were investigated. The filament moved poloidally at a set velocity of 10 km/s. The circular (black line), slightly stretched with 1:2 proportions (red line), and very stretched with 1:4 proportions (green line) filaments are presented. The diameter of the filament was 1.5 cm, and the amplitude was 1% of the density at cutoff of the probing wave (the linear case). The signals showed that, depending on the level of stretching of the filament, the number of signals that reacted to it changed. The more stretched the filament, the more channels formed peaks. This was similar to the strip model, where the signals on all channels were formed. The radially stretched filament could potentially be used to explain the experimentally observed filaments found on only two or three DBS signals.

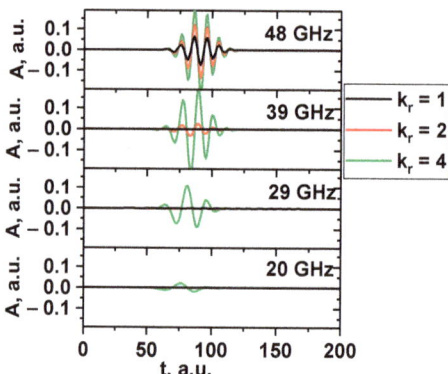

Figure 27. DBS signals for several probing frequencies. The circular (black line), slightly stretched with 1:2 proportions (red line), and very stretched with 1:4 proportions (green line) filaments are presented.

When analyzing the dependency of the signal amplitude, frequency, and velocity on the probing frequency in the linear case (0.1% amplitude) presented in Figure 28, the conclusion was that the radially stretched filament had the same dependencies, and even values, as the circular filament of the same size. The main difference was the signal amplitude for the 3 cm filament (blue line on the left). The amplitude values implied that it could potentially be detected by a lower range of probing frequencies of 20–45 GHz. The velocity on the right was very similar to the circular case, with the stretched filament with a diameter of 0.5 cm having the closest values to the set 10 km/s (olive-green line), and the others being lower in value.

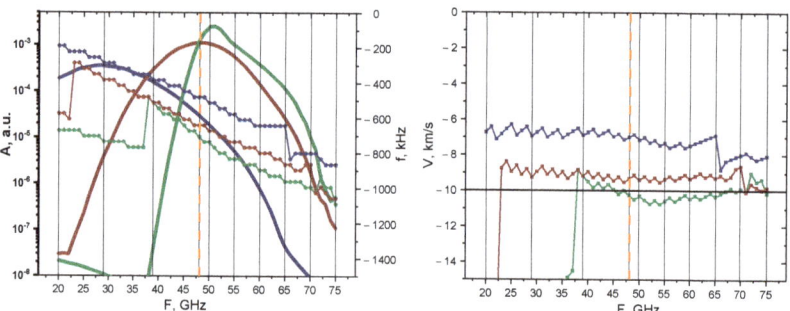

Figure 28. For the radially stretched filament with an amplitude of 0.1%: (**left**) dependency of the DBS signal amplitude (bold lines with scale on the left) and frequency (lines with pentagon shapes with scale on the right) on the probing frequency; (**right**) dependency of the calculated velocity on the probing frequency. The navy-blue lines are the circular filament with a 3 cm diameter, the wine-red are that of 1.5 cm, and the olive-green are that of 0.5 cm. The black vertical lines indicate the probing frequencies available in Globus-M2, and the vertical orange dashed line indicates the position of the filament at cutoff of the 48 GHz probing wave. The horizontal line marks the set 10 km/s filament velocity.

This remains true for the non-linear case with a 50% amplitude of the filament presented in Figure 29. The values of both the signal amplitude, frequency, and velocity were very similar in behavior for all the filament sizes (also observable for the circular filaments). The velocity vales for all the sizes were slightly larger than the expected 10 km/s after the 40 GHz frequency.

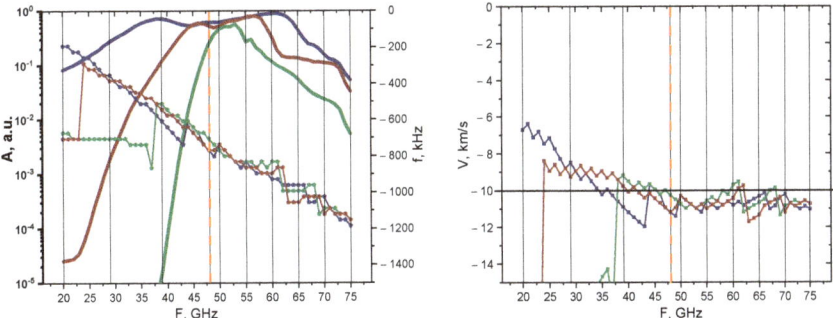

Figure 29. For the radially stretched filament with an amplitude of 50%: (**left**) dependency of the DBS signal amplitude (bold lines with scale on the left) and frequency (lines with pentagon shapes with scale on the right) on the probing frequency; (**right**) dependency of the calculated velocity on the probing frequency. The navy-blue lines are the circular filament with a 3 cm diameter, the wine-red are that of 1.5 cm, and the olive-green are that of 0.5 cm. The black vertical lines indicate the probing frequencies available in Globus-M2, and the vertical orange dashed line indicates the position of the filament at cutoff of the 48 GHz probing wave. The horizontal line marks the set 10 km/s filament velocity.

5.2. Poloidally Stretched Filaments

The poloidally stretched filaments with proportions of 1:2 and 1:4 were compared to the circular one. The DBS signals were calculated for the differently sized filaments and regimes. The filament was positioned at the cut-off radius for the 48 GHz probing frequency, and the signals for this frequency were obtained. It was set to move poloidally at a given velocity of 10 km/s. Figure 30 depicts the obtained signals. The signal amplitude for the poloidally stretched filament (red and green lines) decreased the more it was stretched compared to the circular filament (black line). It can also be said that while in the linear case (top) there was no delay between the signals, in the non-linear case (bottom) a time off-set was clearly visible where the stretched filament signal developed several ms later.

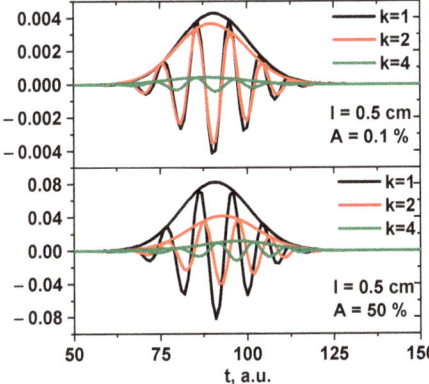

Figure 30. DBS signals for circular (black line), slightly poloidally stretched with 1:2 proportions (red line), and very poloidally stretched with 1:4 proportions (green line) filaments of 0.5 cm diameter: (**top**) amplitude of 0.1% of density at cutoff of the probing wave 48 GHz where the filament was positioned; (**bottom**) amplitude of 50%.

The behavior of the signal amplitude and Doppler frequency is presented in Figure 31. The poloidally stretched filament exhibited behavior that differed from the circular and radially stretched filaments of the same size. Judging by the calculated values of the amplitude of the 3 cm (bold blue line) and 1.5 cm (bold red line) filaments, one can come to the conclusion that no probing frequency signal would be able to detect them; however, for the 45–60 GHz frequencies, it was possible to observe the 0.5 cm poloidally stretched filament (bold green line). The frequencies and velocities were also calculated. Only the values for the 0.5 cm filament were close to the 10 km/s value after the 42 GHz probing frequency, with the 1.5 cm and 3 cm filaments having velocities significantly lower than the expected value.

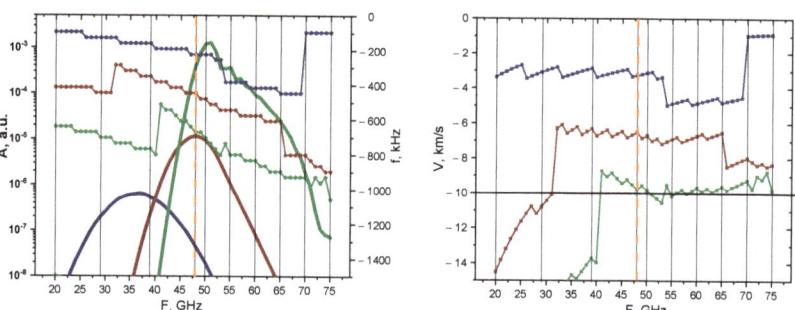

Figure 31. For the poloidally stretched filament with an amplitude of 0.1%: (**left**) dependency of the DBS signal amplitude (bold lines with scale on the left) and frequency (lines with pentagon shapes with scale on the right) on the probing frequency; (**right**) dependency of the calculated velocity on the probing frequency. The navy-blue lines are the circular filament with a 3 cm diameter, the wine-red are that of 1.5 cm, and the olive-green are that of 0.5 cm. The black vertical lines indicate the probing frequencies available in Globus-M2, and the vertical orange dashed line indicates the position of the filament at cutoff of the 48 GHz probing wave. The horizontal line marks the set 10 km/s filament velocity.

The non-linear case yielded different results, which are presented in Figure 32. The signal amplitude values were similar for all the probing frequencies, with the 45–70 GHz channels being able to detect the filament positioned at the 48 GHz cut-off radius. For the smaller probing frequencies, the Doppler shift values for the largest 3 cm filament differed from the 1.5 cm and 0.5 cm ones, which behaved similarly throughout. For probing frequencies below 48 GHz, the velocity had values that differed greatly from the anticipated 10 km/s; however, for channels deeper than 48 GHz, the values remained close to 10 km/s.

The transition from linear (lower than 50%) to non-linear (higher than 50%) was investigated for the poloidally stretched filament, as the DBS signals exhibited interesting characteristics that differed from all the previous cases. Figure 33 depicts the performed calculations and analysis for the 48 GHz probing frequency signal where the filament was positioned. The top figure shows the signal amplitude behavior, while the bottom one shows the signal frequency. The colors of the lines correspond to different filament sizes (blue line—3 cm diameter, red line—1.5 cm, and green line—0.5 cm), and the different symbols represent the degree of stretching (squares—circular filament, pentagon—stretched filament with proportions of 1:2, and stars—stretched filament with proportions of 1:4). In the linear case, the signal amplitude was larger the closer the filament size was to half the wavelength of the probing frequency near the cut-off position. This was in line with the Born approximation prediction. In the non-linear case, everything looked different. For the 1:2 poloidally stretched filament, as well as for the circular filament, the amplitude became larger the more its size increased and was similar in value, despite the poloidal size of the stretched filaments (6 cm) being much larger than half the wavelength. Only for the fourfold stretched filament did the tendency of the signal amplitude to decrease remain after its size changed from the optimal scattering size. The frequency in the linear case was inversely dependent on the size of the poloidal size of the filament; the smaller the length, the higher the frequency. Thus, the minimum frequency was achieved at the maximum size of the filament. This, together with the amplitude behavior of the signal, can be explained by the finite wavenumber resolution of the DBS diagnostics. So, in the presence of plasma fluctuations, the velocity determined using DBS will be significantly underestimated in relation to its true value only for filaments of large scale. In the non-linear case, there is a tendency to align all frequencies close to the frequencies that can be obtained in the linear case where the filament is appropriate in size. However, for very high elongation (1:4), the frequency remained too small even for very large amplitudes of filament. In addition, for the large circular filament, the frequency was generally overestimated.

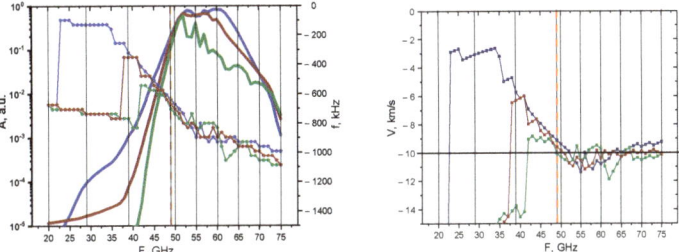

Figure 32. For the poloidally stretched filament with an amplitude of 50%: (**left**) dependency of the DBS signal amplitude (bold lines with scale on the left) and frequency (lines with pentagon shapes with scale on the right) on the probing frequency; (**right**) dependency of the calculated velocity on the probing frequency. The navy-blue lines are the circular filament with a 3 cm diameter, the wine-red are that of 1.5 cm, and the olive-green are that of 0.5 cm. The black vertical lines indicate the probing frequencies available in Globus-M2, and the vertical orange dashed line indicates the position of the filament at cutoff of the 48 GHz probing wave. The horizontal line marks the set 10 km/s filament velocity.

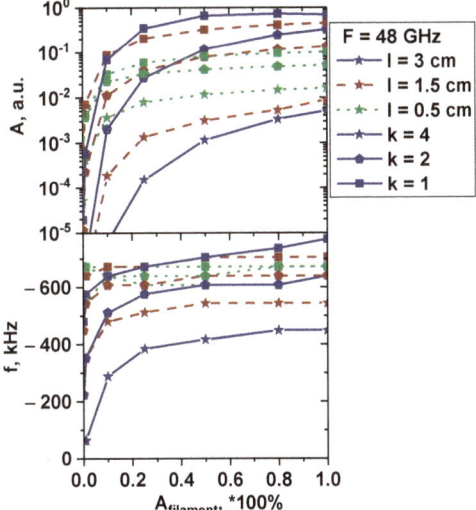

Figure 33. The dependency of the (**top**) signal amplitude and (**bottom**) signal frequency on filament amplitude. The blue lines correspond to 3 cm diameter filaments, the red lines correspond to 1.5 cm, and the green lines correspond to 0.5 cm. The lines with the squares correspond to the circular filament, the lines with pentagon shapes—poloidally stretched filament with proportions of 1:2, and the lines with stars—poloidally stretched filament with proportions of 1:4.

6. Low-Density Profile

This section is devoted to the analysis of DBS signals for the conditions described in previous sections, but for a different electron density profile with density values lower than the one analyzed earlier. The experimental plasma density profile used for the computations is presented in Figure 34 and has values reaching $n_e = 4.4 \times 10^{-19}$ m^{-3} at y = 0.15 m (shot no. 36569 at Globus-M).

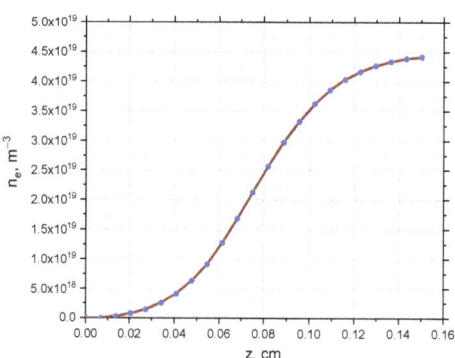

Figure 34. Experimental electron density profile for Globus-M2.

The DBS signals were analyzed to see which channels of the reflectometer would be able to detect the forming filament. The frequency range remained the same, which led to the cut-off being shifted to deeper areas. It turned out that in both the linear (left in Figure 35) and non-linear cases (right in Figure 35), the filaments were visible for a smaller range of probing frequencies. This was indicated by the fact that the solid line profiles (low density profile) were more narrow than the dashed lines for all the filament sizes and forms. In the linear case, this was due to the cut-off positions shifting further away from the filament position compared to the high-density case. For the non-linear case, the observations can be explained by the same effect. All the characteristics of the filament discussed earlier were investigated, and no changes were revealed except for the localization shown.

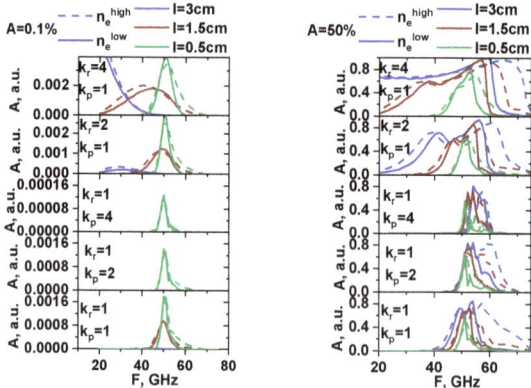

Figure 35. The dependency of the signal amplitude on different probing frequencies for the high-density profile from Figure 1 (dashed lines) and low-density profile from Figure 34 (regular lines): (**left**) filaments with an amplitude of 0.1% of density at cutoff of the probing wave 48 GHz where the filament was positioned; (**right**) filament with an amplitude of 50%. The blue lines correspond to 3 cm diameter filaments, the red lines correspond to 1.5 cm, and the green lines correspond to 0.5 cm. Different proportions of stretching of the filaments are presented in the figure.

7. Real Globus-M2 Geometry

Apart from the modelling performed in slab geometry, which was presented and discussed in detail in all the previous sections, the real Globus-M/M2 geometry was investigated. The question of its effects on the observable phenomena and DBS signals was raised, which is why additional simulations were performed to compare the two scenarios. The circular filament motion was along the magnetic field lines of the Globus-M/M2 tokamak at a constant velocity in the poloidal direction. To obtain this information, the experimental data of magnetic equilibrium (EFIT) and electron density (Thomson diagnostic) were used. A circular filament with a diameter of 3 cm and amplitudes of 1% (linear case) and 50% (non-linear case) was modelled. The filament was positioned at the cut-off of the 20 GHz probing frequency where the filament was generally observed.

The obtained DBS signals are presented in Figure 36. They were obtained for the 20 GHz frequency (left) where the filament was positioned and 48 GHz (right). For better comparison, the maximum signal amplitude was normalized. The signal for the filament with the higher amplitude of 50% developed earlier than the signal for the 1% filament. This was also the case for slab geometry simulations.

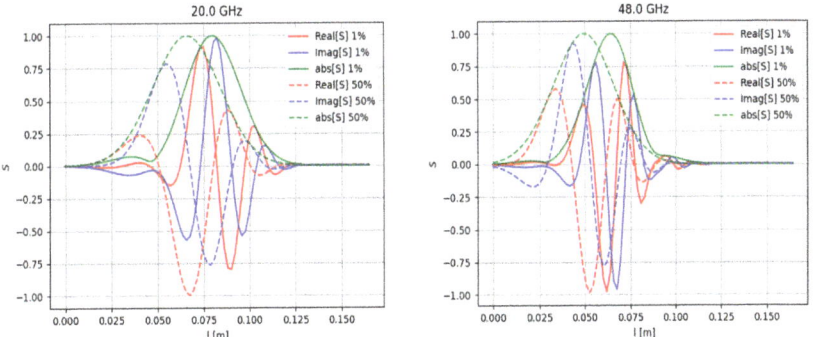

Figure 36. DBS signals calculated for the real Globus-M geometry (**left**) for the 20 GHz probing frequency and (**right**) for 48 GHz frequency. The regular lines correspond to filaments with amplitudes of 1% of density at cutoff of the probing wave, and the dashed lines correspond to filaments with amplitudes of 50% of density at cutoff of the probing wave.

8. Conclusions

Two-dimensional full-wave simulations of backscattering off filaments were performed using the code IPF-FD3D to understand what kind of information about filaments can be extrapolated from signals of Doppler backscattering diagnostics. The DBS systems installed in Globus-M2 were analyzed, and different real experimental scenarios were examined. Computations were performed for a range of probing frequencies from 20 to 75 GHz of electromagnetic waves in O-mode

A circular filament was investigated in slab geometry. Different values of size, amplitude, position, and probing frequencies were analyzed. It was found that after a certain critical size of the circular filament was reached, a significant delay in the formation of the filament between the low and high amplitude case was seen. In addition, the signal frequency significantly increased after the transition to the non-linear regime. In the linear case, DBS was only able to detect the filaments using probing frequencies of 40–55 GHz, and the velocity values were accurate for 35–75 GHz frequencies for small filaments. For the non-linear case, the results indicated that the filaments were observable using a wider range of probing frequencies of 45–65 GHz, and the velocity for all the filament sizes was above the set value for deeper channels. In the linear and non-linear case, the signal frequency coincided with the frequency predicted by the formula for the Doppler shift in the Born approximation when the filament was positioned in the cut off of the given frequency, but it was lower as the position of the filament changed. The delay between different signals can be explained by the stretching and tilt of turbulence eddies or the introduction of the radial velocity.

Filaments in the form of a strip stretching across the whole box were modelled. There was a delay in the signals between the development of the filament in the non-linear regime in comparison to the linear one.

A radially stretched filament was investigated. In the linear case, there was no delay between the signals of the circular and stretched filaments, while in the non-linear case, a time offset was present, as the stretched filament signal developed later. It was found that the more stretched the filament, the more channels formed peaks. This was similar to the strip model. The radially stretched filament could potentially be used to explain the experimentally observed filaments found on only two or three DBS signals. In the linear and non-linear cases, the radially stretched filament had similar dependencies and values as the circular filament of the same size. The filament could potentially be detected by a lower range of probing frequencies of 20–45 GHz.

A poloidally stretched filament was investigated. In the linear case, there was no delay between the signals of the stretched and circular filaments, but in the non-linear case, a time off-set was presented, with the stretched filament signal developing several ms later. No probing frequency signal was able to detect large filaments; however, for the 45–60 GHz frequencies, it was possible to observe small filaments. Only the values for the small filament were close to the set value after the 42 GHz probing frequency, but others had significantly lower velocities. In the non-linear case, the 45–70 GHz channels were able to detect the filament positioned at the 48 GHz cut-off radius. For probing frequencies below 48 GHz, the velocity had values that differed greatly from the anticipated one; however, for channels deeper than 48 GHz, the values remained close to the set value.

For a low-density profile, in the linear and non-linear cases, the filaments were visible for a smaller range of probing frequencies. This was due to the cut-off positions shifting further away from the filament position compared to the high-density case.

The real Globus-M/M2 geometry was investigated. A circular filament moving along the magnetic lines in the poloidal direction was modelled. The signals in this case did not differ significantly from what was obtained in the slab geometry.

The next step in the study and modeling of the behavior of filaments in plasma includes more detailed calculations in the real geometry of the Globus-M2 tokamak to investigate the effect of magnetic field curvature on the resulting signals in the case of more complex filament forms and trajectories. Apart from that, when modelling filaments, background plasma turbulence will be introduced using the codes GENE and GKW. The effect of the tilting of filaments due to the presence of a shear will be studied. In addition, the results will be compared with DBS data collected during a specially designed experiment. Machine learning technologies will be applied to develop the process of recognizing filaments in experimental DBS signals.

Author Contributions: Conceptualization, N.T. and A.Y.; methodology, G.Z.; software, G.Z.; validation, N.T.; formal analysis, A.Y.; investigation, A.Y.; resources, G.Z. and N.T.; data curation, A.Y.; writing—original draft preparation, A.P.; writing—review and editing, A.Y., N.T. and G.Z.; visualization, A.P.; supervision, A.Y.; project administration, A.Y.; funding acquisition, A.Y. All authors have read and agreed to the published version of the manuscript.

Funding: The work is supported by the Russian Science Foundation (Project No. 18-72-10028).

Institutional Review Board Statement: Not applicable.

Informed Consent Statement: Not applicable.

Data Availability Statement: Not applicable.

Acknowledgments: The work is supported by the Russian Science Foundation (Project No. 18-72-10028). The results of the modeling were obtained using computational resources of Peter the Great Saint-Petersburg Polytechnic University Supercomputing Center (www.scc.spbstu.ru, accessed on 1 April 2022). We express our deepest gratitude to Carsten Lechte, the creator of the IPF-FD3D code, for his detailed consulting and assistance in mastering the program.

Conflicts of Interest: The authors declare no conflict of interest.

References

1. Aymar, R. ITER status, design and material objectives. *J. Nucl. Mater.* **2002**, *307–311*, 1–9. [CrossRef]
2. Linke, J.M.; Hirai, T.; Rödig, M.; Singheiser, L.A. Performance of Plasma-Facing Materials Under Intense Thermal Loads in Tokamaks and Stellarators. *Fusion Sci. Technol.* **2004**, *46*, 142–151. [CrossRef]
3. D'Ippolito, D.A.; Myra, J.R.; Zweben, S.J. Convective transport by intermittent blob-filaments: Comparison of theory and experiment. *Phys. Plasmas* **2011**, *18*, 060501. [CrossRef]
4. Fuchert, G.; Birkenmeier, G.; Carralero, D.; Lunt, T.; Manz, P.; Müller, H.W.; Nold, B.; Ramisch, M.; Rohde, V.; Stroth, U.; et al. Blob properties in L- and H-mode from gas-puff imaging in ASDEX upgrade. *Plasma Phys. Control. Fusion* **2014**, *56*, 125001. [CrossRef]
5. Yun, G.S.; Lee, W.; Choi, M.J.; Lee, J.; Park, H.K.; Tobias, B.; Domier, C.W.; Luhmann, N.C., Jr.; Donné, A.J.H.; Team, K. Two-Dimensional Visualization of Growth and Burst of the Edge-Localized Filaments in KSTAR H-Mode Plasmas. *Phys. Rev. Lett.* **2011**, *107*, 045004. [CrossRef]
6. Spolaore, M.; Kovařík, K.; Stöckel, J.; Adamek, J.; Dejarnac, R.; Duran, I.; Komm, M.; Markovic, T.; Martines, E.; Panek, R.; et al. Electromagnetic ELM and inter-ELM filaments detected in the COMPASS Scrape-Off Layer. *Nucl. Mater. Energy* **2017**, *12*, 844. [CrossRef]
7. Zoletnik, S.; Anda, G.; Biedermann, C.; Carralero, A.D.; Cseh, G.; Dunai, D.; Killer, C.; Kocsis, G.; Krämer-Flecken, A.; Otte, M.; et al. Multi-diagnostic analysis of plasma filaments in the island divertor. *Plasma Phys. Control. Fusion* **2020**, *62*, 014017. [CrossRef]
8. Conway, G.D.; Schirmer, J.; Klenge, S.; Suttrop, W.; Holzhauer, E.; ASDEX Upgrade Team. Plasma rotation profile measurements using Doppler reflectometry. *Plasma Phys. Control. Fusion* **2004**, *46*, 951. [CrossRef]
9. Bulanin, V.V.; Varfolomeev, V.I.; Gusev, V.K.; Ivanov, A.E.; Krikunov, S.V.; Kurskiev, G.S.; Larionov, M.M.; Minaev, V.B.; Patrov, M.I.; Petrov, A.V.; et al. Observation of filaments on the Globus-M tokamak by Doppler reflectometry. *Tech. Phys. Lett.* **2011**, *37*, 340–343. [CrossRef]
10. Bulanin, V.V.; Gusev, V.K.; Khromov, N.A.; Kurskiev, G.S.; Minaev, V.B.; Patrov, M.I.; Petrov, A.V.; Petrov, M.A.; Petrov, Y.V.; Prisiazhniuk, D.; et al. The study of filaments by the Doppler backscattering method in the 'Globus-M' tokamak. *Nucl. Fusion* **2019**, *59*, 096026. [CrossRef]
11. Yashin, A.Y.; Bulanin, V.V.; Petrov, A.V.; Gusev, V.K.; Kurskiev, G.S.; Minaev, V.B.; Patrov, M.I.; Petrov, Y.V. Recent Doppler backscattering applications in Globus-M tokamak. *JINST* **2019**, *14*, C10025. [CrossRef]
12. Bulanin, V.V.; Gusev, V.K.; Kurskiev, G.S.; Minaev, V.B.; Patrov, M.I.; Petrov, A.V.; Petrov, Y.V.; Prisyazhnyuk, D.V.; Sakharov, N.V.; Solokha, V.V.; et al. The Effect of Low-Frequency Magnetohydrodynamic Modes on the Development of Filaments in the Globus-M Tokamak. *Tech. Phys. Lett.* **2019**, *45*, 977–980. [CrossRef]
13. Yashin, A.; Bulanin, V.; Petrov, A.; Ponomarenko, A. Review of Advanced Implementation of Doppler Backscattering Method in Globus-M. *Appl. Sci.* **2021**, *11*, 8975. [CrossRef]
14. Trier, E.; Hennequin, P.; Pinzón, J.R.; Hoelzl, M.; Conway, G.D.; Happel, T.; Harrer, G.F.; Mink, F.; Orain, F.; Wolfrum, E.; et al. Studying ELM filaments with Doppler reflectometry in ASDEX Upgrade. In Proceedings of the 45th EPS Conference on Plasma Physics, Prague, Czech Republic, 2–6 July 2018; Available online: http://ocs.ciemat.es/EPS2018PAP/pdf/P1.1023.pdf (accessed on 16 July 2018).
15. Bulanin, V.V.; Gusakov, E.Z.; Gusev, V.K.; Zadvitskiy, G.; Lechte, C.; Heuraux, S.; Minaev, V.B.; Petrov, A.V.; Petrov, Y.V.; Sakharov, N.V.; et al. Full-Wave Modeling of Doppler Backscattering from Filaments. *Plasma Phys. Rep.* **2020**, *46*, 490–495. [CrossRef]
16. Petrov, Y.V.; Gusev, V.K.; Sakharov, N.V.; Minaev, V.B.; Varfolomeev, V.I.; Dyachenko, V.V.; Balachenkov, I.M.; Bakharev, N.N.; Bondarchuk, E.N.; Bulanin, V.V.; et al. Overview of GLOBUS-M2 spherical tokamak results at the enhanced values of magnetic field and plasma current. *Nucl. Fusion* **2022**, *62*, 042009. [CrossRef]
17. Lechte, C. Investigation of the Scattering Efficiency in Doppler Reflectometry by Two-Dimensional Full-Wave Simulations. *IEEE Trans. Plasma Sci.* **2009**, *37*, 1099–1103. [CrossRef]
18. Kirk, A.; Counsell, G.F.; Cunningham, G.; Dowling, J.; Dunstan, M.; Meyer, H.; Price, M.; Saarelma, S.; Scannell, R.; Walsh, M.; et al. Evolution of the pedestal on MAST and the implications for ELM power loadings. *Plasma Phys. Control. Fusion* **2007**, *49*, 1259. [CrossRef]
19. Ben Ayed, N.; Kirk, A.; Dudson, B.; Tallents, S.; Vann, R.G.L.; Wilson, H.R.; the MAST team. Inter-ELM filaments and turbulent transport in the Mega-Amp Spherical Tokamak. *Plasma Phys. Control. Fusion* **2009**, *51*, 035016. [CrossRef]
20. Bulanin, V.V.; Yashin, A.Y.; Petrov, A.V.; Gusev, V.K.; Minaev, V.B.; Patrov, M.I.; Petrov, Y.V.; Prisiazhniuk, D.V.; Varfolomeev, V.I. Doppler backscattering diagnostic with dual homodyne detection on the Globus-M compact spherical tokamak. *Rev. Sci. Instrum.* **2021**, *92*, 033539. [CrossRef]

21. Yashin, A.Y.; Bulanin, V.V.; Gusev, V.K.; Minaev, V.B.; Petrov, A.V.; Petrov, Y.V.; Ponomarenko, A.M.; Varfolomeev, V.I. Doppler backscattering systems on the Globus-M2 tokamak. *JINST* **2022**, *17*, C01023. [CrossRef]
22. Pinzón, J.R.; Estrada, T.; Happel, T.; Hennequin, P.; Blanco, E.; Stroth, U.; the ASDEX Upgrade and TJ-II Teams. Measurement of the tilt angle of turbulent structures in magnetically confined plasmas using Doppler reflectometry. *Plasma Phys. Control. Fusion* **2019**, *61*, 105009. [CrossRef]

Article

VUV to IR Emission Spectroscopy and Interferometry Diagnostics for the European Shock Tube for High-Enthalpy Research

Ricardo Grosso Ferreira [1], Bernardo Brotas Carvalho [1], Luís Lemos Alves [1], Bruno Gonçalves [1], Victor Fernandez Villace [2], Lionel Marraffa [1,2,†] and Mário Lino da Silva [1,*]

[1] Instituto de Plasmas e Fusão Nuclear, Instituto Superior Técnico, Universidade de Lisboa, 1049-001 Lisbon, Portugal; ricardojoaogmferreira@tecnico.ulisboa.pt (R.G.F.); bernardo.carvalho@tecnico.ulisboa.pt (B.B.C.); llalves@tecnico.ulisboa.pt (L.L.A.); bruno@ipfn.tecnico.ulisboa.pt (B.G.); lionel.marraffa@tecnico.ulisboa.pt (L.M.)
[2] European Space Agency—European Space Research and Technology Centre, 2201 AZ Noordwijk, The Netherlands; victor.fernandez.villace@esa.int
* Correspondence: mlinodasilva@tecnico.ulisboa.pt
† Retired.

Abstract: The European Shock Tube for High-Enthalpy Research is a new state-of-the-art facility, tailored for the reproduction of spacecraft planetary entries in support of future European exploration missions, developed by an international consortium led by Instituto de Plasmas e Fusão Nuclear and funded by the European Space Agency. Deployed state-of-the-art diagnostics include vacuum-ultraviolet to ultraviolet, visible, and mid-infrared optical spectroscopy setups, and a microwave interferometry setup. This work examines the specifications and requirements for high-speed flow measurements, and discusses the design choices for the main diagnostics. The spectroscopy setup covers a spectral window between 120 and 5000 nm, and the microwave interferometer can measure electron densities up to 1.5×10^{20} electrons/m^3. The main design drivers and technological choices derived from the requirements are discussed in detail herein.

Keywords: atmospheric entry; shock tube; streak camera; vacuum ultraviolet; visible; infrared; microwave interferometry

1. Introduction

Entry, descent and landing (EDL) is one of the most challenging mission phases for planetary exploration/Earth return spacecrafts, as one needs to ensure the safe and appropriate deceleration of a spacecraft until the soft landing at ground level.

Besides all the specific technological challenges related to the late stages of deceleration (descent and landing), with the definition of appropriate glide systems, parachutes, and landing systems (retropropulsion, inflatables or crushable structures), the atmospheric entry phase remains one of the most complex phases to be tackled. This flight phase occurs at extreme hypersonic speeds with strong deceleration and high heating rates, owing to the so-called *atmospheric entry plasmas*, which are created downstream of strong, detached shock waves typical of hypersonic flow regimes.

These shock waves convert the coherent energy of the flow into thermal agitation energy, impulsively heating it and triggering the internal excitation, dissociation and ionization of the flow species in severe nonequilibrium conditions, ultimately leading to the formation of the entry plasma. This nonequilibrium plasma also radiates significantly besides strongly convecting heat towards the colder spacecraft walls, triggering endothermic surface reactions, which, in general, lead to the ablation of the wall surfaces, which are protected by a thermal protection system (TPS). The precise knowledge of the physical-chemical properties for such plasmas is accordingly key to the efficient design of such TPS systems, among other aspects (aerodynamics, flight stability, blackout issues, etc.).

Ground Testing

A shock tube is a facility designed to create a high temperature gas flow during a short time interval. It is comprised of a high pressure driver section and a low pressure driven section separated by a diaphragm. At a pre-determined pressure, the diaphragm ruptures, and the pressure discontinuity creates a shock wave moving towards the low pressure side. The shock wave will then excite the driven gas, forming a plasma. A simple single-stage shock tube typically cannot generate shocks speeds in excess of 10 km/s, typical of superorbital entries. To overcome this limitation, a double stage shock tube can be deployed, wherein an intermediate section between the driver and the test section is present, acting as a compression/acceleration tube.

Shock tubes are the most faithful facilities for adequately reproducing post-shock conditions for such flows, either directly reproducing the stagnation line flow (behind a normal shock), or indirectly reproducing the other forebody flow regions (behind an oblique shock) by means of straightforward correlations [1]. Such impulsive facilities are complementary with other steady-state plasma wind tunnels, which can, in turn, reproduce the conditions near the spacecraft walls, or afterbody expansions (note that shock tubes may also be deployed as impulsive wind-tunnel facilities, see Ref. [2] for more details). These facilities may also reproduce such flow conditions more approximately, owing to their different plasma excitation mechanism (electromagnetic field), whereas the excitation mechanism in a shock tube is the exact same one as that in flight conditions (shock wave). Figure 1 schematically shows the range of applicability for the different facilities. In short, shock tubes are (among others) key facilities in the planetary entry research ecosystem of space-faring nations.

Akin to the other world's space agencies, the European Space Agency (ESA) supports fundamental and applied research on atmospheric entry plasmas to advance European planetary exploration endeavors. The earliest concerted studies on atmospheric entry flows at European level were carried out in the scope of the HERMES space shuttle development program [3], where two shock tube facilities were developed to support this endeavor: the HEG shock tube at Göttingen, which acted as an aerodynamic shock tube facility [4], and the TCM2 shock tube at Marseilles, which was devoted to fundamental studies on kinetic and radiative shock-induced processes [5]. The latter could only reach velocities of about 8–9 km/s, which was enough for reproducing a return from Earth's orbit, the Huygens mission (which successfully entered Titan at a velocity of 5.15 km/s in 2004) [6–8] or Mars exploration missions [9–11], but not enough for recreating an Earth superorbital entry (11–12 km/s) [12]. However, the renewed ambitions at the beginning of the century, namely the Mars Sample Return mission, required a higher performance facility. In consequence, a competitive tender for the development of a novel facility was launched by ESA in 2009. This competition was won by an international consortium led by the Institute for Plasmas and Nuclear Fusion (IPFN), a research unit of Instituto Superior Técnico (IST) [13,14]. The European Shock Tube for High-Enthalpy Research (ESTHER) was developed in the scope of this contract and is in its final commissioning phase [15].

Such a higher performance shock tube facility requires state-of-the-art instrumentation. Atmospheric entry plasmas are very energetic, and thus radiate very strongly in a broad range of wavelengths. Spectroscopy is therefore a key diagnostic for probing such plasmas, also taking into account that these diagnostics are non-intrusive in nature, hence not disturbing the flow. Measurements of time-dependent emission/absorption in specific wavelength ranges allow probing for specific atomic and molecular quantum transitions, and may indirectly provide information on the time evolution of species concentrations (relative or absolute depending on the setup calibration), as well as flow and species temperatures, providing indication of departures from Boltzmann equilibrium. A direct measurement of the overall time-dependent radiative fluxes (in W cm^{-2}) emitted by the shock wave factors in the TPS design, which needs to withstand the overall convective and radiative fluxes of the flow during the whole entry phase. Measurements carried out in the past were mostly restricted to optical emission spectroscopy from the near-UV to near-IR range, which encompassed most of the spectral features of interest in this velocity region [16–18]. However, at higher velocities, the more prominent radiative features move towards the ultraviolet-vacuum ultraviolet (UV-VUV) region, while on the other side, Mars entries also have peculiar radiative heating characteristics with a prominence in the mid-wave infrared (MWIR) [19]. Accordingly, two companion contracts were awarded to the consortium for the development of additional optical spectroscopy setups, one in the UV-VUV and another in the MWIR region.

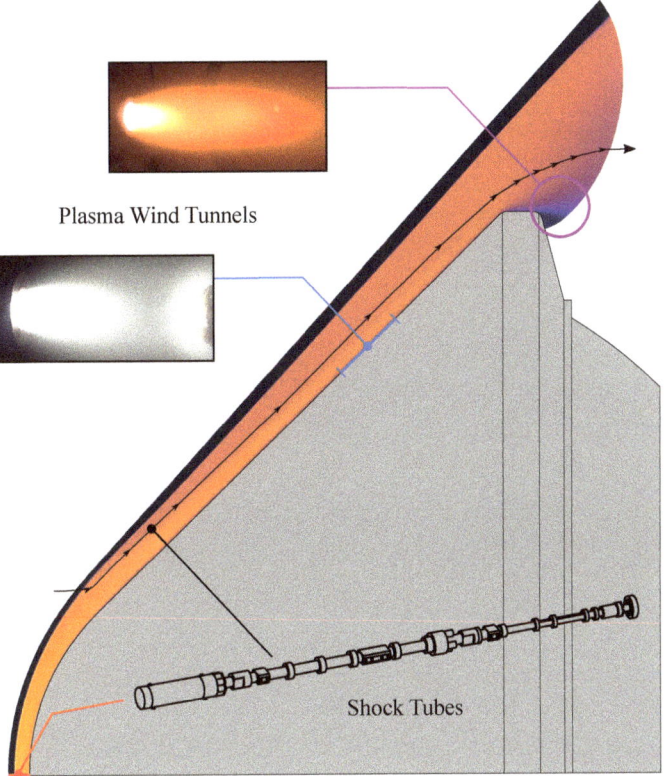

Figure 1. Schematic plot outlining the applicability range of ground test facilities for reproducing the different regions of an atmospheric entry flow. Shock tubes directly reproduce the stagnation streamline region where the heat fluxes are typically peaked (red dot), or other flow streamlines in the forebody (black dot and streamline). Free plasma plumes from wind tunnels are tailored for reproducing expansion regions (large violet circle), or surface ablation if an obstacle is placed inside the plume (blue dot and line). The plasma plume images (with and without obstacles) are taken from the SR5 arcjet plasma wind tunnel [20]. The simulated flowfield is adapted from a previous work reproducing the 1995 Jupiter entry by the Galileo probe [21].

Knowledge on the electronic densities of such plasmas is also a key parameter, as it drives the excitation of radiative states of the flowing species, via electron-impact excitation reactions. Information of heavy species excitation states is also important for the design of TPS [22], as heavy species will heat the wall through convective and radiative heat transfers, possibly endangering the spacecraft if such fluxes exceed the engineering limits [23]. The electron density may be inferred either through spectroscopic methods (Stark broadening techniques [24,25]), or through microwave interferometry techniques. Although these are, to date, not as developed as the formerly discussed optical spectroscopy diagnostics, several proofs of concept for such diagnostics have been demonstrated, and one may expect such techniques to be more vigorously deployed in the near future [26].

2. Materials and Methods

This section discusses the operational requirements of shock facilities in terms of shock speeds and ambient pressure, followed by the general requirements for optical spectroscopy measurements. Then, it discusses existing shock tube facilities for atmospheric entry studies, including our own and its expected performance. It then concludes with a more detailed discussion on the requirements for time-dependent optical spectroscopy and microwave interferometry for radiation and electron density measurements, respectively.

2.1. General Specifications and Requirements

Entry conditions are determined by the spacecraft's orbit into the planet gravitational field. These are then reproduced in key points (typically the peak heating and peak dynamic pressure points), which are obtained through semi-empirical expressions [27,28] on adequate ground test facilities, such as ESTHER. Figure 2 shows the spacecrafts entry conditions in different atmospheres and the corresponding experimental points for different shock tube facilities.

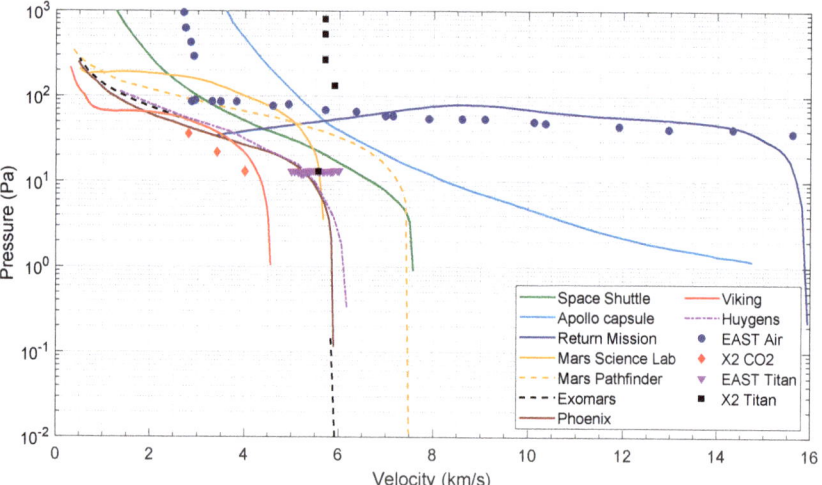

Figure 2. Entry conditions for different spacecrafts. Space Shuttle, Apollo and Return mission in Earth (N_2-O_2); Mars Science Lab, Mars Pathfinder, ExoMars, Phoenix and Viking in Mars (N_2-CO_2); and Huygens in Titan (N_2-CH_4). Adapted from [29].

Once the key trajectory points are known, one may estimate the post-shock temperatures that are reached using simple thermodynamic correlations. Namely, the total stagnation enthalpy h_{stag} of a shock wave system may be expressed as the sum of the internal enthalpy h, and a kinetic term $v^2/2$ due to the flow velocity [30]. The stagnation enthalpy of the system pre- h^0 and post-shock h^1 can be assumed to be conserved, yielding Equations (1) and (2):

$$h_{stag} = h + \frac{v^2}{2} = const. \tag{1}$$

$$h_0 + \frac{v_0^2}{2} = h_1 + \frac{v_1^2}{2} \tag{2}$$

Prior to the arrival of the shock wave, the gas is in chemical equilibrium at a low temperature, around 300 K for a shock tube experiment, where the kinetic energy term is dominant on the left-hand side of Equation (2) $v_0^2/2 \gg h_0$. After the shock, most of the kinetic energy is converted into thermal energy, which both excites the internal energy modes and heats up the gas. Thus, on the right-hand side of Equation (2), the internal energy term is dominant over the kinetic one $h_1 \gg v_1^2/2$. These approximations yield Equation (3), with c_p as the gas specific heat capacity at constant pressure:

$$\frac{v_0^2}{2} \approx h_1 = c_p T_{eq} \quad . \tag{3}$$

The gas total specific heat capacity can be split into two contributions, a constant frozen-gas term c_{p_f} (for an ideal gas), and a contribution from the internal degrees of freedom:

$$c_p = c_{p_f} + \sum_i h_i \left(\frac{\partial c_i}{\partial T} \right)_p \quad . \tag{4}$$

Inserting Equation (4) into (2), and solving for T, we obtain Equation (5a) and both its upper and lower temperature limits, T_{max} (5b) and T_{eq} (5c), respectively.

$$T = \frac{h}{c_{p_f} + \sum_i h_i \left(\frac{\partial c_i}{\partial T} \right)_p} \tag{5a}$$

$$T_{max} = \sqrt{\frac{v_0^2}{2 c_{p_f}}} \tag{5b}$$

$$T_{eq} = \frac{v_0^2}{2 c_p(T_{eq})} = \frac{h}{c_{p_f} + \sum_i h_i \left(\frac{\partial c_i}{\partial T_{eq}} \right)_p} \tag{5c}$$

The expected post-shock temperature range of the gas is shown in Figure 3. The upper limit (full line) is the so-called frozen limit, where all the chemical reactions are ignored, and thus $c_p = c_{p_f}$; the lower limit (dashed line) is the final temperature after chemical equilibrium is reached. Chemical equilibrium temperatures have been computed using our in-house aerothermodynamics code SPARK.

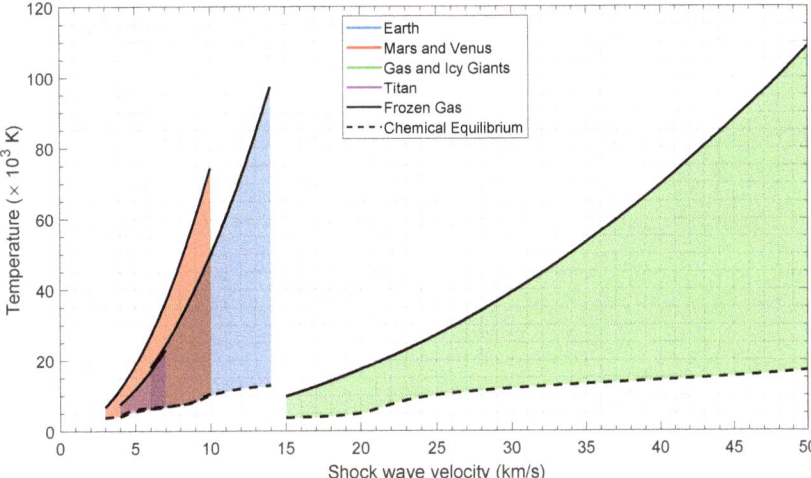

Figure 3. Temperature range for different entry conditions and atmospheres. Upper and lower limits correspond to a calorifically perfect and chemical equilibrium gas, respectively.

Figure 4 depicts the radiation wavelength distribution of a Planck blackbody at different temperatures. As the temperature increases, so does the emitted radiation power with T^4, in line with the law of Stefan–Boltzmann. In addition to this, the peak wavelength of the emitted radiation λ_{peak} moves to the shorter wavelengths following Wien's law $\lambda_{peak} \propto 1/T$ [31].

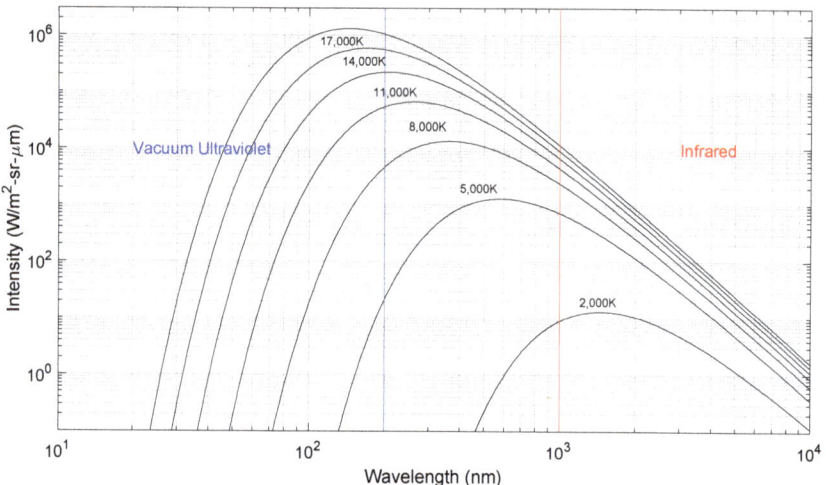

Figure 4. Planck blackbody emission wavelength distribution for various temperatures.

Typically, atmospheric entries occur at a gas pressure low enough that the emitted radiation cannot be assumed to be in equilibrium and treated as a blackbody, yet this assumption provides an upper limit for the emitted radiation over the whole spectral range. In general, the plasma is optically thin, and the discrete radiation spectrum is in strong non-Boltzmann equilibrium. The dominant spectral features will depend on the gas chemical composition and pressure, as well as the temperature (derived from the shock velocity). Entries faster than 7 km/s will usually ionize the gas and thus emit radiation in the ultraviolet region. Slower entries can only excite the molecular internal vibration levels, which radiate in the infrared region. Figure 5 shows the most important emission regions for typical planetary entries, which is based on experimental data from previous shock tube campaigns by different teams all around the world. Data for Earth, Mars and Venus radiative transitions were taken from [17,32], Bose et al. [33], and Cruden et al. [34], respectively. Data for Jupiter were adapted from the work of Cruden and Bogdanoff in [35]. Neptune data were taken from [36,37]. Titan data were reported by Magin et al. [38].

As expected, the radiative features for these different classes of entry flows are quite rich, with a great deal of measured atomic transitions in the VUV region (120–200 nm), the Balmer series of H in the visible, and the O atomic lines in the near-IR region (at 777 nm and at 849 nm, respectively). Molecular radiation features are also very rich, with many emitting systems from C_2, CH, CN and N_2 all over the visible range, and H_2, CO and NO emitting in the VUV range. For the IR, one typically observes emission from the different rovibrational bands of CO and CO_2 from 1.5 µm to over 5 µm. These spectral features may now guide us into selecting the appropriate measurement setup.

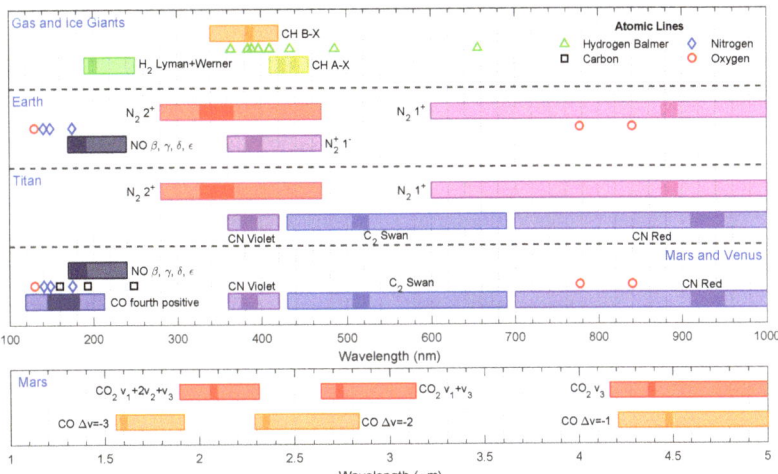

Figure 5. Emission region of different gas species in typical entry conditions for different planets. Observed atomic lines are explicitly presented for their measured wavelengths. Typically, molecular bands have strong bandhead peaks, with more diffuse tails in the neighbouring regions. To provide an estimate of the spectral regions where these bands may be emitting, we ran the in-house spectral line-by-line code SPARK [39,40] with a representative temperature of $T = 5000$ K. We represent the bandheads with a stronger color, and the tails with the same faded-out color. The band heads/tail intensity ratio is taken as a factor of 100.

2.2. Operational Shock Tube Facilities for Fast Atmospheric Entries

Reynier et al. provide a detailed review on hypersonic facilities in [41,42]. Currently, the only ground test facilities capable of achieving superorbital velocities (above 10 km/s) are Ames Electric Arc Shock Tube (EAST—Moffett Field, CA, USA); X2 and X3 expansion tubes (Brisbane, Queensland, Australia); CUBRC LENS XX expansion tube (Buffalo, NY, USA); Hyper Velocity Shock Tube (HVST—Tokyo, Japan); TsAGI ADST shock tube (Moscow, Russia); and the T6 Stalker tunnel (Oxford, UK). Typically, shock tubes use emission spectroscopy to determine the chemical composition of the gas behind the shock wave in its non-equilibrium state. Chemical species concentrations may be determined via absorption spectroscopy. The electron density may be estimated via an indirect method, such as the Stark broadening of some emission lines. Nonetheless, some facilities use microwave interferometry to directly measure the electron density behind a shock wave. Electrostatic (Langmuir) probes are simple diagnostics used for electron density and temperature measurements [43]. However, these are intrusive diagnostics which perturb the flow in the diagnostics region and are damaged by it. Therefore, these drawbacks preclude its use in a shock tube.

EAST [24,44,45] at NASA Ames is equipped with time-of-arrival sensors to have a high-resolution velocity measurement. A long slot optical window is present for shock imaging via spectroscopic instrumentation. A total of four different sets of optics, each with its own spectrometer, perform the imaging at the same axial location. The spectrometers are selected as a function of the region of interest of the electromagnetic spectrum, which itself depends on the shock wave velocity, test gas pressure, and chemical composition. The regions are generally classified as vacuum ultraviolet (120–200 nm), ultraviolet/visible (200–500 nm), visible/near infrared (500–900 nm), near infrared (900–1600 nm) and mid wave infrared (1600–5500 nm). The VUV spectroscopy equipment must operate under vacuum conditions to prevent the ultraviolet radiation to be absorbed along its optical path, and the optical windows must be made of MgF_2 or LiF to minimize absorption.

The University of Queensland hosts three expansion tube facilities, named X1, X2 and X3. The latter two facilities are capable of reaching superorbital velocities and produce VUV and UV radiation. The X2 expansion tube spectroscopy system [46–48] consists of a normal incidence spectrometer and an intensified charge coupled device (iCCD) with a camera of enhanced sensitivity in the VUV spectral range. The system has a theoretical resolution of 0.06 nm/pixel, and a range of 60 nm. The UV spectral region can also be observed using the same setup. An additional visible/NIR spectroscopic system is present, with a wavelength range of 695 to 880 nm and 0.55 nm/pixel resolution. The flow is also monitored with a high-speed camera. Besides the emission spectroscopy diagnostics, X2 has Nd:YAG (355 and 532 nm) interferometry instrumentation equipment capable of measuring density, ionization levels, species concentrations and temperatures [48].

The Japanese HVST [49] located in JAXA's Chofu Aerospace Centre is a free-piston-driven shock tube which operates in both shock and expansion tube modes. A He-Ne laser Schlieren setup is used to detect the shock front and measure the shock wave velocity. Three spectrometers covering the VUV to NIR spectral region are coupled to two CCD arrays [50–53] for radiation emission spectroscopy.

The LENS XX is a hypersonic expansion tube in Buffalo, N.Y., equipped with an emission spectroscopy system [54] in the UV to visible with two gratings (1200 and 150 g/mm) and an iCCD camera. The calibration is performed with a deuterium lamp.

T6 [55] is a Stalker shock tunnel located at the University of Oxford, which may reach velocities up to 18 km/s for light test gases (H_2-He). The emission spectroscopy setup [56] is based on the X2 expansion tube. A series of UV-enhanced aluminum mirrors focus the light into a spectrograph. A 550 nm longpass filter may be applied for measurements in the red and NIR region of the spectrum. The setup can operate in the 350–850 nm range using either gratings of 150 or 1200 g/mm.

2.2.1. The European Shock Tube for High Enthalpy Research

ESTHER is expected to reach shock wave velocities in the range of 6 to 14 km/s in air or above 18 km/s in (H_2-He mixtures [29,57]. Figure 6 depicts a schematic and a photograph overview of ESTHER. The facility is comprised of four sections, separated from each other by diaphragms. These are the combustion chamber driver, the compression tube, the shock tube (also called test section) and the dump tank. A small-scale combustion driver was previously tested in order to de-risk the development of the full-scale driver [14]. Whereas the qualification tests of ESTHER are ongoing [15], the tests of the combustion driver were successfully completed. The driver of ESTHER is a 47-liter cylindrical 200 mm internal diameter combustion chamber, capable of handling $He:H_2:O_2$ or $N_2:H_2:O_2$ mixtures with filling pressures up to 100 bar and post-combustion deflagration pressures of 660 bar. The combustion chamber and its equipment are designed to operate in deflagration (subsonic combustion) mode. Nonetheless, the driver can withstand the detonations (supersonic combustion), which may occasionally occur creating transient pressures up to 1.8 kbar. The ignition of the mixture is attained using a high power Nd:YAG laser [58,59], which fires a 5 ns pulse into the chamber. The compression tube, with an internal diameter of 130 mm, is connected to the driver via a diaphragm designed to open at a predefined pressure. Once filled with helium at pressures between 0.01 and 1 bar, the shock wave moving along the compression tube can reach pressures of up to 70 bar. A second diaphragm divides the compression and the shock tube sections. The shock tube, with a 80 mm internal diameter, is filled with the test gas mixture at pressures between 10 and 100 Pa (0.1 to 1 mbar). The shock wave reaches velocities exceeding 10 km/s in the shock tube, leading to transient pressures reaching up to 20 bar. Pressure sensors and optical detectors are positioned along the shock tube to measure the shock wave velocity and trigger the time-dependent spectroscopic measurements at the test section. Lastly, a 1000 L dump tank, separated from the shock tube by a third diaphragm, recovers the gas flowing in the wake of the shock wave. Following each shot, the liquid phase is drained off, and the remaining contaminated gas mixture is evacuated by the vacuum pumps located in the shock tube section. The tube is then opened for cleaning and diaphragm replacement.

The facility is equipped with 20 ports at 6 different positions along the tube's axial direction, 4 in the compression tube section and 16 in the shock tube section (8 in the test section). There are two measurement stations at the test section, each with four ports located circumferentially, which allow for multiple diagnostics and measurements at the same axial position. These optical ports have a 10 mm diameter cylindrical shape.

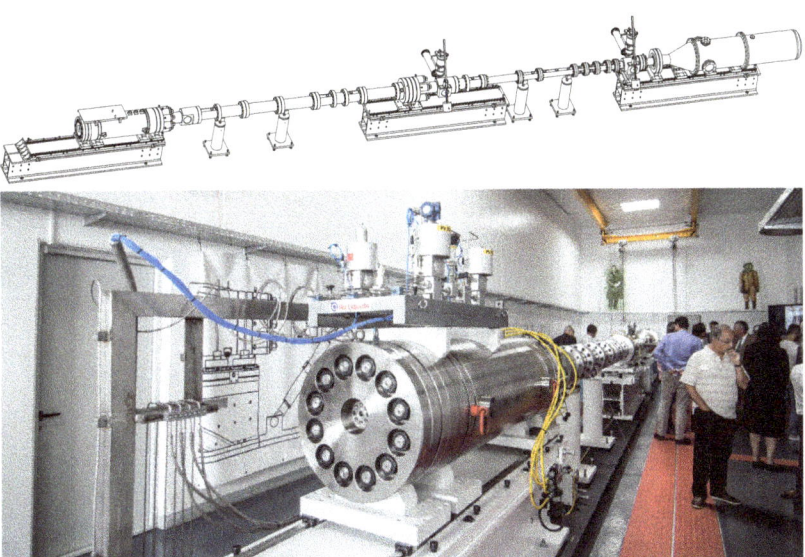

Figure 6. European Shock Tube for High-Enthalpy Research (ESTHER). Driver section to the left, test section on the right side. Shock wave moves from left to right.

2.2.2. ESTHER Performance Map

The ESTHER performance map was predicted using the STAGG code (Shock Tubes and Gas Guns), developed by Fluid Gravity Engineering Ltd. The code solves the set of shock tube equations given the design variables (cross-section areas, length, single/two-stage geometry) and numerical inputs of the chemical mixtures. The numerical model is based on the works of Alpher and White [60], Walenta [61] and Mirels [62]. The gas is assumed to be isentropic and inviscid [63] to compute the shock speed and the pressure along the tube. Further details of this development can be found in [29]. The code was first calibrated using the VUT-1 test data from ESA's CO_2 validation campaign [64]. The STAGG simulations were later re-run following the driver qualification campaign, once the driver combustion performance was assessed [57] in order to have a more realistic performance envelope of the facility.

STAGG simulations were performed in two different modes, optimization and non-optimization. In the first case, the compression tube pressure is adjusted to maximize shock wave velocity at the test section, and in the latter, all input parameters are fixed and the code solves the equations to compute the shock wave pressure, temperature and velocity [63]. Post-combustion temperatures and pressures were obtained during the driver qualification campaign for different gas mixtures and filling pressures [65]. During this campaign, the nominal operational mixtures were tuned to $He:H_2:O_2$ 8:2:1.2–1.4 for the high-velocity experiments (>7 km/s) and $N_2:H_2:O_2$ 8:2:1.4 for the low-velocity experiments (<6 km/s). To generate the performance map, STAGG ran three sets of simulations: Helium driver optimization; Nitrogen driver optimization; and Helium driver non-optimization. These correspond to high (>7 km/s), low (<6 km/s) and medium (6–7 km/s) velocity regions. The input parameters for the different sets of simulations are shown in Table 1. Maximum performance is found when STAGG runs with optimized conditions; however, running it in non-optimized (de-tuned) conditions is also useful to achieve lower shock velocities, extending the operational range. Figure 7 shows ESTHER performance simulation for Earth's (N_2-O_2) atmosphere. The envelope is drawn using the lowest and highest velocity points of each simulation group.

Table 1. STAGG parameters for Earth (N_2-O_2) performance map simulation. Driver input conditions and STAGG running mode (optimization of compression tube pressure or simple computation with fixed conditions).

Mixture Molar Ratios X:H_2:O_2	Driver Pressure Filling-Peak (bar)	Driver Post-Combustion Temperature (K)	STAGG Conditions	Shock Wave Velocity (km/s)
N_2 10:2:1.4	10–38	1250	Optimized	4.0–5.1
N_2 10:2:1.4	30–114	1250	Optimized	4.6–5.5
N_2 10:2:1.4	50–190	1250	Optimized	4.8–5.8
He 8:2:1.6	5–30	2650	De-tuned	5.8–10.0
He 8:2:1.6	10–60	2650	De-tuned	6.7–10.8
He 8:2:1.6	5–30	2650	Optimized	7.3–10.2
He 8:2:1.6	10–60	2650	Optimized	8.0–10.9
He 8:2:1.2	10–61.5	2650	Optimized	8.2–11.0
He 8:2:1.2	50–334	2650	Optimized	10.3–13.2
He 8:2:1.4	100–660	2650	Optimized	11.1–13.8

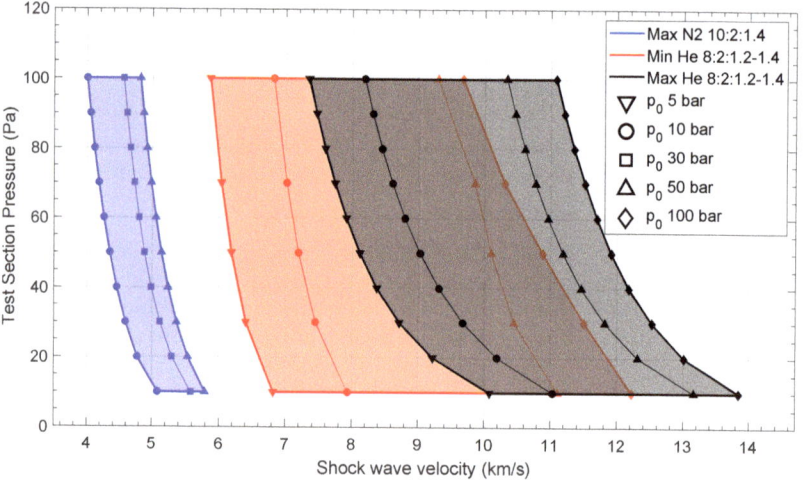

Figure 7. Expected ESTHER performance envelope (test gas pressure vs. shock wave velocity) for air (N_2-O_2) from STAGG code.

2.3. Optical Spectroscopy Specifications and Requirements

Section 2.1 provides us with the operational conditions, pre-shock pressure and target speed, for our setup, and more specifically, Figure 5 provides us with the specific wavelengths of interest for the different gas mixtures considered in the testing. However, the wavelength alone is not sufficient for the specification of the equipment, as one needs to consider as well the shock speeds. These will define the acquisition time, and in turn, the temporal/spatial resolution of the system. The design of the trigger system and of the fast spectroscopy electronics is also influenced by the shock wave speed; the total time should be about 20 μs at worst (for maximum operational speeds).

Another two important parameters are the spectroscopic resolution and the sensitivity of the system. In terms of the spectroscopic resolution, typical spectroscopic systems which work in monochromator mode try to target the best possible resolution, as they are typically deployed in steady-state experiments, where the wavelength can be slowly scanned and acquisitions (either via a photomultiplier or an intensified camera) can be tailored to be long enough so that a good signal-to-noise ratio is achieved. An example for such a setup is described in Ref. [66]. In the case of a shock tube facility, we are collecting light trailing a moving shock wave for a few µs in an experiment that is typically one or few hours in the making. Not only is a spectrograph setup (imaging the full spectral window in one sweep) mandatory, but one needs to ensure that the maximum amount of light is collected during the passage of the shock wave. Therefore, there is a need to ensure that the setup collects the maximum possible amount of light, as well as a need to tailor the spectral window. For the latter, higher spectral resolutions yield lower spectral windows and vice versa. The spectral resolution is determined by the gratings installed in the spectrograph, which can be exchanged. The separation of different wavelengths is called the angular dispersion $\mathcal{D}$ and can be computed as the derivative of the reflection angle θ_m with respect to the wavelength λ:

$$\mathcal{D} = \frac{d\theta_m}{d\lambda} = \frac{k}{a \cos \theta_m} \quad , \qquad (6)$$

where a is the groove distance in the grating (inverse of the groove density). The reflection angles in Equation (6) can be calculated by

$$a(\sin \theta_i + \sin \theta_m) = k\lambda \quad , \qquad (7)$$

where k is the order of diffraction (1, 2, ...). For a set of a and $\theta_{i,m}$, multiple λ satisfy Equation (7). The lowest-order solution, $k = 1$, corresponds to the longer wavelength λ_1, with higher order solutions corresponding to $\lambda_k = \lambda_1/k$. The dispersion of the wavelength at the spectrograph focal plane with focal distance f is computed through

$$\frac{dy}{d\lambda} = f\mathcal{D} = \frac{fk}{a \cos \theta_m} \quad . \qquad (8)$$

A first-order Littrow blaze can be applied to the diffraction grating. It increases the intensity of the refraction order and wavelength of interest by curving the grating surface to direct the light at a preferred angle. These governing equations are useful for defining which gratings will be best suited in terms of apparatus function and spectral window.

Broadly speaking, three different spectral regions (each with its own peculiar characteristics) are identified:

1. The ultraviolet and vacuum ultraviolet regions where most of the radiation for high-speed entries should be emitted (see Figure 4). This region is bounded roughly between 120 nm (below which most windows become opaque) and 300 nm (where the visible region begins).
2. The near-UV to near-IR region in the 300–850 nm range, colloquially referred as the visible range, where many atomic and molecular systems are emissive for moderate entry speeds (see Figure 5).
3. The near-IR to mid-IR region (roughly in the 1–5 µm range) where rovibrational transitions between heteronuclear molecules, such as CO_2, CO, and NO, are strongly emissive. Radiation is observed in this spectral range for low-speed entries in planetary atmospheres with such gases in their composition (mostly Mars, for which the large impact of IR CO_2 radiation was recently assessed [19,67]).

Each of these three spectral regions requires its own setup. This means selecting an adequate spectrograph/camera combo. In terms of spectrographs and beyond the selection of appropriate gratings, one needs to account for the absorption of room air, which encompasses all the spectral ranges below roughly 200 nm, and specific bands in the IR (owing to the trace amounts of CO_2 and water vapor). This means that a VUV setup needs to be held in vacuum, and an IR setup typically needs to be flushed in an inert non-IR absorbing gas, such as Nitrogen. The camera itself needs to be sensitive to the spectral range of interest. Intensified high-speed cameras (iCCD) encompass all the spectral ranges of interest of this work, whereas streak cameras are limited to the UV-VUV and visible ranges. Streak cameras have an advantage over iCCD cameras in the sense that the temporal variations of light intensity at an imaged point are translated to variation in image brightness along the streak direction. In contrast, iCCD cameras sample a line of points where the shock wave evolves over a given acquisition time, hence translating spatial points into time through the shock wave speed. These cameras are subject to motion blur, whereas streak cameras are not. Nonetheless, the deployment of the latter for higher speeds is more arduous. Whereas IR is the only spectral region where streak cameras do not exist, the shock waves are slower and therefore the issue of motion blur is less critical.

The optimal setup selection is therefore schematized in Figure 8. The detailed specifications and requirements for these equipment, as defined by ESA, are presented in Table 2 and can now be discussed in detail. Note that no specifications and requirements have been defined in the visible region, as the optical setup from the previous TCM2 shock tube is reused.

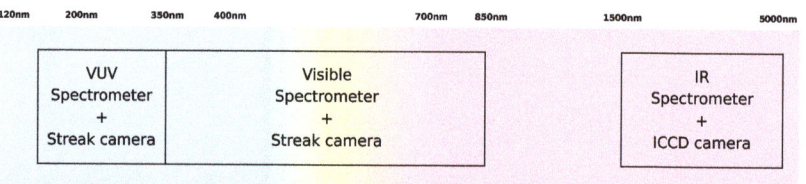

Figure 8. ESTHER instrumentation spectral range coverage.

Table 2. ESTHER spectroscopy specifications and requirements.

UV-VUV Spectroscopy					
Equipment	Spectral Range	Resolution	Accuracy	Integration Time	Signal to Noise Ratio
Spectrometer + Streak Camera	80–350 nm	0.1 nm @ 110 nm	0.01 nm	$\leq 1\ \mu s$	>20
MWIR Spectroscopy					
Equipment	Spectral Range	Resolution	Accuracy	Integration Time	Signal to Noise Ratio
Spectrometer + iCCD camera	1–6 µm	0.5 nm @ 2.7 µm	0.05 nm	$\leq 1\ \mu s$	>20

These specifications and requirements were defined taking into consideration the generic characteristics for spectroscopic setups in their spectral regions. These are understood to be adjustable for a shock tube configuration. In terms of spectral range, resolution and accuracy, most of the commercially available spectrograph equipment are compliant to the specified parameters. In turn, most of the streak and iCCD cameras are capable of achieving integration times lower than 1 µs. The specification of the signal-to-noise ratio will be more experimentally dependent on the amount of radiation emitted by the shocked flow, which then will drive the requirements to minimize stray light and general noise from the acquisition system.

With this said, the assembly and deployment of a setup for detecting short bursts of light typical of shock tube experiments entail a certain number of restrictions, which immediately narrow down the range of compliance to the specifications and requirements of Table 2. Among others, one needs to account for window transmissivity; spectral response, range and sensitivity of the streak/iCCD camera photocathode/CMOS sensor, respectively; bit-rate for the high-speed electronics of the cameras; and availability of fiber optics to connect the spectrograph to the shock tube optical windows. All these narrowed-down specifications and requirements are discussed in detail in Section 3 (taking into account the spectral region, either UV-VUV or IR), except for the window transmissivity, which is discussed here.

VUV light is easily absorbed by gas molecules, and most optical materials, such as quartz or glass. This mandates the VUV system be held in vacuum to prevent the collected light from being absorbed by oxygen and nitrogen molecules. Using an optical fiber cable is also not possible as these have strong attenuation below 300 nm. Alongside, the optical window material should be as transparent as possible in the 100–300 nm range. Figure 9 and Table 3 show the optical transmissivity of different materials in the VUV region. Usually, VUV windows are made of Lithium Fluorite (LiF) or Magnesium Fluorite (MgF_2); however, these materials cannot handle the force of the passing shock wave and will break after two or three experimental runs. For this reason, VUV-graded sapphire was the material chosen for the windows of the spectroscopy system. Regarding the NIR-MWIR spectral region, all the aforementioned optical windows materials are essentially transparent up to 4–5 µm, and there is the additional advantage that infrared optical fibers are available to "transport" the light signal from the shock tube to the spectrometer. These have low attenuation in the infrared region [68] and do not require vacuum like the VUV wavelengths.

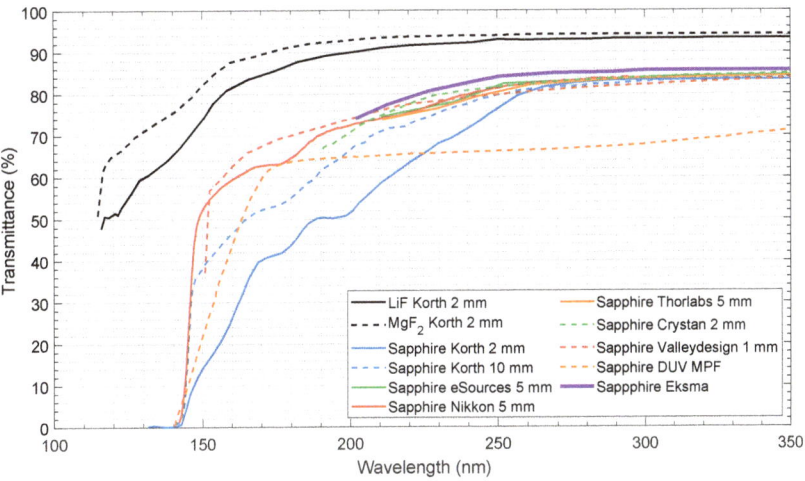

Figure 9. Optical transmissivity for common window materials, data adapted from [69–76].

Table 3. Comparison between different optical window materials for VUV spectroscopy.

Material	Knoop Hardness	Advantages	Disadvantages
Lithium Fluoride	100 Kg/mm^2	Large VUV-IR transmittance	Very brittle, expensive
Magnesium Fluoride	415 Kg/mm^2	Large VUV-IR transmittance	Brittle, expensive
VUV graded Sapphire	1370 Kg/mm^2	Cheap, resistant, some VUV capabilities	Limited transmittance window
Fused Quartz	741 Kg/mm^2	Cheap, resistant	Limited transmittance window

2.4. Microwave Interferometery Specifications & Requirements

Interferometry is an adequate technique for electron density diagnostics in shock tubes since these lack any radial profile (in other words, the shock wave front moves as a "disk" over the tube). Other than a small test campaign carried out at the Moscow Institute of Physics and Technology [41,64,77,78], in the scope of an ESA contract, there are no further direct electron density measurements in hypersonic shock tube facilities using this technique to the authors knowledge. Most plasma interferometry measurements were performed in combustion shock tubes [79] or to measure ionization rates [80]. Interferometry is a common diagnostic in plasma plumes, namely Hall thrusters, because of its non-invasive nature. Examples of this diagnostic include measurements made in Xenon [81–85], and Hydrogen [86]. A recent review on the applicability of this technique for atmospheric entry applications is provided in [26].

The plasma cut-off frequency f_{pe} is the minimum value at which an electromagnetic wave can propagate in a plasma. It relates to the critical electron density $n_{e,p}$ via

Equation (9) [43], where ε_0, e and m_e are the vacuum permittivity, the electron charge and mass, respectively:

$$f_{pe} = \frac{\omega_p}{2\pi} = \frac{1}{2\pi}\sqrt{\frac{n_{e,p}e^2}{m_e\varepsilon_0}} \quad . \tag{9}$$

The cut-off frequency is a key parameter to design any microwave diagnostic. In microwave reflectometry, commonly used in fusion reactors [43,87], the electron density profile is diagnosed using a radar-like technique with microwaves. A probing signal is emitted to the plasma, where it propagates until a layer with n_e equal to the critical value $n_{e,p}$ is found, and is then reflected. The phase difference of the two signals relates to the time of flight of the probe signal and to the plasma refraction index. The latter is associated with the plasma density via the Altar–Appleton equation [88]. Microwave interferometry uses a similar principle to reflectometry; however, the probing signal must completely transverse the plasma. The phase gained by the signal compared to the reference will relate to the average electron density of the plasma. To create a spatial plasma profile, the signal must be de-convoluted via Abel-inversion ("onion-peel") techniques. Both techniques can work monostatically with one antenna for wave emission and reception, or bistatically with one antenna dedicated for emission and another for reception. More details on the specifics for this technique may be found in Ref. [89].

The functional requirements of an interferometer mandate that it should be compact and self-sufficient so it can be assembled and tested in different facilities with ease. Namely, its antennas, emitting and receiving, must be compatible with ESTHER optical plug windows (diameter 10 mm). Its working frequency (f) must be sufficiently high to traverse the plasma without reflecting back, and gain a phase delay [89] significantly large to be detected. The phase shift is given by Equation (10), where D is the plasma thickness, and ω_p and ω are the plasma oscillation and probing angular wave frequencies, respectively:

$$\Delta\phi = \frac{2\pi}{\lambda}\int_0^D (1-\mu)dl = \frac{2\pi}{\lambda}\int_0^D \left(1-\sqrt{1-\omega_p^2/\omega^2}\right)dl \quad . \tag{10}$$

Using Equation (9), $\lambda = c/f$, and ($\omega \gg \omega_p$), Equation (10) can be simplified into

$$\Delta\phi \simeq \frac{e^2}{4\pi c m_e \varepsilon_0 f}\int_0^D n_e dl = \frac{e^2}{4\pi c m_e \varepsilon_0}\frac{\bar{n}_e D}{f} \quad , \tag{11}$$

where $\bar{n}_e$ is the average electron density over the plasma path. This integrated value is a very good approximation to shock tube measurements, as the plasma can be approximated to a disk whose properties change only in the longitudinal direction. Optimally, the working frequency should be high enough to transverse the plasma and guarantee $D/\lambda > 3$. However, if it is too high (small λ), the phase gained may be too small and too difficult to measure. Alongside working as a interferometer, the base equipment base should be convertible to a reflectometer to be mounted on a small spacecraft.

Since not much experimental data are available for direct electron density measurements, a set of CFD (Computational Fluid Dynamics) simulations was performed to estimate the required range for the diagnostics equipment. Shock wave conditions were estimated through the chemical reactive CFD code SPARK (Software Package for Aerothermodynamics, Radiation and Kinetics) [90]. The code is capable of computing the chemical composition behind the shock wave and its respective electron density and emission radiation. A total of six simulation runs were carried out in 1D (post-shock relaxation) conditions. Table 4 shows the CFD simulations initial conditions and references for the chemical–kinetic reactions schemes, as well as for the chosen velocity and pressure conditions. The representative cases are a sample return mission to Earth, and the ExoMars, Huygens and Galileo missions to Mars, Titan and Jupiter, respectively. Simulation conditions for Neptune and Venus were taken from trajectory calculations. The electron densities for these cases are depicted in Figure 10, where the electron density can reach values of 4×10^{23} electrons/m^3

for the case of a Venusian entry. The typical electron density profile has a sharp rise right after the shock front, followed by a slower decay until chemical equilibrium is achieved.

Table 4. CFD simulation conditions and parameters. All simulations 1D post-shock relaxation with initial gas temperature of 300 K.

Planetary Object	Chemical Mixture	Pressure (Pa)	Velocity v_∞ (km/s)	Chemical Model	Reference
Earth	N_2-O_2 (79-21%)	26.66	10.29	[91]	[45,92]
Mars	CO_2-N_2 (95-5%)	57	2.6	[91]	[93]
Venus	CO_2-N_2 (96.5-3.5%)	447	8.9	[91]	[94]
Venus	CO_2-N_2 (96.5-3.5%)	37	10.6	[91]	[94]
Titan	N_2-CH_4 (95-5%)	13.3	5.15	[95]	[94]
Neptune	H_2-He-CH_4 (79.75-18.7-1.54%)	892	18.3	[36]	[36]
Jupiter	H_2-He (89-11%)	27.5	46.7	[96]	[36]

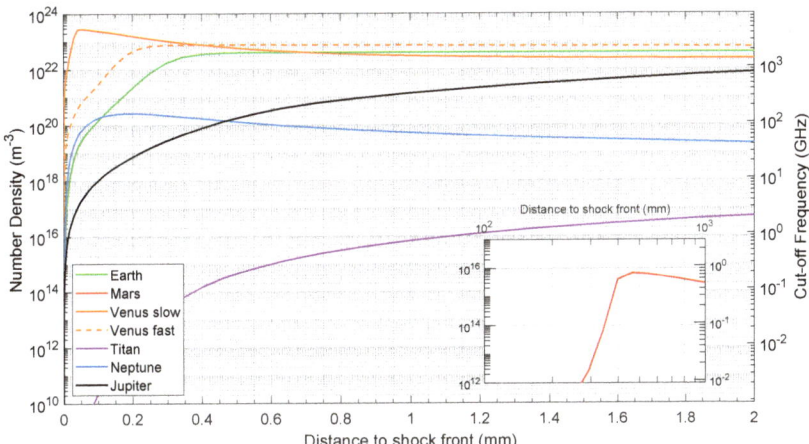

Figure 10. Electron density in shock tube for typical entry conditions in different atmospheres.

As a side note, one needs to point out that Stark broadening measurements of the H-α and H-β lines provide an interesting technique for electron density analysis [24,25]. This may be considered a non-intrusive diagnostic to some extent, as a small percentage of the flow is replaced with Hydrogen (~0.1%). The Hydrogen atom α line at 656 nm and Balmer β line at 486 nm then typically become visible without significantly perturbing the overall spectrum. The observed lines may then be fitted to a Lorentzian curve with all other individual broadening contributions, van der Waals collisional, Doppler and instrument, accounted for. The FWHM (full width at half maximum) for the emission peak may then be compared to tabulated values, such as the ones found in [97], which give an estimation of the electron density. As a caveat, we note that the addition of (even small) quantities of H into the flow may affect wall desorption and slightly alter the flow properties; however, this effect should be relatively limited as shown by Cruden [24].

3. Results and Discussion

The specifications and requirements, as defined in the previous section, led to the final setup for ESTHER's optical spectroscopy setup, which is now discussed in detail:

- A UV-VUV spectroscopy setup covering the 120–350 nm range;
- A visible spectroscopy setup covering the 350–850 nm range;

- A MWIR spectroscopy setup covering the 1–5 μm range.

Here, the spectral regions are bounded by the window transmissivity and grating efficiency of the UV-VUV setup, the streak camera input optics and photocathode sensitivity of the visible setup, and the optical fiber transmissivity of the MWIR setup. For each of these, appropriate blazed gratings are selected, which ideally maximize throughput on a specific wavelength of interest, while maintaining enough transmission efficiency over the whole spectral window. Lastly, finer/coarser gratings are selected to maximize spectral resolution/range over a single measurement, respectively.

In addition to the spectroscopy setups, an interferometer with a 2–18 GHz back end, and a 70.8–112.8 GHz front end was developed for the time-dependent measurement of electron densities. These four setups are discussed in detail in this section.

3.1. UV-VUV Spectroscopy Setup Design and Acceptance Testing

The selected equipment for UV-VUV spectroscopy setup is a McPherson Model 234/302 with an $f/4.5$ aperture and 200 mm focal length, coupled to a UV-VUV-customized Hamamatsu Streak Camera M10913-11, acceptance tested in November 2019. The use of a streak camera presents advantages over a simple iCCD camera. The most relevant is that the signal becomes time resolved since the streak camera creates a time discrimination of the received light. However, increasing the number of elements on the optical path reduces the signal-to-noise ratio. Streak cameras are based on the photoelectric effect, thus they do not work with low energy photons, namely in the infrared region. Therefore, any infrared spectroscopy instrumentation cannot make use of a streak camera.

Figure 11 shows the overall assembly of UV-VUV spectroscopy system and a detailed view of the optical plug, which connects the test section to the UV-VUV spectrometer. The design maximizes the collected light via a 12° light cone throughout the optical path. An optical plug with a VUV graded sapphire window is held in place and vacuum sealed using one viton O-ring, and is connected to a collection optics box by flexible bellows (which allows connecting the shock tube, standing over a seismic slab, to the optics table standing in the experimental hall floor). The optics box is comprised of two mirrors, which collect and focus all the light cone into the spectrograph, hence maximizing the signal throughput. The spectrograph is in turn connected to the streak camera optical relay, which focuses the light into the photocathode that converts the photons into electrons, and deflects them using a saw-tooth signal. The streak camera is coupled to an iCCD camera (model ORCA-Flash4.0 C13440-20CU), which records the final time- and wavelength-dependent signal. Figure 12 depicts a photograph of the UV-VUV spectroscopy setup.

The spectrograph was equipped with two different diffraction gratings of 600 and 1200 g/mm blazed at 150 nm, ensuring an adequate throughput in the 120–350 nm spectral range. Validation tests for the setup were conducted with a mercury vapor lamp and a laser impulse generator. First, the equipment was calibrated with the 1200 g/mm grating using the Hg 253.65 nm line. The center pixel is located at number 628, with a total range of 782 pixels. Using the 1200 g/mm grating, a spectral range of 10 nm can be observed by the streak camera. The spectral dispersion is 10 nm/782 pixels = 0.0128 nm/pixel. The line FWHM at three different points of the photocathode was computed directly by the streak camera computer program to determine the wavelength resolution. Using the coarse grating, the center location was found to be pixel 618. The spectral range is now doubled to 20 nm with 803 pixels, thus giving a spectral dispersion of 20 nm/803 pixel = 0.0249 nm/pixel. The wavelength resolution values are presented in Table 5, and an example for the 1200 g/mm grating is depicted in Figure 13 (top).

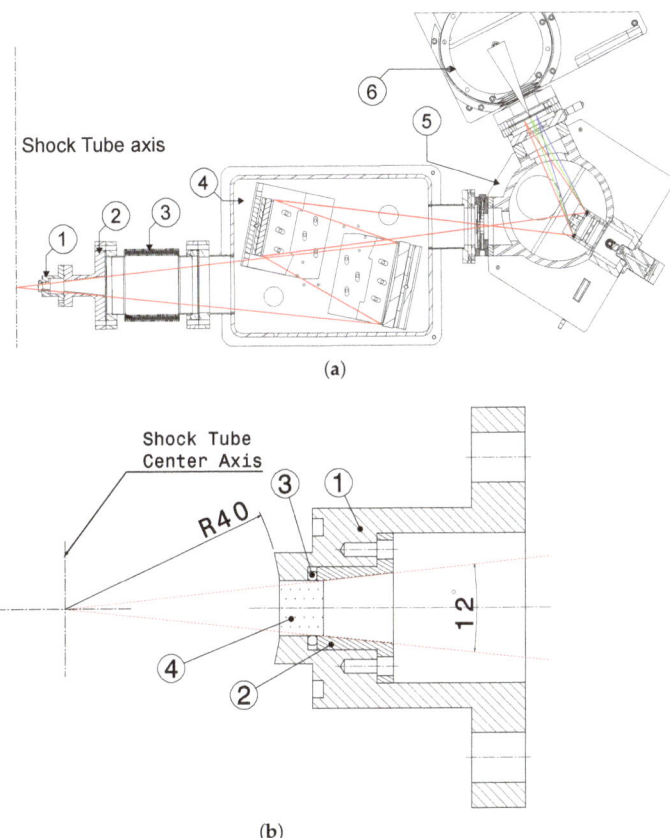

Figure 11. ESTHER UV-VUV spectroscopic setup design schematic. General design (**a**) and optical plug detail (**b**). In (**a**), 1—Shock tube plug; 2—Connection piece; 3—Adjustable Bellows; 4—Optics collection box; 5—Spectrograph with diffraction grating; 6—Connection to streak and CCD camera. In (**b**), 1—Plug; 2—Window tightener piece; 3—Viton O-ring; 4—Sapphire window. Red lines represent the optical path.

Figure 12. ESTHER UV-VUV spectroscopic setup during acceptance testing, ST8 optical port marked as a square in photo.

Table 5. Calibration results for the 1200 g/mm and 600 g/mm gratings, Hg 253.65 nm line, with FWHM in pixels.

1200 g/mm			600 g/mm		
Position	Resolution (Pixel)	Resolution (nm)	Position	Resolution (Pixel)	Resolution (nm)
Left	28	0.36	Left	18.9	0.47
Center	17.8	0.23	Center	24.6	0.61
Right	25.1	0.32	Right	46.8	1.17

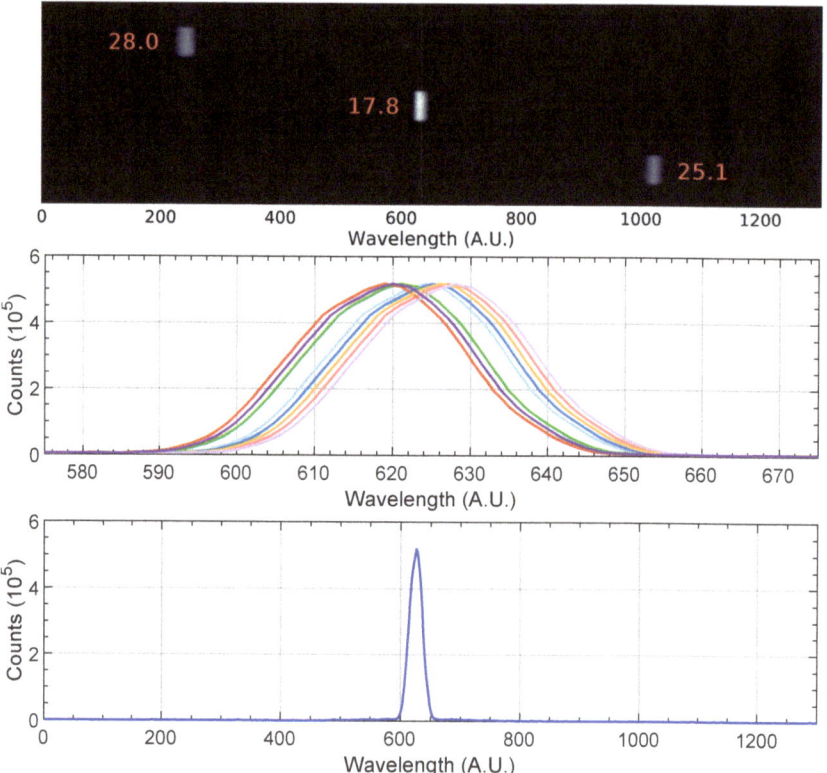

Figure 13. UV-VUV spectroscopic setup acceptance campaign results. Wavelength resolution in pixels for the 1200 g/mm grating (**top**); repeatability and accuracy (**middle**); and signal-to-noise ratio (**bottom**). Each curve in the middle figure represents a different run.

The accuracy and reproducibility were tested by measuring the peak 10 times while sweeping the camera spectral center position. The average center position was found to be at 632.5 ± 4.5 pixels. Lastly, the signal-to-noise ratio was evaluated by the ratio between the peak intensity (513.79) to the noise standard deviation (0.15), yielding a signal-to-noise ratio 513.79/0.15 ≃ 3400. The test values for reproducibility and signal-to-noise are depicted in Figure 13 in the middle and bottom, respectively. The results from the UV-VUV acceptance campaign were in conformity with the requirements. However, the temporary unavailability of the vacuum pumps prevented testing the equipment with the deuterium line at 110 nm or the mercury line at 237.83 nm. The streak camera sampling rate was tested with a picosecond laser. Table 6 presents the outline of the UV-VUV acceptance campaign.

Table 6. ESTHER spectroscopy acceptance campaign results.

-	Resolution 1200 g/mm	Resolution 600 g/mm	Accuracy	Signal-to-Noise Ratio
Specifications and Requirements	0.25 nm	0.50 nm	5 pixels	>100
Test	0.23 nm @ 253.65 nm	0.47 nm @ 253.65 nm	4.5 pixels	3425

3.2. Visible Spectroscopy Setup Description

The visible spectroscopy setup used here was recovered from the previous TCM2 shock tube facility in Marseille, France, and is only be briefly outlined here (see Figure 14). The setup was used previously in a series of spectroscopic studies at the aforementioned facility [5,7,16,98]. It is comprised of a 0.64 m focal length Czerny–Turner Jobin–Yvon HR640 spectrograph $f/5.2$ using the following gratings: two low-resolution large-window 600 and 1200 g/mm gratings blazed at 500 nm, and two high-resolution narrow-window 2400 and 3000 g/mm gratings blazed at 330 nm, and at the 250–550 nm region, respectively. The spectrograph is coupled to a visible-range Hammamatsu streak camera comprised of a M1953 slow speed streak sweep unit and a universal temporal disperser C1587. The streak tube is the YD2369/N1643-01 with spectral range 200–850 nm and input optics model A1975 with spectral range 350–850 nm. A CCD camera (Hamamatsu C4880) is used to record the spectral-time images.

Figure 14. Visible range spectroscopy setup.

3.3. MWIR Spectroscopy Setup Design

The infrared spectroscopy setup is currently in its late definition phase. The preselection of its main components is being carried out, namely, an optical fiber cable transparent in the MWIR, a spectrograph tailored for this spectral region, and a fast IR iCCD camera. The possibility for using an optical fiber cable connected to the shock tube allows for significantly more flexibility in the setup, compared to the UV-VUV system, which requires a bulky, vacuum-purged physical connection between the shock tube and the spectroscopy setup. Figure 15 depicts the optical plug scheme of the aforementioned setup. The optical fiber (4) is tightened to a part (3), which compresses the copper rings and seal the optical window (2), and it is connected on the other end to the spectrometer and the high-speed camera. This design is inspired by the shock detection system of the X2 facility at the University of Queensland, where the optical fiber is connected to a photodetector [99].

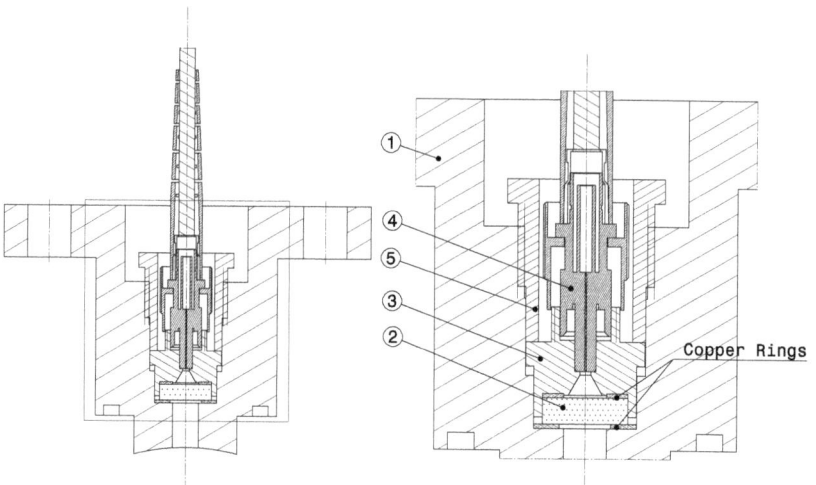

Figure 15. Optical plug for infrared spectroscopy diagnostics. 1—Plug; 2—Sapphire window; 3—Optical fiber connector; 4—Optical fiber cable; 5—Threaded screw tightening the window and optical fiber connector.

Table 7 shows the pre-selected equipment for the infrared spectroscopy setup. Optical fiber cables can either be single or multimode. The former has a higher peak transmittance but a narrow window, while the latter has a lower peak transmittance but broader transmission band. A large core-diameter fiber is also desirable to augment the collected light and provide an adequate signal-to-noise ratio. However, it may distort the signal over time due to possible differences in the travel distance between two photons. All the fiber patches are to be mounted with a ferrule connector on both the spectrograph and shock tube sides. Due to our broad spectral window, the chosen fiber must be a multimode one. For this case, indium fluorite core fibers have a slightly broader transmission band when compared to zirconium fluorite (ZrF_4), despite the latter better transmissivity of the latter in the 2 to 3.6 µm region [68]. Both Thorlabs and Le Verre Fluoré propose fibers made of indium trifluoride (InF_3), which presents good transmittance in the 1 to 5 µm region. The CIR fiber from Art Photonics has a well in the transmittance near 4 µm, and thus the overall signal-to-noise ratio would be lower than for the InF_3 fibers. The selected optical fiber is thus a multimode indium fluorite (InF_3) with a 200 µm core and 2 m of length.

Similarly to the UV-VUV and visible setups, an imaging spectrograph is needed. Table 8 presents popular equipment capable of working in the IR region. These are essentially spectrographs similar to the ones used in the visible range, except they have the possibility of being vacuumed or flushed with inert/dry gases, hence avoiding room air absorption from CO_2/H_2O traces. Longer focal length spectrographs yield larger wavelength separations; however, the spectral window narrows down accordingly. Additionally, wavelength resolution, accuracy, and window will naturally also depend on the selected grating.

The selected imaging spectrograph is a McPherson 2035 equipped with two different diffraction gratings. A finer one with 300 g/mm is blazed at 2.5 µm for higher resolution, and a coarse one with 17.5 µm is blazed at 4.2 µm for a larger spectral window. The setup characteristics for coarse, broadband measurements are very close to the ones of the EAST shock tube at NASA Ames, which provided good-quality data for CO_2 shocked flows in this spectral region for Venus and Mars atmospheric entries [100]. For the high-resolution grating, the specifications were taken considering an optimized spectral region where CO and CO_2 bands may be observed with little interference. The 2–3 µm region is attractive for this purpose as shown by a spectral simulation for the equilibrium radiation of CO and CO_2 at 3000 K equilibrium temperature, with an apparatus function of 0.2 nm, carried out with the SPARK Line-by-Line code [39,40]. Figure 16 shows the spectral features for these bands. CO rotational bands should be well defined and easy to probe in the 2.3–2.6 μm region, whereas CO_2 bands would be prominent in the 2.65–2.9 μm region. The rotational features, allowing determining characteristic temperatures, are evident (see details in Figure 16).

Finally, a group of infrared cameras was also pre-selected for the setup. Table 9 lists the various cameras identified together with the corresponding specifications. A low minimum integration time is desired to avoid excessive smearing of the image in the iCCD since, as previously discussed in Section 2.3, a temporal discrimination of the infrared signal cannot be performed with a streak camera. A short integration time results in minimal smearing of the moving radiation signal; however, short integration times decrease the signal-to-noise ratio. Due to the fast-moving shock wave (up to 6 km/s), a 1 μs integration time is equivalent to a spatial smear of 6 mm in the worst-case scenario. Higher-resolution cameras have more pixels, allowing for a higher spectral resolution without sacrificing the width of the spectral window. Similarly, a higher temporal/spatial resolution can be achieved by increasing the pixel density. A characteristic that is not relevant for our experiment is the acquisition speed/full frame rate. The full frame rate of a fast iCCD camera is limited by the bandwidth of the camera electronics, not by the CCD sensor itself. Faster frame rates may be achieved at the cost of trimming pixels of the camera sensor, thus taking pictures with lower resolution. However, due to the nature of the shock tube experiments, only the first frame is relevant, as it captures the whole useful run time up to the contact wave, and therefore, this is not a concern for our application.

Table 7. Infrared spectroscopy optical fiber pre-selection equipment.

Optical Fiber	Material	Spectral Range (μm)	Core Diam. (μm)	Attenuation dB/m	Transmissivity
Thorlabs MF22L2	InF_3	0.310–5.5	200	<0.25 [2–4 μm]	88%
Thorlabs MZ22L2	ZrF_4	0.285–5.5	200	<0.2 [2–3.5 μm]	91%
Le Verre Fluoré IFG MM 200/260	InF_3	0.310–5.5	200	<0.01 @ 3.5 μm	>81%
Guiding Photonics Mid-IR	Glass	2–16	200	4	45%
Art Photonics CIR250/300	Chalcogenide	1–5.5	250	0.3 [1–4 μm]	42%

Table 8. Infrared spectroscopy spectrograph pre-selection equipment.

Spectrograph	Focal Length	Aperture Number	Grating [Blazing]	Resolution (nm)
McPherson 2035	350 mm	$f/4.8$	300 g/mm [2 μm]	0.2 @ 312.6 nm
			20 g/mm [3.7 μm]	3 @ 312.6 nm
McPherson 207	670 mm	$f/4.7$	300 g/mm [2 μm]	0.16 @ 312.6 nm
			20 g/mm [3.7 μm]	2.04 @ 312.6 nm
McPherson 2061	1000 mm	$f/7$	300 g/mm [2 μm]	0.07 @ 312.6 nm
			20 g/mm [3.7 μm]	1.05 @ 312.6 nm
McPherson 209	1330 mm	$f/4.7$	300 g/mm [2 μm]	0.04 @ 312.6 nm
			20 g/mm [3.7 μm]	0.6 @ 312.6 nm
Princeton HRS-300	300 mm	$f/3.9$	300 g/mm [2 μm]	0.4 @ 500 nm
			50 g/mm [0.6 μm]	2.4 @ 500 nm
Princeton HRS-750	750 mm	$f/9.7$	300 g/mm [2 μm]	0.16 @ 500 nm
			50 g/mm [0.6 μm]	0.96 @ 500 nm

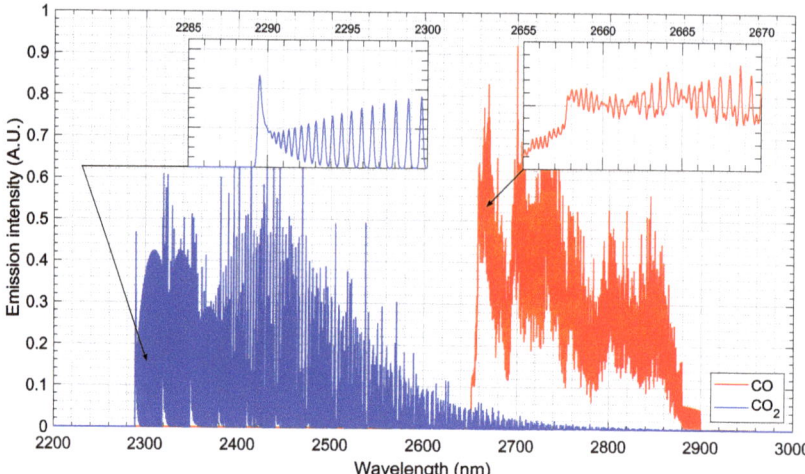

Figure 16. Synthetic spectrum for CO (blue) and CO_2 (red) radiation in the IR. The spectrum is normalized for both radiative systems and assumes an equilibrium temperature of 3000 K and an apparatus function of 0.2 nm.

The spectral cameras of interest exhibit similar characteristics, with the cost driven primarily by the capability of the equipment electronics to achieve high-resolution, high-frequency frame rates. Since this is of no concern for our application, the final selection criteria will be likely driven by equipment cost.

Table 9. Infrared spectroscopy fast iCCD cameras characteristics.

Fast Infrared Camera	Aperture Number	Resolution	Min. Int. Time (µs)	NETD (mK) *
Flir X8580	$f/2.5$	1280 × 1024	0.27	⩽30
Flir X6980	$f/2.5$	640 × 512	0.27	⩽30
Infratec Image IR 9400	$f/2.2$	640 × 512	0.10	<30 @ 30 °C
Telops FAST M1K	$f/2.5$	640 × 512	0.27	⩽25
Telops FAST M200hd	$f/3$	1280 × 1024	0.50	⩽20
Telops FAST M100hd	$f/3$	1280 × 1024	0.50	⩽20
Tigris 640 InSb BB	$f/3$	640 × 512	NA	⩽25

* The noise equivalent temperature different (NETD) is the parameter that regulates the signal photon resolution.

3.4. Microwave Interferometry

As discussed in Section 2.4 and shown in Figure 10, the working frequency of a microwave diagnostic will depend on the electron densities of interest. These can fall into the microwave (3–30 GHz), the millimeter-wave (30–300 GHz), or the sub-millimeter-wave (>300 GHz) region. Accordingly, an interferometry system needs to be flexible to encompass several frequency-band segments and switch among them to scan a specific range, or alternatively, it may consist of a stack-up of simpler instruments running in parallel. This latter approach allows the development of very compact instruments and does not limit future extensions of the bands to cover. A bespoke equipment [26], depicted in Figure 17, was developed according to this design philosophy, which is composed of two sections: a back-end signal generator and a front-end frequency multiplier and antenna assembly.

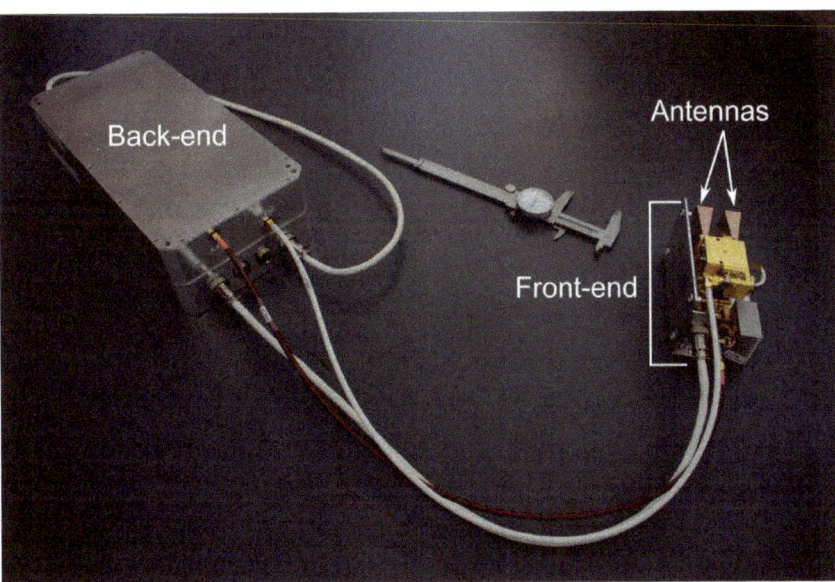

Figure 17. 70.8–112.8 GHz interferometer.

The back end is a complete, digitally controlled, signal-swept generator operating at a convenient microwave frequency range, such as 8–12 GHz or 12–18 GHz (or 10–20 GHz). The fast-swept signal is generated inside the back end by ramping the control voltage of a VCO (voltage controlled oscillator). There are several VCOs available that cover a variety of frequency ranges, but none of them, without exception, exhibit a linear relation of control voltage versus frequency. Therefore, the proper sweep of a specific frequency range requires pre-distorted voltage ramps to obtain a linearized frequency sweep at the output. This is achieved by a digital arbitrary waveform generator, which is integrated in the back end. Additionally, to allow the easy control of several units, each back-end unit is controlled/configured over a TCP/IP connection.

The front end comprises an active frequency multiplier, which produces the actual frequency range that probes the plasma, the antenna and signal detection device, which interface the plasma, and either a mixer or a single-end detector. We chose to multiply the back-end frequencies by 6 since this brings us to the 70.8–112.8 GHz frequencies (NATO W bands). The choice for this specific front-end was driven by the legacy of the successful application of this sampling technique in the VUT-1 shock tube [26] for low-ionization shocked flows. Selecting higher frequencies might be more appropriate for typical atmospheric entry conditions (see Figure 10), as the 70.8–112.8 GHz band is likely to be below the cutoff frequency of most shocked flows of interest. On the other hand, front-end developments for higher frequencies incur steeply increasing costs. Moreover, these high-frequency front ends are more difficult to test with steady-state plasma sources since very-high-density plasmas are required. The development of higher-frequency front ends is therefore contingent on the results obtained using the selected frequency band. We note that this equipment could also be considered for probing integrated electronic densities in steady-state plasma wind tunnels. Here, the issue lies in our probing frequencies being significantly above the plasma cutoff frequency, for which the measurements of the phase shift (see Equation (10)) will be less accurate (with phase shifts of a few degrees at most). To offset this issue, a lower-frequency front end working in the 4–18 GHz range (essentially with no frequency multiplication) is under development.

With this architectural choice, a standard compact back end that may connect to different front-end units will enable quick instrument re-configuration to probe different plasma scenarios. Ultimately, several back-end units, along with corresponding front ends, may be run in parallel for probing vaster plasma density regions, simultaneously. This type of modular design shares the hardware components with a reflectometer and can be modified into one if needed.

3.5. Proof-of-Concept Application to a Steady-State Plasma Source

A fluorescent lamp was used as a plasma source for demonstrating the interferometry concept. The lamp was placed between the emitter/receptor antennas and a metallic plate, where the electromagnetic waves are reflected. With the lamp in the off position, the received signal has a given phase corresponding to the path that the electromagnetic wave travels over the air and inside the lamp, then back inside the lamp and the air upon bouncing on the metallic plate. Once the lamp is turned on, the electromagnetic wave is further delayed, owing to the presence of free electrons inside the lamp, leading to a phase shift in the signal, according to Equation (10). The setup is shown in Figure 18. Data acquisition is performed using a digital oscilloscope.

The signal frequency was set to 70.8 GHz, corresponding to a wavelength of 4.24 mm. The plasma thickness D inside the lamp tube was calculated, taking the tube diameter D_{tube} and subtracting its thickness t_{tube}. Since the microwave crosses N lamp tubes, the result was multiplied by the $2 \times N$ to account for the reflection at the metallic wall and round trip as shown in Equation (12):

$$D = 2N(D_{tube} - 2 \times t_{tube}) = 2 \times 4 \times (11.4 - 2 \times 1.4) = 68.8 \text{ mm}. \qquad (12)$$

To convert the voltage signal into an angular phase signal, first the full amplitude (2π rad) of the phase signal needs to be known. This is estimated by performing a frequency sweep on the setup and measuring the full amplitude of the signal, yielding 2π rad $\equiv$ 670 mV.

Figure 19 shows the phase difference in relation to base value and the corresponding average electron density. At $t \approx -0.02$ s, the lamp is turned on, leading to a sharp rise in the voltage signal, and turned off at $t \approx 0.42$ s. Once the lamp is turned on, the plasma density starts to increase until a steady state is reached at $t \approx 0.1$ s. During lamp operation, an oscillation of the signal is present, which corresponds to the electric grid outlet frequency of 50 Hz.

Computing the phase difference in radians, and then applying Equation (11), since we verify the condition $D/\lambda > 3$, we obtain an average electron density of 1.2×10^{17} m^{-3}. The results are on the same order of magnitude as the ones observed by Liu in [101], where the author performed a similar experiment to measure the influence of electric frequency on the electron density of the fluorescent lamps.

This simplified experiment highlights the simplicity and flexibility of this non-intrusive technique for measuring plasma electron densities. The tested configuration with the emitting and receiving antenna placed side by side allow for seamless use of the equipment in any arbitrary plasma source, provided it has two facing windows with the front end placed in front of one and a metallic plate placed in front of the other. Obviously, this requires that a careful alignment is performed to ensure that the electromagnetic waves are properly reflected in the metallic plate and not elsewhere. An alternative would be to decouple the receiving antenna, which would be placed facing the emitting one; however, this requires a bespoke configuration adapted for each plasma source (a specific configuration of this type will be used for measurements in ESTHER). Another limitation is that a plasma source needs to have a size of the same order of magnitude or bigger than the horn antennas to ensure that most of the emitted electromagnetic waves effectively cross the plasma. Lastly, the range of frequencies preferentially needs to be "compatible" with the application as discussed in Section 2.4. The frequency should be higher than the plasma cutoff frequency, as otherwise the wave will not propagate, yet it should not be so high that the phase shift becomes as small as the background noise. With this said, this small experiment shows that reasonable measurements of plasma densities may still be achieved with just a few degrees of phase shift.

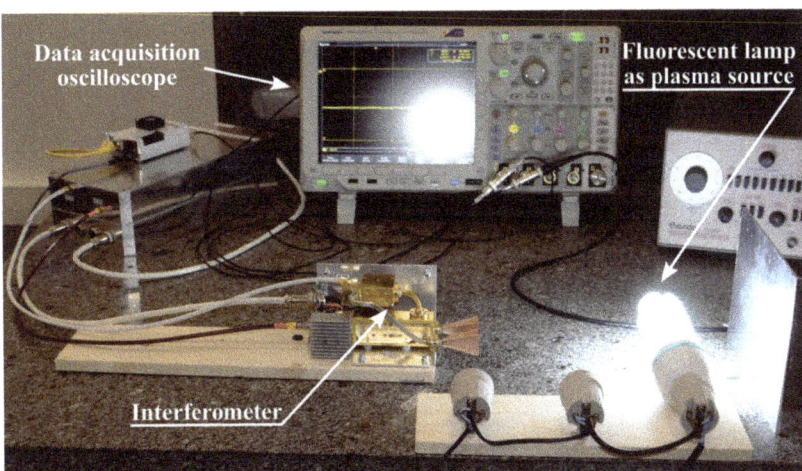

Figure 18. Setup for proof-of-concept experiment.

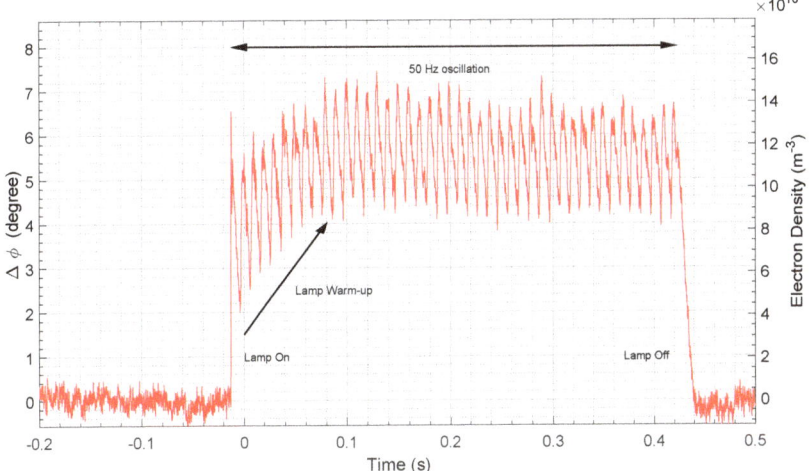

Figure 19. Interferometer phase difference and electron density for the fluorescent lamp plasma source setup.

Microwave electron density measurements may be cross checked with other techniques, specifically Stark broadening, which is the only other diagnostic which may be straightforwardly applied to shock tubes. To properly assess the applicability range of Stark-broadening techniques, two 1D shock wave simulations were ran, yielding the chemical conditions behind the shock wave for a high-ionization case (Jupiter entry) and a low-ionization case (Mars entry). The broadening of the H-α line was then computed from the post-shock conditions for peak electron density. Table 10 presents the expected broadening of the H-α spectral line caused by different broadening mechanisms. A more detailed description of these is found in [20,39,40]. The total spectral broadening may be modeled as a convolution of the Lorentzian and the Doppler broadening. For the Jupiter entry case, the high entry velocity leads to a post-shock temperature in the vicinity of 42,000 K, and a corresponding electron density of about 1.40×10^{21} electron/m^3. In the case of a Mars entry, the peak electron density is 1.74×10^{17} electrons/m^3 at a temperature of around 3200 K. In the former case, the Lorentz component of the broadening is dominated by the Stark effect, whereas in the latter, the collisional term is dominant. As expected, for lower ionization flows, the Stark line broadening effects are not dominant, meaning interferometry measurements are more advantageous. For highly ionized flows, where the plasma cutoff frequencies require THz-rated diagnostics, Stark-broadening techniques might be more straightforward.

Table 10. Discrimination of the different spectral broadening effects for shock tube experiments on the $H-\alpha$ line (656.46 nm). Data were computed from the simulations in Table and using SPARK-LbL code [39].

Case	FWHM Broadening (cm^{-1})					
	Collisional	van der Waals	Resonance	Stark	Lorentz (All)	Doppler
Jupiter	0.056	0.011	0.173	0.522	0.762	2.228
Mars *	0.551	0.068	0.006	0.004	0.630	0.713

* Mars conditions adapted from VUT-1 shock tube test case [64]: pure CO_2, $v_\infty = 3.4$ km/s, $p_\infty = 826.5$ Pa.

4. Conclusions

High-speed events, such as planetary entry shock waves, are very challenging to examine in shock tube facilities, owing to their very short timescales (in the order of the µs), hence mandating the deployment of fast diagnostic techniques. ESTHER is a new state-of-the-art facility designed to reproduce and characterize high-speed entry flows (>10 km/s) by means of spectroscopy and microwave interferometry. The importance of examining the different spectral regions lies in characterizing the physical and chemical processes governing the behavior of the entry plasmas to perfect the numerical models. The instrumentation setups of ESTHER for each of the spectral regions of UV-VUV, visible and NIR/MWIR are discussed. Following a successful application of this concept in the older VUT-1 shock tube, a microwave interferometry setup was developed to enable time-resolved electron-density measurements in ESTHER. An initial proof-of-concept test was conducted, and its results show good agreement with the bibliography review.

As ESTHER is now finally coming to its first operational tests, after a long-winded development cycle of about a decade, it is expected that the array of diagnostics described here will act as an enabler for high-level science and technology obtained in the facility (and other analogues), allowing to shed light on one of the final frontiers of fluid mechanics: aerothermodynamics.

Author Contributions: Conceptualization, R.G.F. and M.L.d.S.; methodology, M.L.d.S.; software, R.G.F., B.B.C. and M.L.d.S.; validation, R.G.F. and M.L.d.S.; formal analysis, R.G.F. and M.L.d.S.; investigation, R.G.F. and M.L.d.S.; resources, M.L.d.S., L.L.A. and B.G.; data curation, R.G.F., B.B.C. and M.L.d.S.; writing—original draft preparation, R.G.F. and M.L.d.S.; writing—review and editing, All; visualization, R.G.F. and M.L.d.S.; supervision, M.L.d.S., L.L.A., B.G., L.M. and V.F.V.; project administration, M.L.d.S., L.L.A., B.G., L.M. and V.F.V.; funding acquisition, M.L.d.S., L.L.A., B.G., L.M. and V.F.V. All authors have read and agreed to the published version of the manuscript.

Funding: This work was conducted under a program of, and funded by, the European Space Agency (contract nos. 4200023086, 4000111557 and 4000118059). The views expressed herein can in no way be taken to reflect the official opinion of the European Space Agency. IST/IPFN manpower was funded by Fundação para a Ciência e Tecnologia under Projects UIDB/50010/2020 and UIDP/50010/2020 and grant PD/BD/114325/2016 (PD-F APPLAuSE).

Institutional Review Board Statement: Not applicable.

Informed Consent Statement: Not applicable.

Data Availability Statement: Not applicable.

Acknowledgments: The authors would like to acknowledge Christopher James, University of Queensland, for useful information regarding the design of optical fiber connections to a shock tube test section. These designs were adapted for our IR setup. The authors would also like to acknowledge Miguel Levy from the University of Lisbon for assistance in image post-processing.

Conflicts of Interest: The authors declare no conflict of interest.

Abbreviations

The following abbreviations are used in this manuscript:

CFD	Computational Fluid Dynamics
EAST	Electric Arc Shock Tube
ESA	European Space Agency
ESTHER	European Shock Tube for High Enthalpy Research
FWHM	Full Width at Half Maximum
HVST	Hyper Velocity Shock Tube
iCCD	Intensified Charge Coupled Device
IPFN	Instituto de Plasmas e Fusão Nuclear
IST	Instituto Superior Técnico
MWIR	Mid-Wavelength Infrared
NETD	Noise Equivalent Temperature Different
NIR	Near Infrared
SPARK	Software Package for Aerothermodynamics Radiation and Kinetics
STAGG	Shock Tube And Gas Gun
UV	Ultraviolet
VCO	Voltage Controlled Oscillator
VUV	Vacuum Ultraviolet

References

1. Johnston, C.O. Evaluating Shock-Tube Informed Biases for Shock-Layer Radiative Heating Simulations. *J. Thermophys. Heat Transf.* **2021**, *35*, 349–361. [CrossRef]
2. Łukasiewicz, J. *Experimental Methods of Hypersonics*; Gasdynamics Series; Marcel Dekker Incorporated: New York, NY, USA, 1973; Volume 3.
3. Delahais, M.; Nerault, M. The Hermes system: Programme status and technology aspects. *Acta Astronaut.* **1991**, *25*, 11–22. [CrossRef]
4. Hannemann, K. High Enthalpy Flows in the HEG Shock Tunnel: Experiment and Numerical Rebuilding (Invited). In Proceedings of the 41st Aerospace Sciences Meeting and Exhibit, Reno, NV, USA, 6–9 January 2003; pp. 1–18. [CrossRef]
5. Rond, C.; Boubert, P.; Félio, J.M.; Chikhaoui, A. Nonequilibrium radiation behind a strong shock wave in CO_2–N_2. *Chem. Phys.* **2007**, *340*, 93–104. [CrossRef]
6. Ramjaun, D.; Dumitrescu, M.; Brun, R. Kinetics of Free Radicals Behind Strong Shock Waves. *J. Thermophys. Heat Transf.* **1999**, *13*, 219–225. [CrossRef]
7. Rond, C.; Boubert, P.; Félio, J.M.; Chikhaoui, A. Radiation Measurements in a Shock Tube for Titan Mixtures. *J. Thermophys. Heat Transf.* **2007**, *21*, 638–646. [CrossRef]
8. Herdrich, G.; Fertig, M.; Lohle, S. Experimental Simulation of High Enthalpy Planetary Entries. *Open Plasma Phys. J.* **2009**, *2*, 150–164. [CrossRef]
9. Ramjaun, D. Cinétique des Radicaux Libres à L'aval D'ondes de Choc dans des Atmosphères Planétaires: Expérimentation en Tube à Choc. Ph.D. Thesis, Université de Provence Marseille, Marseille, France, 1998.
10. Boubert, P.; Rond, C. Nonequilibrium Radiation in Shocked Martian Mixtures. *J. Thermophys. Heat Transf.* **2010**, *24*, 40–49. [CrossRef]
11. Brandis, A. Experimental Study and Modelling of Non-Equilibrium Radiation During Titan and Martian Entry. Ph.D. Thesis, The University of Queensland, Brisbane, Australia, 2009.
12. Desai, P.N.; Lyons, D.T.; Tooley, J.; Kangas, J. Entry, Descent, and Landing Operations Analysis for the Stardust Entry Capsule. *J. Spacecr. Rocket.* **2008**, *45*, 1262–1268. [CrossRef]
13. Lino da Silva, M.; Chikhaoui, A.; Smith, A.; Dudeck, M. *Performance Design and Development of the New European Shock-Tube Facility ESTHER*; Ouwehand, L., Ed.; ESA Special Publication; ESA: Noordwijk, The Netherlands, 2012; Volume 714, p. 20.
14. Lino da Silva, M.; Brotas de Carvalho, B.; Smith, A.; Marraffa, L. High-pressure $H_2/He/O_2$ Combustion Experiments for the Design of the ESTHER Shock-tube Driver. In Proceedings of the 46th AIAA Thermophysics Conference, AIAA Aviation Forum, Washington, DC, USA, 13–17 June 2016; pp. 1–14. [CrossRef]
15. Lino da Silva, M.; Grosso Ferreira, R.; Rodrigues, R.; Alves, L.; Gonçalves, B.; Smith, A.; Merrifield, J.; Villace, V.; Marraffa, L. Qualification of the European Shock-Tube for High Enthalpy Research. In Proceedings of the AIAA Scitech 2020 Forum, American Institute of Aeronautics and Astronautics, Orlando, FL, USA, 6–10 January 2020. [CrossRef]
16. Rond, C.; Boubert, P.; Félio, J.M.; Chikhaoui, A. Studies of the radiative emission behind a shock wave in Titan like mixture on TCM2 shock tube. In Proceedings of the Radiation of High Temperature Gases in Atmospheric Entry, Rome, Italy, 6–8 September 2005; Volume 583, p. 33.

17. Brandis, A.M.; Johnston, C.O.; Cruden, B.A. Investigation of Non-equilibrium Radiation for Earth Entry. In Proceedings of the 46th AIAA Thermophysics Conference, American Institute of Aeronautics and Astronautics, Washington, DC, USA, 13–17 June 2016. [CrossRef]
18. Boubert, P.; Chaix, A.; Chikhaoui, A.; Robin, L.; Vervisch, P. Aerodynamic calibration of TCM2 facility and study of a bow shock layer by emission and laser spectroscopy. *Shock Waves* **2002**, *11*, 341–351. [CrossRef]
19. Lino da Silva, M.; Beck, J. Contribution of CO_2 IR Radiation to Martian Entries Radiative Wall Fluxes. In Proceedings of the 49th AIAA Aerospace Sciences Meeting including the New Horizons Forum and Aerospace Exposition, American Institute of Aeronautics and Astronautics, Orlando, FL, USA, 4–7 January 2011. [CrossRef]
20. Lino da Silva, M. Simulation des Propriétés Radiatives du Plasma Entourant un Véhicule Traversant une Atmosphère Planétaire à Vitesse Hypersonique—Application à la Planète Mars. Ph.D. Thesis, Université d'Orléans, Orleans, France, 2004.
21. Santos Fernandes, L.; Lopez, B.; Lino da Silva, M. Computational fluid radiative dynamics of the Galileo Jupiter entry. *Phys. Fluids* **2019**, *31*, 106104. [CrossRef]
22. Kleb, B.; Johnston, C. Uncertainty analysis of air radiation for lunar return shock layers. In Proceedings of the AIAA Atmospheric Flight Mechanics Conference and Exhibit, Honolulu, HI, USA, 18–21 August 2008; p. 6388.
23. Grinstead, J.; Wilder, M.; Olejniczak, J.; Bogdanoff, D.; Allen, G.; Dang, K.; Forrest, M. Shock-heated air radiation measurements at lunar return conditions. In Proceedings of the 46th AIAA Aerospace Sciences Meeting and Exhibit, Reno, NV, USA, 7–10 January 2008; p. 1244.
24. Cruden, B.; Martinez, R.; Grinstead, J.; Olejniczak, J. Simultaneous vacuum-ultraviolet through near-IR absolute radiation measurement with spatiotemporal resolution in an electric arc shock tube. In Proceedings of the 41st AIAA Thermophysics Conference, San Antonio, TX, USA, 22–25 June 2009; p. 4240. [CrossRef]
25. Cruden, B.A. Electron density measurement in reentry shocks for lunar return. *J. Thermophys. Heat Transf.* **2012**, *26*, 222–230. [CrossRef]
26. Grosso Ferreira, R.; Vicente, J.; Silva, F.; Gonçalves, B.; Cupido, L.; Lino da Silva, M. Reflectometry diagnostics for atmospheric entry applications: State-of-the-art and new developments. *CEAS Space J.* **2023**. [CrossRef]
27. Tauber, M.E. A review of high-speed, convective, heat-transfer computation methods. In *Technical Report TP-2914*; National Aeronautics and Space Administration, NASA Ames Research Center Moffett Field: Silicon Valley, CA, USA, 1989.
28. Tauber, M.E.; Sutton, K. Stagnation-point radiative heating relations for Earth and Mars entries. *J. Spacecr. Rocket.* **1991**, *28*, 40–42. [CrossRef]
29. Luís, D. Performance Design of Hypervelocity Shock Tube Facilities. Master's Thesis, Instituto Superior Técnico, Lisbon, Portugal, 2018.
30. Anderson, J.D., Jr. *Hypersonic and High Temperature Gas Dynamics*; American Institute of Aeronautics & Astronautics: Reston, VA, USA, 2000.
31. Blundell, S.J.; Blundell, K.M. *Concepts in Thermal Physics*, 2nd ed.; Oxford University Press: London, UK, 2009.
32. Davis, D.; Braun, W. Intense vacuum ultraviolet atomic line sources. *Appl. Opt.* **1968**, *7*, 2071–2074. [CrossRef] [PubMed]
33. Bose, D.; Grinstead, J.H.; Bogdanoff, D.W.; Wright, M.J. Shock layer radiation measurements and analysis for mars entry. In Proceedings of the 3rd International Workshop on Radiation of High Temperature Gases in Atmospheric Entry, Heraklion, Greece, 30 September–3 October 2008.
34. Cruden, B.A.; Prabhu, D.; Martinez, R. Absolute radiation measurement in venus and mars entry conditions. *J. Spacecr. Rocket.* **2012**, *49*, 1069–1079. [CrossRef]
35. Cruden, B.A.; Bogdanoff, D.W. Shock radiation tests for Saturn and Uranus entry probes. *J. Spacecr. Rocket.* **2017**, *54*, 1246–1257. [CrossRef]
36. Coelho, J.; Lino da Silva, M. Aerothermodynamic analysis of Neptune ballistic entry and aerocapture flows. *Adv. Space Res.* **2023**, *71*, 3408–3432. [CrossRef]
37. Poloni, E.; Leiser, D.; Ravichandran, R.; Delahaie, S.; Hufgard, F.; Eberhart, M.; Grigat, F.; Sautière, Q.; Loehle, S. Emission Spectroscopy of Plasma Flows for Ice Giant Entry. In Proceedings of the 9th International Workshop on Radiation of High Temperature Gases for Space Missions, Santa Maria, Portugal, 12–16 September 2022.
38. Magin, T.E.; Caillault, L.; Bourdon, A.; Laux, C.O. Nonequilibrium radiative heat flux modeling for the Huygens entry probe. *J. Geophys. Res. Planets* **2006**, *111*, E7. [CrossRef]
39. Lino da Silva, M. The SPARK Line-by-Line Radiative Code, v.3.0. 2021. Available online: http://esther.ist.utl.pt/sparklbl/ (accessed on 24 February 2023).
40. Lino da Silva, M. An adaptive line-by-line—statistical model for fast and accurate spectral simulations in low-pressure plasmas. *J. Quant. Spectrosc. Radiat. Transf.* **2007**, *108*, 106–125. [CrossRef]
41. Bugel, M.; Reynier, P.; Smith, A. Survey of European and Major ISC Facilities for Supporting Mars and Sample Return Mission Aerothermodynamics and Tests Required for Thermal Protection System and Dynamic Stability. *Int. J. Aerosp. Eng.* **2011**, *2011*, 937629. [CrossRef]
42. Reynier, P. Survey of high-enthalpy shock facilities in the perspective of radiation and chemical kinetics investigations. *Prog. Aerosp. Sci.* **2016**, *85*, 1–32. [CrossRef]
43. Hutchinson, I.H. *Principles of Plasma Diagnostics*; Cambridge University Press: Cambridge, UK, 2002.

54. Sharma, S.P.; Park, C. Operating characteristics of a 60-and 10-cm electric arc-driven shock tube. I-The driver. II-The driven section. *J. Thermophys. Heat Transf.* **1990**, *4*, 259–265. [CrossRef]
45. Cruden, B.A. Absolute Radiation Measurements in Earth and Mars Entry Conditions. NATO RTO Lecture Series RTO-EN-AVT-218. 2014. Available online: https://ntrs.nasa.gov/citations/20140008609 (accessed on 5 May 2023).
46. Hermann, T.; Löhle, S.; Bauder, U.; Morgan, R.; Wei, H.; Fasoulas, S. Quantitative Emission Spectroscopy for Superorbital Reentry in Expansion Tube X2. *J. Thermophys. Heat Transf.* **2017**, *31*, 257–268. [CrossRef]
47. Sheikh, U.A.; Morgan, R.G.; McIntyre, T.J. Vacuum Ultraviolet Spectral Measurements for Superorbital Earth Entry in X2 Expansion Tube. *AIAA J.* **2015**, *53*, 3589–3602. [CrossRef]
48. Morgan, R.; McIntyre, T.; Buttsworth, D.; Jacobs, P.; Potter, D.; Brandis, A.; Gollan, R.; Jacobs, C.; Capra, B.; McGilvary, M.; et al. *Shock and Expansion Tube Facilities For the Study of Radiating Flows*; University of Queensland: Brisbane, Australia, 2008.
49. Yamada, G.; Suzuk, T.; Takayanagi, H.; Fujita, K. Development of Shock Tube for Ground Testing Reentry Aerothermodynamics. *Trans. Jpn. Soc. Aeronaut. Space Sci.* **2011**, *54*, 51–61. [CrossRef]
50. Lemal, A.; Nishimura, S.; Nomura, S.; Takayanagi, H.; Matsuyama, S.; Fujita, K. Analysis of VUV radiation measurements from high temperature air mixtures. In Proceedings of the 54th AIAA Aerospace Sciences Meeting, American Institute of Aeronautics and Astronautics, San Diego, CA, USA, 4–8 January 2016. [CrossRef]
51. Takayanagi, H.; Fujita, K. Absolute Radiation Measurements Behind Strong Shock Wave In Carbon Dioxide Flow for Mars Aerocapture Missions. In Proceedings of the 43rd AIAA Thermophysics Conference, American Institute of Aeronautics and Astronautics, New Orleans, LA, USA, 25–28 June 2012. [CrossRef]
52. Takayanagi, H.; Fujita, K. Infrared Radiation Measurement behind Shock Wave in Mars Simulant Gas for Aerocapture Missions. In Proceedings of the 44th AIAA Thermophysics Conference, American Institute of Aeronautics and Astronautics, San Diego, CA, USA, 24–27 June 2013. [CrossRef]
53. Takayanagi, H.; Nomura, S.; Fujita, K. Emission Intensity Measurements around Mars Entry Capsule with a Free-Piston-Driven Expansion Tube. In Proceedings of the 52nd Aerospace Sciences Meeting, American Institute of Aeronautics and Astronautics, National Harbor, MD, USA, 13–17 January 2014. [CrossRef]
54. Parker, R.; MacLean, M.; Dufrene, A.; Holden, M.; Desjardin, P.; Weisberger, J.; Levin, D. Emission Measurements from High Enthalpy Flow on a Cylinder in the LENS-XX Hypervelocity Expansion Tunnel. In Proceedings of the 51st AIAA Aerospace Sciences Meeting including the New Horizons Forum and Aerospace Exposition, American Institute of Aeronautics and Astronautics, Grapevine, TX, USA, 7–10 January 2013. [CrossRef]
55. McGilvray, M.; Doherty, L.J.; Morgan, R.G.; Gildfind, D.; Jacobs, P.; Ireland, P. T6: The Oxford University Stalker Tunnel. In Proceedings of the 20th AIAA International Space Planes and Hypersonic Systems and Technologies Conference, American Institute of Aeronautics and Astronautics, Glasgow, UK, 6–9 July 2015. [CrossRef]
56. Collen, P.; Doherty, L.J.; McGilvray, M. Measurements of radiating hypervelocity air shock layers in the T6 Free-Piston Driven Shock Tube. In Proceedings of the ESA Conference Bureau. ESA Conference Bureau, Louisville, KY, USA, 16 August 2019.
57. Oliveira, B. High-Pressure He/H_2/O_2 Mixtures Combustion on the ESTHER Driver: Experiment and Modeling. Master's Thesis, Instituto Superior Técnico, Lisbon, Portugal, 2021.
58. Grosso Ferreira, R. Laser Ignition of a High-Pressure H_2/He/O_2 Combustible Mixture. Master's Thesis, Instituto Superior Técnico, Lisbon, Portugal, 2017.
59. Grosso Ferreira, R.; Brotas Carvalho, B.; Rodrigues, J.; Rodrigues, R.; Smith, A.; Marraffa, L.; Lino da Silva, M. Unfocused laser ignition of high-pressure He-H_2-O_2 combustible mixtures. *arXiv* **2022**, arXiv:2203.04278v1. [CrossRef]
60. Alpher, R.A.; White, D.R. Flow in shock tubes with area change at the diaphragm section. *J. Fluid Mech.* **1958**, *3*, 457. [CrossRef]
61. Walenta, Z.A. Optimization of the parameters of a double-diaphragm shock tube. *Arch. Mech. Stosow.* **1967**, *19*, 665–685.
62. Mirels, H. Test Time in Low-Pressure Shock Tubes. *Phys. Fluids* **1963**, *6*, 1201. [CrossRef]
63. Billet, S.J.; Smith, A.J. *STAGG User Manual*; Confidential and Property Product of Fluid Gravity Engineering Ltd.: St Andrews, UK, 2013.
64. Chikhaoui, A.; Lino da Silva, M.; Mota, S.; Resendes, D. *Support during the MIPT Shock Tube Calibration*; Technical Report; European Space Agency: Paris, France, 2008.
65. Grosso Ferreira, R.; Lino da Silva, M.; Brotas de Carvalho, B.; Rodrigues, R. High-pressure Combustion and Qualification of ESTHER Shock-tube Driver. In Proceedings of the 9th International Workshop on Radiation of High Temperature Gases for Space Missions, Santa Maria, Portugal, 12–16 September 2022.
66. Espinho, S.; Felizardo, E.; Tatarova, E.; Dias, F.M.; Ferreira, C.M. Vacuum ultraviolet emission from microwave Ar–H_2 plasmas. *Appl. Phys. Lett.* **2013**, *102*, 114101. [CrossRef]
67. Vargas, J.; Lopez, B.; Lino da Silva, M. CDSDv: A compact database for the modeling of high-temperature CO_2 radiation. *J. Quant. Spectrosc. Radiat. Transf.* **2020**, *245*, 106848. [CrossRef]
68. Thorlabs Inc. Multimode Fluoride Fiber Optic Patch Cables. Available online: https://www.thorlabs.com/newgrouppage9.cfm?objectgroup_id=7840 (accessed on 10 December 2022).
69. Korth Kristalle GmbH. Our Materials. Available online: https://www.korth.de/en/materials (accessed on 10 December 2022).
70. Crystran Ltd. Optical Materials. Available online: https://www.crystran.co.uk/optical-materials/sapphire-al2o3 (accessed on 10 December 2022).

71. EKSMA Optics. Sapphire (Al_2O_3) Windows. Available online: https://eksmaoptics.com/optical-components/uv-and-ir-optics/sapphire-al2o3-windows/ (accessed on 10 December 2022).
72. eSource Optics. VUV-UV Optical Material Properties. Available online: https://www.esourceoptics.com/vuv_material_properties.html (accessed on 10 December 2022).
73. MPF Products, Inc. DUV Grade Sapphire Transmission Graph DUV (Laser Grade) Grade Sapphire Transmission Chart. Available online: https://mpfpi.com/duv-sapphire-transmission-chart/ (accessed on 10 December 2022).
74. Nikon Corporation. Optical Components & Materials. Available online: https://www.nikon.com/products/components/downloads/ (accessed on 10 December 2022).
75. Thorlabs Inc. Sapphire Windows. Available online: https://www.thorlabs.com/NewGroupPage9.cfm?ObjectGroup_ID=3982 (accessed on 10 December 2022).
76. Valley Design Corp. Sapphire Optical Properties and Sapphire Optical Transmission. Available online: https://valleydesign.com/sapppic.htm (accessed on 10 December 2022).
77. Beck, J. *CFD Validation in a CO_2 Environment: Synthesis Report*; Technical Report, Fluid Gravity Report CR012/08; CFD: Emsworth, UK, 2008.
78. Anokhin, E.M. Physical and Chemical Relaxation Behind Strong Shock Waves in CO_2-N_2 Mixtures. Ph.D. Thesis, Moscow Institute of Physics and Technology (MIPT), Moscow, Russia, 2005.
79. Toujani, N.; Alquaity, A.B.S.; Farooq, A. Electron density measurements in shock tube using microwave interferometry. *Rev. Sci. Instrum.* **2019**, *90*, 054706. [CrossRef]
80. Schneider, K.P.; Park, C. Shock tube study of ionization rates of NaCl-contaminated argon. *Phys. Fluids* **1975**, *18*, 969. [CrossRef]
81. Ohler, S.; Gilchrist, B.; Gallimore, A. Microwave plume measurements of an SPT-100 using xenon and a laboratory model SPT using krypton. In Proceedings of the 31st Joint Propulsion Conference and Exhibit, American Institute of Aeronautics and Astronautics, San Diego, CA, USA, 10–12 July 1995. [CrossRef]
82. Ohler, S.; Gilchrist, B.E.; Gallimore, A. Microwave Plume Measurements of a Closed Drift Hall Thruster. *J. Propuls. Power* **1998**, *14*, 1016–1021. [CrossRef]
83. Cappelli, M.A.; Gascon, N.; Hargus, W.A. Millimetre wave plasma interferometry in the near field of a Hall plasma accelerator. *J. Phys. Appl. Phys.* **2006**, *39*, 4582–4588. [CrossRef]
84. Reed, G.; Hargus, W.; Cappelli, M. Microwave Interferometry (90 GHz) for Hall Thruster Plume Density Characterization. In Proceedings of the 41st AIAA/ASME/SAE/ASEE Joint Propulsion Conference & Exhibit, American Institute of Aeronautics and Astronautics, Tucson, AZ, USA, 10–13 July 2005. [CrossRef]
85. Kuwabara, N.; Chono, M.; Yamamoto, N.; Kuwahara, D. Electron Density Measurement Inside a Hall Thruster Using Microwave Interferometry. *J. Propuls. Power* **2021**, *37*, 491–494. [CrossRef]
86. Ohler, S.G.; Gilchrist, B.E.; Gallimore, A.D. Nonintrusive electron number density measurements in the plume of a 1 kW arcjet using a modern microwave interferometer. *IEEE Trans. Plasma Sci.* **1995**, *23*, 428–435. [CrossRef]
87. Estrada, T.; Nagasaki, K.; Blanco, E.; Perez, G.; Tribaldos, V. Microwave Reflectometry Diagnostics: Present Day Systems and Challenges for Future Devices. *Plasma Fusion Res.* **2012**, *7*, 2502055. [CrossRef]
88. Bachynski, M.P. Electromagnetic wave penetration of reentry plasma sheaths. *J. Res. Natl. Bur. Stand. Sect. Radio Sci.* **1965**, *69*, 147. [CrossRef]
89. Tudisco, O.; Lucca Fabris, A.; Falcetta, C.; Accatino, L.; De Angelis, R.; Manente, M.; Ferri, F.; Florean, M.; Neri, C.; Mazzotta, C.; et al. A microwave interferometer for small and tenuous plasma density measurements. *Rev. Sci. Instrum.* **2013**, *84*, 033505. [CrossRef] [PubMed]
90. Lopez, B.; Lino Da Silva, M. SPARK: A Software Package for Aerodynamics, Radiation and Kinetics. In Proceedings of the 46th AIAA Thermophysics Conference, American Institute of Aeronautics and Astronautics, Washington, DC, USA, 13–17 June 2016. [CrossRef]
91. Park, C. A review of reaction rates in high temperature air. In Proceedings of the 24th AIAA Thermophysics Conference, American Institute of Aeronautics and Astronautics, Buffalo, NY, USA, 12–14 June 1989. [CrossRef]
92. Brandis, A.M.; Cruden, B.A. Benchmark Shock Tube Experiments of Radiative Heating Relevant to Earth Re-entry. In Proceedings of the 55th AIAA Aerospace Sciences Meeting, American Institute of Aeronautics and Astronautics, Grapevine, TX, USA, 9–13 January 2017. [CrossRef]
93. Gülhan, A.; Thiele, T.; Siebe, F.; Kronen, R.; Schleutker, T. Aerothermal Measurements from the ExoMars Schiaparelli Capsule Entry. *J. Spacecr. Rocket.* **2019**, *56*, 68–81. [CrossRef]
94. Marraffa, L.; Santovincenzo, A.; Roumeas, R.; Huot, J.P.; Scoon, G.; Smith, A. Aerothermodynamics aspects of Venus sample return mission. In Proceedings of the 3rd European Symposium on Aerothermodynamics for Space Vehicles, ESTEC, Noordwijk, The Netherlands, 24–26 November 1998; Volume 426, p. 139.
95. Lino da Silva, M.; Vargas, J. A Physically-Consistent Chemical Dataset for the Simulation of N_2-CH_4 Shocked Flows Up to T = 100,000 K. *arXiv* **2022**, arXiv:2212.09911. [CrossRef]
96. Leibowitz, L.P.; Kuo, T.J. Ionizational Nonequilibrium Heating During Outer Planetary Entries. *AIAA J.* **1976**, *14*, 1324–1329. [CrossRef]
97. Gigosos, M.A.; Cardenoso, V. New plasma diagnosis tables of hydrogen Stark broadening including ion dynamics. *J. Phys. At. Mol. Opt. Phys.* **1996**, *29*, 4795. [CrossRef]

98. Rond, C.; Boubert, P. Chemical Kinetic and Radiative Simulations for Titan Atmospheric Entry. *J. Thermophys. Heat Transf.* **2009**, *23*, 72–82. [CrossRef]
99. James, C.M. Radiation from Simulated Atmospheric Entry into the Gas Giants. Ph.D. Thesis, The University of Queensland, Brisbane, Australia, 2018.
100. Cruden, B.A.; Brandis, A.M.; Prabhu, D.K. Measurement and Characterization of Mid-wave Infrared Radiation in CO_2 Shocks. In Proceedings of the 11th AIAA/ASME Joint Thermophysics and Heat Transfer Conference, Atlanta, GA, USA, 16–20 June 2014. [CrossRef]
101. Liu, Y.; Hou, Z. Diagnostics of plasma electron density and collision frequency of fluorescent lamp using microwave transmission diagnostics. *J. Phys. Conf. Ser.* **2019**, *1324*, 012073. [CrossRef]

Disclaimer/Publisher's Note: The statements, opinions and data contained in all publications are solely those of the individual author(s) and contributor(s) and not of MDPI and/or the editor(s). MDPI and/or the editor(s) disclaim responsibility for any injury to people or property resulting from any ideas, methods, instructions or products referred to in the content.

Article

Angular-Resolved Thomson Parabola Spectrometer for Laser-Driven Ion Accelerators

Carlos Salgado-López *, Jon Imanol Apiñaniz, José Luis Henares, José Antonio Pérez-Hernández, Diego de Luis, Luca Volpe and Giancarlo Gatti

Centro de Láseres Pulsados (CLPU), Edificio M5, Parque Científico USAL, C/Adaja, 8, 37185 Villamayor, Salamanca, Spain; japinaniz@clpu.es (J.I.A.); jlhenares@clpu.es (J.L.H.); japerez@clpu.es (J.A.P.-H.); ddeluis@clpu.es (D.d.L.); lvolpe@clpu.es (L.V.); ggatti@clpu.es (G.G.)
* Correspondence: csalgado@clpu.es

Abstract: This article reports the development, construction, and experimental test of an angle-resolved Thomson parabola (TP) spectrometer for laser-accelerated multi-MeV ion beams in order to distinguish between ionic species with different charge-to-mass ratio. High repetition rate (HHR) compatibility is guaranteed by the use of a microchannel plate (MCP) as active particle detector. The angular resolving power, which is achieved due to an array of entrance pinholes, can be simply adjusted by modifying the geometry of the experiment and/or the pinhole array itself. The analysis procedure allows for different ion traces to cross on the detector plane, which greatly enhances the flexibility and capabilities of the detector. A full characterization of the TP magnetic field is implemented into a relativistic code developed for the trajectory calculation of each pinhole beamlet. We describe the first test of the spectrometer at the 1 PW VEGA 3 laser facility at CLPU, Salamanca (Spain), where up to 15 MeV protons and carbon ions from a 3 μm laser-irradiated Al foil are detected.

Keywords: plama diagnostics; charged-particle spectroscopy; ion beams; instrumentation

Citation: Salgado-López, C.;
Apiñaniz, J.I.; Henares, J.L.;
Pérez-Hernández, J.A.; de Luis, D.;
Volpe, L.; Gatti, G. Angular-Resolved
Thomson Parabola Spectrometer for
Laser-Driven Ion Accelerators.
Sensors 2022, 22, 3239. https://doi.org/10.3390/s22093239

Academic Editor: Bruno Goncalves

Received: 29 March 2022
Accepted: 21 April 2022
Published: 22 April 2022

Publisher's Note: MDPI stays neutral with regard to jurisdictional claims in published maps and institutional affiliations.

Copyright: © 2022 by the authors. Licensee MDPI, Basel, Switzerland. This article is an open access article distributed under the terms and conditions of the Creative Commons Attribution (CC BY) license (https://creativecommons.org/licenses/by/4.0/).

1. Introduction

Since the advent of the chirped pulse amplification (CPA) technology [1], the range of accessible intensities on focus for ultra-bright, short-pulse lasers has increased significantly, reaching current values above 10^{22} W/cm^2 [2]. Such enhancement has paved the way for laser-based particle accelerators, mainly for ions [3,4] and electrons [5]; however, acceleration schemes have been also demonstrated for positrons [6] and neutrons [7].

The range of applications of the accelerated beams is quite rich, profiting from the low-emittance and ultrashort duration (and high peak current) of the generated particle beams, well-fitted characteristics for applications. Specifically, since the demonstration of collimation and monochromatisation of laser-driven multi-MeV ion beams [8,9], their potential employments, such as ultrafast proton probing [10–13], isochoric heating of dense plasmas [14], fast ignition of inertial confinement fusion reactions [15], material science [16], and medical purposes [17,18], have attained plenty of attention [19].

The compactness and costs of high-power laser facilities are important advantages when compared to conventional radio frequency acceleration facilities, as well as the reduced size of the radio protection requirements. Most of the potential industrial applications of these sources require a high time-averaged particle flux, which stresses the importance of developing high-repetition-rate (HRR) laser sources, targetry instrumentation [20], and diagnostics.

One of the key diagnostics for ion acceleration investigation are the Thomson parabola spectrometers, first developed by Thomson in 1907 [21], i.e., in-line diagnostics which can rate the particles depending on their energy, momentum, and charge-to-mass ratio [22]. The element of the spectrometer sensible to particles can be either a passive detector—for instance, imaging plates (IPs) or a CR39 nuclear track detector [23], which require post-processing to retrieve data, or an active one -microchannel-plate [24]—or plastic scintillators [25], well fitted for HRR operation, due to their ability to perform on-line measurements for every single laser shot. The main drawback of an ordinary TP is the incapability of deconvolving the angular distribution of the measured beam, as only a particular angle of the beam (with an insignificant angular spread) is measured, as the particles measured have to cross a pinhole. This fact also makes this detector specially sensitive to alignment.

Tracing the angular-resolved spectrum of the ions is a vital milestone in the study of the beam properties. For instance, from this kind of work, we have learnt that the most widely used laser-driven ion acceleration mechanism, target normal sheath acceleration (TNSA) [3,4], normally achieved by the ultra-bright laser irradiation of thin metallic films, is able to emit extraordinarily laminar, low-emittance beams from the rear surface of the target [26], coming from a source with a diameter size as big as a few hundred micrometers [10] and a total beam divergence angle around 20°. Potential applications benefit from the transport properties of these laminar beams, which have proven to be suitable when focused to millimetre-sized spots [27,28]. While increasing the resolution of the diagnostics responsible for measuring the ion phase space, new beam features have been discovered, such as the beam pointing deviation from the target normal for certain laser and target conditions [29–31]. Tomography-like measurements have also shown that there is a different source size and divergence for each ion energy [32–36]. Non-laminar ion acceleration has also been demonstrated when triggering plasma instabilities under specific circumstances, for instance due to the generation of a preplasma prior to the laser interaction with the target [37–39] or the use of ultrathin (nanometric-thick) film targets in the radiation pressure acceleration scheme (RPA) [40,41].

In order to retrieve angular-resolved spectral information about the beam, radiochromic film (RCF) or scintillator stacks are practical diagnostic tools [25,42], typically yielding a discretized spectrum of $\Delta E \approx 1$ MeV, which is much coarser when compared to the continuous spectral TP resolution [22]. Moreover, these diagnostics cannot discriminate between different q/m ionic species. This fact is compensated in some experimental layouts by the combination of perforated RCF stacks and TPs [43]; thus, part of the beam reaches the latter. Despite yielding complementary information, this method is limited in spectral resolution for most parts of the beam.

In this work, we present a multi-pinhole Thomson parabola spectrometer, which combines sharp spectral/angular precision, besides the ionic species sorting capability. Furthermore, the use of a MCP detector device allows for single-shot HRR acquisition. Section 2 describes the basic operation principle of the detector and depicts its physical properties and parameters. The experimental layout where the detector was tested and the analysed results are presented in Section 3.

2. Materials and Methods

2.1. Thomson Parabola Design and Operation

The Thomson parabola works according to magnetic and electric sector spectrometer principles. The entrance pinhole selects a beamlet composed by ions with a specific charge-to-mass ratio q/m, with $q = Ze$. The ion charge is deflected by parallel (or antiparallel) magnetic B and electric E fields of length l_2. The initial velocity of the particles is perpendicular to the field lines; thus, the deflection directions of the two fields are mutually orthogonal, allowing species separation (by electric field, in x-axis) and energy separation (mostly by magnetic field, in y-axis) after some propagation distance l_3. The ions are measured on a two-dimensional spatially resolved particle-sensitive detector; in our case, a MCP. In the small deflection approximation $\sin(\theta) \approx \theta$, considering perfectly sharp and homogeneous fields and nonrelativistic particle energies, the deviation coordinates at the detector plane caused by the Lorentz force are given by

$$x = \frac{qEl_2l_3}{2E_{kin}}, \qquad (1)$$

$$y = \frac{qBl_2l_3}{\sqrt{2mE_{kin}}}, \qquad (2)$$

where E_{kin} is the kinetic energy of the ion. When combining (1) and (2), we obtain the parabolic equation

$$y^2 = \frac{q}{m}\frac{B^2 l_2 l_3}{E} x. \qquad (3)$$

Ions with the same charge-to-mass ratio will reach the same parabolic trace on the detector plane; meanwhile, their position along the trace will define their energy. Photons are not deflected by the fields and travel straightly through the source–pinhole axis and impact the MCP, providing the zero deflection reference (used for spectrum data interpretation).

The ultimate energy resolution of the TP depends on the spatial separation of the different energies at the active area of the detector (which depends on the magnetic field strength B, its length l_2 and the distance to MCP l_3) and on the capability of the system to resolve this separation (which depends on the magnification of the imaging system collecting MCP signal and on the trace thickness δ). In the considered approximation, δ (which is inversely proportional to the spectrometer resolution) is given by the setup geometry and the pinhole diameter d, similarly as in a pinhole camera, as $\delta = d + (s+d)L'/L$, where L is the distance from source to pinhole, $L' = l_1 + l_2 + l_3$ is the distance from pinhole to detector, and s is the source size [44].

Several concept modifications have been proposed to improve the basic functioning of TPs, such as a tunable magnetic dipole [45] or electromagnets [24] for adaptable energy resolution, exotic electrode geometry (trapezoidal or wedged) [45–47] for extended retrieval of lower part of the spectrum, transient electric field for time-gated measurement of the beam [48], designs with two in-line entrance pinholes for spatially resolved measurements of the ion source [36], or simultaneous measurements of ion and electron [49] or plasma-emitted extreme ultraviolet radiation spectra [50].

2.2. Multi-Pinhole Thomson Parabola Spectrometer

Here, we propose a modification of the basic TP design, consisting on the substitution of the pinhole by a horizontal array of pinholes. This array chops the incoming cone of particles in several beamlets which are simultaneously detected. In this way, we can measure the different angles adding angularly resolved spectral information. Similar strategies were already proposed [32–36,51–55] in most of the cases, dismissing the electric field for charge-to-mass ratio inspections, as the authors claim to accelerate a single ion species (protons). Some references [56] showed the possibility of joining different ion diagnostics in order to have extended information about the beam, but lacking ion discrimination capability at different subtended angles. A few works have proposed [45] or demonstrated [57] an absolute capability of angular spectral-q/m resolution, but with strong limitations in the available species to be investigated, as well as their subtended angles, due to data analysis constraints. In order to facilitate the spectrum retrieval, they managed to avoid crossing traces from neighbouring beamlets at the detector plane.

We propose a more general multi-pinhole TP spectrometer, including the use of electric and magnetic fields for identifying different q/m ions, angular selection of beamlets, and a more generic post-processing method, which does not limit the available ion species to be investigated.

In the case shown here, three pinholes of $d = 200\,\mu$m, separated by $a = 3$ mm and aligned along the x-axis (electric deflection direction, see Figure 1), were drilled into a W 1 mm-thick 25 mm-diameter plate. A copper nosepiece was designed to easily exchange between these substrates, with different pinhole array combinations with pre-set alignment orientations. A $B = 0.4$ T, $l_2 = 75$ mm long permanent dipole magnet is located $l_1 = 13.5$ mm after the pinholes. The direction of the magnetic field $(-x)$ deflects the ions upwards $(+y)$. Magnets are attached to an iron yoke, leaving a gap between poles of 16 mm, on-axis with respect to the central pinhole. Two thin copper electrodes are placed over the magnet poles. A variable voltage difference can be applied between electrodes, up to

$U = 10$ kV, which deviates the particles parallel to the field lines. After a propagation distance $l_3 = 135$ mm a 80×30 mm^2, a single-stage MCP with attached phosphor screen (Hamamatsu F283-12P, 12 µm channel diameter) converts the two-dimensional ion traces into visible photons at the rear side of the TP. This signal is acquired by a properly calibrated image recording system, in order to retrieve particle data for every laser shot. All the components are light-tight covered by a shielding made of overlapped layers of Al, polyethylene, and Pb, whose goal is to protect the MCP from secondary radiation sources. The full assembly is 10 kg heavy and $130 \times 255 \times 175$ mm^3 in volume, which makes it relatively easy to set and align inside an experimental vacuum chamber.

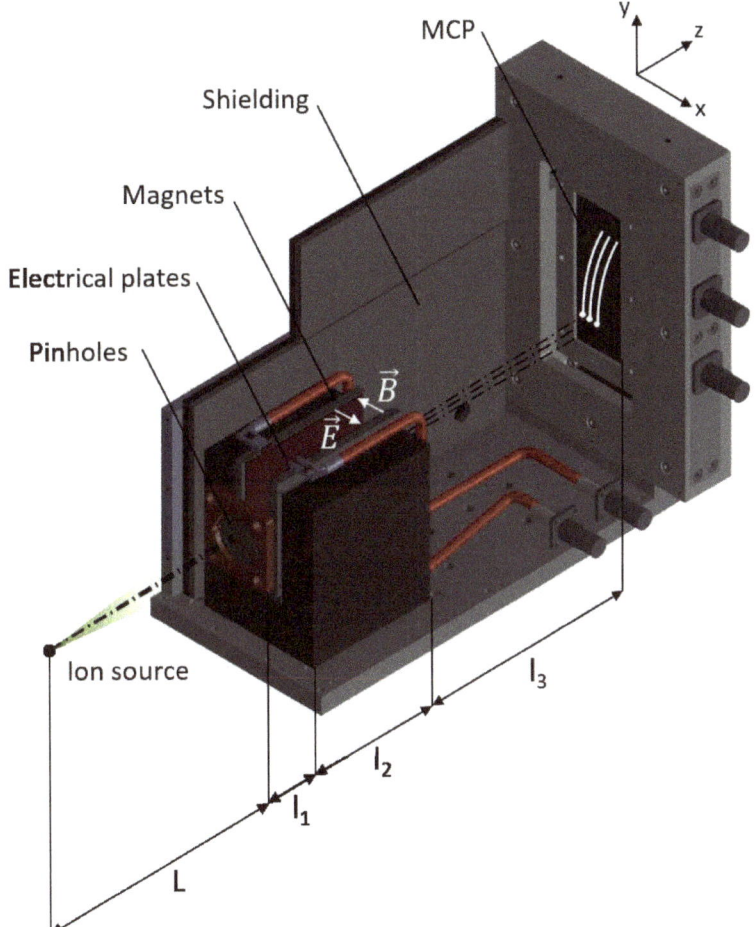

Figure 1. Multi-pinhole TP spectrometer design.

A 3D numerical solver was developed for trajectory simulation to provide expected traces in MCP and energy–position relation for all the required charged species. It consisted of a second-order Verlet algorithm and the numerical error was estimated by comparison to analytical results on 0.23% at 20 MeV. The magnetic field was mapped by a Hall probe in the central vertical plane (between dipole magnets) and implemented in the code for realistic trajectory calculation (see Figure 2). The electric field could not be measured; therefore, it is implemented in the code as a perfectly sharp and homogeneous field between plates, ignoring edge effects. The numerical simulation provided precise energy position calibrations, accounting for the exact trajectory of particles coming from each pinhole of the array.

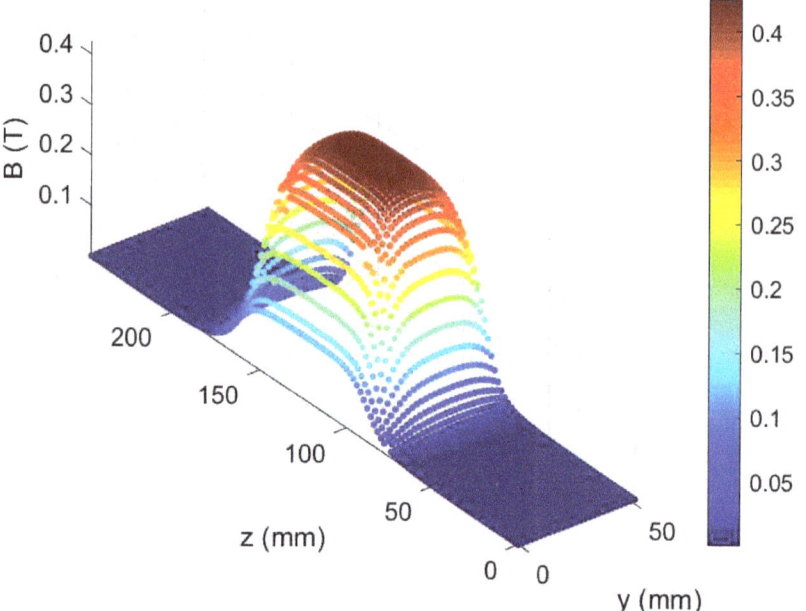

Figure 2. Measured magnetic field distribution. Ions propagate in z-direction through the field towards the MCP, deflected on the y-direction by the Lorentz force.

In the presented design, proton energies between 300 keV and 25 MeV are accessible. The ultimate energy resolution of the TP spectrometer is determined by the energy separation at the MCP position ($\Delta E/\Delta x$) and the trace width. For instance, at 20 MeV, $\Delta E/\Delta x = 4.33$ MeV/mm, with a measured trace width of 0.30 mm, the minimum uncertainty is 4.33 MeV/mm × 0.3 mm = 1.3 MeV.

In this geometrical configuration, the crossing of traces from different pinholes are abundant and need to be processed (see Figure 3). Crossings produce peak artifacts in the spectrum readout due to intersections. To avoid the artifacts, we proceeded in identifying the species involved on each crossing event. If one works in the pixel position vs the pixel value space, the peaks produced by intersections are aligned in position for all the involved species and can be treated in the same basis. The intersection peaks were identified, located, and eliminated from the raw data to later perform a linear interpolation between both edges of the gap left by extraction. All the intersections were identified, but only the ones producing significant distortion were eliminated. We eliminated the peaks that were still observable after performing a Gaussian smoothing over the data in the pixel space. The convolved Gaussian radius was set to 10 points as Fourier analysis of raw data revealed peaks of size structures from 5 to 20 points. In this way, both statistical variations and small structures up to 20 points were considered as noise and contribute to the root mean square deviation (RMSD) value. The interpolated segments were considered to follow the trend within variations according to this RMSD.

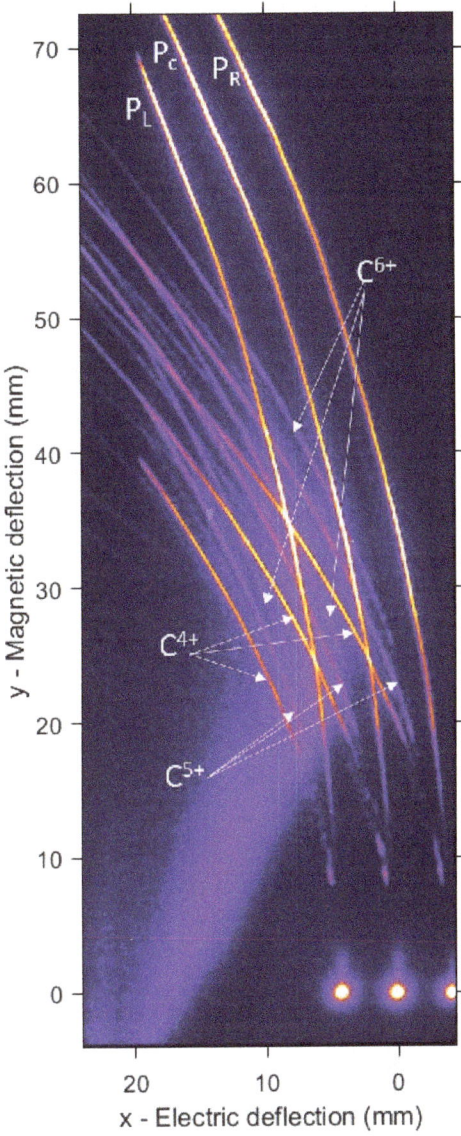

Figure 3. Multi-pinhole Thomson parabola traces obtained from a single laser shot at VEGA 3. The origin of the coordinate system corresponds to the zero-deflection point of the central beamlet. P_L, P_C, and P_R corresponds to the tracks related to the left, central, and right beamlets, respectively. C^{n+} indicates the three n-charged ion traces. The halo at the bottom left corner is originated by a leak of white light from plasma emission, which reaches the MCP through the junction of MCP structure and shielding.

3. Experimental Test and Results

The detector was tested at the VEGA 3 petawatt laser facility at Centro de Láseres Pulsados (CLPU, Spain). The target, a 3 μm-thick Al planar foil, was irradiated at 10° to form the normal target in the horizontal plane by the VEGA 3 laser. For the case shown, a single 0.8 μm linearly p-polarized pulse of 10.9 J (on target) and a 180 fs full width at a half-maximum (FWHM) duration was focalized by an F/11 off-axis parabolic mirror onto a 9.7 µm FWHM spot, containing 20% of the pulse energy, yielding an averaged intensity of 1.7×10^{19} W/cm^2 inside the FWHM and resulting in the acceleration of ions from the contamination layer on the target rear surface, rich in H and C atoms, by the TNSA mechanism. As a side effect, ultraviolet and X-rays photons are generated during the laser–plasma interaction, which can define the non-deflected axis of the detector.

The TP spectrometer was carefully aligned so the central pinhole was exactly facing the interaction point in the normal direction of the target surface at a distance $L = 508$ mm. The electrode plates were fed with a voltage difference of $U = 10,000$ V. The photons from the phosphor screen were collected by an imaging system consisting of an objective (Nikon AF-S NIKKOR 18–105 mm 1:3.5–5.6G ED) with adaptable focal length range of 18–105 mm and a Blackfly PGE-23S6M-C CMOS optical camera of 1920×1200 pixels with a laterial size of 5.86 µm. The system was set up to an ultimate magnification of 0.0854 at the detector plane (14.57 pixels/mm), which was calibrated thanks to a laser machined pattern in the object plane, next to the MCP. For the detection distance L, the choice of pinhole separation $a = 3$ mm keeps a reasonably separation of the three traces group, occupying as much as possible the MCP surface. Figure 3 shows the raw traces acquired in a single laser shot.

It is important to note that the source-to-detector distance L in the setup prepared is much larger than the source size s (typically a few hundred of micrometers in diameter). The magnification of the pinhole camera effect on the detector plane is barely enough to distinguish spatial effects of the source, therefore considering a point-like source for analysis means. The examined angles in this case are 0 and $\pm \alpha$, being $\alpha = \tan^{-1}(a/L) = 0.3°$.

A single beamlet proton spectrum is plotted in Figure 4, together with the traces intersections signal peaks, which are considered measurement artifacts and removed. Figure 5 shows the three spectra retrieved from the three analysed proton traces. As expected, due to the small difference of the angles probed when compared to the typical TNSA beam divergence (around 20°), all spectra are similar. This is also true for the carbon ions (see Figure 6). For sake of comparison, different ion species from the same pinhole are plotted in Figure 7. The RMSD was calculated from raw data with respect to smoothed data and some example values are 8310 (a.u.) for left trace protons and 655 (a.u.) for C^{4+} left.

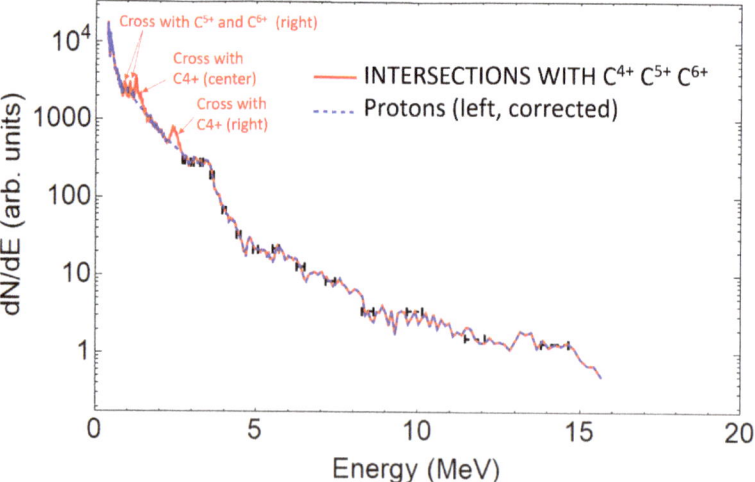

Figure 4. Left trace proton spectrum in logarithmic scale. Blue: corrected spectrum. Red: same spectrum showing the peaks subtracted, corresponding to trace crosses. Several representative error bars are plotted, picturing the estimated energy resolution.

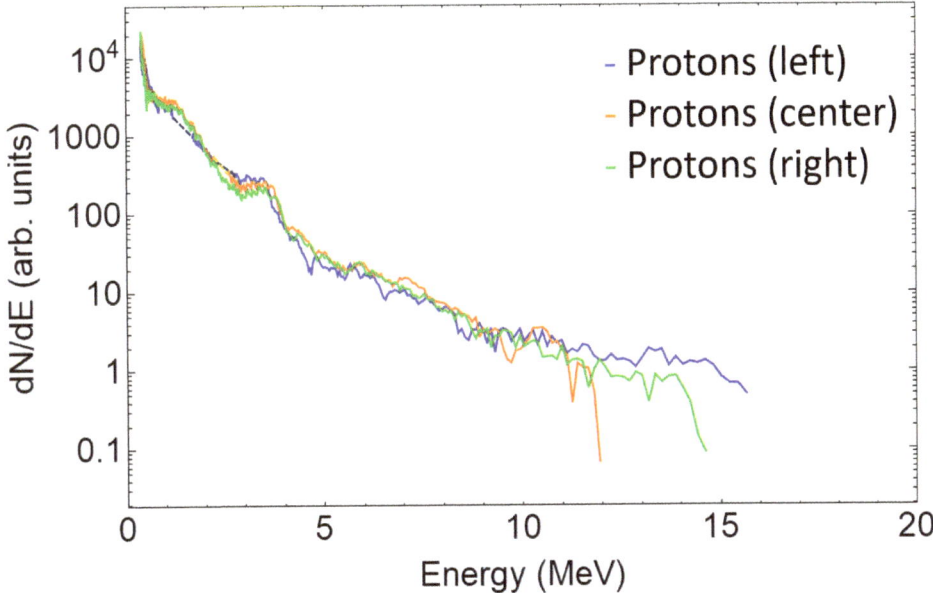

Figure 5. Left, center, and right proton beamlets spectra in the logarithmic scale. Dashed lines show the reconstructed spectrum range. It is important to note that the crosstalk between traces lays at different energy locations for each beamlet. This effect is clear from Figure 3; the relative horizontal position of each proton trace makes the crosses appear at different energetic levels. Note for instance that the proton right trace does not suffer any crosstalk; thus, no correction is applied.

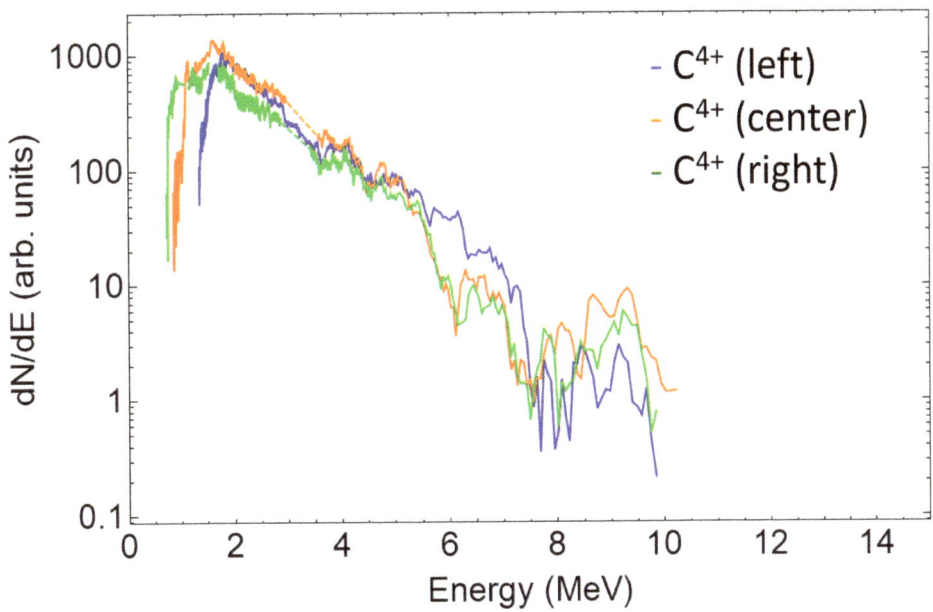

Figure 6. Left, center, and right C^{4+} beamlets spectra in the logarithmic scale. Dashed lines show the reconstructed spectrum range.

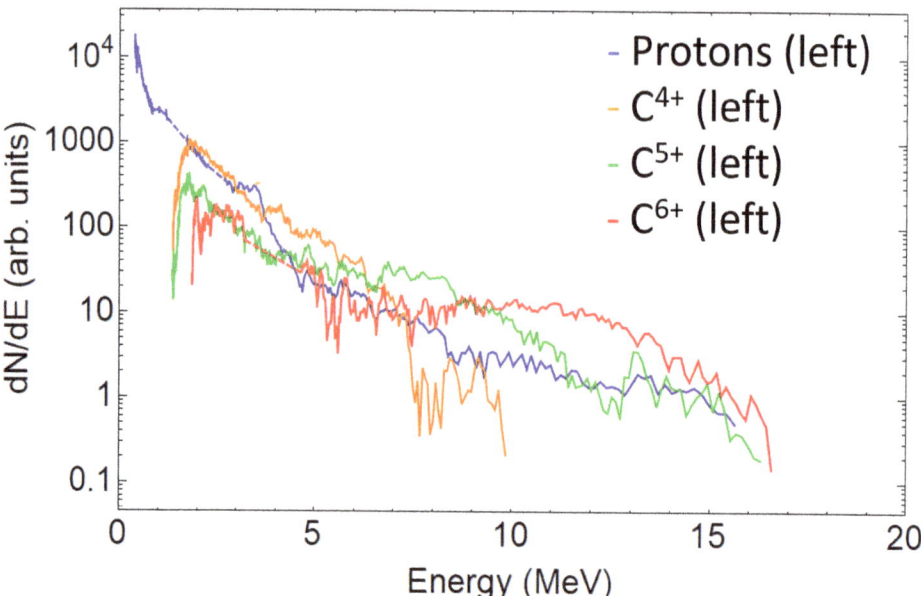

Figure 7. Left beamlets spectra for protons and C^{4+}, C^{5+}, and C^{6+} ions in the logarithmic scale. Dashed lines show the reconstructed spectrum range.

4. Discussion

We designed, built, and tested a Thomson parabola which can measure the spectra of discretized beamlets with different emission angles from a laser–plasma interaction experiment at a high repetition rate, simultaneously sorting the ion species by their charge-to-mass ratio.

A novel analysis method, which can examine the crossing parabolic traces on the detector plane, grants access to several variables parallelly. The peak artifacts due to trace crossing are corrected by subtraction of the contribution of the non-desired traces by means of a fitted gaussian intersection function in the pixel position space, ensuring the graph continuity of the analyzed trace.

Future improvements of the detector are considered. An absolute charge calibration can be used a next step for detector refinement, by combining the use of the MCP with CR39 or RCF stacks [24], or possibly by testing the TP in a charge-controlled electrostatic particle accelerator. From the results presented, it also became clear that one of the limitations of the system is the parallel configuration of the electric plates. The detector can be improved by adapting it to a wedged geometry (separation between the electrodes increases with ion path) [45–47], thus preventing low energy ions from colliding with the plates, specially pronounced for the ions from the less favourable (left) beamlet (see for instance "left" carbons spectra in Figures 6 and 7), as also demonstrated in Figure 3. Additionally, an increased size of MCP (or use of other larger 2-D active detector) can be beneficial. The conversion of the TP into an ion wide angle spectrometer [55] (iWASP) is possible by switching off the electric potential of the electrodes, maintaining only the magnetic deflection, and replacing the pinhole array by an horizontal micrometer-wide slit.

Finally, new pinhole array combinations are foreseen to explore different and more numerous probing angles. A larger pinhole imaging magnification can be implemented by placing the detector closer to the target [30], followed by a proper dimensioning of the pinhole array configuration, implying clearer variations detected between traces. For instance, a line of three pinholes with separation $a = 500\,\mu m$ can set the TP as close as $L = 12.5$ mm by avoiding particle collisions with the electrodes, yielding probing angles of 0 and $\pm \alpha \approx \pm 4.5°$, where the differences in spectra among the beamlets will be presumably noticeable. Moreover, each pinhole can apply an imaging magnification of $L'/L \approx 18$, which will considerably enhance the spatial resolution. Considering high laminar beams, this high magnification imaging mode should be sufficient to resolve changes on the target emission coordinates and beam pointing variations as a function of energy for each beamlet [30,33–35], therefore granting a beam emittance characterization. This increased magnification can cause a cost of loss of energy resolution. An estimation of the resolution in the worst case scenario of a pure non-laminar beam provides a value of ± 3.5 MeV at 15 MeV. However, a large percentage of laminarity is expected at these energies to reduce the trace thickness and increase the resolution. In any case, the observation of how laminarity depends on energy is itself an important achievement, and this type of setup can provide a direct measurement of it. On the other hand, it seems possible to apply the multi-pinhole TP to diagnose energy (and species) laser-generated ion beamlines. In such a case, the detector field configuration can be adapted in order to comprise an adequate range of energies, therefore improving the energy resolution considerably. All these possibilities show the potential interest of such kind of detectors for beam analysis of novel acceleration mechanisms under research, such as collisionless shock acceleration (CSA) [58] or RPA [40,41,59], as well as for measurements of transported ion beamlines, well fitted for applications.

Author Contributions: Conceptualization, C.S.-L., J.I.A., J.L.H. and G.G.; methodology, C.S.-L., J.I.A., J.L.H. and J.A.P.-H.; software, C.S.-L. and J.I.A.; validation, C.S.-L., J.I.A., J.L.H., J.A.P.-H. and L.V.; formal analysis, C.S.-L., J.I.A., J.L.H. and J.A.P.-H.; investigation, C.S.-L., J.I.A. and G.G.; resources, D.d.L.; data curation, C.S.-L.; writing—original draft preparation, C.S.-L.; writing—review and editing, C.S.-L., J.I.A., J.L.H. and L.V; visualization, C.S.-L., J.I.A. and D.d.L.; supervision, L.V. and G.G. All authors have read and agreed to the published version of the manuscript.

Funding: The research leading to these results has received funding from LASERLAB-EUROPE V (Grant Agreement No. 871124, European Union Horizon 2020 research and innovation program), and from IMPULSE (Grant Agreement No. 871161, European Union Horizon 2020 research and innovation program). Support from Equipment Grant No. EQC2018-005230-P, Unidad de Investigación Consolidada 167 from Junta de Castilla y León and Junta de Castilla y León (Grant No. CLP263P20), are acknowledged.

Institutional Review Board Statement: Not applicable.

Informed Consent Statement: Not applicable.

Data Availability Statement: Not applicable.

Acknowledgments: We thank all the staff of CLPU involved in the experiment where the spectrometer was tested, including the technical area, the radio protection department, and the engineering section. Special mention should be made for the scientific division personnel collaborating in the experimental campaign and D. Arana who machined several pieces of the detector at CLPU workshop.

Conflicts of Interest: The authors declare no conflict of interest.

Abbreviations

The following abbreviations are used in this manuscript:

TP	Thomson parabola
HHR	High repetition rate
MCP	Microchannel plate
CPA	Chirped pulse amplification
IP	Imaging plate
TNSA	Target normal sheath acceleration
RPA	Radiation pressure acceleration
RCF	Radiochromic film
FWHM	Full width at half-maximum
iWASP	Ion-wide angle spectrometer
CSA	Collisionless shock acceleration

References

1. Strickland, D.; Mourou, G. Compression of amplified chirped optical pulses. *Opt. Commun.* **1985**, *55*, 447–449. [CrossRef]
2. Yoon, J.W.; Kim, Y.G.; Choi, I.W.; Sung, J.H.; Lee, H.W.; Lee, S.K.; Nam, C.H. Realization of laser intensity over 10^{23}W/cm^2. *Optica* **2021**, *8*, 630–635. [CrossRef]
3. Daido, H.; Nishiuchi, M.; Pirozhkov, A.S. Review of laser-driven ion sources and their applications. *Rep. Prog. Phys.* **2012**, *75*, 056401. [CrossRef] [PubMed]
4. Macchi, A.; Borghesi, M.; Passoni, M. Ion acceleration by superintense laser-plasma interaction. *Rev. Mod. Phys.* **2013**, *85*, 751–793. [CrossRef]
5. Esarey, E.; Schroeder, C.B.; Leemans, W.P. Physics of laser-driven plasma-based electron accelerators. *Rev. Mod. Phys.* **2009**, *81*, 1229–1285. [CrossRef]
6. Chen, H.; Wilks, S.C.; Meyerhofer, D.D.; Bonlie, J.; Chen, C.D.; Chen, S.N.; Courtois, C.; Elberson, L.; Gregori, G.; Kruer, W.; et al. Relativistic Quasimonoenergetic Positron Jets from Intense Laser-Solid Interactions. *Phys. Rev. Lett.* **2010**, *105*, 015003. [CrossRef]
7. Roth, M.; Jung, D.; Falk, K.; Guler, N.; Deppert, O.; Devlin, M.; Favalli, A.; Fernandez, J.; Gautier, D.; Geissel, M.; et al. Bright Laser-Driven Neutron Source Based on the Relativistic Transparency of Solids. *Phys. Rev. Lett.* **2013**, *110*, 044802. [CrossRef]
8. Ter-Avetisyan, S.; Schnürer, M.; Polster, R.; Nickles, P.; Sandner, W. First demonstration of collimation and monochromatisation of a laser accelerated proton burst. *Laser Part. Beams* **2008**, *26*, 637–642. [CrossRef]
9. Toncian, T.; Borghesi, M.; Fuchs, J.; d'Humières, E.; Antici, P.; Audebert, P.; Brambrink, E.; Cecchetti, C.A.; Pipahl, A.; Romagnani, L.; et al. Ultrafast Laser-Driven Microlens to Focus and Energy-Select Mega-Electron Volt Protons. *Science* **2006**, *312*, 410–413. [CrossRef]
10. Borghesi, M.; Mackinnon, A.J.; Campbell, D.H.; Hicks, D.G.; Kar, S.; Patel, P.K.; Price, D.; Romagnani, L.; Schiavi, A.; Willi, O. Multi-MeV Proton Source Investigations in Ultraintense Laser-Foil Interactions. *Phys. Rev. Lett.* **2004**, *92*, 055003. [CrossRef]
11. Santos, J.J.; Bailly-Grandvaux, M.; Giuffrida, L.; Forestier-Colleoni, P.; Fujioka, S.; Zhang, Z.; Korneev, P.; Bouillaud, R.; Dorard, S.; Batani, D.; et al. Laser-driven platform for generation and characterization of strong quasi-static magnetic fields. *New J. Phys.* **2015**, *17*, 083051. [CrossRef]
12. Apiñaniz, J.I.; Malko, S.; Fedosejevs, R.; Cayzac, W.; Vaisseau, X.; De Luis, D.; Gatti, G.; McGuffey, C.; Bailly-Grandvaux, M.; Bhutwala, K.; et al. A quasi-monoenergetic short time duration compact proton source for probing high energy density states of matter. *Sci. Rep.* **2021**, *11*, 6881. [CrossRef] [PubMed]
13. Malko, S.; Cayzac, W.; Ospina-Bohorquez, V.; Bhutwala, K.; Bailly-Grandvaux, M.; McGuffey, C.; Fedosejevs, R.; Vaisseau, X.; Tauschwitz, A.; Aginako, J.A.; et al. Proton stopping measurements at low velocity in warm dense carbon. *Nat. Commun.* **2017**. [CrossRef]
14. Patel, P.K.; Mackinnon, A.J.; Key, M.H.; Cowan, T.E.; Foord, M.E.; Allen, M.; Price, D.F.; Ruhl, H.; Springer, P.T.; Stephens, R. Isochoric Heating of Solid-Density Matter with an Ultrafast Proton Beam. *Phys. Rev. Lett.* **2003**, *91*, 125004. [CrossRef]
15. Roth, M.; Cowan, T.E.; Key, M.H.; Hatchett, S.P.; Brown, C.; Fountain, W.; Johnson, J.; Pennington, D.M.; Snavely, R.A.; Wilks, S.C.; et al. Fast Ignition by Intense Laser-Accelerated Proton Beams. *Phys. Rev. Lett.* **2001**, *86*, 436–439. [CrossRef]
16. Mirani, F.; Maffini, A.; Casamichiela, F.; Pazzaglia, A.; Formenti, A.; Dellasega, D.; Russo, V.; Vavassori, D.; Bortot, D.; Huault, M.; et al. Integrated quantitative PIXE analysis and EDX spectroscopy using a laser-driven particle source. *Sci. Adv.* **2021**, *7*, eabc8660. [CrossRef]
17. Spencer, I.; Ledingham, K.; Singhal, R.; McCanny, T.; McKenna, P.; Clark, E.; Krushelnick, K.; Zepf, M.; Beg, F.; Tatarakis, M.; et al. Laser generation of proton beams for the production of short-lived positron emitting radioisotopes. *Nucl. Instrum. Methods Phys. Res. Sect. B* **2001**, *183*, 449–458. [CrossRef]
18. Ledingham, K.W.; Bolton, P.R.; Shikazono, N.; Ma, C.M. Towards Laser Driven Hadron Cancer Radiotherapy: A Review of Progress. *Appl. Sci.* **2014**, *4*, 402–443. [CrossRef]
19. Volpe, L.; Fedosejevs, R.; Gatti, G.; Pérez-Hernández, J.A.; Méndez, C.; Apiñaniz, J.; Vaisseau, X.; Salgado, C.; Huault, M.; Malko, S.; et al. Generation of high energy laser-driven electron and proton sources with the 200 TW system VEGA 2 at the Centro de Laseres Pulsados. *High Power Laser Sci. Eng.* **2019**, *7*, e25. [CrossRef]
20. Puyuelo Valdes, P.; de Luis, D.; Hernandez, J.; Apiñaniz, J.; Curcio, A.; Henares, J.L.; Huault, M.; Perez-Hernandez, J.A.; Roso, L.; Gatti, G.; et al. Implementation of a thin, flat water target capable of high-repetition-rate MeV-range proton acceleration in a high-power laser at the CLPU. *Plasma Phys. Control. Fusion* **2022**, accepted. [CrossRef]
21. Thomson, J. XLVII. On rays of positive electricity. *Lond. Edinb. Dublin Philos. Mag. Sci.* **1907**, *13*, 561–575. [CrossRef]
22. Bolton, P.R.; Borghesi, M.; Brenner, C.; Carroll, D.C.; De Martinis, C.; Fiorini, F.; Flacco, A.; Floquet, V.; Fuchs, J.; Gallegos, P.; et al. Instrumentation for diagnostics and control of laser-accelerated proton (ion) beams. *Phys. Medica* **2014**, *30*, 255–270. [CrossRef] [PubMed]
23. Cobble, J.A.; Flippo, K.A.; Offermann, D.T.; Lopez, F.E.; Oertel, J.A.; Mastrosimone, D.; Letzring, S.A.; Sinenian, N. High-resolution Thomson parabola for ion analysis. *Rev. Sci. Instrum.* **2011**, *82*, 113504. [CrossRef] [PubMed]
24. Harres, K.; Schollmeier, M.; Brambrink, E.; Audebert, P.; Blažević, A.; Flippo, K.; Gautier, D.C.; Geißel, M.; Hegelich, B.M.; Nürnberg, F.; et al. Development and calibration of a Thomson parabola with microchannel plate for the detection of laser-accelerated MeV ions. *Rev. Sci. Instrum.* **2008**, *79*, 093306. [CrossRef] [PubMed]

25. Huault, M.; De Luis, D.; Apiñaniz, J.I.; De Marco, M.; Salgado, C.; Gordillo, N.; Gutiérrez Neira, C.; Pérez-Hernández, J.A.; Fedosejevs, R.; Gatti, G.; et al. A 2D scintillator-based proton detector for high repetition rate experiments. *High Power Laser Sci. Eng.* **2019**, *7*, e60. [CrossRef]
26. Cowan, T.E.; Fuchs, J.; Ruhl, H.; Kemp, A.; Audebert, P.; Roth, M.; Stephens, R.; Barton, I.; Blazevic, A.; Brambrink, E.; et al. Ultralow Emittance, Multi-MeV Proton Beams from a Laser Virtual-Cathode Plasma Accelerator. *Phys. Rev. Lett.* **2004**, *92*, 204801. [CrossRef]
27. Brack, F.E.; Kroll, F.; Gaus, L.; Bernert, C.; Beyreuther, E.; Cowan, T.E.; Karsch, L.; Kraft, S.; Kunz-Schughart, L.A.; Lessmann, E.; et al. Spectral and spatial shaping of laser-driven proton beams using a pulsed high-field magnet beamline. *Sci. Rep.* **2020**, *10*, 9118. [CrossRef]
28. Milluzzo, G.; Petringa, G.; Catalano, R.; Cirrone, G.A. Handling and dosimetry of laser-driven ion beams for applications. *Eur. Phys. J. Plus.* **2021**, *136*, 1170. [CrossRef]
29. Lindau, F.; Lundh, O.; Persson, A.; McKenna, P.; Osvay, K.; Batani, D.; Wahlström, C.G. Laser-Accelerated Protons with Energy-Dependent Beam Direction. *Phys. Rev. Lett.* **2005**, *95*, 175002. [CrossRef]
30. Schreiber, J.; Ter-Avetisyan, S.; Risse, E.; Kalachnikov, M.P.; Nickles, P.V.; Sandner, W.; Schramm, U.; Habs, D.; Witte, J.; Schnürer, M. Pointing of laser-accelerated proton beams. *Phys. Plasmas* **2006**, *13*, 033111. [CrossRef]
31. Nakamura, T.; Mima, K.; Ter-Avetisyan, S.; Schnürer, M.; Sokollik, T.; Nickles, P.V.; Sandner, W. Lateral movement of a laser-accelerated proton source on the target's rear surface. *Phys. Rev. E* **2008**, *77*, 036407. [CrossRef]
32. Chen, H.; Hazi, A.U.; van Maren, R.; Chen, S.N.; Fuchs, J.; Gauthier, M.; Le Pape, S.; Rygg, J.R.; Shepherd, R. An imaging proton spectrometer for short-pulse laser plasma experiments. *Rev. Sci. Instrum.* **2010**, *81*, 10D314. [CrossRef] [PubMed]
33. Ter-Avetisyan, S.; Schnürer, M.; Nickles, P.V.; Sandner, W.; Nakamura, T.; Mima, K. Correlation of spectral, spatial, and angular characteristics of an ultrashort laser driven proton source. *Phys. Plasmas* **2009**, *16*, 043108. [CrossRef]
34. Ter-Avetisyan, S.; Borghesi, M.; Schnürer, M.; Nickles, P.V.; Sandner, W.; Andreev, A.A.; Nakamura, T.; Mima, K. Characterization and control of ion sources from ultra-short high-intensity laser–foil interaction. *Plasma Phys. Control. Fusion* **2009**, *51*, 124046. [CrossRef]
35. Ter-Avetisyan, S.; Schnürer, M.; Nickles, P.V.; Sandner, W.; Borghesi, M.; Nakamura, T.; Mima, K. Tomography of an ultrafast laser driven proton source. *Phys. Plasmas* **2010**, *17*, 063101. [CrossRef]
36. Ter-Avetisyan, S.; Romagnani, L.; Borghesi, M.; Schnürer, M.; Nickles, P. Ion diagnostics for laser plasma experiments. *Nucl. Instrum. Methods Phys. Res. Sect. A Accel. Spectrometers Detect. Assoc. Equip.* **2010**, *623*, 709–711. [CrossRef]
37. Göde, S.; Rödel, C.; Zeil, K.; Mishra, R.; Gauthier, M.; Brack, F.E.; Kluge, T.; MacDonald, M.J.; Metzkes, J.; Obst, L.; et al. Relativistic Electron Streaming Instabilities Modulate Proton Beams Accelerated in Laser-Plasma Interactions. *Phys. Rev. Lett.* **2017**, *118*, 194801. [CrossRef]
38. Scott, G.G.; Brenner, C.M.; Bagnoud, V.; Clarke, R.J.; Gonzalez-Izquierdo, B.; Green, J.S.; Heathcote, R.I.; Powell, H.W.; Rusby, D.R.; Zielbauer, B.; et al. Diagnosis of Weibel instability evolution in the rear surface density scale lengths of laser solid interactions via proton acceleration. *New J. Phys.* **2017**, *19*, 043010. [CrossRef]
39. Qin, C.Y.; Zhang, H.; Li, S.; Zhai, S.H.; Li, A.X.; Qian, J.Y.; Gui, J.Y.; Wu, F.X.; Zhang, Z.X.; Xu, Y.; et al. Mapping non-laminar proton acceleration in laser-driven target normal sheath field. *High Power Laser Sci. Eng.* **2022**, *10*, e2. [CrossRef]
40. Palmer, C.A.J.; Schreiber, J.; Nagel, S.R.; Dover, N.P.; Bellei, C.; Beg, F.N.; Bott, S.; Clarke, R.J.; Dangor, A.E.; Hassan, S.M.; et al. Rayleigh-Taylor Instability of an Ultrathin Foil Accelerated by the Radiation Pressure of an Intense Laser. *Phys. Rev. Lett.* **2012**, *108*, 225002. [CrossRef]
41. Gonzalez-Izquierdo, B.; King, M.; Gray, R.; et al.. Towards optical polarization control of laser-driven proton acceleration in foils undergoing relativistic transparency. *Nat. Commun.* **2016**, *7*, 12891. [CrossRef] [PubMed]
42. Nürnberg, F.; Schollmeier, M.; Brambrink, E.; Blažević, A.; Carroll, D.C.; Flippo, K.; Gautier, D.C.; Geißel, M.; Harres, K.; Hegelich, B.M.; et al. Radiochromic film imaging spectroscopy of laser-accelerated proton beams. *Rev. Sci. Instrum.* **2009**, *80*, 033301. [CrossRef] [PubMed]
43. Flippo, K.; Hegelich, B.; Albright, B.; Yin, L.; Gautier, D.; Letzring, S.; Schollmeier, M.; Schreiber, J.; Schulze, R.; Fernandez, J.; et al. Laser-driven ion accelerators: Spectral control, monoenergetic ions and new acceleration mechanisms. *Laser Part. Beams* **2007**, *25*, 3–8. [CrossRef]
44. Rajeev, R.; Rishad, K.P.M.; Trivikram, T.M.; Narayanan, V.; Krishnamurthy, M. A Thomson parabola ion imaging spectrometer designed to probe relativistic intensity ionization dynamics of nanoclusters. *Rev. Sci. Instrum.* **2011**, *82*, 083303. [CrossRef]
45. Kojima, S.; Inoue, S.; Dinh, T.H.; Hasegawa, N.; Mori, M.; Sakaki, H.; Yamamoto, Y.; Sasaki, T.; Shiokawa, K.; Kondo, K.; et al. Compact Thomson parabola spectrometer with variability of energy range and measurability of angular distribution for low-energy laser-driven accelerated ions. *Rev. Sci. Instrum.* **2020**, *91*, 053305. [CrossRef]
46. Carroll, D.; Brummitt, P.; Neely, D.; Lindau, F.; Lundh, O.; Wahlström, C.G.; McKenna, P. A modified Thomson parabola spectrometer for high resolution multi-MeV ion measurements—Application to laser-driven ion acceleration. *Nucl. Instrum. Methods Phys. Res. Sect. A Accel. Spectrometers Detect. Assoc. Equip.* **2010**, *620*, 23–27. [CrossRef]
47. Gwynne, D.; Kar, S.; Doria, D.; Ahmed, H.; Cerchez, M.; Fernandez, J.; Gray, R.J.; Green, J.S.; Hanton, F.; MacLellan, D.A.; et al. Modified Thomson spectrometer design for high energy, multi-species ion sources. *Rev. Sci. Instrum.* **2014**, *85*, 033304. [CrossRef]
48. Ter-Avetisyan, S.; Schnürer, M.; Nickles, P.V. Time resolved corpuscular diagnostics of plasmas produced with high-intensity femtosecond laser pulses. *J. Phys. Appl. Phys.* **2005**, *38*, 863–867. [CrossRef]

49. Ter-Avetisyan, S.; Schnürer, M.; Busch, S.; Risse, E.; Nickles, P.V.; Sandner, W. Spectral Dips in Ion Emission Emerging from Ultrashort Laser-Driven Plasmas. *Phys. Rev. Lett.* **2004**, *93*, 155006. [CrossRef]
50. Ter-Avetisyan, S.; Ramakrishna, B.; Doria, D.; Sarri, G.; Zepf, M.; Borghesi, M.; Ehrentraut, L.; Stiel, H.; Steinke, S.; Priebe, G.; et al. Complementary ion and extreme ultra-violet spectrometer for laser-plasma diagnosis. *Rev. Sci. Instrum.* **2009**, *80*, 103302. [CrossRef]
51. Sokollik, T.; Schnürer, M.; Ter-Avetisyan, S.; Nickles, P.V.; Risse, E.; Kalashnikov, M.; Sandner, W.; Priebe, G.; Amin, M.; Toncian, T.; et al. Transient electric fields in laser plasmas observed by proton streak deflectometry. *Appl. Phys. Lett.* **2008**, *92*, 091503. [CrossRef]
52. Ter–Avetisyan, S.; Schnürer, M.; Nickles, P.V.; Sokollik, T.; Risse, E.; Kalashnikov, M.; Sandner, W.; Priebe, G. The Thomson deflectometer: A novel use of the Thomson spectrometer as a transient field and plasma diagnostic. *Rev. Sci. Instrum.* **2008**, *79*, 033303. [CrossRef] [PubMed]
53. Zheng, Y.; Su, L.N.; Liu, M.; Liu, B.C.; Shen, Z.W.; Fan, H.T.; Li, Y.T.; Chen, L.M.; Lu, X.; Ma, J.L.; et al. Note: A new angle-resolved proton energy spectrometer. *Rev. Sci. Instrum.* **2013**, *84*, 096103. [CrossRef]
54. Yang, S.; Yuan, X.; Fang, Y.; Ge, X.; Deng, Y.; Wei, W.; Gao, J.; Fu, F.; Jiang, T.; Liao, G.; et al. A two-dimensional angular-resolved proton spectrometer. *Rev. Sci. Instrum.* **2016**, *87*, 103301. [CrossRef]
55. Jung, D.; Hörlein, R.; Gautier, D.C.; Letzring, S.; Kiefer, D.; Allinger, K.; Albright, B.J.; Shah, R.; Palaniyappan, S.; Yin, L.; et al. A novel high resolution ion wide angle spectrometer. *Rev. Sci. Instrum.* **2011**, *82*, 043301. [CrossRef] [PubMed]
56. Senje, L.; Yeung, M.; Aurand, B.; Kuschel, S.; Rödel, C.; Wagner, F.; Li, K.; Dromey, B.; Bagnoud, V.; Neumayer, P.; et al. Diagnostics for studies of novel laser ion acceleration mechanisms. *Rev. Sci. Instrum.* **2014**, *85*, 113302. [CrossRef] [PubMed]
57. Zhang, Y.; Zhang, Z.; Zhu, B.; Jiang, W.; Cheng, L.; Zhao, L.; Zhang, X.; Zhao, X.; Yuan, X.; Tong, B.; et al. An angular-resolved multi-channel Thomson parabola spectrometer for laser-driven ion measurement. *Rev. Sci. Instrum.* **2018**, *89*, 093302. [CrossRef]
58. Fiuza, F.; Stockem, A.; Boella, E.; Fonseca, R.A.; Silva, L.O.; Haberberger, D.; Tochitsky, S.; Gong, C.; Mori, W.B.; Joshi, C. Laser-Driven Shock Acceleration of Monoenergetic Ion Beams. *Phys. Rev. Lett.* **2012**, *109*, 215001. [CrossRef]
59. Macchi, A.; Cattani, F.; Liseykina, T.V.; Cornolti, F. Laser Acceleration of Ion Bunches at the Front Surface of Overdense Plasmas. *Phys. Rev. Lett.* **2005**, *94*, 165003. [CrossRef]

Article

Development of the Measurement of Lateral Electron Density (MOLE) Probe Applicable to Low-Pressure Plasma Diagnostics

Si-jun Kim [1], Sang-ho Lee [1,2], Ye-bin You [1], Young-seok Lee [1], In-ho Seong [1], Chul-hee Cho [1], Jang-jae Lee [3] and Shin-jae You [1,4,*]

1. Applied Physics Lab for PLasma Engineering (APPLE), Department of Physics, Chungnam National University, Daejeon 34134, South Korea; sjk@o.cnu.ac.kr (S.-j.K.); esangho35@kimm.re.kr (S.-h.L.); 201500963@o.cnu.ac.kr (Y.-b.Y.); lerounsukre@o.cnu.ac.kr (Y.-s.L.); showing123@o.cnu.ac.kr (I.-h.S.); paulati@o.cnu.ac.kr (C.-h.C.)
2. Department of Plasma Engineering, Korea Institute of Machinery and Materials (KIMM), Daejeon 34104, South Korea
3. Samsung Electronics, Hwaseong-si 18448, South Korea; jangjae2.lee@samsung.com
4. Institute of Quantum Systems (IQS), Chungnam National University, Daejeon 34134, South Korea
* Correspondence: sjyou@cnu.ac.kr

Abstract: As the importance of measuring electron density has become more significant in the material fabrication industry, various related plasma monitoring tools have been introduced. In this paper, the development of a microwave probe, called the measurement of lateral electron density (MOLE) probe, is reported. The basic properties of the MOLE probe are analyzed via three-dimensional electromagnetic wave simulation, with simulation results showing that the probe estimates electron density by measuring the surface wave resonance frequency from the reflection microwave frequency spectrum (S_{11}). Furthermore, an experimental demonstration on a chamber wall measuring lateral electron density is conducted by comparing the developed probe with the cutoff probe, a precise electron density measurement tool. Based on both simulation and experiment results, the MOLE probe is shown to be a useful instrument to monitor lateral electron density.

Keywords: plasma diagnostics; non-invasive electron density measurement; planar microwave probes; plasma monitoring

Citation: Kim, S.-j.; Lee, S.-h.; You, Y.-b.; Lee, Y.-s.; Seong, I.-h.; Cho, C.-h.; Lee, J.-j.; You, S.-j. Development of the Measurement of Lateral Electron Density (MOLE) Probe Applicable to Low-Pressure Plasma Diagnostics. *Sensors* **2022**, *22*, 5487. https://doi.org/10.3390/s22155487

Academic Editor: Bruno Goncalves

Received: 21 June 2022
Accepted: 18 July 2022
Published: 22 July 2022

Publisher's Note: MDPI stays neutral with regard to jurisdictional claims in published maps and institutional affiliations.

Copyright: © 2022 by the authors. Licensee MDPI, Basel, Switzerland. This article is an open access article distributed under the terms and conditions of the Creative Commons Attribution (CC BY) license (https://creativecommons.org/licenses/by/4.0/).

1. Introduction

Plasma applications in modern technologies cover numerous fields such as material fabrication, nuclear fusion, medical treatment, agriculture, and catalysis [1–4]. Among them, plasma has played a key role in the state-of-the-art processing of semiconductor fabrication, particularly in ultra-high-aspect ratio etching, atomic layer etching, and deposition, since plasma has chemically and physically active species [2,5].

Recently, to improve throughput and productivity, process monitoring technology has become significant because the price of a patterned wafer has steadily grown and loss by unstable process is no longer negligible [6–9]. Particularly, advanced process control (APC), which is a process-tuning technology based on real-time signals from monitoring devices, has attracted strong interest from industry [10,11]. The APC gathers monitoring data in real-time and diagnoses whether plasma processing steps are normal or not based on post-process algorithms [12–14]. Non-invasive measurement devices create tremendous monitoring data such as optical emission spectra, voltages applied and current flowing through the discharge electrode or antenna, capacitor positions in an impedance matcher, throttle valve positions, gas flow rates, and plasma parameters such as electron density and temperature [13,15–18].

Electron density is a significant factor among these monitoring parameters since it is believed to be directly related to processing time and quality [2,5,10,11,17,18]. To measure electron density non-invasively, various diagnostic methods have been developed, including actinometry and the line ratio method by analyzing optical emission spectra [6,19,20], laser Thomson scattering by measuring scattered laser light [20,21], and planar microwave probes by analyzing reflected or transmitted microwave signals [18,22,23]. They all have been frequently employed in the research field but some optical and laser methods have limitations for industrial application, as follows. First, the actinometry and line ratio approaches are only applicable in a narrow processing window, while the laser Thomson scattering method requires a large and stable space to generate the laser and fine-tune its sensitivity. Second, both optical and laser methods are strongly affected by any contamination of the viewport [24]. On the other hand, planar microwave probes are not affected by contamination of the microwave antenna [25] and only slightly perturb the processing plasma [26]. Hence, the planar microwave probe is seen as a promising electron density monitoring device, as evidenced by numerous probe designs as follows.

Based on the plasma cutoff phenomenon, which was firstly employed in a microwave interferometry measuring phase shift due to the plasma cutoff for measuring the line-integrated electron density [27,28], You et al. developed the planar cutoff probe (PCP), which measures the cutoff frequency in the transmission microwave frequency spectrum (S_{21}), which is defined as transmitted power of port 2 over radiated power from port 1 in a frequency spectrum [10,11,18,25]. They proposed the planar cutoff probe [11], developed it as a real-time monitoring instrument [17], and further optimized and analyzed the PCP [10,18]. Sugai et al. developed the planar curling probe (CP), based on the quarter-wave resonance (QWR) on a curling antenna, that measures the shift of QWR frequency induced by plasma in the reflection microwave frequency spectrum (S_{11}), which is defined as reflected power of port 1 over radiated power from port 1 in a frequency spectrum [22,29–31]. They deeply analyzed the CP [29–31], developed it as a real-time monitoring instrument for electron density as well as film thickness [22]. With similar principle of the CP, Beckers et al. have developed microwave cavity resonance spectroscopy (MCRS), that measures cavity resonance frequency shift by plasma in S11 spectrum [32–34]. In the recent attempt, they demonstrated operation of the MCRS with the wall-mounted invasive antenna. Brinkmann et al. developed the planar multipole resonance probe (pMRP) based on dipole resonance; this probe measures multipole resonance frequencies in S_{11} [23,35–38]. They introduced the pMRP [36], developed it as a monitoring sensor [37], and further investigated the pMRP [23,38]. Among those probes, the MCRS and microwave interferometry show wide measurement range and short time resolution, whereas there is lack of study for measurement limitations of the PCP, the CP, and the pMRP; those planar probes would be deeply investigated in terms of its detection limitations. Although these planar probes all show high measurement accuracy, the probe module size and design are bulky and complex, respectively.

A small and simple probe size is desirable for the following reasons. First, a small probe design lowers plasma perturbations as well as thermal damage on the probe antenna from the plasma. Recently, high ion fluxes and energy have been generated with high average power (10–30 kW), so thermal damage issues should be considered [39,40]. Second, a simple probe design expands its applicability, such as enabling installation into an electrostatic chuck (ESC) as well as on vacuum chamber walls, with improved assembly tolerance. It should be noted that the capability to be installed onto chamber walls in place of an ESC or powered radio-frequency (rf) electrode has various advantages: addressing rf noise issues induced by high power application, minimizing process perturbation (with no requirement for any modification of an ESC or rf electrode before probe insertion) [17,18], and lowering the thermal damage from high-energy ions and neutrals created near the rf electrode.

To realize such advantages, this paper proposes the measurement of lateral electron density (MOLE) probe. As the probe can be made by merely cutting an RG-401 coaxial cable, it represents the smallest and simplest design among planar microwave probes. For the principle, the MOLE probe is based on the surface wave resonance (SWR) at the plasma–sheath interface and measures the SWR frequency in S_{11}. Here, a sheath is an ion space charge region either covering materials immersed into plasma or plasma itself and is caused by mobility difference of electron and ions. In fact, the MOLE probe shares the same operation physics as the plasma absorption probe (PAP) [41] and the plasma transmission probe (PTP) [42]. The difference compared with the PAP will be discussed in the next section. Here, we briefly explain the difference between the PTP and the MOLE probe: the former measures the SWR frequency in S_{21}, but the latter does so in S_{11} [42]. This fact brings about two effects: first, the PTP requires a receiving antenna but the MOLE probe does not. Second, the sensitivity of the MOLE probe is higher than that of the PTP; the background signal level is high enough to be comparable to the SWR signal since the radiating and receiving antennae of the PTP are adjacent through a dielectric medium. Furthermore, the SWR frequency is closer to the ideal SWR frequency for the MOLE probe than for the PTP, and thus the MOLE probe can more precisely measure the SWR frequency.

The remainder of this paper is as follows. In the second section, the simulation method to verify the operation physics of the proposed probe is discussed. In the third section, an experimental demonstration of the MOLE probe is described and analyzed. Finally, in the conclusion section, a summary of this paper is provided.

2. Three-Dimensional Electromagnetic Wave Simulation Analysis

2.1. Simulation Details

To analyze the basic properties of the MOLE probe, the high-frequency time-domain solver provided by CST Microwave Suite was utilized. It solves Maxwell's equations in three-dimensional (3D) space via the finite-difference time-domain method for the typical microwave range, from several MHz to several tens of GHz. This simulation approach has been widely used to analyze and optimize microwave probes such as the planar cutoff probe, planar curling probe, and planar multipole resonance probe; related simulation details can be found in [10,23,31,35].

Figure 1 shows a schematic diagram for the simulation configuration. A coaxial cable is partially immersed into a rectangular plasma of dimensions $100 \times 100 \times 300$ mm^3. The plasma was assumed as a dispersive dielectric material, or more specifically the Drude model provided by the solver. One can find details on the Drude model in [10]. Briefly, based on the Drude model, the plasma dielectric constant can be calculated by the electron density (n_e), collision frequency (ν_m), and microwave frequency (f). Here, ν_m is calculated with the same assumptions as in [10], and the sheath is assumed as a vacuum (the relative dielectric constant of the sheath is unity). In fact, the sheath is an ion-space charge region (rigorously not a vacuum) but in terms of microwave, of which frequency is much higher than ion plasma oscillation frequency, ions are immobile and therefore the sheath can be assumed as a vacuum. In this simulation condition, n_e, ν_m, and sheath width are variables.

The benchmark dimension of the coaxial cable is a RG-401 cable used in this experiment; not exactly the same but similar dimensions for simple numeral. The coaxial cable consists of a cylindrical copper core 2.0 mm in diameter, a polytetrafluoroethylene (PTFE) tube 6.7 mm in outer diameter and 2.0 mm in inner diameter, a relative dielectric constant of 2.1, and an aluminum tube 8.0 mm in outer diameter and 6.7 mm in inner diameter.

A nanosecond Gaussian pulse ($P_{in}(t)$) including microwaves from 0 to 10 GHz is applied at the waveguide port, as shown in Figure 1. Then, reflected signals ($P_{ref}(t)$) come from the end of the probe, sheath, and plasma in the time domain. As we interested in analysis on the reflected signals in the frequency domain, by using fast Fourier transform, both $P_{in}(t)$ and $P_{ref}(t)$ are converted into $P_{in}(f)$ and $P_{ref}(f)$, which are input and reflected powers, respectively, in the frequency domain. Then, S_{11} is calculated as $10\log_{10}(P_{in}(f)/P_{ref}(f))$.

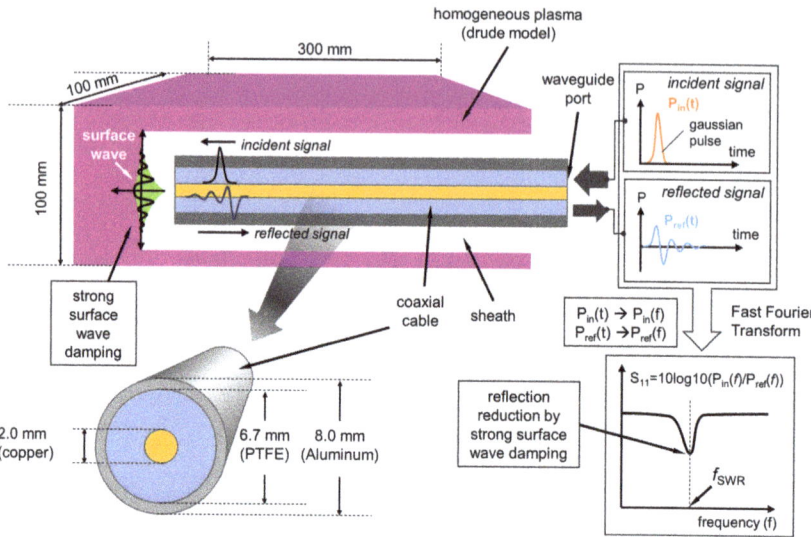

Figure 1. Schematic diagram for the simulation configuration.

2.2. Validation of Surface Wave Resonance

To analyze the basic characteristics of the SWR, a simple simulation model was employed with a thick ground shield 300 mm in diameter and other parameters the same as those in Figure 1. Then, the plasma–sheath interface near the coaxial cable becomes flat and the surface wave propagation on the interface is clearly visualized as shown in Figure 2a–c, which show the magnitude of the absolute electric field with a cross-sectional view at $\pi/2$ phase and different frequencies. Based on Figure 2a,b, a surface wave is generated from the plasma–sheath interface. It can be noted that the magnitude of the electric field at a frequency of 1.40 GHz is significantly large, indicating that the surface wave at this frequency is strongly localized at the interface and absorbed by the plasma. This minimizes the amount of the reflected waves. Such a result corresponds to the simulation result that the minimum S_{11} is at 1.40 GHz, as shown in Figure 2d. The surface wave disappears at higher frequencies (>1.40 GHz), as shown in Figure 2c, which implies that the dispersion relation of the surface wave no longer exists above 1.40 GHz and, hence, this frequency indicates the SWR frequency [43]. Furthermore, this frequency is significantly close to the SWR frequency theoretically defined as $f_{SWR} = f_{pe}/\sqrt{2}$ [5,43] at an infinite boundary and without collisions, where f_{pe} is the plasma oscillation frequency ($f_{pe} = 8980\sqrt{n_e}$); this theoretical SWR frequency is about 1.42 GHz. It can therefore be concluded that the resonance frequency at which the S_{11} value is minimum is the SWR frequency.

Moreover, the fact that the electric field strongly localizes at the plasma–sheath interface means that the surface wave resonance frequency includes information on the plasma, the dimensions of which are nearly the same as the diameter of the probe. This implies that the MOLE probe can measure the lateral electron density when it is installed on a chamber wall.

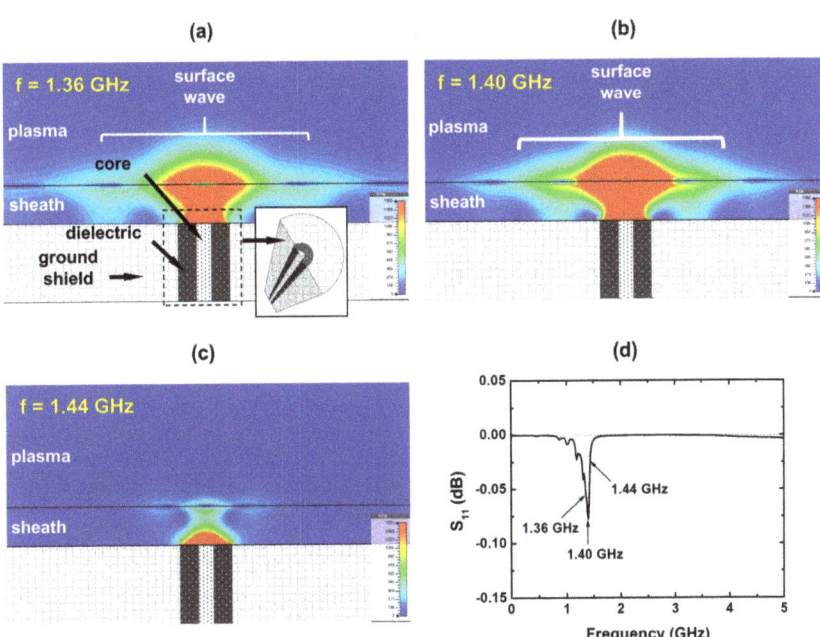

Figure 2. Magnitude of the absolute electric field ($E^2 = E_x^2 + E_y^2 + E_z^2$) at different frequencies: (**a**) 1.36 GHz, (**b**) 1.40 GHz, and (**c**) 1.44 GHz, and (**d**) S_{11} spectrum The electron density is 5×10^{10} cm^{-3}, pressure 100 mTorr, and sheath width 5.0 mm. Here, the resonance frequency is 1.40 GHz.

2.3. Simulation Results and Discussion

Through the last section, it can be verified that the SWR induces the minimum S_{11} value. Now, this section further analyzes the SWR frequency based on the configuration shown in Figure 1. Figure 3 plots the S_{11} spectra from 0 to 10 GHz with various input electron densities and pressures at a constant sheath width of 0.234 mm. At a low input n_e of 5.0×10^9 cm^{-3}, one can find a sharp resonance peak near 0.3 GHz and multiple peaks above 1 GHz. The former is believed to be the SWR peak and the others due to standing waves formed inside the coaxial cable based on an electric field analysis, which is not included in this paper as it is considered out of scope. As shown in Figure 3a–c, both the SWR frequency and the Q-factor (sharpness of the peak) increase with increasing input n_e. As mentioned previously, at the SWR condition, the surface wave is localized at the plasma–sheath interface and absorbed by the plasma, leading to an abrupt drop in S_{11} at the SWR frequency. These wave confinement and absorption effects increase at higher electron densities, with the signal reduction at the SWR frequency becoming stronger.

Furthermore, the Q-factor abruptly decreases with increasing pressure, as shown in Figure 3. At high-pressure conditions, electrons barely interact with electromagnetic (E/M) waves due to frequent collisions between argon atoms and electrons, and consequently the wave absorption by the plasma becomes inefficient. As a result, the wave confinement effect decreases and the S_{11} reduction at the SWR frequency becomes weaker. Figure 4 plots the calculated f_{SWR} over input n_e at various pressures and sheath widths. The dashed line in each plot indicates the ideal SWR frequency as defined by $f_{SWR} = 6349\sqrt{n_e}$. At a thin sheath width (Figure 4a), there is a large difference between the ideal SWR and the calculated f_{SWR}. The maximum discrepancy (defined as $f_{idealSWR} - f_{SWR}/f_{idealSWR}$) is 36.5% at an n_e of 1.0×10^{12} cm^{-3}. This is because of the boundary effect, which is not included in the ideal SWR model. At a thin sheath width, the E/M wave damped along the plasma–sheath interface interacts with the plasma as well as with the ground aluminum tube of the coaxial cable. As the sheath width increases, interaction between the E/M wave and the plasma becomes dominant and f_{SWR} corresponds to the ideal SWR frequency for most pressure conditions, as shown in Figure 4d. Through the E/M wave simulation, it is verified that the sharp resonance peak in the S_{11} spectrum of the MOLE probe originates from the SWR and values of n_e can be inferred by measuring the SWR frequency by the relation $f_{SWR} = 6349\sqrt{n_e}$. However, depending on the sheath width, the measurement discrepancy may increase up to 36.5%.

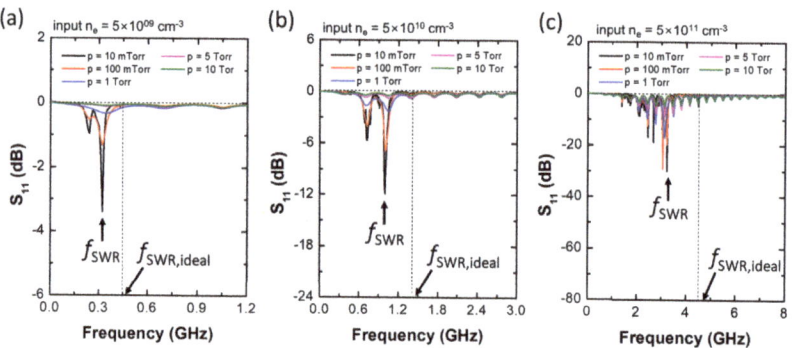

Figure 3. S_{11} spectra with various pressures (p) at a constant sheath width of 0.234 mm and electron densities (n_e) of (**a**) 5×10^9 cm^{-3}, (**b**) 5×10^{10} cm^{-3}, and (**c**) 5×10^{11} cm^{-3}.

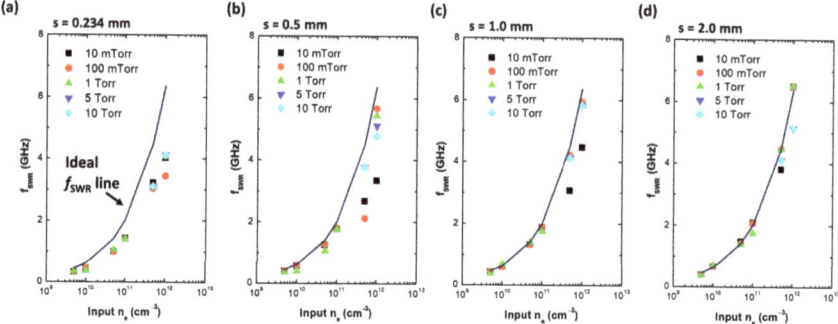

Figure 4. Surface wave resonance frequencies (f_{SWR}) with various pressures and input electron densities at sheath widths of (**a**) 0.234 mm, (**b**) 0.5 mm, (**c**) 1.0 mm, and (**d**) 2.0 mm.

Those simulation results indicate that both detection limit and accuracy largely depend on the sheath width. In the case of floating sheath, as electron density lowered sheath becomes thicker, which leads to a reduction in signal-to-noise ratio since most electromagnetic waves are already reflected on the open termination of the probe edge at thick sheath condition. Meanwhile, at high electron density, the sheath becomes thin, which leads to decreasing accuracy.

2.4. Difference between the MOLE Probe and the PAP

As mentioned in the *Introduction* section, the principle of the MOLE probe is the same as the PAP [41,44,45], except for probe configuration: the PAP has a radiating tip whereas the MOLE probe is tipless, as shown in Figure 1. This difference seems to be trivial, just either *presence* of the tip or *not*, but we found that there is a significant difference in terms of practical use. In general, the S_{11} spectrum of the PAP shows multiple resonance peaks caused by higher resonance modes as in [41,45] whereas the MOLE probe exhibits just *double* peaks. In fact, these multiple peaks are already addressed in [44] where the sensitive PAP was proposed to minimize higher mode peaks. We found, however, that the sensitive PAP is not effective under different conditions. Figure 5 shows simulated S_{11}

spectra with various tip lengths at different sheath widths of 0.234 mm and 3.0 mm. As shown in Figure 5a, multiple peaks caused by higher mode resonances [45] show higher S_{11} values compared with the fundamental peak (f_{SWR}). In practical situations where one determines the f_{SWR} in a measured S_{11} spectrum, higher mode peaks would result in a large measurement discrepancy of measurement or require additional analytic processes. The higher mode peaks are however suppressed with lessening the tip length and at the tipless condition, S_{11} spectrum shows clear double peaks or a single peak, depending on conditions. Similarly, at the thick sheath condition (Figure 5b), one can find a complicated S_{11} spectrum as well as peaks with the probe tip whereas, at the tipless condition, clear double peaks are observed. The simulation result ensures that the MOLE probe exhibits a clear and simple S_{11} spectrum and as a result, easy analysis to infer electron density (n_e) just by using the equation, $f_{SWR} = 6349\sqrt{n_e}$. Furthermore, as the MOLE probe has a flat antenna, it is applicable for plasma monitoring installed on the vacuum chamber wall, which is the main purpose of the MOLE probe.

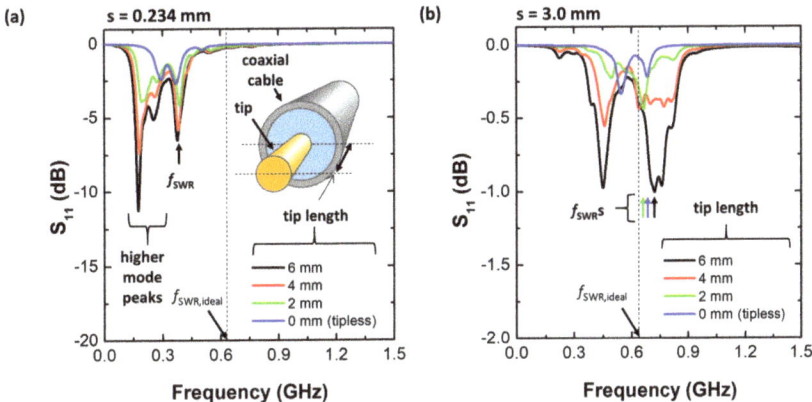

Figure 5. S_{11} spectra with various tip lengths at a sheath width of (**a**) 0.234 mm and (**b**) 3.0 mm, electron density of 1×10^{10} cm^{-3}, and pressure of 100 mTorr.

3. Experimental Demonstration

3.1. Experimental Setup

Figure 6 shows the experimental setup for a demonstration of the MOLE probe. The probe was made by cutting an RG-401 cable, a common coaxial cable. The diameter of the core conductor (copper) was 1.63 mm and the inner and outer diameters of the jacket conductor (copper) were 5.31 and 6.35 mm, respectively. The dielectric material was PTFE, of which the inner and outer diameters were 1.63 and 5.31 mm, respectively. The cutoff frequency of the RG-401 cable was 20 GHz, which is much higher than the frequency range used in this experiment, and the characteristic impedance was 50 Ω.

Inductively coupled plasma was generated by applying rf power from an rf generator (YSR-06MF, YongSin RF Inc., Hanam-si, Korea) to an in-house one-turn antenna through an rf matcher (YongSin RF Matcher, YongSin RF Inc., Gyonggi-do, Korea). A ceramic plate was used to both sustain the vacuum and propagate rf power into the gas inside the chamber. Argon gas (99.999% purity) was injected through a mass flow controller (MFC, LineTech Inc., Daejeon, Korea). A rotary pump (DS102, Agilent Inc., Santa Clara, CA, USA) was used to generate a vacuum in the chamber via pumping port, with a resultant base pressure of 0.6 mTorr.

The MOLE probe was inserted into the chamber via a NW-40 port with distance of 140 mm from the ceramic plate. One end of the probe was located in the center of the chamber (Figure 6a) and the other end was connected to a commercial SMA connector, which itself was connected to vector network analyzer 1 (E5071B 300 kHz–8.5 GHz, Agilent

Inc., Santa Clara, CA, USA)) via SMA cable, as shown in Figure 6. To remove the signal loss by the SMA cable, calibration was conducted at the end of the cable with a calibration kit (85033D, Agilent Inc.).

(a)

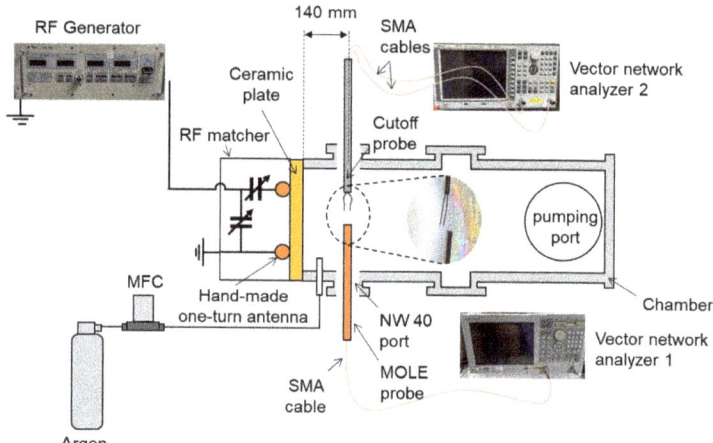

(b)

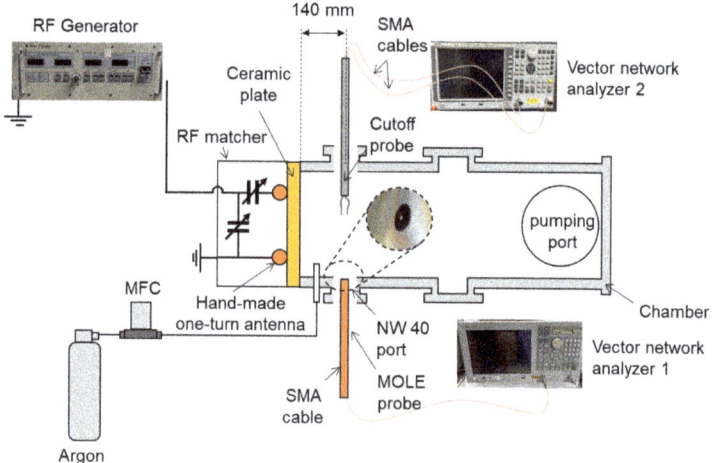

Figure 6. Schematic diagrams of the experimental setup for a demonstration of the MOLE probe positioned (**a**) in the center of the chamber and (**b**) on the chamber wall.

To demonstrate the operation of the MOLE probe, a cutoff probe was inserted into the chamber via a NW-40 port opposite from the MOLE probe. The cutoff probe was connected to vector network analyzer 2 (S3601B 100 kHz–8.5 GHz, SALUKI Inc., Taipei, Taiwan) via two SMA cables. The cutoff probe can precisely measure the electron density by measuring the cutoff frequency in S_{21}; details can be found elsewhere [9,46–49]. In the next section, operation of the MOLE probe at the center of the chamber is first demonstrated compared to the cutoff probe. Operation on the chamber wall is then considered, with the cutoff probe still located at the center of the chamber.

3.2. Experimental Results and Discussion

Figure 7 shows the S_{11} spectra of the MOLE probe with various rf powers at 5.0 mTorr. A peak appears near 0.5 GHz with the probe at various rf powers (201–593 W) and pressures (5.0–18.2 mTorr) depending on the rf power; based on the simulation analysis, this peak is believed to be the SWR peak, with the other peaks above 1 GHz thought to originate from the standing wave resonance formed inside the coaxial cable. Since these peaks complicate the analysis, converted S_{11} spectra [defined as S_{11}(plasma) − S_{11}(vacuum)] are introduced in Figure 8a–c. With these, the SWR resonance can be more clearly observed. The SWR frequency is proportional to the rf power, and as the pressure increases, the Q-factor of the peak abruptly decreases. As shown in Figure 8d–g, the cutoff frequency ($f_{cutoff} = 8980\sqrt{n_e}$) in S_{21}, which is marked with an arrow at each rf power condition, is proportional to the rf power [47–49]. This implies that the electron density is also proportional to the rf power. Combining the findings of the converted S_{11} analysis and S_{21} analysis, the SWR frequency as well as the signal reduction are proportional to the electron density. Moreover, the Q-factor of the SWR peak decreases with increasing pressure. These trends are exactly the same as those in the simulation analysis. The dependence of the SWR frequency as well as the cutoff frequency on the rf power is further analyzed by considering the wall-positioned MOLE probe results, as follows.

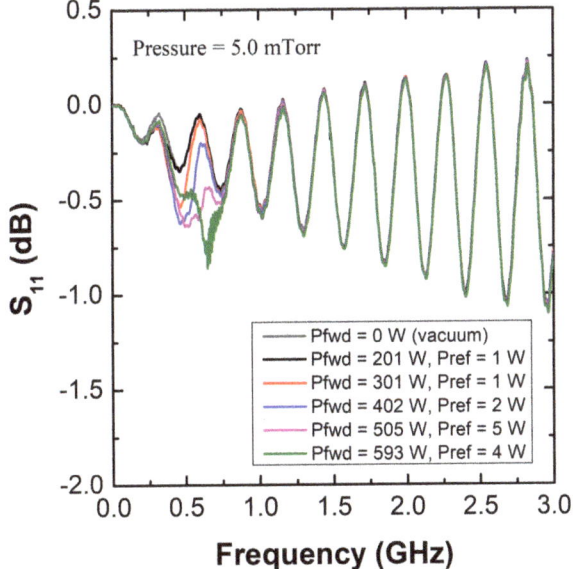

Figure 7. S_{11} spectra of the MOLE probe with various rf powers at a pressure of 5.0 mTorr.

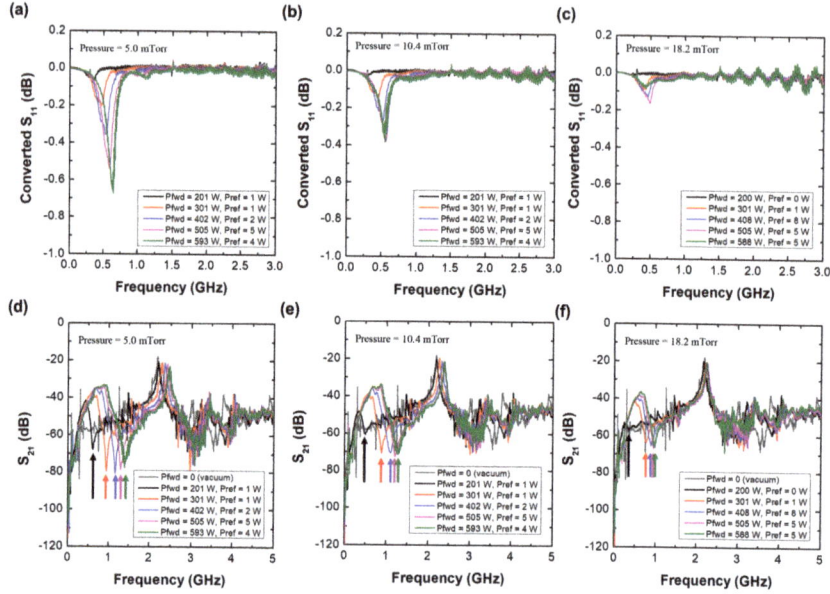

Figure 8. (a–c) Converted S_{11} spectra [defined as S_{11}(plasma)–S_{11}(vacuum)] of the MOLE probe. (d–f) S_{21} spectra of the cutoff probe at various rf powers (201–593 W) and pressures (5.0–18.2 mTorr).

Finally, to demonstrate the operation of the MOLE probe in non-invasively measuring lateral electron density, the probe was positioned on the chamber wall as shown in Figure 6b. Here, lateral means the plasma–sheath edge near the chamber wall. Figure 9a,b show the S_{11} and converted S_{11} spectra of the wall-positioned MOLE probe at 5 mTorr. Compared to Figure 7, the SWR peak as well as the signal reduction in the S_{11} spectra are unclear since the electron density near the chamber wall is, in general, lower than that in the center of the chamber [5]. Although the SWR peak is unclear in S_{11}, there is a distinct SWR peak in the converted S_{11} spectra, and this peak is proportional to the rf power. Figure 9c shows that the cutoff frequency is also proportional to the rf power. Once again, by combining these two facts, the f_{SWR} from the wall-positioned MOLE probe is found to depend on the electron density at the center.

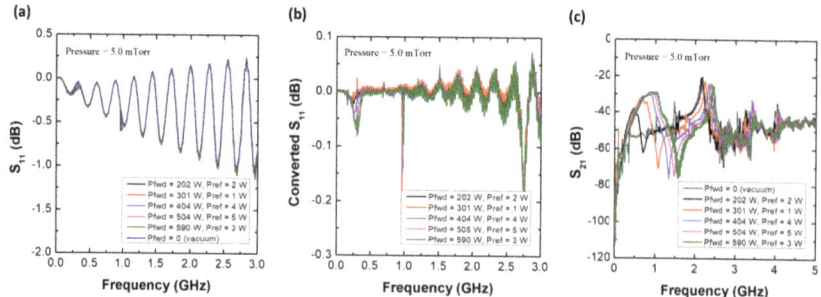

Figure 9. (a) S_{11} spectra and (b) converted S_{11} spectra [defined as S_{11}(plasma)–S_{11}(vacuum)] of the MOLE probe. (c) S_{21} spectra of the cutoff probe with various rf powers (202–590 W) at a pressure of 5 mTorr.

Figure 10 exhibits estimated electron densities from the MOLE probe at bulk over densities from the cutoff probe at various pressures RF powers based on the measurement results of Figures 8 and 9. In terms of the trend, the $n_{e,\text{MOLEprobe}}$ measured at the bulk represents a good linearity with the $n_{e,\text{cutoffprobe}}$, which means that the MOLE probe catches well the variation of the bulk plasma. For the wall mount MOLE probe, the $n_{e,\text{MOLEprobe}}$ also follows the bulk variation as shown in Figure 10. Hence, through experimental demonstration with the cutoff probe, we find that the MOLE probe on the wall works well.

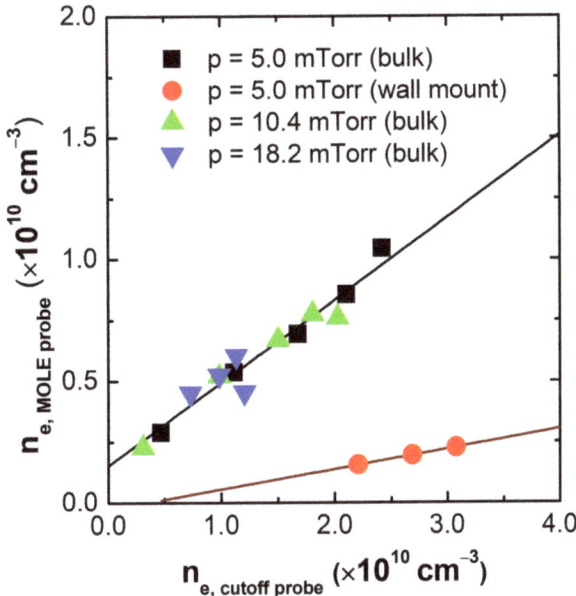

Figure 10. Estimated electron density from the MOLE probe ($n_{e,\text{MOLEprobe}}$) at the bulk and on the wall over densities from the cutoff probe ($n_{e,\text{cutoffprobe}}$) at various pressures.

The MOLE probe however measures the electron density several times lower for the bulk and 10 times lower for the wall mount than the cutoff probe. This is because of thin sheath condition and the plasma non-uniformity [17,50]. Based on the simulation result (Figure 4), at a thin sheath width the f_{SWR} is lower than the ideal one, which means the MOLE probe estimates lower density. In fact, sheath width in the inductively coupled plasma source is known as floating sheath, which is five times the Debye length ($5 \times \lambda_{\text{De}}$) [51], where $\lambda_{\text{De}} (= 740\sqrt{T_e/n_e})$ is the Debye length in cm and T_e and n_e are the electron temperature in eV and electron density in cm^{-3}, respectively. Assuming the n_e as 2×10^{10} cm^{-3} based on the cutoff probe and T_e as a few eV[5], the estimated sheath width is the order of 0.1 mm and it surely causes the underestimation of the SWR frequency. Moreover, at low pressure, the inductively coupled plasma source has a non-uniformity as in [17] where the measured density near wall is 10 times lower than the bulk due to ambipolar potential. Hence, with considering the non-uniformity effect, the electron density by the MOLE probe on the wall is reasonable based on Figure 10.

4. Conclusions

The development of a microwave probe called the measurement of lateral electron density (MOLE) probe was described in this work with simulation study and experimental demonstration. Lateral as used here refers to the plasma–sheath edge near the chamber wall. Results of a 3D E/M wave simulation analysis established that the MOLE probe can measure the SWR frequency ($f_{SWR} = 6349\sqrt{n_e}$), with a theoretical measurement discrepancy up to 36.5% depending on the sheath width; especially, the discrepancy increases as sheath width decreases. Experimental demonstrations including a cutoff probe for comparison exhibits that the MOLE probe represents a good linearity with the cutoff probe at bulk as well as on the chamber wall, which means that MOLE probe can measure the lateral electron density. In conclusion, the MOLE probe positioned on the chamber wall is believed to be a useful tool to monitor lateral electron density.

Author Contributions: Conceptualization, S.-j.Y., S.-h.L.; validation, S.-j.K., I.-h.S., J.-j.L., Y.-s.L., C.-h.C.; formal analysis, S.-j.K., Y.-b.Y.; writing—original draft preparation, S.-j.K.; writing—review and editing, S.-j.y., Y.-s.L., C.-h.C.; supervision, S.-j.Y. All authors have read and agreed to the published version of the manuscript.

Funding: This research was supported by the National Research Council of Science & Technology (NST) grant by the Korea government (MSIP) (No. CAP-17-02-NFRI, CRF-20-01-NFRI), by Next-generation Intelligence semiconductor R&D Program through the Korea Evaluation Institute of Industrial Technology(KEIT) funded by the Korea government (MOTIE), by the Korea Institute of Energy Technology Evaluation and Planning (KETEP) and the MOTIE of the Republic of Korea (20202010100020), by the MOTIE (20009818, 20010420) and KSRC (Korea Semiconductor Research Consortium) support program for the development of the future semiconductor device, by the Korea Institute for Advancement of Technology (KIAT) grant funded by the Korea Government (MOTIE)(P0008458, HRD Program for Industrial Innovation) by a Basic Science Research Program through the National Research Foundation of Korea (NRF) funded by the Ministry of Education (NRF-2020R1A6A1A03047771), and by KIMM Institutional Program (NK236F) and NST/KIMM.

Data Availability Statement: The data presented in this study are available on request from the corresponding author.

Conflicts of Interest: The authors declare no conflict of interest.

References

1. Bogaerts, A.; Tu, X.; Whitehead, J.C.; Centi, G.; Lefferts, L.; Guaitella, O.; Azzolina-Jury, F.; Kim, H.H.; Murphy, A.B.; Schneider, W.F.; et al. The 2020 plasma catalysis roadmap. *J. Phys. Appl. Phys.* **2020**, *53*, 443001. [CrossRef]
2. Adamovich, I.; Baalrud, S.; Bogaerts, A.; Bruggeman, P.; Cappelli, M.; Colombo, V.; Czarnetzki, U.; Ebert, U.; Eden, J.; Favia, P.; et al. The 2017 Plasma Roadmap: Low temperature plasma science and technology. *J. Phys. Appl. Phys.* **2017**, *50*, 323001. [CrossRef]
3. Laroussi, M.; Bekeschus, S.; Keidar, M.; Bogaerts, A.; Fridman, A.; Lu, X.; Ostrikov, K.; Hori, M.; Stapelmann, K.; Miller, V.; et al. Low-temperature plasma for biology, hygiene, and medicine: perspective and roadmap. *IEEE Trans. Radiat. Plasma Med. Sci.* **2022**, *6*, 127–157. [CrossRef]
4. Attri, P.; Ishikawa, K.; Okumura, T.; Koga, K.; Shiratani, M. Plasma agriculture from laboratory to farm: A review. *Processes* **2020**, *8*, 1002. [CrossRef]
5. Lieberman, M.A.; Lichtenberg, A.J. *Principles of Plasma Discharges and Materials Processing*; John Wiley & Sons: Hoboken, NJ, USA, 2005; pp. 1–22.
6. Yang, R.; Chen, R. Real-time plasma process condition sensing and abnormal process detection. *Sensors* **2010**, *10*, 5703–5723. [CrossRef]
7. Abbas, M.A.; Dijk, L.V.; Jahromi, K.E.; Nematollahi, M.; Harren, F.J.; Khodabakhsh, A. Broadband time-resolved absorption and dispersion spectroscopy of methane and ethane in a plasma using a mid-infrared dual-comb spectrometer. *Sensors* **2020**, *20*, 6831. [CrossRef]
8. Wang, J.; Ji, W.; Du, Q.; Xing, Z.; Xie, X.; Zhang, Q. A Long Short-Term Memory Network for Plasma Diagnosis from Langmuir Probe Data. *Sensors* **2022**, *22*, 4281. [CrossRef]
9. Kim, S.J.; Lee, J.J.; Lee, Y.S.; Cho, C.H.; You, S.J. Crossing Frequency Method Applicable to Intermediate Pressure Plasma Diagnostics Using the Cutoff Probe. *Sensors* **2022**, *22*, 1291. [CrossRef]
10. Kim, S.J.; Lee, J.J.; Lee, Y.S.; Yeom, H.J.; Lee, H.C.; Kim, J.H.; You, S.J. Computational Characterization of Microwave Planar Cutoff Probes for Non-Invasive Electron Density Measurement in Low-Temperature Plasma: Ring-and Bar-Type Cutoff Probes. *Appl. Sci.* **2020**, *10*, 7066. [CrossRef]
11. Kim, D.; You, S.; Kim, S.; Kim, J.; Lee, J.; Kang, W.; Hur, M. Planar cutoff probe for measuring the electron density of low-pressure plasmas. *Plasma Sources Sci. Technol.* **2019**, *28*, 015004. [CrossRef]
12. Lynn, S.; Ringwood, J.; Ragnoli, E.; McLoone, S.; MacGearailty, N. Virtual metrology for plasma etch using tool variables. In Proceedings of the 2009 IEEE/SEMI Advanced Semiconductor Manufacturing Conference, Berlin, Germany, 10–12 May 2009; pp. 143–148.

3. Jang, Y.C.; Park, S.H.; Jeong, S.M.; Ryu, S.W.; Kim, G.H. Role of features in plasma information based virtual metrology (PI-VM) for SiO2 etching depth. *J. Semicond. Disp. Technol.* **2019**, *18*, 30–34.
4. Kim, B.; Bae, J.K.; Hong, W.S. Plasma control using neural network and optical emission spectroscopy. *J. Vac. Sci. Technol. A* **2005**, *23*, 355–358. [CrossRef]
5. Kwon, J.W.; Ryu, S.; Park, J.; Lee, H.; Jang, Y.; Park, S.; Kim, G.H. Development of virtual metrology using plasma information variables to predict Si etch profile processed by SF6/O2/Ar capacitively coupled plasma. *Materials* **2021**, *14*, 3005. [CrossRef] [PubMed]
6. Lee, H.K.; Baek, K.H.; Shin, K. Resolving critical dimension drift over time in plasma etching through virtual metrology based wafer-to-wafer control. *Jpn. J. Appl. Phys.* **2017**, *56*, 066502. [CrossRef]
7. Yeom, H.; Kim, J.; Choi, D.; Choi, E.; Yoon, M.; Seong, D.; You, S.J.; Lee, H.C. Flat cutoff probe for real-time electron density measurement in industrial plasma processing. *Plasma Sources Sci. Technol.* **2020**, *29*, 035016. [CrossRef]
8. Kim, S.; Lee, J.; Lee, Y.; Kim, D.; You, S. Finding the optimum design of the planar cutoff probe through a computational study. *AIP Adv.* **2021**, *11*, 025241. [CrossRef]
9. Fantz, U. Basics of plasma spectroscopy. *Plasma Sources Sci. Technol.* **2006**, *15*, S137. [CrossRef]
20. Engeln, R.; Klarenaar, B.; Guaitella, O. Foundations of optical diagnostics in low-temperature plasmas. *Plasma Sources Sci. Technol.* **2020**, *29*, 063001. [CrossRef]
21. Seo, B.; Kim, D.W.; Kim, J.H.; You, S. Investigation of reliability of the cutoff probe by a comparison with Thomson scattering in high density processing plasmas. *Phys. Plasmas* **2017**, *24*, 123502. [CrossRef]
22. Ogawa, D.; Nakamura, K.; Sugai, H. Experimental validity of double-curling probe method in film-depositing plasma. *Plasma Sources Sci. Technol.* **2021**, *30*, 085009. [CrossRef]
23. Friedrichs, M.; Oberrath, J. The planar Multipole Resonance Probe: A functional analytic approach. *EPJ Tech. Instrum.* **2018**, *5*, 7. [CrossRef]
24. Kim, J.; Lee, K.I.; Jeong, H.Y.; Lee, J.H.; Choi, Y.S. Anti-contamination SMART (Spectrum Monitoring Apparatus with Roll-to-roll Transparent film) window for optical diagnostics of plasma systems. *Rev. Sci. Instrum.* **2021**, *92*, 013507. [CrossRef] [PubMed]
25. Kim, J.H.; Choi, S.C.; Shin, Y.H.; Chung, K.H. Wave cutoff method to measure absolute electron density in cold plasma. *Rev. Sci. Instrum.* **2004**, *75*, 2706–2710. [CrossRef]
26. You, K.; You, S.; Kim, D.W.; Na, B.; Seo, B.; Kim, J.; Seong, D.; Chang, H. Measurement of electron density using reactance cutoff probe. *Phys. Plasmas* **2016**, *23*, 053515. [CrossRef]
27. Tudisco, O.; Lucca Fabris, A.; Falcetta, C.; Accatino, L.; De Angelis, R.; Manente, M.; Ferri, F.; Florean, M.; Neri, C.; Mazzotta, C.; et al. A microwave interferometer for small and tenuous plasma density measurements. *Rev. Sci. Instrum.* **2013**, *84*, 033505. [CrossRef]
28. Neumann, G.; Bänziger, U.; Kammeyer, M.; Lange, M. Plasma-density measurements by microwave interferometry and Langmuir probes in an rf discharge. *Rev. Sci. Instrum.* **1993**, *64*, 19–25. [CrossRef]
29. Arshadi, A.; Brinkmann, R.P.; Hotta, M.; Nakamura, K. A simple and straightforward expression for curling probe electron density diagnosis in reactive plasmas. *Plasma Sources Sci. Technol.* **2017**, *26*, 045013. doi: 10.1088/1361-6595/aa60f2. [CrossRef]
30. Arshadi, A.; Brinkmann, R.P. Analytical investigation of microwave resonances of a curling probe for low and high-pressure plasma diagnostics. *Plasma Sources Sci. Technol.* **2016**, *26*, 015011. [CrossRef]
31. Liang, I.; Nakamura, K.; Sugai, H. Modeling microwave resonance of curling probe for density measurements in reactive plasmas. *Appl. Phys. Express* **2011**, *4*, 066101. [CrossRef]
32. van Ninhuijs, M.A.; Daamen, K.; Beckers, J.; Luiten, O. Design and characterization of a resonant microwave cavity as a diagnostic for ultracold plasmas. *Rev. Sci. Instrum.* **2021**, *92*, 013506. [CrossRef]
33. Staps, T.; Platier, B.; Mihailova, D.; Meijaard, P.; Beckers, J. Numerical profile correction of microwave cavity resonance spectroscopy measurements of the electron density in low-pressure discharges. *Rev. Sci. Instrum.* **2021**, *92*, 093504. [CrossRef] [PubMed]
34. Beckers, J.; Van De Wetering, F.; Platier, B.; Van Ninhuijs, M.; Brussaard, G.; Banine, V.; Luiten, O. Mapping electron dynamics in highly transient EUV photon-induced plasmas: A novel diagnostic approach using multi-mode microwave cavity resonance spectroscopy. *J. Phys. D Appl. Phys.* **2018**, *52*, 034004. [CrossRef]
35. Lapke, M.; Oberrath, J.; Schulz, C.; Storch, R.; Styrnoll, T.; Zietz, C.; Awakowicz, P.; Brinkmann, R.P.; Musch, T.; Mussenbrock, T.; et al. The multipole resonance probe: Characterization of a prototype. *Plasma Sources Sci. Technol.* **2011**, *20*, 042001. [CrossRef]
36. Schulz, C.; Styrnoll, T.; Awakowicz, P.; Rolfes, I. The planar multipole resonance probe: Challenges and prospects of a planar plasma sensor. *IEEE Trans. Instrum. Meas.* **2014**, *64*, 857–864. [CrossRef]
37. Pohle, D.; Schulz, C.; Oberberg, M.; Awakowicz, P.; Rolfes, I. The planar multipole resonance probe: A minimally invasive monitoring concept for plasma-assisted dielectric deposition processes. *IEEE Trans. Microw. Theory Tech.* **2020**, *68*, 2067–2079. [CrossRef]
38. Wang, C.; Friedrichs, M.; Oberrath, J.; Brinkmann, R.P. Kinetic investigation of the planar multipole resonance probe in the low-pressure plasma. *Plasma Sources Sci. Technol.* **2021**, *30*, 105011. [CrossRef]
39. Huang, S.; Huard, C.; Shim, S.; Nam, S.K.; Song, I.C.; Lu, S.; Kushner, M.J. Plasma etching of high aspect ratio features in SiO2 using Ar/C4F8/O2 mixtures: A computational investigation. *J. Vac. Sci. Technol. A* **2019**, *37*, 031304. [CrossRef]

40. Ishikawa, K.; Karahashi, K.; Ishijima, T.; Cho, S.I.; Elliott, S.; Hausmann, D.; Mocuta, D.; Wilson, A.; Kinoshita, K. Progress in nanoscale dry processes for fabrication of high-aspect-ratio features: How can we control critical dimension uniformity at the bottom? *Jpn. J. Appl. Phys.* **2018**, *57*, 06JA01. [CrossRef]
41. Kokura, H.; Nakamura, K.; Ghanashev, I.P.; Sugai, H. Plasma absorption probe for measuring electron density in an environment soiled with processing plasmas. *Jpn. J. Appl. Phys.* **1999**, *38*, 5262. [CrossRef]
42. Dine, S.; Booth, J.P.; Curley, G.A.; Corr, C.; Jolly, J.; Guillon, J. A novel technique for plasma density measurement using surface-wave transmission spectra. *Plasma Sources Sci. Technol.* **2005**, *14*, 777. [CrossRef]
43. Anwar, R.S.; Ning, H.; Mao, L. Recent advancements in surface plasmon polaritons-plasmonics in subwavelength structures in microwave and terahertz regimes. *Digit. Commun. Netw.* **2018**, *4*, 244–257. [CrossRef]
44. Nakamura, K.; Ohata, M.; Sugai, H. Highly sensitive plasma absorption probe for measuring low-density high-pressure plasmas. *J. Vac. Sci. Technol. A* **2003**, *21*, 325–331. [CrossRef]
45. Lapke, M.; Mussenbrock, T.; Brinkmann, R.; Scharwitz, C.; Böke, M.; Winter, J. Modeling and simulation of the plasma absorption probe. *Appl. Phys. Lett.* **2007**, *90*, 121502. [CrossRef]
46. Kim, S.; Lee, J.; Kim, D.; Kim, J.; You, S. A transmission line model of the cutoff probe. *Plasma Sources Sci. Technol.* **2019**, *28*, 055014. [CrossRef]
47. Kim, S.; Lee, J.; Lee, Y.; Kim, D.; You, S. Effect of an inhomogeneous electron density profile on the transmission microwave frequency spectrum of the cutoff probe. *Plasma Sources Sci. Technol.* **2020**, *29*, 125014. [CrossRef]
48. Kim, D.W.; You, S.; Kwon, J.; You, K.; Seo, B.; Kim, J.; Yoon, J.S.; Oh, W.Y. Reproducibility of the cutoff probe for the measurement of electron density. *Phys. Plasmas* **2016**, *23*, 063501. [CrossRef]
49. Kim, D.W.; You, S.; Kim, J.; Chang, H.; Oh, W.Y. Computational comparative study of microwave probes for plasma density measurement. *Plasma Sources Sci. Technol.* **2016**, *25*, 035026. [CrossRef]
50. cheol Kim, Y.; Lee, H.C.; Chung, C.W. Study on Plasma Uniformity Using 2-D Measurement Method in Argon Inductively Coupled Plasmas. *IEEE Trans. Plasma Sci.* **2014**, *42*, 2858–2859.
51. Kim, D.W.; You, S.J.; Kim, J.H.; Seong, D.J.; Chang, H.Y.; Oh, W.Y. Computational study on reliability of sheath width measurement by the cutoff probe in low pressure plasmas. *J. Instrum.* **2015**, *10*, T11001. [CrossRef]

Article

Development of a High-Linearity Voltage and Current Probe with a Floating Toroidal Coil: Principle, Demonstration, Design Optimization, and Evaluation

Si-jun Kim [1], In-ho Seong [1], Young-seok Lee [1], Chul-hee Cho [1], Won-nyoung Jeong [1], Ye-bin You [1], Jang-jae Lee [2] and Shin-jae You [1,3,*]

1 Applied Physics Lab for PLasma Engineering (APPLE), Department of Physics, Chungnam National University, Daejeon 34134, Korea
2 Samsung Electronics, Hwaseong-si 18448, Korea
3 Institute of Quantum Systems (IQS), Chungnam National University, Daejeon 34134, Korea
* Correspondence: sjyou@cnu.ac.kr

Abstract: As the conventional voltage and current (VI) probes widely used in plasma diagnostics have separate voltage and current sensors, crosstalk between the sensors leads to degradation of measurement linearity, which is related to practical accuracy. Here, we propose a VI probe with a floating toroidal coil that plays both roles of a voltage and current sensor and is thus free from crosstalk. The operation principle and optimization conditions of the VI probe are demonstrated and established via three-dimensional electromagnetic wave simulation. Based on the optimization results, the proposed VI probe is fabricated and calibrated for the root-mean-square (RMS) voltage and current with a high-voltage probe and a vector network analyzer. Then, it is evaluated through a comparison with a commercial VI probe, with the results demonstrating that the fabricated VI probe achieved a slightly higher linearity than the commercial probe: R^2 of 0.9967 and 0.9938 for RMS voltage and current, respectively. The proposed VI probe is believed to be applicable to plasma diagnostics as well as process monitoring with higher accuracy.

Keywords: plasma diagnostics; plasma monitoring; voltage and current (VI) probe; floating toroidal coil; simulation optimization; VI probe calibration

Citation: Kim, S.-j.; Seong, I-h.; Lee, Y.-s.; Cho, C.-h.; Jeong, W.-n.; You, Y.-b.; Lee, J.-j.; You, S.-j. Development of a High-Linearity Voltage and Current Probe with a Floating Toroidal Coil: Principle, Demonstration, Design Optimization, and Evaluation. *Sensors* **2022**, *22*, 5871. https://doi.org/10.3390/s22155871

Academic Editor: Bruno Goncalves

Received: 24 June 2022
Accepted: 4 August 2022
Published: 5 August 2022

Publisher's Note: MDPI stays neutral with regard to jurisdictional claims in published maps and institutional affiliations.

Copyright: © 2022 by the authors. Licensee MDPI, Basel, Switzerland. This article is an open access article distributed under the terms and conditions of the Creative Commons Attribution (CC BY) license (https://creativecommons.org/licenses/by/4.0/).

1. Introduction

Plasma, called the fourth state of matter, consists of physically energetic charged particles (electrons, positive ions, negative ions) and chemically reactive neutral particles (radicals) [1]. Due to their high physical energy and chemical reactivity, plasma has been widely used in various fields such as semiconductor fabrication, medical and environmental industries, aerospace, bio, and nuclear fusion science [2,3]. In particular, in semiconductor fabrication, plasmas significantly influence the plasma etching [4–7], ashing [8,9], and deposition [10,11] processes to realize feature sizes on the nanoscale. As feature sizes continue to shrink towards a few nanometers with improved levels of integration, process abnormalities such as arcing and leakage that reduce productivity have been regarded as serious problems [12–14].

To improve process productivity, process monitoring techniques based on plasma diagnostic methods have garnered much attention [15] since key process parameters such as etching and deposition rates are related to the plasma parameters [16–20]. Plasma diagnostic methods employ an analysis of (i) the current-voltage characteristics of plasma using the Langmuir probe [21–23], (ii) the response characteristics of plasma to microwaves using resonators (microwave probes) [24–32], (iii) the optical emission characteristics of plasma using an optical emission spectrometer (OES) [33,34], and (iv) the voltage and current (VI) waveforms on a powered electrode using VI probes with circuit modeling [35–37].

These diagnostic techniques have been well studied and are commonly used in various research fields. Some of them, however, especially the Langmuir probe and microwave probes, are not suitable for plasma process monitoring, since they are invasive and as a result would distort and be perturbed by the processing. Recently, low-frequency modulation technology and non-invasive types have been proposed and are still under development [27,31,32,38,39]. Commonly implemented plasma process monitoring tools are the OES and VI probe; they are non-invasive and easy to install in the process equipment [40–44]. In general, an OES measures the optical emission from plasma via an optical window and is used for gas composition analysis and anomalous behavior detection. Despite their convenience, however, OESs have limitations in the following three aspects: optical window contamination, narrow spaces of process facilities, and complicated analysis. Process gases such as CF_4, C_4F_8, CHF_3, and SiH_4 cause optical window contamination that either degrades the emission intensity or cuts off some spectral bands [45], issues for which several techniques have been developed [45–47]. Moreover, some process chambers have no optical window since it would perturb process uniformity. Finally, the optical spectra of process gases are highly complicated and pose challenges to analysis since the atomic and molecular spectra overlap, and in certain cases there are no fundamental spectral data for some gases and their compounds [48,49].

The VI probe, in general, measures the voltage and current of the electrode (or antenna) used to generate plasma [35,43,44] and is employed for plasma parameter analysis with some circuit modeling and sensitive detection of anomalous behaviors, especially arcing. As VI probes can be conveniently installed between the electrode (or antenna) and an impedance matcher, they are free from contamination. Nevertheless, since traditional VI probes have separate voltage and current sensors, crosstalk, which is defined as capacitive coupling between the sensors, leads to a degradation of measurement linearity, or in other words, accuracy. To minimize crosstalk, one commonly employed technique is to separate the voltage and current sensors by inserting a metal shield (called a Faraday shield) between them. Lafleur et al. [50] invented a coaxial-type VI probe named the Vigilant probe, where the voltage sensor (called the D-dot antenna) has a conical shape and the current sensor has an axisymmetric groove. Since the current sensor is embedded into external grounded metal and is separated from the voltage sensor, crosstalk can be minimized . In another example, Plasmart Inc. (Daejeon, Korea) [51] developed a printed circuit board (PCB)-type VI probe with a Faraday shield located between the voltage and current sensors to block crosstalk through the inside of the PCB. Despite the Faraday shield, however, crosstalk passing over the PCB still exists. To remove crosstalk completely, Kim et al. [52] developed a VI probe with double walls designed to prohibit the crosstalk passing over as well as through the inside of the PCB . However, in a high power environment, crosstalk can penetrate the Faraday shield, and conventional blocking methods are not effective.

Here, we propose a VI probe with a floating toroidal coil (FTC). Since the FTC plays a role in both voltage and current sensing, the VI probe is free from crosstalk. Through three-dimensional (3D) electromagnetic wave simulation, we first demonstrate the operation principle and establish optimization conditions. Then, based on the optimization results, we fabricated the VI probe and evaluated it with a comparison to a commercial VI probe. The results demonstrate that the fabricated VI sensor has a higher linearity than the commercial probe.

The rest of this paper is organized as follows. The Section 2 provides an explanation and demonstration of the operation principle of the FTC with 3D electromagnetic wave simulation. Design optimization procedures through simulation, and the resultant optimum conditions are also presented. In the Section 3, calibration and evaluation of the fabricated VI probe are investigated. Then, in the Section 4, we summarize the significant results of this paper.

2. Principle, Demonstration, and Design Optimization of the VI Probe

2.1. Principle of a Floating Toroidal Coil as a Voltage and Current Sensor

In this section, the operation principle of the FTC is qualitatively explored. Figure 1a presents a schematic diagram of the FTC with a cross-sectional view of the signal rod connected to a radio frequency (RF) generator. When RF power is applied to the signal rod, RF voltage is created and RF current flows through the signal rod. For easy understanding, we initially assume two ideal cases: (i) only RF voltage (V_{RF}), and (ii) only RF current (I_{RF}). For the former case, voltage on the FTC is induced by capacitive coupling between the FTC and ground through a time-varying electric field, depicted with green arrows in Figure 1a. Here, capacitive coupling means that the FTC plays a role as a counter-electrode with respect to the rod like a capacitor. Since the RF wavelength is much longer than the dimensions of the FTC, the FTC voltage (V_{coil}) is uniformly distributed between points a and b (Figure 1a) at any RF phase, as shown in Figure 1b; the uniform V_{coil}, therefore, sinusoidally oscillates with time. For the latter RF current-only case, a voltage difference between the FTC ends (a and b, Figure 1a) is induced by inductive coupling between the FTC and the rod through a time-varying magnetic field. Inductive coupling here follows Faraday's law of induction: an electromotive force is induced to disturb the time-varying magnetic field created by I_{RF}. As shown in Figure 1c, the V_{coil} is non-uniformly distributed. Note that the center of the FTC acts as a ground and the ends show push/pull characteristics during RF oscillation.

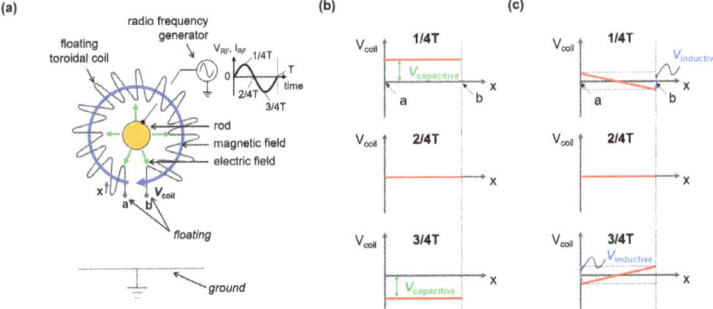

Figure 1. (a) Schematic of a floating toroidal coil (FTC). (b,c) Voltage of the FTC (V_{coil}) when only radio frequency (RF) voltage is applied (b) and when only RF current flows through the rod (c) at different RF phases (1/4T, 2/4T, and 3/4T), where the T is the period of the RF signal.

Considering a realistic situation, V_{RF} and I_{RF} simultaneously exist. This means that V_{coil} is induced by a combination of both capacitive and inductive coupling effects. Provided that these effects can be linearly combined (as proved in the next section), the spatiotemporal behavior of V_{coil} becomes the sum of Figure 1b,c. Therefore, the center V_{coil} and the different V_{coil} between the ends represent $V_{capacitive}$ and $V_{inductive}$, respectively. Here, $V_{capacitive}$ and $V_{inductive}$ mean the magnitude of their couplings, as shown in Figure 1b,c.

Practical use of the FTC to estimate V_{RF} and I_{RF} is as follows. We assume that from two points a to b the FTC is symmetric in terms of its center, as shown in Figure 1a. Then, the center V_{coil} is the same as $V_{capacitive}$, since $V_{inductive}$ is zero during RF oscillation at that position (see Figure 1c). Provided that V_{coil} is symmetrically distributed throughout the FTC, the average value can be the arithmetic mean of the voltages at the ends; hence, $V_{capacitive}$ is defined as

$$V_{capacitive} = V_{coil}^{avg} = \frac{V_{coil}^a + V_{coil}^b}{2}, \quad (1)$$

where V_{coil}^a and V_{coil}^b are the voltages of the FTC at each end (a and b shown in Figure 1a). As $V_{capacitive}$ results from capacitive coupling, it is noted that the summation of V_{coil}^a and

V_{coil}^b can be proportional to V_{RF} and thus a good indicator to measure V_{RF} with a coefficient, α, as

$$V_{coil}^a + V_{coil}^b = \alpha V_{RF}. \tag{2}$$

With a similar perspective, measuring I_{RF} can be explained as follows. Regarding that the voltage difference of V_{coil} at the ends originates from inductive coupling, $V_{inductive}$ is defined as

$$V_{inductive} = V_{coil}^a - V_{coil}^b. \tag{3}$$

Similar to the above, it is worthwhile to note that here, the subtraction of V_{coil}^b from V_{coil}^a can be proportional to I_{RF} and thus is a good indicator to measure I_{RF} with a coefficient, β, as

$$V_{coil}^a - V_{coil}^b = \beta I_{RF}. \tag{4}$$

Equations (2) and (4) imply that by measuring V_{coil}^a and V_{coil}^b, V_{RF} and I_{RF} can be assessed, provided that calibration factors α and β are known.

2.2. Simulation Demonstration

In this section, we demonstrate the principle introduced in the previous section via 3D electromagnetic wave simulation, CST Microwave Suite [53]. Figure 2a–c show schematic diagrams of three simulation cases: (i) capacitive and inductive coupling (with no shields), (ii) capacitive coupling only (with an inductive coupling shield), and (iii) inductive coupling only (with a capacitive coupling shield). For these three cases, the common configurations are the FTC, the coaxial cables, and the rod, as shown in Figure 2d,g. This apparatus is covered by a rectangular case that is electrically grounded (not depicted in the figure for clarity). The dimensions are listed in Table 1.

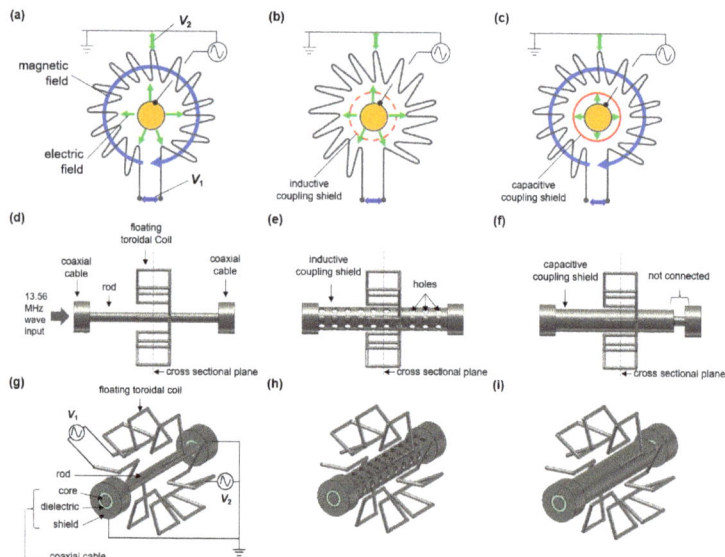

Figure 2. (**a**–**c**) Schematic diagrams of the simulation configurations and (**d**–**i**) corresponding three-dimensional images. Capacitive and inductive coupling is present in (**a,d,g**); only capacitive coupling is present in (**b,e,h**); and only inductive coupling is present in (**c,f,i**). Here, V_1 means the voltage difference between the ends of the floating toroidal coil, and V_2 is the voltage difference between the center of the coil and the grounded case.

Table 1. Dimensions used in the three-dimensional electromagnetic wave simulation. PEC: perfect electric conductor.

Coaxial cable	Outer diameter of core	6 mm
	Conductance of core	infinity (PEC)
	Outer diameter of dielectric	9 mm
	Relative dielectric constant of dielectric	2.1
	Outer diameter of shield	19 mm
	Conductance of shield	infinity (PEC)
	Length	10 mm
Rod	Diameter	6 mm
	Length	80 mm
	Conductance	infinity (PEC)
Floating toroidal coil	Inner diameter	30 mm
	Outer diameter	60 mm
	Width	20 mm
	Wire diameter	2 mm
	Turns	9
	Conductance	infinity (PEC)
Inductive coupling shield (ICS)	Inner diameter	12 mm
	Outer diameter	14 mm
	Hole diameter	4 mm
	Length	80 mm
	Conductance	infinity (PEC)
Capacitive coupling shield (CCS)	Inner diameter	12 mm
	Outer diameter	14 mm
	Length	73 mm
	Conductance	infinity (PEC)
Rectangular case	Volume	$100 \times 100 \times 100$ mm^3
	Thickness	5 mm
	Conductance	infinity (PEC)

The coaxial cables play a role as input and output ports for voltage and current waves. Incident waves from the input port are carried via the rod and induce V_{coil} on the FTC. In this simulation, a voltage monitor function, which integrates the electric field along a given line, is used to calculate the voltage difference. Here, the voltage monitors V_1 and V_2 shown in Figure 2g, respectively, mean the voltage difference between the ends of the FTC, that is $V_{inductive}$, and between the center of the FTC and the rectangular case, that is $V_{capacitive}$.

A brief explanation about the role of the inductive coupling shield (ICS) and the capacitive coupling shield (CCS) is as follows. As shown in Figure 2e,h, since the ICS is connected to the coaxial cable shields, which are electrically grounded, a closed current loop from the rod to the ICS forms. Based on Ampere's law, no net current source exists outside the ICS, since the current in the rod and the shield have the same magnitude but the opposite direction. As a result, no magnetic field outside the ICS can exist, meaning that inductive coupling is blocked. Capacitive coupling in this case exists between the rod and the FTC through the holes in the ICS, as shown in Figure 2e,h. As for the CCS shown in Figure 2f,i, this shield is connected to only one of the coaxial cable shields. In this configuration, no closed current loop can form, meaning that capacitive coupling is blocked while inductive coupling is not.

Simulation results are summarized as follows. Figure 3a–f show the magnetic field vectors and magnitude of the electric field on the cross-sectional plane, respectively, at the phases where their values are maximum. Since magnetic and electric fields form with rotational and diverse directions, respectively, different figure plots (vector and contour) are used for clarity. As for simulation case (i) involving both capacitive and inductive coupling, a rotating magnetic field by RF current in the rod forms inside the FTC, as shown in Figure 3a, demonstrating that the inductive coupling is effective. Furthermore, an electric field strongly forms between the rod and the inner side of the FTC, as shown in Figure 3d, demonstrating that the capacitive coupling is also effective. Since both couplings are effective, the voltage monitors $V_1 (= V_{\text{inductive}})$ and $V_2 (= V_{\text{capacitive}})$ show a sinusoidal waveform signal (Figure 3g). For case (ii) with only capacitive coupling, no magnetic fields are created inside the FTC, since the currents in the rod and in the ICS are opposite (Figure 3b), as explained in the previous paragraph. As shown in Figure 3e, small electric fields escape through the holes (see the green area), which render capacitive coupling effective despite its small magnitude. Furthermore, it is noted that V_1 is extremely small but V_2 shows a sinusoidal waveform (Figure 3h), meaning that only capacitive coupling is present. Combining these results, we note that V_2 can be an indicator of inductive coupling, that is $V_{\text{inductive}}$. As for case (iii) with only inductive coupling, Figure 3c shows that a magnetic field is well produced inside the FTC, similar to Figure 3a, while Figure 3f shows that no electric field forms between the rod and the inner side of the FTC (as electric fields are blocked inside the CCS). This implies that inductive coupling is effective but capacitive coupling is blocked by the CCS. Notably, V_1 shows a sinusoidal waveform and is much larger than V_2, as shown in Figure 3i. Hence, V_1 can be an indicator of $V_{\text{capacitive}}$.

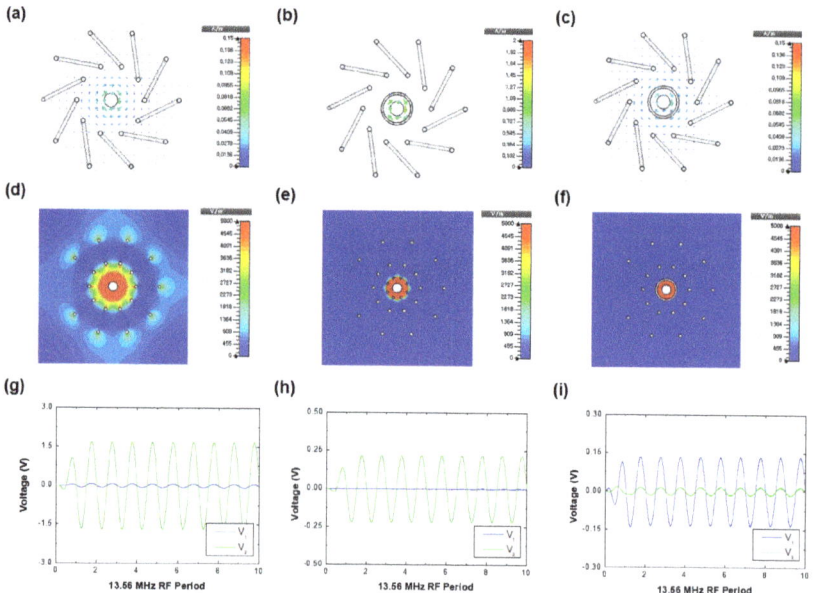

Figure 3. Magnetic field vectors $\vec{B} = \vec{B}_x + \vec{B}_y + \vec{B}_z$ (top row), magnitude of electric fields $|E| = \sqrt{E_x^2 + E_y^2 + E_z^2}$ (middle row), and voltage waveforms of V_1 and V_2 (bottom row) for (**a**,**d**,**g**) capacitive and inductive coupling, (**b**,**e**,**h**) capacitive coupling only, and (**c**,**f**,**i**) inductive coupling only. In the figure, V_1 means the voltage difference between the ends of the floating toroidal coil, and V_2 is the voltage difference between the center of the coil and the grounded case.

2.3. Design Optimization through Simulation

We demonstrated the workings of the FTC in the previous section via simulation. Before fabrication of the proposed sensor for a practical demonstration, it is highly useful to find the optimum conditions to achieve the highest sensitivity also through computer simulation rather than practical trials to minimize development costs. For this, the best method may be to examine all simulation cases for optimization, but this is not recommended due to the simulation cost. Instead, the following procedure is believed to be reasonable [52]. Assuming there are three parameters a, b, and c for optimization, the first step is to sweep the a parameter while fixing the other parameters at arbitrary values to find the optimum condition of a. The second step sweeps the b parameter with the optimized a and finds the optimum condition of b. The next trial sweeps the c parameter with the optimized a and b and finds the optimum condition of c. This process represents one sweeping cycle. By performing several cycles, provided that the optimized conditions of a, b, and c are the same as those of prior sweeping cycles, the final values are the optimum ones.

Figure 4a shows the simulation configuration of the proposed VI probe and each component: the FTC, U-cut printed circuit board (PCB), signal output lines, rod, dielectric holder, case, and coaxial cables, as well as the parameters for optimization: the number of turns, coil distance, and coil length. The dimensions are listed in Table 2. Here, each signal output line is connected to the two ends and the center of the FTC. The three lines terminate at the end of the U-cut PCB. Three voltage monitors calculate the voltage difference between the case (grounded) and each end of the signal output lines. Based on Equation (1), the center voltage monitor (V_{CTR}) represents $V_{capacitive}$, and based on Equation (3), the difference between the end voltage monitors (V_{end}s) is $V_{inductive}$. We introduce the center signal line for an exact measurement of $V_{capacitive}$. Hence, in this optimization procedure, the optimum condition is defined in terms of the highest signal amplitude of V_{CTR} and V_{end} for the fabrication of sensitive VI probe. If their maximum condition is different, the optimum condition is selected with an alternative way: at first, analyzing the tendency of V_{CTR} and V_{end} with optimization parameters and then finding the condition where either V_{CTR} or V_{end} is the highest value.

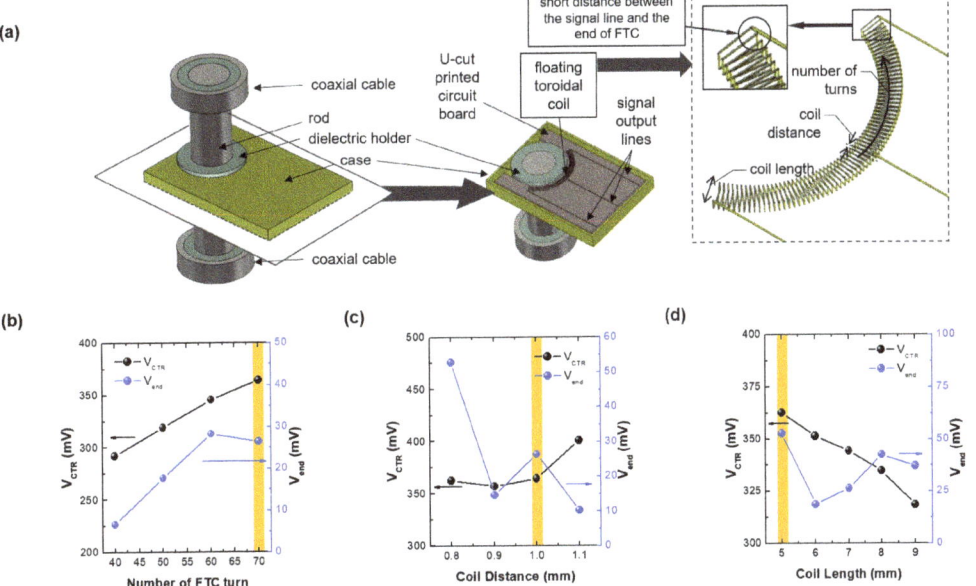

Figure 4. (a) Schematic of the VI probe showing the FTC embedded in the PCB, rod, and coaxial lines. The parameters for optimization are illustrated in the dashed box. Magnitude of the voltage difference between the ends of the FTC (V_{end}) and between the center of the coil and ground (V_{CTR}) by number of toroidal coil turns (**b**), coil distance (**c**), and coil length (**d**). The yellow bars highlight the optimum conditions.

Table 2. Dimensions used in the optimization simulation. PEC: perfect electric conductor.

Coaxial cable	Outer diameter of core	30 mm
	Conductance of core	infinity (PEC)
	Outer diameter of dielectric	45 mm
	Relative dielectric constant of dielectric	2.1
	Outer diameter of shield	55 mm
	Conductance of shield	infinity (PEC)
	Length	15 mm
Rod	Diameter	30 mm
	Length	120 mm
	Conductance	infinity (PEC)
Dielectric holder	Inner diameter	30 mm
	Outer diameter	50
	Length	18 mm
	Relative dielectric constant	2.1
Floating toroidal coil	Inner diameter	30 mm
	Outer diameter	60 mm
	Width	20 mm
	Wire diameter	2 mm
	Turns	9
	Conductance	infinity (PEC)
Printed circuit board	Board volume	75.8 × 110 × 2.60 mm^3
	Pattern thickness	0.07 mm
	Pattern width	0.2 mm
	Pattern conductance	5.96 × 0^7 S/m (copper)
Rectangular case	Volume	122 × 86 × 15 mm^3
	Thickness	2 mm
	Conductance	infinity (PEC)

Figure 4b shows the amplitude of the V_{CTR} and V_{end} waveforms from 40 to 70 turns of the FTC with a coil distance of 1.0 mm and a coil length of 5.0 mm, which are arbitrarily selected. As their maximum conditions are different, the optimum condition is selected with the alternative way. As the number of turns increases, V_{CTR} monotonically increases since the capacitive coupling area enlarges. On the other hand, V_{end} is saturated at 60 turns because the effective inductive coupling area inside the FTC becomes saturated. At 70 turns, the signal lines connected to the FTC ends are close to each other, as shown in Figure 4a, while above 70 turns, they are overlapped. Accordingly, the effective number of turns is saturated, and as a result, the optimum condition is 70 turns.

Figure 4c shows the optimization result for the coil distance at the optimized number of turns (70) and a coil length of 5.0 mm. Again, as their maximum conditions are different, the optimum condition is selected with the alternative way. As the coil distance increases, V_{CTR} gradually increases because the capacitive coupling area is slightly enlarged. Conversely, V_{end} decreases, except for at a coil distance of 1.0 mm, which results from the decrease in the number of turns per unit length. The opposite trends of V_{CTR} and V_{end} imply that the optimum condition is from 0.9 to 1.0 mm. Hence, we choose 1.0 mm as the optimum coil distance since the associated V_{end} is higher, although the spike of V_{end} at the 1.0 mm coil distance is not yet well understood.

In the final procedure in one cycle with two optimum conditions (70 turns and 1.0 mm coil distance), as shown in Figure 4d, as the coil length increases, V_{CTR} decreases while V_{end} abruptly decreases and then gradually rises. In this case, their maximum conditions are the same, the optimum condition is selected with the highest values of them. Since the outer edges of the FTC get farther away from the rod with increasing coil length, the effective capacitive coupling area decreases, which results in the decrease in V_{CTR}. The abrupt drop of V_{end} can be explained with the decrease in the number of turns per unit length since the outer arc length increases. The increase in coil length also results in an enlarged area inside the FTC, leading to an enhancement of inductive coupling, which causes the increase in V_{end}. Based on this analysis, while reducing the coil length may seem beneficial, doing so would lead to an overlap of the signal lines at the FTC ends. Hence, the optimum coil length is 5.0 mm.

It is noted that the initial conditions of 1.0 mm coil distance and 5.0 mm coil length at the initial optimization procedure (sweeping the number of turns) are the same as the results from the final optimization procedure. Accordingly, the optimization process is terminated despite the single cycle, and the final conditions are 70 turns, 1.0 mm coil distance, and 5.0 mm coil length. More detailed specifications are listed in Table 3.

Table 3. Dimensions of the optimized floating toroidal coil.

Optimized floating toroidal coil		
	Inner diameter	27 mm
	Outer diameter	32 mm
	Coil length	5.0 mm
	Coil distance	1.0 mm
	Turns	70
	Pattern width	0.2 mm
	Pattern height	0.07 mm

3. Experiment Results and Discussion

3.1. Fabrication

The fabricated PCB including the FTC, signal lines, and huge ground pads is shown in Figure 5. In the device, we removed the center signal line to minimize the number of signal ports; in fact, $V_{capacitive}$ can be estimated by measuring the voltages at the FTC ends based on Equation (2). It is important for the VI probe to have high sensitivity, so to minimize RF noise effects, a large grounded pad is attached near the FTC and signal lines. Furthermore, parallel capacitors are installed as a high frequency pass filter, and the signal lines are fabricated as microstrip lines with a characteristic impedance of 50 Ω. Each end of the signal lines is connected with an SMA connector that acts as a signal port.

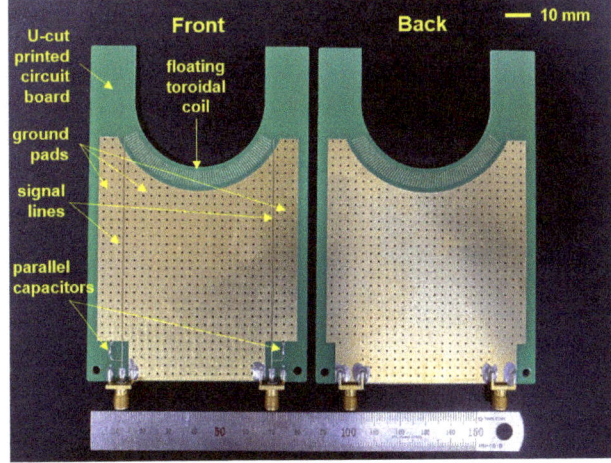

Figure 5. Photograph of the fabricated FTC embedded in a PCB showing both front and back sides.

Figure 6 shows the components of the fabricated VI probe: N-type connectors, mounts, cases (top and bottom), rod, dielectric holder, and printed circuit board. The N-type connectors coupled with the rod play a role as the input and output ports of the fabricated VI probe. The assembly procedure is described in Figure 7. As shown in Figures 6 and 7, the fabricated VI probe is both easy to assemble and robust.

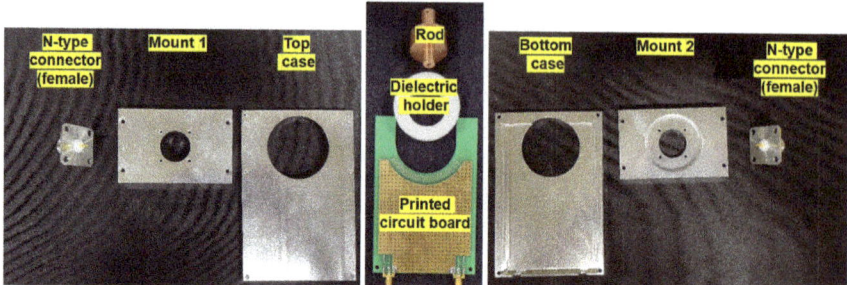

Figure 6. Photographs of the FTC-based VI probe components: N-type connector, mounts, cases, rod, dielectric holder, and PCB.

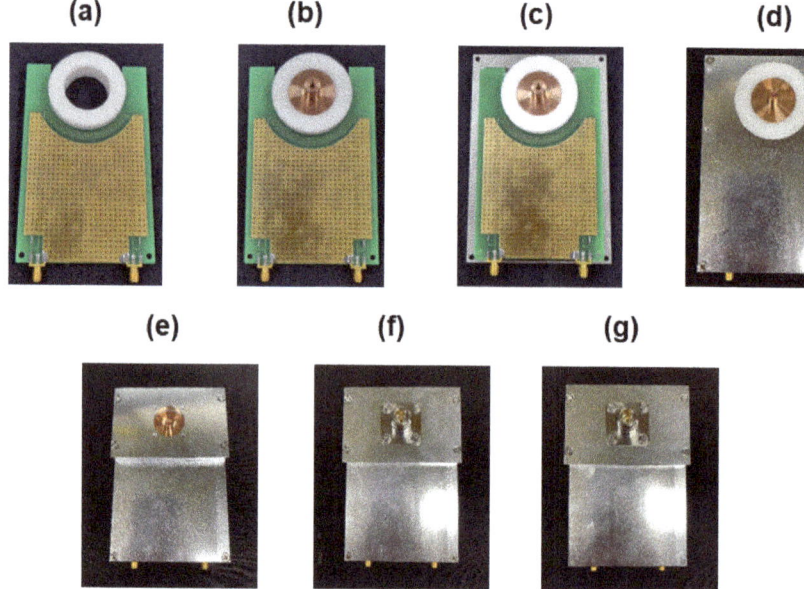

Figure 7. Photographs of the probe assembly procedure. (**a**) The U-cut printed circuit board (PCB) is inserted in the groove of a dielectric holder. (**b**) The rod is placed in the hole inside the dielectric holder. (**c**) The module is mounted on the bottom case. (**d**) The module is covered with the top case. (**e**) Mount 1 is installed. (**f**) An N-type connector is installed. (**g**) Mount 2 and an N-type connector are installed on the back.

3.2. Calibration

The experimental setup to identify the coefficients α and β from Equations (2) and (4) is shown in Figure 8. Details of this setup are also described in [26]. For high power calibration, a cylindrical vacuum chamber with a turbomolecular pump (D-35614 Asslar, Pfeiffer Vacuum, Inc., Asslar, Germany) and a rotary pump (GHP-800K, KODIVAC Ltd., Gyeongsan-si, Korea) are employed as the dummy load in this calibration system. The pressure of the vacuum chamber, measured by a vacuum gauge (Baraton, MKS Instruments

Inc., Andover, MA, USA), is maintained below 1 mTorr to suppress vacuum discharge causing impedance variation during the calibration procedure; here, the chamber pressure is lower than the minimum measurable range of the vacuum gauge. A cylindrical electrode with a diameter of 150 mm connected with an RF matcher (PathFinder, Plasmart Inc., Daejeon, Korea) is inserted into the vacuum chamber. To minimize impedance variation by thermal effects, coolant flows through the electrode. The fabricated VI probe is installed on the input port of the RF matcher with an N-type Tee adaptor. The two signal ports of the fabricated VI probe are connected to channel 1 and 2 of an oscilloscope (TDS3054B, Tektronix Inc., Beaverton, OR, USA) through coaxial cables with BNC-SMA adaptors. A high-voltage probe (P5100, Tektronix Inc., Beaverton, OR, USA) along with the oscilloscope measures the voltage of the open (left) port of the tee adaptor.

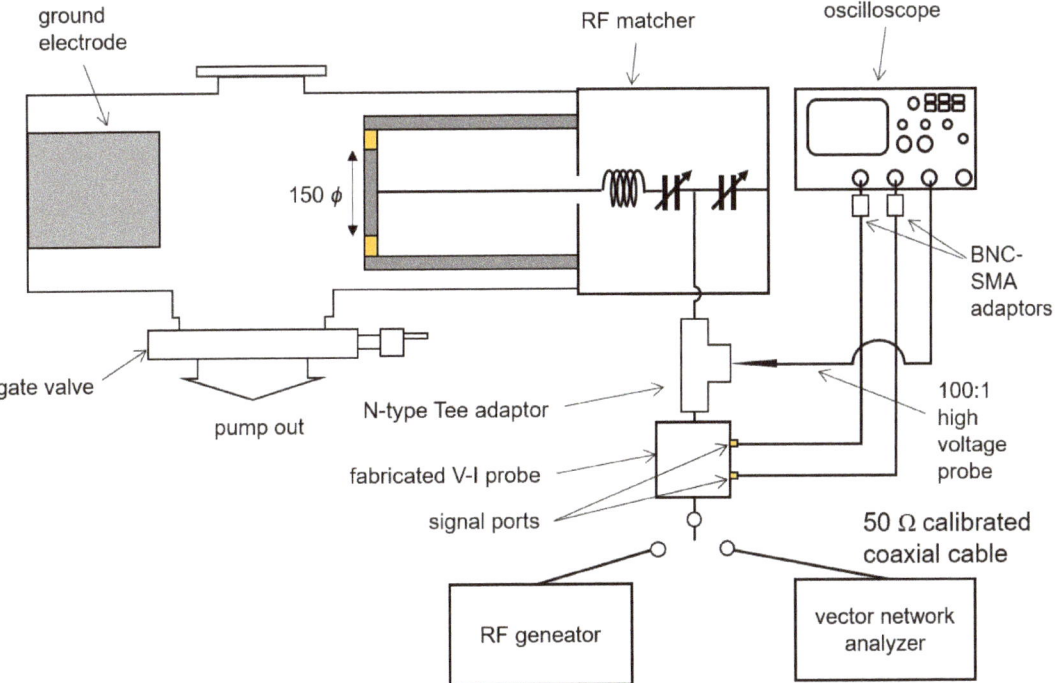

Figure 8. Schematic diagram of the calibration setup for the fabricated FTC-based VI probe. The fabricated VI probe is installed on the input port of the RF matcher with the N-type Tee adaptor. The two signal ports of the fabricated VI probe are connected to channel 1 and 2 of an oscilloscope through coaxial cables with BNC-SMA adaptors.

The calibration procedure is as follows. First, we connect a vector network analyzer (E5071B, Agilent Inc., Santa Clara, CA, USA) to the input port of the fabricated VI probe with a coaxial cable with the end calibrated with a kit (SAV20201B, Saluki Technology Inc., Taipei, Taiwan) as shown in Figure 8. Then, the RF matcher is manually manipulated to match the input impedance (Z_{input}) as 50 Ω while the vector network analyzer measures the input impedance. Second, provided that the impedance matching is terminated, the vector network analyzer is replaced with an RF generator (YSR-06MF, Yongshin RF Inc., Hanam-si, Korea). While 13.56 MHz power from 50 W to 300 W is applied to the electrode, the reference voltage (V_{RF}) and current (I_{RF}) are measured by the high-voltage probe and calculated by $I_{RF} = V_{RF}/|Z_{input}|$, respectively. Each measurement is carried out 20 times.

Figure 9 shows the root-mean-square (RMS) values of the voltage and current signals from the fabricated VI probe, $V_{voltage,rms}$ and $V_{current,rms}$, over the RMS reference voltage and current, $V_{RF,rms}$ and $I_{RF,rms}$, respectively. Here, $V_{voltage,probe}$ is calculated from the RMS value of $(V_{ch1} + V_{ch2})/2$, where V_{ch1} and V_{ch2} are the voltage waveforms recorded from channel 1 and 2 of the oscilloscope, respectively. Similarly, $V_{current,probe}$ is from the RMS value of $V_{ch1} - V_{ch2}$.

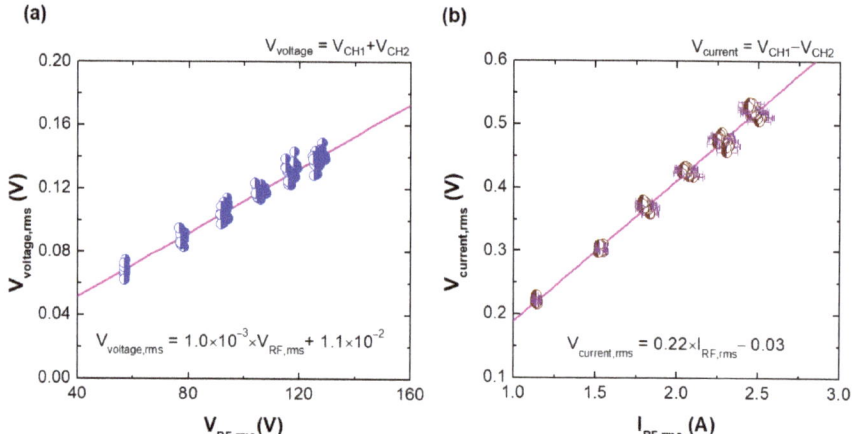

Figure 9. Calibration results of the (**a**) voltage and (**b**) current along increasing RF input voltage and current. To avoid impedance variation by plasma formation during the calibration procedure, the pressure of the vacuum chamber is maintained below 1 mTorr (lower than the minimum measurable range of the vacuum gauge).

Since RF power is dissipated as heat by each component, such as the electrode, RF matcher, etc., the impedance changes, and this affects the accuracy of calibration. To assess the impedance variance by thermal effects during the calibration procedure, Z_{input} was measured again after the procedure. The impedance variance is considered to calculate $I_{RF,rms}$ as the min-max value, represented in Figure 9b as error bars on the x-axis.

3.3. Comparison with a Commercial VI Probe

For an evaluation of the fabricated VI probe via comparison with a commercial VI probe, the experimental setup is slightly changed, as shown in Figure 10. A commercial VI probe (Octive poly, Impedans Ltd., Dublin, Ireland) is installed between the RF generator and the fabricated VI probe for the comparison. A mass flow controller (MFC, TN280, SMTEK Co., Ltd., Seongnam-si, Korea) maintains the flow rate of argon gas at 100 sccm into the vacuum chamber to maintain the chamber pressure at 20 mTorr. The RF generator applies power to the electrode and argon plasma is generated.

Since the RF matcher maintains the source impedance at 50 Ω while the plasma is sustained, the relationships of V_{RF} and I_{RF} to the RF power (P_{RF}) are $P_{RF} = V_{RF}^2/50$ and $P_{RF} = 50 I_{RF}^2$, respectively. Figure 11a plots the square of the RMS voltage measured by the fabricated VI probe, the commercial VI probe, and the high-voltage probe with the oscilloscope over input RF power. As the input RF power increases, all probes show a linear increase. Among them, the fabricated VI probe shows a higher R^2 of 0.9967 for linear fitting than that of the commercial probe. As shown in Figure 11b, the squares of the RMS currents by the fabricated and commercial VI probes also show a linear increase. The fabricated VI probe again shows a higher R^2 of 0.9938 for the current compared to the commercial probe. In summary, the fabricated VI probe demonstrates a good linearity for both voltage and current, at slightly higher levels than the commercial VI probe.

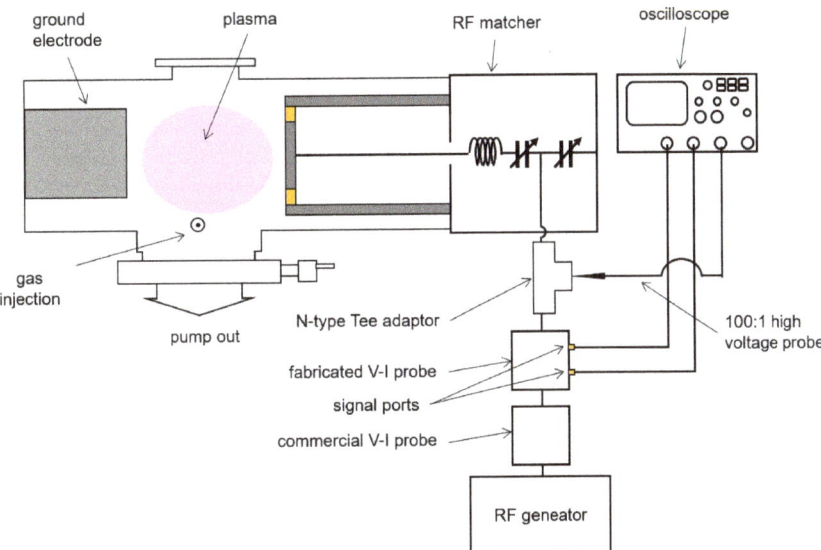

Figure 10. Experimental setup for a comparison of the fabricated VI probe with a commercial VI probe. A commercial VI probe is installed between the RF generator and the fabricated VI probe for the comparison. The RF generator applies power to the electrode and argon plasma is generated.

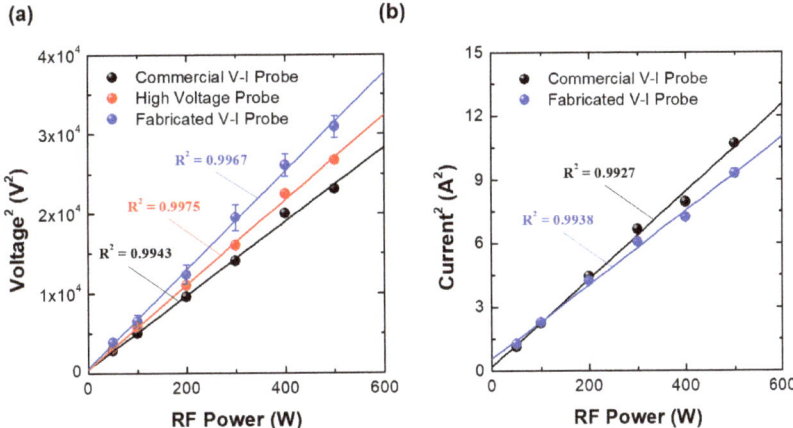

Figure 11. Square of the root-mean-square (RMS) (**a**) voltage and (**b**) current measured by the three probes over RF input power at an argon gas injection rate of 100 sccm, pressure of 20 mTorr, and linearity factors (R-squared values (R^2)).

Here, the squares of the RMS currents from the high-voltage measurement with the oscilloscope is excluded in Figure 11b since it requires the impedance information during plasma discharge. While the RF power is applied, the impedance cannot be measured with the vector network analyzer since the internal impedance of the VNA is 50 Ω and the applied voltage is beyond the measurement limitation of the vector network analyzer.

It should be noted that the voltage level of $V_{voltage,rms}$ is much lower than $V_{current,rms}$ based on Figure 9. Traditional VI probes show the opposite characteristic, where the capacitive signal is much larger than the inductive signal as in [52]. This results from the small area of capacitive coupling; traditional voltage sensors use a large area electrode, whereas the FTC consists of wire-type electrodes and naturally has a much smaller coupling area. Further development of the proposed VI probe is therefore important to enhance the capacitive coupling, such as by using other dielectric holders with higher dielectric constants, increasing (decreasing) the radius of the rod (FTC), etc.

The evaluation result for RMS voltage and current does not mean the performance of the proposed probe is better than the commercial probe. The data acquisitions of ten times for each RF power condition in the evaluation process is not enough to exactly compare the proposed VI probe with the commercial probe. Nevertheless, this evaluation result means the successful operation of the prototype. Furthermore, the proposed probe is not fully optimized based on various practical tests; the simulation plays a role in bringing the probe design to near optimized conditions. There are still several practical-test-based optimizations. Later, practical optimization to enhance its performance and exact comparison with the commercial ones will be reported.

4. Conclusions

In this paper, we proposed a VI sensor based on a floating toroidal coil. The operation principle of the FTC was demonstrated and its optimum design was established through 3D electromagnetic wave simulation. Here, optimization parameters of the FTC on a printed-circuit board are the number of turns, the coil distance, and the coil length. The resultant optimum conditions are 70 turns, coil distance of 1.0 mm, and coil length of 5.0 mm. Based on the optimum conditions, the proposed VI probe with FTC was fabricated and calibrated based on the high-voltage probe measurement for voltage and the vector network analyzer measurement for the current. During calibration procedure, impedance change by plasma formation and thermal expansion of electrode are suppressed by maintaining pressure of the vacuum chamber below 1 mTorr and flowing coolant through the electrode, respectively. Then, it was evaluated by comparison with a commercial VI probe. The results demonstrated that the FTC-based probe achieved a slightly higher linearity than the commercial one, with an R^2 of 0.9967 for RMS voltage and 0.9938 for RMS current.

Author Contributions: Conceptualization , S.-j.Y. and Y.-b.Y.; validation, S.-j.K., I.-h.S., J.-j.L., Y.-s.L. and C.-h.C.; formal analysis, S.-j.K. and W.-n.J.; writing—original draft preparation, S.-j.K.; writing—review and editing, S.-j.Y., Y.-s.L. and C.-h.C.; supervision, S.-j.Y. All authors have read and agreed to the published version of the manuscript.

Funding: This research was supported by a National Research Council of Science & Technology (NST) grant by the Korean government (MSIP) (No. CAP-17-02-NFRI, CRF-20-01-NFRI), by the Next-generation Intelligent Semiconductor R&D Program through the Korea Evaluation Institute of Industrial Technology (KEIT) funded by the Korean government (MOTIE), by the Korea Institute of Energy Technology Evaluation and Planning (KETEP) and the MOTIE of the Republic of Korea (20202010100020), by the MOTIE (20009818, 20010420) and KSRC (Korea Semiconductor Research Consortium) support program for the development of future semiconductor devices, by a Korea Institute for Advancement of Technology (KIAT) grant funded by the Korean Government (MOTIE) (P0008458, HRD Program for Industrial Innovation), by the Basic Science Research Program through the National Research Foundation of Korea (NRF) funded by the Ministry of Education (NRF-2020R1A6A1A03047771), and by the Korea Institute of Machinery & Materials (KIMM) Institutional Program (NK236F) and NST/KIMM.

Data Availability Statement: The data presented in this study are available on request from the corresponding author.

Conflicts of Interest: The authors declare no conflict of interest.

References

1. Lieberman, M.A.; Lichtenberg, A.J. *Principles of Plasma Discharges and Materials Processing*, 2nd ed.; Wiley&Sons. Inc.: Hobken, NJ, USA, 2005; pp. 1–22.
2. Adamovich, I.; Baalrud, S.D.; Bogaerts, A.; Bruggeman, P.J.; Cappelli, M.; Colombo, V.; Czarnetzki, U.; Ebert, U.; Eden, J.G.; Favia, P.; et al. The 2017 Plasma Roadmap: Low temperature plasma science and technology. *J. Phys. D Appl. Phys.* **2017**, *50*, 323001. [CrossRef]
3. Bogaerts, A.; Tu, X.; Whitehead, J.C.; Centi, G.; Lefferts, L.; Guaitella, O.; Azzolina-Jury, F.; Kim, H.-H.; Murphy, A.B.; Schneider, W.F. The 2020 plasma catalysis roadmap. *J. Phys. D Appl. Phys.* **2020**, *53*, 443001. [CrossRef]
4. Ishikawa, K.; Karahashi, K.; Ishijima, T.; Cho, S.I.I.; Elliott, S.; Hausmann, D.; Mocuta, D.; Wilson, A.; Kinoshita, K. Progress in nanoscale dry processes for fabrication of high-aspect-ratio features: How can we control critical dimension uniformity at the bottom? *Jpn. J. Appl. Phys.* **2018**, *57*, 06JA01. [CrossRef]
5. Vincent, M. Donnelly and Avinoam Kornblit, Plasma etching: Yesterday, today, and tomorrow. *J. Vac. Sci. Technol.* **2013**, *31*, 050825.
6. Seong, I.; Lee, J.; Cho, C.; Lee, Y.; Kim, S.; You, S. Characterization of SiO2 Over Poly-Si Mask Etching in Ar/C4F8 Capacitively Coupled Plasma. *Appl. Sci. Converg. Technol.* **2021**, *30*, 176–182. [CrossRef]
7. Lee, Y.; Seong, I.; Lee, J.; Lee, S.; Cho, C.; Kim, S.; You, S. Various evolution trends of sample thickness in fluorocarbon film deposition on SiO_2. *J. Vac. Sci. Technol. A* **2022**, *40*, 013001. [CrossRef]
8. Lee, H.; Chung, C. Electron heating and control of electron energy distribution for the enhancement of the plasma ashing processing. *Plasma Sources Sci. Technol.* **2015**, *24*, 024001. [CrossRef]
9. Susa, Y.; Ohtake, H.; Jianping, Z.; Chen, L.; Nozawa, T. Characterization of CO_2 plasma ashing for less low-dielectric-constant film damage. *J. Vac. Sci. Technol. A* **2015**, *33*, 061307. [CrossRef]
10. Hamedani, Y.; Macha, P.; Bunning, T.J.; Naik, R.R.; Vasudev, M.C. Plasma-Enhanced Chemical Vapor Deposition: Where we are and the Outlook for the Future. In *Chemical Vapor Deposition, Recent Advances and Applications in Optical, Solar Cells and Solid State Devices*; Neralla, S., Ed.; IntechOpen: London, UK, 2016; pp. 247–251.
11. Vasudev, M.C.; Anderson, K.D.; Bunning, T.J.; Tsukruk, V.V.; Naik, R.R. Exploration of Plasma-Enhanced Chemical Vapor Deposition as a Method for Thin-Film Fabrication with Biological Applications. *ASC Appl. Mater. Interfaces* **2013**, *5*, 3983–3994. [CrossRef]
12. Chung, Y.; Lung, C.; Chiu, Y.; Lee, H.; Lian, N.; Yang, T.; Chen, K.; Lu, C. Study of Plasma Arcing Mechanism in High Aspect Ratio Slit Trench Etching. In Proceedings of the 2019 30th Annual SEMI Advanced Semiconductor MAnufacturing Conference (ASMC), Saratoga Springs, NY, USA, 6–9 May 2019.
13. Carter, D.; Walde, H.; Nauman, K. Managing arcs in large area sputtering applications. *Thin Solid Films* **2012**, *520*, 4199–4202. [CrossRef]
14. Lee, H.J.; Seo, D.S.; May, G.S.; Hong, S.J. Use of In-Situ Optical Emission Spectroscopy for Leak Fault Detection and Classification in Plasma Etching. *J. Semicond. Technol. Sci.* **2013**, *13*, 4. [CrossRef]
15. Marchack, N.; Buzi, L.; Farmer, D.B.; Miyazoe, H.; Papalia, J.M.; Yan, H.; Totir, G.; Engelmann, S.U. Plasma processing for advanced microelectronics beyond CMOS. *J. Appl. Phys.* **2021**, *130*, 080901. [CrossRef]
16. Seman, M.; Wolden, C.A. Investigation of the role of plasma conditions on the deposition rate and electrochromic performance of tungsten oxide thin films. *J. Vac. Sci. Technol. A* **2003**, *21*, 6. [CrossRef]
17. Gopikishan, S.; Banerjee, I.; Bogle, K.A.; Das, A.K.; Pathak, A.P.; Mahapatra, S.K. Paschen curve approach to investigate electron density and deposition rate of Cu in magnetron sputtering system. *Radiat. Eff. Deffects Solids* **2016**, *171*, 999–1005. [CrossRef]
18. Cho, C.; You, K.; Kim, S.; Lee, Y.; Lee, J.; You, S. Characterization of SiO2 Etching Profiles in Pulse-Modulated Capacitively Coupled Plasmas. *Materials* **2021**, *14*, 5036. [CrossRef]
19. Hopkins, M.B.; Lawler, J.F. Plasma diagnostics in industry. *Plasma Phys. Control. Fusion* **2000**, *42*, B189–B197. [CrossRef]
20. Baek, K.H.; Jung, Y.; Min, G.J.; Kang, C.; Cho, H.K.; Moon, J.T. Chamber maintenance and fault detection technique for a gate etch process via self-excited electron resonance spectroscopy. *J. Vac. Sci. Technol. B* **2005**, *23*, 125–129. [CrossRef]
21. Lobbia, R.B.; Beal, B.E. Recommended Practice for Use of Langmuir Probes in Electric Propulsion Testing. *J. Propuls. Power* **2017**, *33*, 3. [CrossRef]
22. Chen, F.F. Lecture Notes on Langmuir Probe Diagnostics. In Proceedings of the 30th International Conference on Plasma Science, Jeju, Korea, 2–5 June 2003.
23. Godyak, V. RF discharge diagnostics: Some problems and their resolution. *J. Appl. Phys.* **2021**, *129*, 041101. [CrossRef]
24. Sugai, H.; Nakamura, K. Recent innovations in microwave probes for reactive plasma diagnostics. *Jpn. J. Appl. Phys.* **2019**, *58*, 060101. [CrossRef]
25. Kim, S.J.; Lee, J.J.; Kim, D.W.; Kim, J.H.; You, S.J. A transmission line model of the cutoff probe. *Plasma Sources Sci. Technol.* **2019**, *28*, 055014. [CrossRef]
26. Kim, S.; Lee, J.; Lee, Y.; Cho, C.; You, S. Crossing Frequency Method Applicable to Intermediate Pressure Plasma Diagnostics Using the Cutoff Probe. *Sensors* **2022**, *22*, 1291. [CrossRef]
27. Kim, S.J.; Lee, J.J.; Lee, Y.S.; Kim, D.W.; You, S.J. Finding the optimum design of the planar cutoff probe through a computational study. *AIP Adv.* **2021**, *11*, 025241. [CrossRef]

28. Piejak, R.B.; Godyak, V.A.; Garner, R.; Alexandrovich, B.M.; Sternber, N. The hairpin resonator: A plasma density measuring technique revisited. *J. Appl. Phys.* **2004**, *95*, 7. [CrossRef]
29. Dine, S.; Booth, J.-P.; Curley, G.A.; Corr, C.S.; Jolly, J.; Guillon, J. A novel technique for plasma density measurement using surface-wave transmission spectra. *Plasma Sources Sci. Technol.* **2005**, *14*, 777–786. [CrossRef]
30. Styrnoll, T.; Harhausen, J.; Lapke, M.; Storch, R.; Brinkmann, R.P.; Foest, R. A Ohl and P Awakowicz, Process diagnostics and monitoring using the multipole resonance probe in an inhomogeneous plasma for ion-assisted deposition of optical coatings. *Plasma Sources Sci. Technol.* **2013**, *22*, 045008. [CrossRef]
31. Wang, C.; Friedrichs, M.; Oberrath, J.; Brinkmann, R.P. Kinetic investigation of the planar multipole resonance probe in the low-pressure plasma. *Plasma Sources Sci. Technol.* **2021**, *30*, 105011. [CrossRef]
32. Ogawa, D.; Nakamura, K.; Sugai, H. Experimental validity of double-curling probe method in film-depositing plasma. *Plasma Sources Sci. Technol.* **2021**, *30*, 085009. [CrossRef]
33. Mackus, A.J.M.; Heil, S.B.S.; Langereis, E.; Knoops, H.C.M.; Sanden, M.C.M.V.; Kessels, W.M.M. Optical emission spectroscopy as a tool for studying, optimizing, and monitoring plasma-assisted atomic layer deposition processes. *J. Vac. Sci. Technol. A* **2010**, *28*, 77. [CrossRef]
34. Engeln, R.; Klarenaar, B.; Guaitella, O. Foundations of optical diagnostics in low-temperature plasmas. *Plasma Sources Sci. Technol.* **2020**, *29*, 063001. [CrossRef]
35. Sobolewski, M.A. Electrical characterization of radio-frequency discharges in the Gaseous Electronics Conference Reference Cell. *J. Vac. Sci. Technol. A* **1992**, *10*, 3550–3562. [CrossRef]
36. Dewan, N.A. Analysis and Modelling of the Impact of Plasma RF Harmonics in Semiconductor Plasma Processing. Ph.D. Thesis, Dublin City University, Dublin, Ireland, 2001.
37. Sezemsky, P.; Stranak, V.; Kratochvil, J.; Cada, M.; Hippler, R.; Hrabovsky, M.; Hubicka, Z. Modified high frequency probe approach for diagnostics of highly reactive plasma. *Plasma Sources Sci. Technol.* **2019**, *28*, 115009. [CrossRef]
38. Lee, M.; Jang, S.; Chung, C. Floating probe for electron temperature and ion density measurement applicable to processing plasmas. *J. Appl. Phys.* **2007**, *101*, 033305. [CrossRef]
39. Zhang, A.; Kwon, D.; Chung, C. A method for measuring negative ion density distribution using harmonic currents in a low-pressure oxygen plasma. *Plasma Sources Sci. Technol.* **2020**, *29*, 065017. [CrossRef]
40. Yang, R.; Chen, R. Real-Time Plasma Process Condition Sensing and Abnormal Process Detection. *Sensors* **2010**, *10*, 5703–5723. [CrossRef] [PubMed]
41. Yang, R.; Chen, R. Real-Time Fault Classification for Plasma Processes. *Sensors* **2011**, *11*, 7037–7054. [CrossRef]
42. Yang, J.; McArdle, C.; Daniels, S. Dimension Reduction of Multivariable Optical Emission Spectrometer Datasets for Industrial Plasma Processes. *Sensors* **2014**, *14*, 52–67. [CrossRef]
43. Motomura, T.; Kasashima, Y.; Uesugi, F.; Kurita, H.; Kimura, N.A. Real-time characteristic impedance monitoring for end-point and anomaly detection in the plasma etching process. *Jpn. J. Appl. Phys.* **2014**, *53*, 03DC03. [CrossRef]
44. Kang, G.; An, S.; Kim, K.; Hong, S. An in situ monitoring method for PECVD process equipment condition. *Plasma Sci. Technol.* **2019**, *21*, 064003. [CrossRef]
45. Kim, J.; Lee, K.; Jeong, H.; Lee, J.; Choi, Y.S. Anti-contamination SMART (Spectrum Monitoring Apparatus with Roll-to-roll Transparent film) window for optical diagnostics of plasma systems. *Rev. Sci. Instrum.* **2021**, *92*, 013507. [CrossRef]
46. Kim, N. Self-Plasma Optical Emission Spectroscopy Having Active Contamination Preventing Equipment and Method of Preventing Contaminaion of Plasma Chamber. KR 101273922B1, 11 June 2013.
47. Cha, D. Window Contaminating Delay Apparatus for Semiconductor Process. KR 101410296B1, 20 June 2014.
48. Huang, X.; Xin, Y.; Yang, L.; Ye, C.; Yuan, Q.; Ning, Z. Analysis of optical emission spectroscopy in a dual-frequency capacitively coupled CHF3 plasma. *Phys. Plasmas* **2009**, *16*, 043509. [CrossRef]
49. Nakano, T.; Samukawa, S. Effects of Ar dilution on the optical emission spectra of fluorocarbon ultrahigh-frequency plasmas: C4F8 vs CF4. *J. Vac. Sci. Technol. A* **1999**, *17*, 686. [CrossRef]
50. Lafleur, T.; Delattre, P.A.; Booth, J.P.; Johnson, E.V.; Dine, S. Radio frequency current-voltage probe for impedance and power measurements in multi-frequency unmatched loads. *Rev. Sci. Instrum.* **2013**, *84*, 015001. [CrossRef] [PubMed]
51. Lee, S.; Kim, J. Sensor for measuring electrical characteristics. KR 10-2011-0024791, 29 December 2011
52. Kim, K.K.; Lee, J.J.; Kim, S.J.; Cho, C.H.; Yoo, S.W.; You, S.J. Development of High-precision RF Sensor. *Appl. Sci. Converg. Technol.* **2019**, *28*, 88–92. [CrossRef]
53. CST Studio Suite. Available online: https://www.3ds.com/ (accessed on 16 November 2018).

Article

Development of a Noninvasive Real-Time Ion Energy Distribution Monitoring System Applicable to Collisional Plasma Sheath

Inho Seong [1], Sijun Kim [1], Youngseok Lee [1], Chulhee Cho [1], Jangjae Lee [2], Wonnyoung Jeong [1], Yebin You [1] and Shinjae You [1,3,*]

1. Applied Physics Lab for PLasma Engineering (APPLE), Department of Physics, Chungnam National University, Daejeon 34134, Korea
2. Samsung Electronics, Hwaseong-si 18448, Korea
3. Institute of Quantum Systems (IQS), Chungnam National University, Daejeon 34134, Korea
* Correspondence: sjyou@cnu.ac.kr

Abstract: As the importance of ion-assisted surface processing based on low-temperature plasma increases, the monitoring of ion energy impinging into wafer surfaces becomes important. Monitoring methods that are noninvasive, real-time, and comprise ion collision in the sheath have received much research attention. However, in spite of this fact, most research was performed in invasive, not real-time, and collisionless ion sheath conditions. In this paper, we develop a noninvasive real-time IED monitoring system based on an ion trajectory simulation where the Monte Carlo collision method and an electrical model are adopted to describe collisions in sheaths. We technically, theoretically, and experimentally investigate the IED measurement with the proposed method, and compared it with the result of IEDs measured via a quadrupole mass spectrometer under various conditions. The comparison results show that there was no major change in the IEDs as radio-frequency power increased or the IED gradually became broad as gas pressure increased, which was in a good agreement with the results of the mass spectrometer.

Keywords: plasma; ion energy distribution (IED); real time; monitoring; noninvasive

1. Introduction

Plasma processing in modern semiconductor manufacturing has attracted enormous interest in fabricating fine structures on wafers. For nanoscale electronic devices, conventional plasma technology is challenging to meet the demanded performance for the device. Therefore, advanced low-temperature plasma technology is required for high-level processes such as atomic-layer and high-aspect-ratio etching, which may satisfy zero-defect etching results and deep-and-narrow contact hole profiles [1–7]. To attain high-level processes, key process parameters such as etching and deposition rate, which determines the surface treatment results, were investigated to find their correlations to plasma internal parameters, namely, electron density, electron temperature, and ion energy [5,6,8–12]. Ion energy and its bombardment are important parts of physical etching such as sputtering and improving chemical surface reactions because they directly transfer energy to surfaces [13–16]. Ion energy thus needs be carefully optimized and controlled to meet the desired conditions. However, to monitor and precisely control ion energy distributions (IEDs) is very difficult; thus, this leads to the process drift or abnormalities causing low-productivity fabrication [17–19]. Therefore, a reliable method of IED control is needed for the next-generation process [20].

In academic areas such as universities, invasive diagnostics to measure the IED are retarding field energy analyzers (RFEAs) [21,22] and quadrupole mass spectrometer (QMSs) [23–26]. An RFEA measures ion fluxes overcoming grid potentials that repel ions of lower energy than that of their potential. IEDs are inferred by the first derivative of ion fluxes as a function of the grid potential [22]. A QMS comprises an energy analyzer, a quadrupole mass filter, and an ion detector, and measures ion fluxes with a specific energy that can pass through the analyzer. An IED is the measured ion fluxes by sweeping the potential of the energy analyzer [24]. To measure IEDs, such diagnostics are mounted onto an electrode surface or inserted into a reactor via a port. However, the semiconductor industry does not prefer to use such a diagnostic tool for the following several reasons. First, reactors used in the industry have few view ports because they can be a source of asymmetry in plasma uniformity. Second, an invasive method such as a QMS can perturb the plasma. To avoid this perturbation, a QMS can be placed onto the radio-frequency (RF) powered electrode, but a proper measurement is not allowed due to huge RF voltage interference to the QMS.

To overcome such issues from invasive diagnostics, an alternative is an ion energy monitoring system that measures and analyzes voltage waveforms on a powered electrode. Normally, voltage and current (V–I) probes used for arcing signal, particles, and etch end point detection are used by installing them behind a powered electrode [22,27–34]. The first attempt of the V–I probe as a diagnostic method is to estimate the plasma density and total ion current through simple analytical approaches from the measured electrode voltage and current waveform [34,35]. With continuous development, various plasma parameters, including an IED, can be obtained through a numerical sheath model that analyzes the measured voltage signal.

Sobolewski studied V–I waveforms between a powered electrode and impedance matching network, and developed models to estimate various plasma parameters with waveforms for a few decades [34–38]. He reported an ion flux and energy monitoring system with a collisionless sheath model [36,38]. Initially, the proposed technique was demonstrated with a bare metallic electrode without a wafer. Then, for practical use in plasma etching, the improved monitoring system with the high-speed data acquisition method was validated in a wafer-loaded environment in extremely low pressure [37]. However, the collisionless sheath model is not applicable for film deposition processing where the reactor pressure ranges several hundred mTorr. In this pressure range, the ion mean free path is shorter than the sheath thickness; as a result, the collision effect is included in the motion of ions in the sheath.

Recently, Bogdanova et al. have reported the feasibility of a virtual IED sensor based on the fast calculation of the ion trajectory simulation coupled to the Monte Carlo collision (MCC) method [39–41]. As the MCC increases the simulation time, to decrease the calculation time, the authors introduced a simple analytical model of a time-varying linear electric field and measured sheath voltage waveform that was used in the input data and defined as the subtraction of the powered-electrode voltage waveform from the plasma potential waveform. In this work, for rapid calculations, the plasma potential was implemented, and the following calculation time is investigated and optimized as a function of the number of injection particles. The sheath voltage waveform, which is a significant parameter to determine the shape of IEDs, is calculated as the difference between the potentials of the plasma and the powered electrode. Additionally, the simple analytical model of a time-varying linear electric field equation was used. Since actual plasma processing chambers do not allow for any invasive measurement, Bogdanova et al. investigated IEDs with a change in the plasma potential and the powered electrode voltage type: (i) invasive measurement of both voltage waveforms, (ii) assumption of plasma potential as sine, and (iii) assumption of both voltage waveforms as sine. Results show that, as assumptions are added, the accuracy decreases despite this being a noninvasive method.

In previous works, IED monitoring systems had the limitations of, for example, neglecting collisions, measuring invasively, and assuming the plasma potential to be a sine wave. To overcome these limitations, we propose a real-time IED monitoring system with an MCC model and an electrical sheath model. Instead of plasma potential measurement, the proposed system calculates it through the sheath model with a powered-electrode voltage waveform. To take into account collisions and give real-time feedback, we used the standard null collision method and simplified cross-sectional method for various collisions. We validated the resulting IEDs from our monitoring system by comparing them to the IEDs obtained with a commercial QMS in argon plasma at various conditions (RF powers, pressure). The result shows that the IED measured with the proposed monitoring system was in a good agreement with the QMS results, which shows the possibility of measuring accurate IEDs according to changes in various discharge conditions. The details of the methods and experimental setup of our IED monitoring system are described in the following section. The comparison results are shown and discussed.

2. Experiment
2.1. Vacuum and Plasma Conditions

To validate the concept of the monitoring system, experiments were conducted with an asymmetric capacitively coupled plasma (CCP) source to which 13.56 MHz RF power was applied. Figure 1 shows a schematic of the CCP source where the diagnosis and determination of the plasma parameters such as a voltage waveform and IEDs were performed. The CCP chamber had a diameter of 200 mm and contained parallel-plate electrodes separated by 120 mm. The bottom electrode was surrounded by a ceramic ring for electrical insulation from the ground, and connected to a 13.56 MHz RF generator via an L-type matching network. The top electrode, which had a shower head to uniformly inject gases into the processing chamber, was grounded. A QMS (PSM, Hiden Analytic, Warrington, Cheshire, UK) that could measure IEDs by collecting energetic ions escaping from plasma was mounted to the chamber sidewall via a port. The vacuum system in this processing chamber consisted of a rotary pump and turbo molecular pump that arweree equipped to produce base pressure in the order of 10^{-6} Torr. We injected 100 sccm Ar gas through the shower head, and a variable-conductance valve between the processing chamber and the turbo molecular pump controlled the pressure.

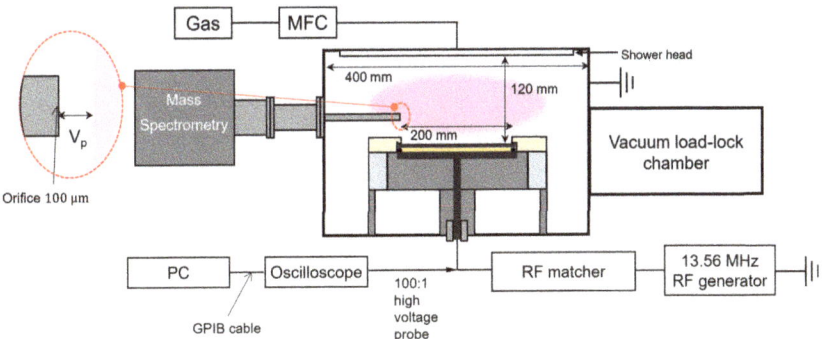

Figure 1. Schematic diagram of the experimental setup of the noninvasive IED monitoring system.

Since various internal and external parameters, electrode voltage, and pressure can affect the profile of IEDs, the experiments were carried out by changing the applied RF power from 100 to 500 W, and the pressure from 10 to 50 mTorr to validate our noninvasive IED monitoring system. The ions were accelerated with a time-varying electric field in the sheath near the powered and grounded electrode, and the chamber wall. IEDs mostly strongly depend on the electric field in the sheath in several types of acceleration and collisions.

2.2. Diagnostics

IEDs were measured with a QMS mounted on the processing chamber sidewall for a comparative study [23,26]. This spectrometer comprised the inline energy analyzer *Bessel box*, a triple-filter quadrupole analyzer, and an ion counting detector with a differential pump housing composed of a rotary pump and turbo molecular pump that maintained the base pressure kn the order of 10^{-7} Torr for ions to transit into the QMS without collisions to the detector [24]. Ions accelerated in the sheath flowed into the chamber of the QMS through a 100 μm orifice located at the end of a grounded sampling probe, but since the commercial sampling probe that contained the orifice could not sample ions in this system, the sampling probe had to be extended and be exposed to plasma. Ions that flowed into the chamber of the QMS through the grounded sampling probe were focused by the electrical lens system to obtain high transmission before reaching the entrance hole of the energy analyzer that was composed of a cylindrical vessel and two circular plates connected to the different voltage sources. Sweeping the voltage sources that are fixing the potential distribution in the analyzer, IEDs could be measured. Ions resolved from the energy analyzer flowed into the quadrupole analyzer where four parallel rods equidistant from the central axis were biased with RF and direct current (DC) voltages, with the opposite rods being at the same potential. Since the RF and DC voltages determine which mass of ions would pass through the quadrupole analyzer, they were set to measure the energy distribution of argon ions in these experiments. Filtered ions need to be transformed from the current into readable signals, which is carried out in the ion detector.

2.3. Validation

To test the validity of our proposed concept, the QMS result and our result were compared by comparing both IEDs on the grounded electrode (or wall) rather than powered electrode, because the commercial QMS of which the sampling probe was grounded could only measure the IED on the grounded electrode. Our IED monitoring system, which can obtain IEDs on a both powered and grounded electrodes, needs a voltage waveform on a powered electrode as an input parameter provided by high-voltage probe measurement with oscilloscopes.

3. Noninvasive Monitoring System

Since the concept of our IED monitoring system is noninvasive, it is combination of hardware for the measurements of electrode voltage as an input parameter and model-based software for analyzing voltage measured by the hardware and describing ion motion. This system was based on a flow chart for input-parameter acquisition, electrical-model calculation, ion-trajectory simulation, and IED data acquisition as shown in Figure 2. Since another concept of this system is real-time monitoring, various methods, such as simplifying the cross-section of ion-neutral collisions, were used for cost calculation. Details on the hardware, software, and speed-up are outlined later.

3.1. Input Parameter

Since IEDs strongly depend on the sheath voltage waveform, it is important to take exact waveforms of the sheath voltage, which can be obtained by subtracting the electrode voltage from the plasma potential. In previous research, the plasma potential was measured via an RF antenna inserted into plasma. Because this method is not applicable to our noninvasive one, the powered electrode voltage was measured instead of it [40].

The electrode voltage waveform was measured with a commercial high-voltage probe (P5100, Tektronix Inc., Beaverton, OR, USA) that was readily available, and mounted between the electrode and the matching network as shown in Figure 1. The waveform was digitized with an oscilloscope (TDS3054B, Tektronix Inc., Beaverton, OR, USA) that was linked to the computer with lab-produced software (MATLAB2018, Natick, MA, USA) via a general-purpose interface bus (GPIB) cable.

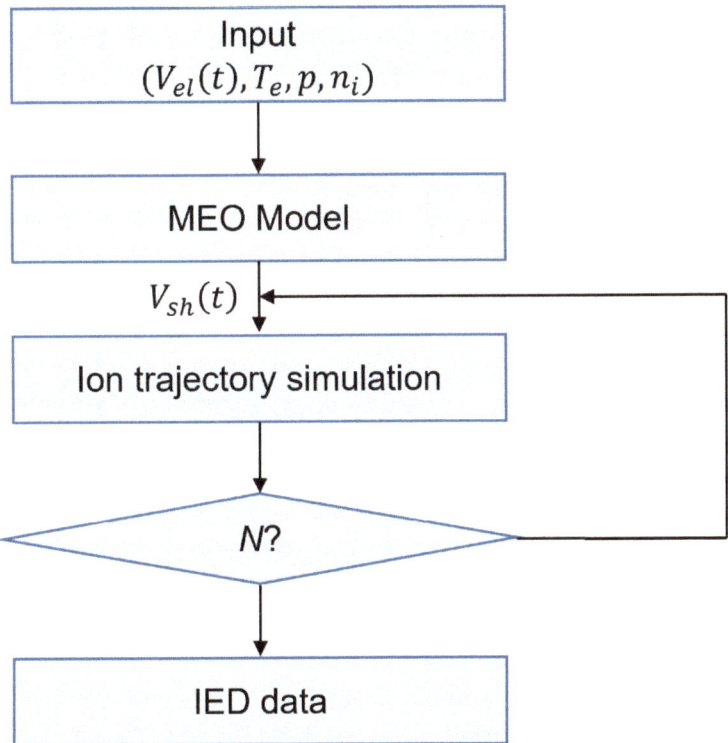

Figure 2. Flowchart of the simulation algorithm in the noninvasive IED monitoring system.

The analysis of electrode voltage allows for obtaining the plasma potential via electrical sheath model Metze, Erine, and Oskam (MEO) [42]. The details of this model are described in Section 3.2.1. Thus, this model was used to obtain the plasma potential without any invasive diagnostics in our system, which requires the electrode voltage waveform as an input parameter. The sheath voltage waveform was obtained by subtracting the calculated plasma potential from the measured electrode voltage.

3.2. Model
3.2.1. Electrical Sheath Model

The model of Metze et al. was applicable to calculating the plasma potential in a fast feedback concept, although this model becomes more accurate as the applied field approaches the low-frequency regime where ions and electrons both respond instantaneously to an imposed time variation of the sheath voltage [42]. In this section, the MEO model adjusted in our system is explored. A schematic of the equivalent electric circuit model for the plasma reactor is shown in Figure 3. Here, V_{rf} is the voltage of the applied RF signal from the matched RF generator, V_{el} is the voltage applied on a powered electrode through a blocking capacitor of which capacitance is represented as C_B, and V_p is the plasma potential. The sheath capacitance and conduction current adjacent to the powered electrode are represented as C_{el}^{sh} and I_{el}, respectively, while the corresponding values adjacent to the grounded wall are C_w^{sh} and I_w. V_{el}^{sh} and V_w^{sh} represent the voltages across the grounded wall-plasma sheath and the powered electrode-plasma sheath, respectively, but those are not shown in Figure 3. Since the details of the derivation and analytic expressions for the sheath capacitance and conduction current were clearly described in [42], they are omitted in this work.

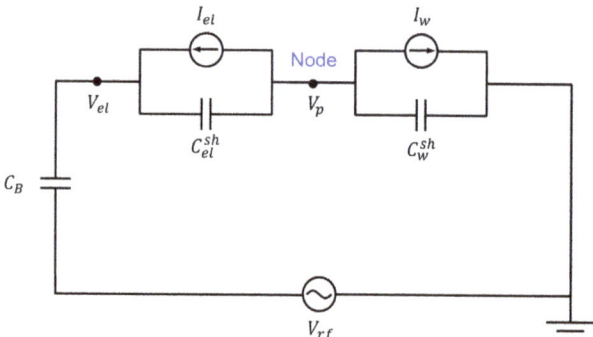

Figure 3. Schematic diagram of the equivalent circuit for the RF sheath model.

The purpose of our system was to obtain a time-varying plasma potential from the powered electrode voltage measured via a voltage probe through this electric circuit. From this equivalent circuit and the current conservation law, as shown in Figure 3, it can be expressed as follows:

$$C_w^{sh}\frac{\partial V_p}{\partial t} + I_w + C_{el}^{sh}\frac{\partial(V_p - V_{el})}{\partial t} + I_{el} = 0. \tag{1}$$

If the analytic expressions for C^{sh} and I of both sheaths from [42] are used in Equation (1), time-varying plasma potential V_p can be numerically obtained. To solve this partial differential equation, the Runge–Kutta fourth-order method was used. Figure 4 shows the calculated waveforms for electrode voltage V_{el}, plasma potential V_p, sheath voltage V_{el}^{sh} adjacent to the powered electrode, and magnified plasma potential in Ar plasma at a pressure of 20 mTorr and RF power of 300 W. The calculated plasma potential was similar except for the phase when the electrode voltage was positive. The sheath voltage always had negative potential with respect to the plasma, which was in a good agreement with measurements by Bruce [43].

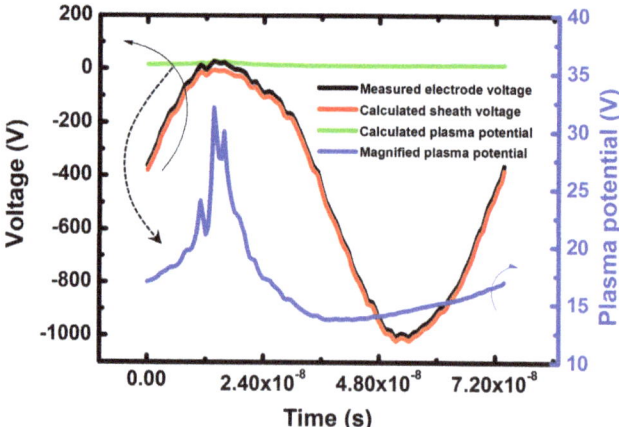

Figure 4. Measurements of electrode voltage, sheath voltage adjacent to the powered electrode, plasma potential, and magnified plasma potential.

Since plasma processing such as surface modification and etching is performed on a powered electrode, it is greatly meaningful to obtain and analyze the sheath voltage waveform adjacent to that electrode. Nevertheless, since the QMS can only measure IEDs at the grounded sheath due to the grounded sampling probe, we obtained the sheath voltage waveform adjacent to the grounded electrode for the IED comparison between our monitoring system and the commercial QMS, as mentioned in Section 2.3.

3.2.2. Ion Trajectory Simulation

In order to obtain the IED on the grounded or powered electrode, the sheath was simulated on the basis of a simple matrix sheath model, with sheath voltage waveforms obtained from the measured voltage adjacent to the powered electrode with the MCC algorithm for ion-neutral collisions [44,45]. From the voltage adjacent to the grounded electrode, which was equal to the plasma potential, shown in Figure 4, an electron sheath edge $s(t)$ was calculated as follows:

$$s(t) = \sqrt{\frac{\epsilon_0 V_p(t)}{2en_0}}, \qquad (2)$$

where ϵ_0 is permittivity in vacuum, $V_p(t)$ is the time-varying plasma potential, e is an electron charge, and n_0 is a constant ion density. If the sheath adjacent to the powered electrode is considered, $V_p(t)$ should be replaced with the absolute value of the sheath voltage on the side of powered electrode, which is always negative with respect to the plasma. Spatially linear electric field $E(x,t)$ in the sheath was derived from Poisson's equation as below.

$$E(x,t) = \begin{cases} \frac{en_0}{\epsilon_0}(x - s(t)), & x < s(t) \\ 0, & x \geq s(t). \end{cases} \qquad (3)$$

To solve the ion trajectory simulation, boundary conditions such as injected ion velocities into the sheath and the initial plasma phase for each ion at $t = 0$ for consideration on one RF period should be defined. Here, ion incident velocities at the ion sheath edge where quasineutrality is always satisfied during discharge are given by

$$u_B = \sqrt{\frac{eT_e}{m_i}}, \qquad (4)$$

where T_e is the electron temperature, and m_i is ion mass. This velocity is technically termed as Bohm velocity, and u_B is a criterion to keep the quasineutrality of plasma [46]. Because the motion of ions can vary according to the plasma phase as they pass through the ion sheath edge, in this simulation, ions were launched from all phases in one RF period.

The model consisted of three coupled modules, namely, the field solver, particle mover, and MCC model. First, the electric field was calculated by the field solver for two cases: (i) if ions in the plasma phase, and the electric field being zero; and (ii) if ions in the sheath phase, and the electric field following the equation specified above. The second is the particle mover modulus in which the motion of the ions is determined by the force equation with the electric field calculated in the field solver. Third, in order to take collisions into account, we used the MCC method from which the probability of the elastic and charge exchange collisions is calculated. Electron impact collisions such as ionization and dissociation were assumed to be negligible in this monitoring system. In thermal equilibrium, the density of electrons satisfies the Boltzmann relation because electrons are highly mobile. Therefore, electron-neutral collisions can be neglected because the electron density was small in the sheath, of which the voltage was much higher than the electron temperature. The distance was calculated from the particle mover in the sheath during the time interval. A comparison between the probability for each collision and a random number between 0 and 1 determinded whether the electrons collide with neutral gas or not. These steps were repeated until the ion reached the electrode.

3.3. Speed Up

One of the key properties of our monitoring system is real-time measurement. The calculation time of a conventional IED monitoring system is a few minutes, however, which suggests that it is not likely that the industry would adopt the monitoring system in their processing. Therefore, the calculation time of our monitoring system had to be shorter to meet the industrial demand while the data quality should not deteriorate. We introduce the three methods to reduce the calculation time for one IED. The IED calculation times were investigated through a simple ion trajectory simulation where arbitrary sinusoidal sheath voltage waveforms were entered with amplitude of 300 V and pressure of 20 mTorr.

First, the number of calculated particles was optimized to minimize the calculation time without the loss of information on the IED profiles. As shown in Figure 5, in the case of an IED with 2×10^3 particles, there was noise that blurred which signals were reasonable over the entire energy range. On the other hand, the noise was greatly reduced in the case of 5×10^3 (B) and 8×10^3 (A) particles, especially at the lowest energy and near 50 eV. Figure 5a shows that the calculation time of 8×10^3 particles was longer than that of the 5×10^3 particle case, even though the IED profiles were similar to each other, as seen in Figure 5c. Thus, we used the particle number of 5×10^3 that guaranteed good IED quality and the shortest calculation time. Second, the cross-section of collisions was simplified to be constant. The function of fitting curves is well-known for the experimental cross-sectional data of argon ion-neutral collisions [47]. Conventionally, the fitting function of the cross-section was calculated as a function of the energy of the ions in the sheath. An investigation of the calculation time for each code provided by MATLAB confirmed that the cross-sectional calculation takes a long time. Therefore, the cross-section was simplified to a constant with the following steps. Before calculating one IED, we (1) acquired the maximal sheath voltage measured at the target electrode (powered or grounded), and (2) averaged the cross-section from zero to the maximal sheath voltage. The collision cross-sections obtained in the following steps were employed for all collisions in the ion trajectory simulation. Third, we used the well-known null collision method [48] calculates the number of ions involved in a collision using the maximal probability involved in a collision at single time step and randomly selects ions from among all ions. Figure 5b shows the calculated IEDs according to additional speed-up methods such as the constant cross-section of collisions (C) and null collision (D), including the optimization of calculated particles (B). The profiles of the IEDs according to the additional methods were almost the same, but the calculation time was shorter than 10 s, as shown in Figure 5c. However, since 10 s is also a long time to give fast feedback in current industrial process systems, we improved this simulation by using the vectorization of the code and loading a time-varying electron sheath edge from a database instead of calculation; we achieved a calculation time of a few seconds.

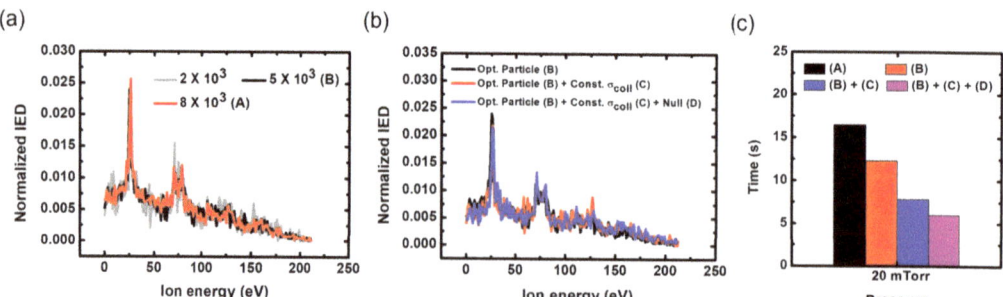

Figure 5. (**a**) Calculated IEDs according to the number of ions; (**b**) calculated IEDs; (**c**) corresponding calculation times with additional speed-up methods. (A) Nonoptimized particles 8×10^3 without speed-up methods; (B) optimized particles of 5×10^3 without speed-up methods; (C) constant cross-section method; (D) null collision method.

4. Results and Discussion

The accuracy of our noninvasive IED monitoring system was determined by how the system could produce the IED profile and the IED measured via commercial QMS depending on the RF powers and pressures. The y-axis unit of the IED in the commercial system and our system was different before data processing. In the case of our system, the y axis was the number of ions for that energy; in the case of QMS, it was the number of ions counted per second. Therefore, since the integration of an IED is a parameter representing the total number of ions, we normalized the IED to the integration value of each IED for the comparison.

Figure 6 shows the comparison between the IEDs measured via our noninvasive monitoring system and the commercial QMS with increasing RF power at a fixed pressure of 20 mTorr in Ar plasma. Figure 6 shows that the measured IEDs via the commercial QMS had a single energy peak because, at the frequency of this input RF power higher than the ion frequency, the ions could respond to a time-averaged sheath potential that was equal to a time-averaged plasma potential at the grounded electrode. As the input RF power increased, the energy peak also slightly changed from 12 to 13 eV because it did not significantly affect the plasma potential, but RF and DC self-bias voltages applied on the powered electrode increased [49–51]. The IED calculated with our monitoring system showed similar changes according to the power variation measured via the commercial QMS, which implies that IEDs of both our and commercial diagnostics were in good agreement with each other.

Figure 7 plots the IEDs of both diagnostics as a function of pressure ranging from 10 to 50 mTorr at a fixed input RF power of 300 W. The IEDs obtained from the commercial QMS with increasing pressure had similar trends of a single energy peak, and that changed with power variation. From the aspect of a single energy peak, there was a slight change because as chamber pressure changed from 10 to 50 mTorr, the electron temperature barely changed, which implied that the plasma potential remained almost unchanged. With increasing pressure, the ion energy distribution caused by the collision is broadened [45,52,53]. Figure 7 shows that the IED measured via the commercial QMS became broader as chamber pressure increased, which also appeared in the IED calculated with our monitoring system.

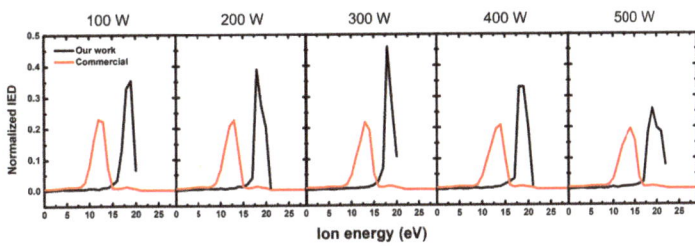

Figure 6. Normalized IEDs measured via our noninvasive monitoring system and by a commercial QMS with an increase in RF power from 100 to 500 W at a fixed pressure of 20 mTorr.

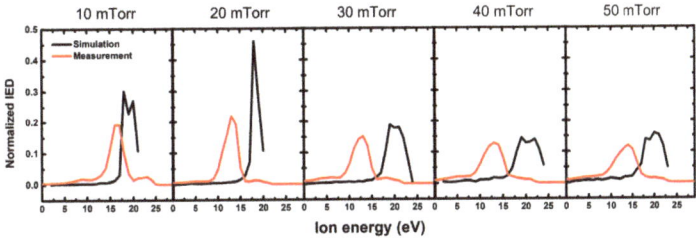

Figure 7. Normalized IEDs measured via our noninvasive monitoring system and by a commercial QMS with an increase in pressure from 10 to 50 mTorr at a fixed RF power of 300 W.

Nevertheless, while the amplitude of the peak in the IED measured with the commercial QMS did not change with the change in external conditions, the amplitude of our system tended to change, which may be attributed to two limitations of our system. First, the error of the voltage measurement at the powered electrode could have caused the difference between both spectra in Figures 6 and 7. In the case of our monitoring system, the IED greatly depended on the sheath voltage at a ground electrode, which was obtained from the measured voltage applied to the powered electrode through the model. This obtained sheath voltage was small compared with the measured voltage. and was easily affected by the noise of the measured voltage. For this reason, the variation in their amplitude may have been caused as a function of RF input power due to the error of the measured voltage or the limitation of the measuring equipment. Second, since the electrical sheath model of our system was mainly used for low-frequency analysis, there was a limit to the application of the high-frequency regime. The displacement current, defined as $\frac{\partial}{\partial t}(C_{sh}(t)V_{sh}(t))$, where $C_{sh}(t)$ and $V_{sh}(t)$ are time-varying sheath capacitance and voltage, respectively, created the harmonics of the sheath oscillation, which resulted in the plasma potential oscillation. The displacement current rarely influences the total discharge current in a low-frequency regime, and the plasma potential waveform becomes not sinusoidal but highly nonlinear, as depicted in Figure 4. This potential waveform resulted in a narrow IED, as most ions were accelerated in the flat potential region. On the other hand, since the displacement current dominates in a high-frequency regime, the plasma potential became sinusoidal. However, since the time-varying sheath capacitance is ignored in the MEO model, this high-frequency effect worked weakly on calculating the waveform of the plasma potential in our monitoring system. If a model is used that considers the time-varying sheath capacitance, the sheath voltage becomes a more sinusoidal waveform. In other words, the IED became broader than the IED provided by our monitoring system, and the amplitude of their peak decreases because ions were energized by sinusoidally oscillating waveform of the sheath voltage. An improvement in the electrical sheath model is our next challenge for high accuracy in high-frequency sheaths.

Figure 8 shows the calculation time of one IED as a function of pressure and power. This monitoring system includes the methods presented in the Section 3.3, and the vectorization of the code and loading method from the database of electron sheath edge motion, taking a calculation time of under 2 s overall RF power and pressure.

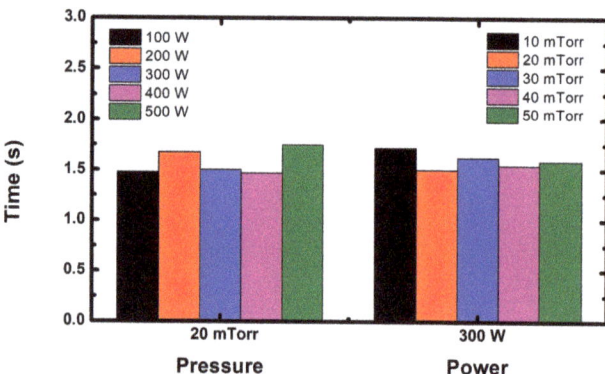

Figure 8. Calculation time for one IED with different pressure and RF power levels.

Unfortunately, it was impossible to compare the IEDs of both diagnostics adjacent to the powered electrode, and the IEDs under the condition that there was a dramatic change in sheath voltage waveforms because the commercial QMS could measure the IED adjacent to the grounded electrode due to the grounded sampling probe of the QMS, while our system had no constraint, whether powered or grounded. It is necessary to investigate further whether the IED was measured even under various changes in plasma conditions. As shown in Figures 6 and 7, the IEDs calculated with our system were higher than those of the commercial QMS under all conditions. In order to find out the cause of this difference, it was necessary to measure the voltage waveform at the electrode and compare the difference with the voltage waveform at the point outside the cell where the probes were located. However, it was difficult to accurately measure the voltage waveform at the powered electrode due to plasma perturbation. Accordingly, although this difference was not clearly identified, it may have been caused by a stray impedance between the electrode and location of the probe [34]. Additionally, in actual plasma processing, fluorocarbon gas can produce numerous ion and neutral species in plasma, so improving this monitoring system that works even in this type of gas is our next work. Nevertheless, the results of this section show that this monitoring system is capable of noninvasively real-time monitoring the IED.

5. Conclusions

The invasive diagnostics of IEDs adjacent to a grounded or powered electrode during plasma processing are difficult for several reasons. Therefore, numerous studies have been conducted for noninvasive and real-time monitoring systems. Nevertheless, there are imperfect factors such as collision or noninvasive measurements.

In this paper, a noninvasive and real-time IED monitoring system was proposed and validated in an asymmetric RF CCP discharge in Ar plasma. Results show that changes in the IED measured via our system according to the input RF power and pressure variation were in good agreement with the results from a commercial QMS. This proposed monitoring system is capable of the noninvasive and real-time monitoring of IEDs in actual processing.

Author Contributions: Conceptualization, S.Y.; validation, I.S., J.L., S.K., Y.L., C.C., W.J. and Y.Y.; formal analysis, I.S.; writing—original draft preparation, I.S.; writing—review and editing, S.Y., S.K., Y.L. and C.C.; supervision, S.Y. All authors have read and agreed to the published version of the manuscript.

Funding: This research was supported by a National Research Council of Science and Technology (NST) grant from the Korean Government (MSIP) (no. CAP-17-02-NFRI, CRF-20-01-NFRI), the Next-Generation Intelligence Semiconductor R&D Program through the Korean Evaluation Institute of Industrial Technology (KEIT) funded by the Korean Government (MOTIE), by the Korea Institute of Energy Technology Evaluation and Planning (KETEP) and the MOTIE of the Republic of Korea (20202010100020), by the MOTIE (20009818, 20010420) and Korean Semiconductor Research Consortium (KSRC) support program for the development of the future semiconductor device, by the Korean Institute for Advancement of Technology (KIAT) grant funded by the Korean Government (MOTIE) (P0008458, HRD Program for Industrial Innovation), by a Basic Science Research Program through the National Research Foundation of Korea (NRF) funded by the Ministry of Education (NRF-2020R1A6A1A03047771), and by the KIMM Institutional Program (NK236F) and NST/KIMM.

Institutional Review Board Statement: Not applicable.

Informed Consent Statement: Not applicable.

Data Availability Statement: The data presented in this study are available on request from the corresponding author.

Conflicts of Interest: The authors declare no conflict of interest.

References

1. Adamovich, I. The 2017 Plasma Roadmap: Low temperature plasma science and technology. *ESC J. Phys. D Appl. Phys.* **2017**, *50*, 323001. [CrossRef]
2. Oehrlein, G.S.; Metzler, D.; Li, C. Atomic Etching at the Tipping Point: An Overview. *ESC J. Solid State Sci. Technol.* **2015**, *4*, N5041. [CrossRef]
3. Kaler, S.S.; Lou, Q.; Donnelly, V.M.; Economou, D.J. Atomic layer etching of silicon dioxide using alternating C_4F_8 and energetic Ar^+ plasma beams. *J. Phys. D Appl. Phys.* **2017**, *50*, 234001. [CrossRef]

4. Lee, Y.S.; Kim, S.J.; Lee, J.J.; Cho, C.H.; Seong, I.H.; You, S.J. Purgeless atomic layer etching of SiO_2. *J. Phys. D Appl. Phys.* **2022**, *55*, 365203. [CrossRef]
5. Suto, S.; Hayasaka, N.; Okano, H. Highly Selective Etching of Si_3N_4 to SiO_2 Employing Fluorine and Chlorine Atoms Generated by Microwave Discharge. *J. Electrochem. Soc.* **1989**, *136*, 2032. [CrossRef]
6. Hayashi, H.; Kurihara, K.; Sekine, M. Characterization of Highly Selective SiO_2/Si_3N_4 Etching of high-Aspect-Ratio Holes. *Jpn. J. Appl. Phys.* **1996**, *35*, 2488. [CrossRef]
7. Kasternmeier, B.E.E.; Matsuo, P.H.; Oehrlein, G.S. Highly selective etching of silicon nitride over silicon and silicon dioxide. *J. Vac. Sci. Techno. A* **1999**, *17*, 3179–3184. [CrossRef]
8. Seman, M.; Wolden, C.A. Investigation of the role of plasma conditions on the deposition rate and electrochromic performance of tungsten oxide thin films. *J. Vac. Sci. Technol. A* **2003**, *21*, 6. [CrossRef]
9. Radjenovic, B.M.; Radmilovic-Radjenovic, M.D.; Petrovicm, Z.L. Dynamics of the Profile Charging During SiO_2 Etching in Plasma for High Aspect Ratio Trenches. *IEEE Trans Plasma Sci.* **2008**, *36*, 874. [CrossRef]
10. Brichon, P.; Pujo, E.D.; Mourey, O.; Joubert, O. Key plasma parameters for nanometric precision etching of Si films in chlorine discharge. *J. Appl. Phys.* **2015**, *118*, 053303. [CrossRef]
11. Gopikishan, S.; Banerjee, L.; Bigkem, K.A.; Das, A.K.; Pathak, A.P.; Mahapatra, S.K. Paschen curve approach to investigate electron density and deposition rate of Cu in magnetron sputtering system. *Radiat. Eff. Deffects Solids* **2016**, *171*, 999. [CrossRef]
12. Cho, C.H.; You, K.H.; Kim, S.J.; Lee, Y.S.; Lee, J.J.; You, S.J. Characterization of SiO_2 Etching Profiles in Pulse-Modulated Capacitively Coupled Plasmas. *Materials* **2021**, *14*, 5036. [CrossRef] [PubMed]
13. Kanarik, K.J.; Tan, S.; Gottscho, R.A. Atomic layer etching: Rethinking the art of etch. *J. Phys. Chem. Lett.* **2018**, *16*, 4814. [CrossRef] [PubMed]
14. Faraz, T.; Arts, K.; Karwal, S.; Knoops, H.C.M.; Kessels, W.M.M. Energetic ions during plasma-enhanced atomic layer deposition and their role in tailoring material properties. *Plasma Source Sci. Technol.* **2019**, *28*, 024002. [CrossRef]
15. Chang, W.S.; Yook, Y.G.; You, H.S.; Park, J.H.; Kwon, D.C.; Song, M.Y.; Yoon, J.S.; Kim, D.W.; You, S.J.; Yu, D.H.; et al. A unified semi-global surface reaction model of polymer deposition and SiO_2 etching in fluorocarbon plasma. *Appl. Surf. Sci.* **2020**, *515*, 145975. [CrossRef]
16. Seong, I.H.; Lee, J.J.; Cho, C.H.; Lee, Y.S.; Kim, S.J.; You, S.J. Characterization of SiO_2 over poly-*Si* mask etching in and Ar/C_4F_8 capacitively coupled plasma. *Appl. Sci. Converg. Technol.* **2021**, *30*, 6. [CrossRef]
17. Chung, Y.A.; Lung, C.Y.; Chiu, Y.C.; Lee, H.J.; Lian, N.T.; Yang, T.; Chen, K.C.; Lu, C.Y. Study of Plasma Arcing Mechanism in High Aspect Ratio Slit Trench Etching. In Proceedings of the 2019 30th Annual SEMI Advanced Semiconductor Manufacturing Conference (ASMC), Saratoga Springs, NY, USA, 6–9 May 2019.
18. Carter, D.; Walde, H.; Nauman, K. Managing arcs in large area sputtering applications. *Thin Solid Film* **2012**, *520*, 4199–4202. [CrossRef]
19. Lee, H.J.; Seo, D.S.; May, G.S.; Hong, S.J. Use of In-Situ Optical Emission Spectroscopy for Leak Fault Detection and Classification in Plasma Etching. *J. Semicond. Technol. Sci.* **2013**, *13*, 4. [CrossRef]
20. Marchack, N.; Buzi, L.; Farmer, D.B.; Miyazoe, H.; Papalia, J.M.; Yan, H.; Totir, G.; Engelmann, S.U. Plasma prcessing for advanced microelectronics beyond CMOS. *J. Appl. Phys.* **2021**, *130*, 080901. [CrossRef]
21. Simpson, J.A. Design of Retarding Field Energy Analyzers. *Rev. Sci. Instrum.* **1961**, *32*, 1283. [CrossRef]
22. Gahan, D.; Dolinaj, B.; Hopkins, M.B. Retarding field analyzer for ion energy distribution measurements at a radio-frequency biased electrode. *Rev. Sci. Instrum.* **2008**, *79*, 033502. [CrossRef] [PubMed]
23. Kreul, S.G.; Hübner, S.; Schneider, S.; Ellerweg, D.; Keudell, A.V.; Matejčík, S.; Benedikt, J. Mass spectrometry of atmospheric pressure plasmas. *Plasma Sources Sci. Technol.* **2015**, *24*, 044008. [CrossRef]
24. Benedikt, J.; Hecimovic, A.; Ellerweg, D.; Keudell, A.V. Quadrupole mass spectrometry of reactive plasmas. *J. Phys. D Appl. Phys* **2012**, *45*, 403001. [CrossRef]
25. Singh, H.; Coburn, J.W.; Graves, D.B. Appearance Potential Mass Spectrometry: Discrimination of Dissociative Ionization Products. *J. Vac. Sci. Technol. A Vacuum Surfaces Film.* **2000**, *18*, 299–305. [CrossRef]
26. Bohlmark, J.; Lattemann, M.; Gudmundsson, J.T.; Ehiasarian, A.P.; Gonzalvo, Y.A.; Brenning, N.; Helmersson, U. The ion energy distributions and ion flux composition from a high power impulse magnetron sputtering discharge. *Thin Solid Films* **2006**, *515*, 1522–1526. [CrossRef]
27. Ghidini, R.; Groothuis, C.H.J.M.; Sorokin, M.; Kroesen, G.M.W.; Stoffels, W.W. Electrical and optical characterization of particle formation in an argon-silane capacitively coupled radio-frequency discharge. *Plasma Sources Sci. Technol.* **2004**, *13*, 143. [CrossRef]
28. Schauer, J.C.; Hong, S.; Winter, J. Electrical measurements in dusty plasmas as a detection method for the early phase of particle formation. *Plasma Sources Sci. Technol.* **2004**, *13*, 636. [CrossRef]
29. Hong, S.; Berndt, J.; Winter, J. Growth precursors and dynamics of dust particle formation in the Ar/CH_4 and Ar/C_2H_2 plasmas. *Plasma Sources Sci. Technol.* **2002**, *12*, 46. [CrossRef]
30. Shen, Z.; Kortshagen, U. Experimental study of the influence of nanoparticle generation on the electrical characteristics argon-silane capacitive radio-frequency plasmas. *J. Vac. Sci. Technol. A* **2002**, *20*, 153. [CrossRef]
31. Boufendi, L.; Gaudin, J.; Huet, S.; Viera, G.; Dudemaine, M. Detection of particles of less than 5 nm in diameter formed in an argon-silane capacitively coupled radio-frequency discharge. *Appl. Phys. Lett.* **2001**, *79*, 4301. [CrossRef]

32. Sezemsky, P.; Stranak, V.; Kratochvil, J.; Cada, M.; Hippler, R.; Hrabovsky, M.; Hubicka, Z. Modified high frequency probe approach for diagnostics of highly reactive plasma. *Plasma Sources Sci. Technol.* **2019**, *28*, 115009. [CrossRef]
33. Dewan, N.A. Analysis and Modelling of the Impact of Plasma RF Harmonics in Semiconductor Plasma Processing. Ph.D. Thesis, Dublin City University, Dublin, Ireland, 2001.
34. Sobolewski, M.A. Electrical characterization of radio-frequency discharges in the Gaseous Electronics Conference Reference Cell. *J. Vac. Sci. Technol. A* **1992**, *10*, 3550–3562. [CrossRef]
35. Sobolewski, M.A. Measuring the ion current in high-density plasmas using radio-frequency current and voltage measurements. *J. Appl. Phys.* **2001**, *90*, 2660. [CrossRef]
36. Sobolewski, M.A. Monitoring sheath voltages and ion energies in high-density plasmas using noninvasive radio-frequency current and voltage measurements. *J. Appl. Phys.* **2004**, *95*, 4593. [CrossRef]
37. Sobolewski, M.A. Real-time, noninvasive monitoring of ion energy and ion current at a wafer surface during plasma etching. *J. Vac. Sci. Technol. A* **2006**, *24*, 1892. [CrossRef]
38. Sobolewski, M.A. Noninvasive monitoring of ion energy drift in an inductively coupled plasma reactor. *J. Appl. Phys.* **2005**, *97*, 033301. [CrossRef]
39. Bogdanova, M.A.; Lopaev, D.; Zyryanov, S.M.; Rakhimov, A.T. "Virtual IED sensor" at an rf-biased electrode in low-pressure plasma. *Phys. Plasmas* **2016**, *23*, 073510. [CrossRef]
40. Bogdanova, M.A.; Lopaev, D.; Rakhimov, A.T.; Zotovich, A.; Zyryanov, S.M. 'Virtual IED sensor' for df rf CCP discharges. *Plasma Sources Sci. Technol.* **2021**, *30*, 075020. [CrossRef]
41. Bogdanova, M.A.; Lopaev, D.; Zyryanov, S.M.; Voloshin, D.; Rakhimov, A.T. Ion composition of rf CCP in Ar/H_2 mixtures. *Plasma Sources Sci. Technol.* **2019**, *28*, 095017. [CrossRef]
42. Metze, A.; Erine, D.W.; Oskam, H.J. Application of the physics of plasma sheaths to the modeling of rf plasma reactors. *J. Appl. Phys.* **1986**, *60*, 3081. [CrossRef]
43. Bruce, R.H. Ion response to plasma excitation frequency. *J. Appl. Phys.* **1981**, *52*, 7064. [CrossRef]
44. Vahedi, V.; Surendra, M. A Monte Carlo collision model for the particle-in-cell method: Applications to argon and oxygen discharges. *Comput. Phys. Commun.* **1995**, *87*, 179–198. [CrossRef]
45. Lieberman, M.A.; Lichtenberg, A.J. *Principles of Plasma Discharges and Materials Processing*, 2nd ed.; Wiley & Sons. Inc.: Hobken, NJ, USA, 2005; pp. 1–22.
46. Bohm, D. *The Characteristics of Electrical Discharges in Magnetic Fields*; McGraw-Hill: New York, NY, USA, 1949.
47. Dai, Z.L.; Wang, Y.N. Simulations of ion transport in a collisional radio-frequency plasma sheath. *Phys. Rev. E* **2004**, *69*, 036403. [CrossRef] [PubMed]
48. Skullerud, H.R. The stochastic computer simulation of ion motion in a gas subjected to a constant electric field. *J. Phys. D Appl. Phys.* **1968**, *1*, 1567. [CrossRef]
49. Köhler, K.; Coburn, J.W.; Horne, D.E.; Kay, E.; Keller, J.H. Plasma potentials of 13.56 MHz rf argon glow discharges in a planar system. *J. Appl. Phys.* **1985**, *57*, 59. [CrossRef]
50. Edelberg, E.A.; Perry, A.; Benjamin, N.; Aydil, E.S. Energy distribution of ions bombarding biased electrodes in high density plasma reactors. *J. Vac. Sci. Technol. A.* **1999**, *17*, 506. [CrossRef]
51. Rusu, I.A.; Popa, G.; Sullivan, J.L. Electron plasma parameters and ion energy measurement at the grounded electrode in an rf discharge. *J. Phys. D Appl. Phys.* **2002**, *35*, 2808–2814. [CrossRef]
52. Chen, F.F. *Introduction to Plasma Physics and Controlled Fusion*, 3rd ed.; Springer: Berlin/Heidelberg, Germany, 2018.
53. Chabert, P.; Braithwaite, N. *Physics of Radio-Frequency Plasmas*; Cambridge University Press: Cambridge, UK, 2011; pp. 101–110.

Article

Refined Appearance Potential Mass Spectrometry for High Precision Radical Density Quantification in Plasma

Chulhee Cho [1], Sijun Kim [1], Youngseok Lee [1], Wonnyoung Jeong [1], Inho Seong [1], Jangjae Lee [2], Minsu Choi [1], Yebin You [1], Sangho Lee [1,3], Jinho Lee [1] and Shinjae You [1,4,*]

1. Department of Physics, Chungnam National University, 99 Daehak-ro, Yuseong-gu, Daejeon 34134, Korea
2. Samsung Electronics, Samsungjeonja-ro, Hwaseong-si 18448, Korea
3. Korea Institute of Machinery & Materials, 156 Gajeongbuk-ro, Yuseong-gu, Daejeon 34103, Korea
4. Institute of Quantum System (IQS), Chungnam National University, Daejeon 34134, Korea
* Correspondence: sjyou@cnu.ac.kr

Abstract: As the analysis of complicated reaction chemistry in bulk plasma has become more important, especially in plasma processing, quantifying radical density is now in focus. For this work, appearance potential mass spectrometry (APMS) is widely used; however, the original APMS can produce large errors depending on the fitting process, as the fitting range is not exactly defined. In this research, to reduce errors resulting from the fitting process of the original method, a new APMS approach that eliminates the fitting process is suggested. Comparing the neutral densities in *He* plasma between the conventional method and the new method, along with the real neutral density obtained using the ideal gas equation, confirmed that the proposed quantification approach can provide more accurate results. This research will contribute to improving the precision of plasma diagnosis and help elucidate the plasma etching process.

Keywords: plasma; quadrupole mass spectrometer; radical density; quantification

1. Introduction

Low-temperature plasma is generated by an electric field in a space. Energetic electrons from the electric field hit neutrals, which divides them into reactive neutrals called radicals and charged particles [1,2]. These particles play individual roles in industry; charged particles physically affect a surface, while radicals chemically react with a surface. Such properties have led to the widespread use of plasma in semiconductor fabrication [3–11], agriculture [12–14], and medical treatment [15–17]. In particular, the role of plasma in semiconductor processing has gained importance in recent years along with the rapid growth of semiconductor performance. The current trend, therefore, is to seek high precision and refinement of the plasma process. However, due to the complexity of plasmas and inconsistencies in plasma analysis and processing from inaccurate plasma diagnosis, development remains based on trial and error. As the complexity of the plasma process increases, the trial-and-error approaches are reaching their limit, which has led to intense research to achieve more accurate plasma diagnostic methods.

The purpose of plasma diagnostics is typically to obtain particle densities and energy distributions. In particular, particle quantification has been extensively investigated [18–24]. As a representative example, the Langmuir probe is widely used to measure various plasma properties, such as electron energy distribution, electron density, plasma potential, and ion density [25]. Further, microwave probes, such as the cutoff probe, are used for electron quantification [26]. However, while electron quantification methods are diverse and accurate, the quantification of neutral species is difficult due to the low accuracy and high complexity of the current approaches. Since actinometry—a neutral quantification method used in optical emission spectrometry (OES)—requires a small amount of additional gas, it is difficult to use in the highly precise and refined semiconductor process [27].

Alternatively, the quadrupole mass spectrometer (QMS) is another tool that is widely used to obtain the density of plasma-neutral species [28–30]. A QMS measures the partial pressure of the neutrals without any extra gas in the chamber, providing an advantage over OES. Since the 1990s, steady efforts have been made by international standardization organizations to establish a standardized calibration procedure for QMSs, but as this is still in the first stage, quantification with a QMS remains difficult, especially in achieving high accuracy [31].

The main issue with neutral quantification via QMS is that additional neutral products are generated in the QMS. The QMS consists of three main components: an ionizer, a mass filter, and a detector. Electrons emitted by a filament in the ionizer are accelerated by the electric field between the positively biased anode grid and the negatively biased repeller; therefore, the electron energy is set to the voltage difference between the anode grid and the repeller. Neutrals are ionized by the accelerated electrons, with two cases of ionization by electron impact: direct ionization and dissociative ionization. Direct ionization means that molecules colliding with electrons lose their own electrons without decomposition. Direct ionization can be expressed as

$$A + e^- \rightarrow A^+ + 2e^-,$$

where A is any neutral atom and e^- is an electron. The direct ionization cross-section is zero below the threshold energy and rises linearly just above the electron energy. The maximum cross-section is usually reached at 50–100 eV [32]. Otherwise, dissociative ionization means ionization with the decomposition of molecules, which can be expressed as

$$AB + e^- \rightarrow A^+ + B + 2e^-,$$

where AB is any molecule consisting of A and B atoms. As dissociative ionization can occur from the ions resulting from the direct ionization of other molecules, crucial errors can appear in the quantification of neutrals. Therefore, the separation of direct and dissociative ionization is one of the most important challenges to obtaining high precision neutral densities.

To separate the direct and dissociative ion signals, appearance potential mass spectrometry (APMS) is often used. APMS signals from a QMS, which sweeps the electron energy in the ionizer, are obtained by increasing the electron energy from below the ionization threshold energy to above the threshold energy and linearly fitting the signal to distinguish the direct and dissociative ionization signals. The energy dependence of the signals is eliminated by the linear fitting, allowing APMS to be used simply and briefly. However, the APMS signals curve up near the threshold energy, so no fitting standard has been confirmed. Therefore, the slope of the linear fitting values is changed with the fitting standard; consequently, neutral quantification with inaccurate fitting values can lead to enormous errors. The specifics of this method are described in Section 3.1 of this paper.

In this work, a refined APMS approach without a fitting process is suggested. This work compares the same neutral density in three different ways, as follows. First, the neutral density is calculated with the ideal gas equation for use as a reference. Second, the neutral density is obtained by the original APMS approach from the QMS signals gathered by sweeping the electron energies and conducting the fitting process. Third, the neutral density is acquired through the refined approach, which uses the same signals as the original APMS but with no fitting process.

The rest of this paper is organized as follows. Details of the experimental setup for APMS are described in Section 2. The specific methods and results for the original APMS method are given in Section 3.1, and the results of the refined APMS method are discussed in Section 3.2. Section 4 concludes the work.

2. Experimental Setup

A schematic outline of the vacuum chamber is shown in Figure 1. Two mass flow controllers (LineTech Inc., Daejeon, Korea) injected 99.999% purity *Ar* and *He* gas at a maximum flow rate of 50 standard cubic centimeters per minute (sccm). The chamber was pumped out by a 99-L-per-min (L/min) rotary pump (Varian, Crawley, UK) that was connected to the center of the chamber body. The base pressure of the chamber was 5.3 Pascal (Pa), which was read by a Baratron gauge (MKS, Andover, MA, USA).

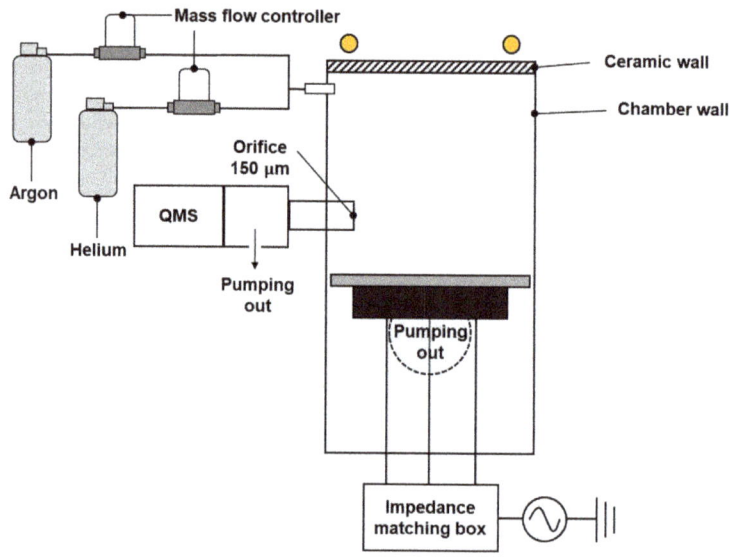

Figure 1. Experimental setup for *He* density measurement with a quadrupole mass spectrometer.

A QMS (Hiden Analytical, Warrington, UK, PSM) was installed in the chamber via a 150-μm orifice that separated the QMS chamber to create a high-level vacuum to remove neutral–neutral collisions in the QMS. The base pressure of the QMS chamber was below 1.3×10^{-5} Pa, which was read by a compact cold cathode vacuum gauge (Pfeiffer IKR 251, Pfeiffer Vacuum, Asslar, Germany). This high vacuum was sustained by a 300 L/s single turbo molecular pump (Edwards, Burgess Hill, UK) and a 200 L/min rotary pump (WSA, Gunpo, Korea).

When comparing the APMS approaches, a neutral density reference should first be obtained. In a vacuum chamber filled with inert gas, such as *Ar* and *He*, the density of the gas can be calculated from the pressure of the chamber using the ideal gas law because of the non-reaction of inert gas. The ideal gas law is expressed as [33]

$$n = \frac{P}{kT}, \qquad (1)$$

where n is the neutral density in the main chamber, P is the main chamber pressure, k is the Boltzmann constant, and T is the gas temperature. Here, the gas temperature was approximated to be room temperature (300 K). In this work, neutral densities for two cases were calculated. The conditions of the first case were 5 sccm *Ar* injection, 10.13 Pa main chamber pressure, and 2.45×10^9 m^{-3} *Ar* density. The conditions of the second case were 20 sccm *He* injection, 31.6 Pa chamber pressure, and 7.64×10^9 m^{-3} *He* density. This *He* density was used for the theoretical value of the density. The QMS signals used in the APMS approaches were obtained under the same conditions as above. In other words, the QMS signals were obtained from two conditions, one with the chamber filled with 20 sccm *He* and the other with the chamber filled with 5 sccm *Ar*. Before the experiments, the QMS filament was heated for over 2 h, and the pressure of the chamber was kept at vacuum for more than 1 day. The QMS signals were obtained over 30 cycles for reproducibility, and the average values of the signals were used. To calibrate the QMS system, the built-in QMS calibration software RGAtune was used.

3. Results and Discussion

3.1. Original APMS Approach

The QMS displays signals by detecting the current generated by ions. Therefore, the expression of the QMS signal from the neutral species is given by [33]

$$S = \beta t\left(\frac{m}{e}\right)\theta\left(\frac{m}{e}\right) l_{cage} I_e \sigma n_{ionizer},\tag{2}$$

where S is the QMS signal intensity, β is the extraction efficiency of the ions from the ionizer, t is the transmission efficiency of the quadrupole mass filter (which is a function of the ratio of mass to charge), θ is the detection coefficient of the channeltron detector, l_{cage} is the ionizer length, I_e is the electron current in the ionizer, σ is the cross-section of ionization, and $n_{ionizer}$ is the density of neutrals in the ionizer. As Equation (2) shows, neutrals can be detected in the QMS when becoming ions. Figure 2a shows the measured QMS signal of Ar at 10.13 Pa chamber pressure and He at 31.6 Pa chamber pressure, and Figure 2b shows the Ar and He direct ionization cross-section [32]. The signals are obtained by sweeping the electron energy of the ionizer from 10 eV to 40 eV. Referring to the cross-section, Ar becomes $Ar+$ after the electron energy exceeds 15.8 eV, and He becomes $He+$ after the electron energy exceeds 24.7 eV. Therefore, the signals appear after the electron energies increase above each threshold energy of neutral ionization. The signal curves up near the threshold energy, and after that, it linearly increases with electron energy. When the electron energy is higher than the threshold energy by 5–10 eV, the increase rate of the signal slightly declines. These signals are used for calculating the density via APMS.

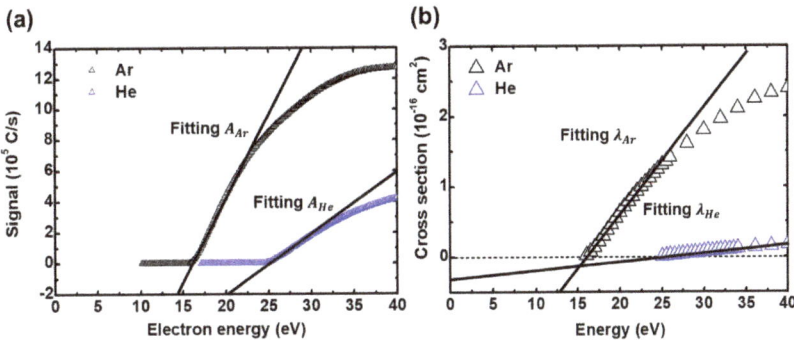

Figure 2. (a) Ar and He signals from the QMS. (b) Ar and He direct ionization cross-section.

The original APMS method is well known for measuring the radical species from plasma [33]. Singh et al. suggested the radical density quantification of X in plasma using Equation (2) as shown below [33]

$$n_X^{on} = \left(\frac{A^{X\to X^+}}{A^{Ar\to Ar^+}}\right)\left(\frac{\lambda^{Ar\to Ar^+}}{\lambda^{X\to X^+}}\right)\left[\frac{t(m_{Ar^+})\theta(m_{Ar^+})}{t(m_{X^+})\theta(m_{X^+})}\right] n_{Ar}^{off}.\tag{3}$$

Here, as the purpose of this work was to compare He density, X is He, and A and λ were the linear fitting slope of the QMS signal and cross-section, as shown in Figure 2. The superscripts *on* and *off* of the density in Equation (3) indicate whether plasma is generated or not; note these superscripts in Equation (3) are excluded in this article since the QMS signal measurements were conducted not in a plasma environment but in a vacuum. We also note that Singh measured the oxygen atom density generated from plasma, while in the current work He density is unchanged with the plasma operating condition (*on* or *off*), so the plasma is not operated and n_X^{on} is transformed to n_{He}^{off}. In addition, Singh stated that $t(m)\theta(m)$ for singly charged ions can be expressed as being proportional to m^{-1}, but also that it can change by instrument characteristics. Lee et al. recommended a calculation method for $t(m)\theta(m)$ using a standard gas, namely a gas mixture of various noble gas species [34]. Following this recommendation, $f(m)$ as a function of $t(m)\theta(m)$, is given as

$$f(m) = 1.05 \times 10^7 e^{-0.028m}. \tag{4}$$

With this, Equation (3) can be rewritten as below,

$$n_{He} = \left(\frac{A^{He \rightarrow He^+}}{A^{Ar \rightarrow Ar^+}} \right) \left(\frac{\lambda^{Ar \rightarrow Ar^+}}{\lambda^{He \rightarrow He^+}} \right) \left[\frac{f(m_{Ar})}{f(m_{He})} \right] n_{Ar}. \tag{5}$$

Here, A and λ, which are suggested by Singh, should be calculated near the threshold energy. However, since QMS signals are not linear right above the threshold, the linear fitting results can differ significantly according to their fitting ranges. Figure 3a,b show different fitting lines with different fitting ranges of Ar and He, respectively. Each fitting range was determined to be from the threshold energy to some energy higher than the threshold by an arbitrarily chosen amount. The fitting ranges and resulting slopes obtained from the Ar and He signals are listed in Table 1. Ranges 1–5 correspond to fitting ranges of 0.3, 0.5, 1, 3, and 5, respectively. In Figure 3, it can readily be seen that the different fitting ranges cause various slopes of the fitting lines, leading to inaccurate density calculations. In addition, it should be noted that decisions on which range is the most suitable can also be altered by the axis scales of the plots, as shown in the insets of Figure 3a,b. This may possibly generate considerable errors in density quantifications from QMS signals as well. Table 2 lists the fitting slopes of the cross-section, obtained from only one range since the cross-sections linearly increase near the threshold energy.

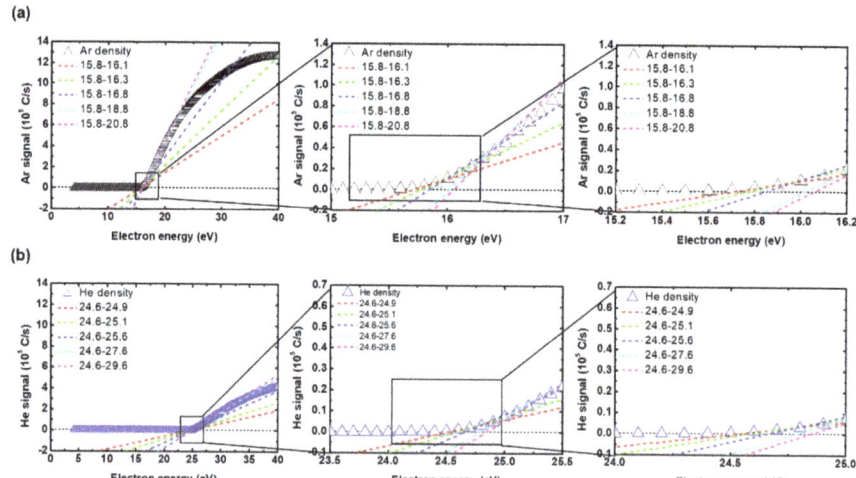

Figure 3. Different fitting results with different fitting ranges of the QMS signals for (**a**) Ar and (**b**) He.

Table 1. Linear fitting slope of *Ar* and *He* signals from the QMS with different ranges.

Range	A_{Ar}	A_{He}
Range 1	35,420	12,427
Range 2	52,751	17,423
Range 3	73,067	24,875
Range 4	108,389	33,386
Range 5	113,050	35,307

Table 2. Linear fitting slope of the *Ar* and *He* direct ionization cross-section.

Range	σ_{Ar}	σ_{He}
Range 1	0.16181	0.01249

By substituting the fitting results in Tables 1 and 2 for A and λ, respectively, in Equation (5), and the *Ar* density obtained by the ideal gas law in Section 2 for n_{Ar} in Equation (5), the *He* densities resulting from the different fitting ranges can be calculated. Figure 4 plots the calculated *He* densities from Equation (5). Note that $A^{Ar \rightarrow Ar^+}$ and $A^{He \rightarrow He^+}$ in Equation (5) each had five different values, as listed in Tables 1 and 2, which resulted in a matrix multiplication of 25 total densities. As shown in Figure 4, *He* density increased as the fitting range for *He* was increased, while for *Ar* density decreased with an increase in fitting range. The average of the calculated *He* densities was 4.53×10^9 m^{-3}, which is similar to the theoretical *He* density calculated from the ideal gas law. The *He* densities calculated from the same *Ar* and *He* ranges (i.e., 1-1, 2-2, 3-3, etc.) were closer to the average value. This phenomenon occurs because the rates of change of $A^{He \rightarrow He^+}$ and $A^{Ar \rightarrow Ar^+}$ are similar.

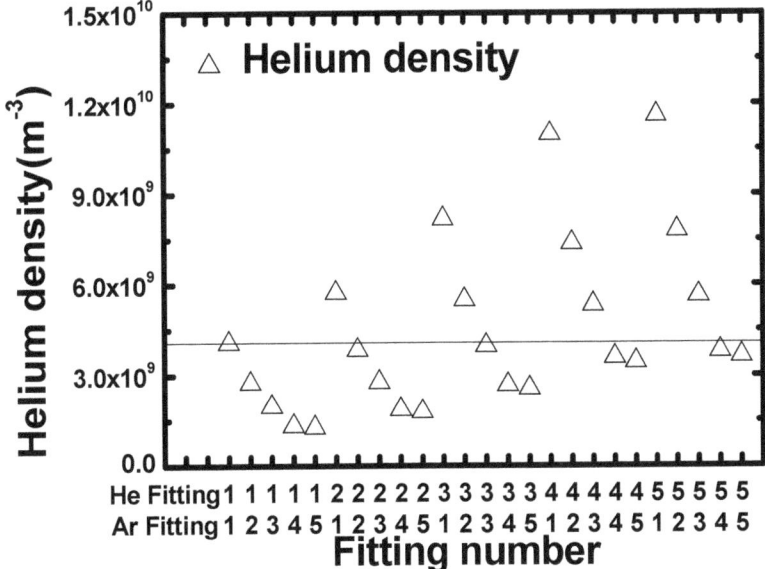

Figure 4. Plot of *He* densities calculated with each *He* and *Ar* fitting range. The horizontal line is the average value of the density.

3.2. Refined APMS

The *He* densities from the original APMS approach have a huge deviation, leading to insufficient precision. Because of the curvature of the signal near the threshold energy, it is hard to choose the correct fitting range, and this can cause errors. Therefore, a quantification method without the fitting process is suggested. Singh et al. [33]. stated that the QMS signal and direct ionization cross-section can be expressed from the linear fitting slope with energy dependence as below,

$$S(E) = A\left(E - E_i^{X \to X^+}\right), \quad (6)$$

$$\sigma(E) = \lambda\left(E - E_i^{X \to X^+}\right), \quad (7)$$

where E is the electron energy and E_i is the direct ionization threshold energy. Therefore, to remove the fitting process, the refined quantification method neglects the slope through the fitting and directly uses the signal and cross-section with energy dependence. According to Equations (6) and (7), A and λ in Equation (5) can be substituted with S and σ. The refined equation is then expressed as below,

$$n_{He}(E) = \left(\frac{S^{He \to He^+}(E)}{S^{Ar \to Ar^+}(E)}\right)\left(\frac{\sigma^{Ar \to Ar^+}(E)}{\sigma^{He \to He^+}(E)}\right)\left[\frac{f(m_{Ar})}{f(m_{He})}\right]n_{Ar}. \quad (8)$$

Here, S is the QMS signal and σ is the direct ionization cross-section. The measured QMS signals are 0.1 eV apart, and both Ar and He signals are calculated at the same energy. The cross-section is calculated by changing the interval of 1 eV in the low energy region to an interval of 0.1 eV through data preprocessing. Through the above formula, it is possible to calculate the density of He according to electron energy. Since the density of He is the value in the main chamber, there should be no change with electron energy sweeping.

Figure 5 plots the He densities calculated with the refined APMS approach. As for the calculated densities, the signal appears beyond the electron energy of 24.7 eV (the threshold energy of He direct ionization) and it can be confirmed that it increases slightly as the energy increases, eventually saturating at around 27 eV. The initial value, which is the smallest one, appears to be approximately half of the saturated value. The measured He density can be obtained by averaging the densities in the region with the smallest change. Here, even if the result of the initial value with the largest error is included, there is no significant change in the average value because the region between the lowest point and the saturation point of He density is local compared to the saturation region. Therefore, He density from the refined APMS approach is obtained by averaging the entire electron energy area.

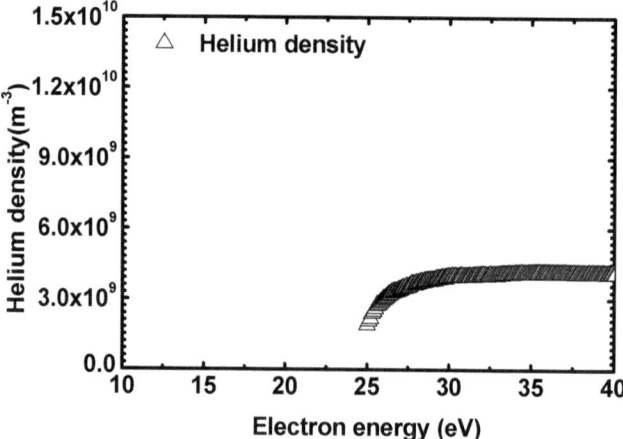

Figure 5. Plot of He densities obtained through the refined APMS method.

In Figure 6, the *He* densities from each quantification method are shown. To check the maximum error, the maximum deviation was used. Therefore, the lower limits of the error bars were the minimum values in Figures 4 and 5, while the upper limits were the maximum values of *He* density in Figures 4 and 5. The density from the refined APMS method was 3.83×10^9 m^{-3}, a value that was further from the theoretical *He* density from the ideal gas law (7.64×10^9 m^{-3}) compared to the value from the original APMS method (4.53×10^9 m^{-3}). However, the original APMS results have both a larger maximum (3.25×10^9 m^{-3}) and standard (2.79×10^9 m^{-3}) deviation of *He* density due to fitting error compared to the refined APMS results with a smaller maximum (1.93×10^9 m^{-3}) and standard (4.47×10^8 m^{-3}) deviation. This demonstrates that the refined APMS approach provides more precise quantification results from QMS measurements than the original approach. Moreover, the accuracy of the refined APMS method can be further improved by considering the pressure dependence of $f(m)$. The black and blue circles in Figure 7 are the $f(m)$ values used at different pressures, where $f(m_{Ar})$ and $f(m_{He})$ were taken as averages. These values seem to be pressure-dependent, so the dashed lines in Figure 7 were obtained by fitting the circle values. Using the dashed lines, we obtained $f(m_{Ar}, 76$ mTorr$)$ and $f(m_{He}, 237$ mTorr$)$. Then the *He* density, which was re-calculated with $f(m_{Ar}, 76$ mTorr$)$ and $f(m_{He}, 237$ mTorr$)$, increased compared to before; the refined result, (1.35×10^{10} m^{-3}), was closer to the ideal result than that from the original approach (1.59×10^{10} m^{-3}). This is illustrated in Figure 6 with the solid plots. Therefore, the possibility of achieving higher accuracy with the refined APMS method exists.

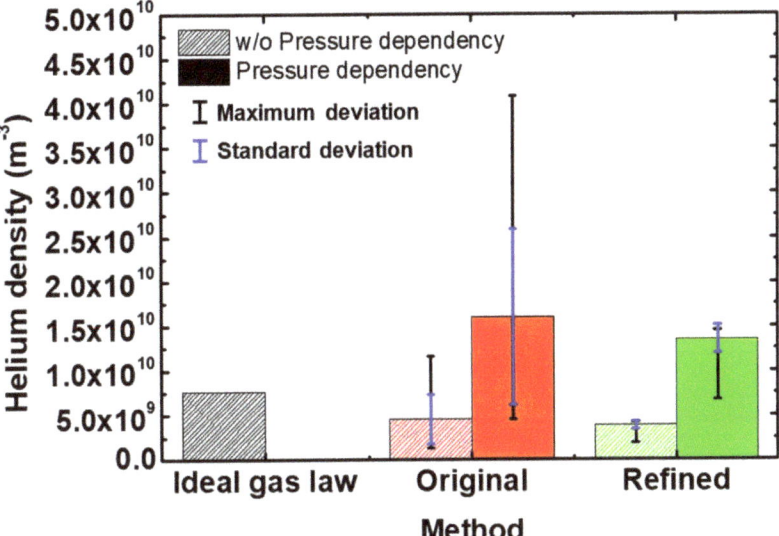

Figure 6. *He* densities from each quantification method. The maximum deviation is used.

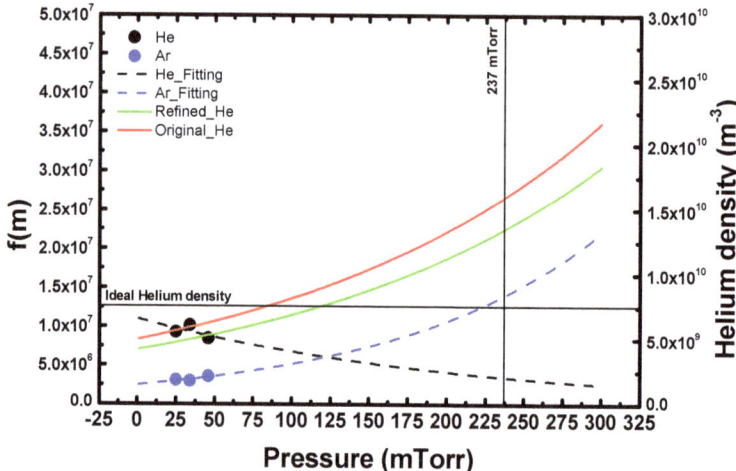

Figure 7. *He* density and f(m) with varying pressure. Black and blue circles are f(m), black and blue dashed lines are the fitted f(m), and green and red lines are the *He* density from the refined and original APMS methods, respectively.

4. Conclusions

In this paper, we proposed a refined APMS approach for quantifying radical densities in plasma. With a simple modification of the original APMS approach, the fitting process was eliminated, and the *He* density was obtained over the entire electron energy ranges. After averaging the *He* density of each electron energy, the *He* density was compared to the density from the ideal gas law. As a result, while the *He* density from the refined APMS method, 3.83×10^9 m^{-3}, was further from the density of the ideal gas law, (7.64×10^9 m^{-3}), than that from the original approach, (4.53×10^9 m^{-3}), the deviation of the density from the refined method was much smaller than that from the original APMS method. Based on this, the refined APMS approach is believed to be applicable to neutral density quantification with higher precision. This method has some limitations. To obtain correct signals from the QMS, the QMS chamber pressure should be below 1.3×10^{-5} Pa, and highly reactive radicals, such as fluorine and chlorine, should be avoided to prevent chemical reactions with the filament. Additionally, to use either APMS method, a cross-section is required. If an algorithm that can be applied to existing equipment is developed in future research, such as APMS, it is expected to enable more precise and accurate measurement, which could lead to adoption in various fields that require precise plasma diagnosis, such as the semiconductor etching process.

Author Contributions: Conceptualization, S.Y.; validation, C.C., S.K. and S.L.; formal analysis, C.C., J.L. (Jangjae Lee), W.J., Y.Y., M.C., J.L. (Jinho Lee) and S.L.; writing—original draft preparation, C.C.; writing—review and editing, S.Y., S.K., Y.L. and I.S.; supervision, S.Y. All authors have read and agreed to the published version of the manuscript.

Funding: This research was supported by National Research Council of Science & Technology (NST) grants provided by the Korean government (MSIP) (No. CAP-17-02-NFRI, CRF-20-01-NFRI); the Next-Generation Intelligent Semiconductor R&D program through the Korea Evaluation Institute of Industrial Technology (KEIT) funded by the Korean government (MOTIE); the Korea Institute of Energy Technology Evaluation and Planning (KETEP) and the MOTIE of the Republic of Korea (20202010100020); the MOTIE (20009818, 20010420) and the KSRC (Korea Semiconductor Research Consortium) support program for the development of future semiconductor devices; a Korea Institute for Advancement of Technology (KIAT) grant funded by the Korean government (MOTIE) (P0008458, HRD Program for Industrial Innovation); the Basic Science Research Program through the National Research Foundation of Korea (NRF) funded by the Ministry of Education (NRF-2020R1A6A1A03047771); and the KIMM Institutional Program (NK236F) and NST/KIMM.

Institutional Review Board Statement: Not applicable.

Informed Consent Statement: Not applicable.

Data Availability Statement: The data presented in this study are available upon request from the corresponding author.

Conflicts of Interest: The authors declare no conflict of interest.

References

1. Chen, F.F. *Introduction to Plasma Physics and Controlled Fusion*, 3rd ed.; Springer: Cham, Switzerland, 2018.
2. Lieberman, M.A.; Lichtenberg, A.J. *Principles of Plasma Discharges and Materials Processing*, 2nd ed.; John Wiley & Sons: Hoboken, NJ, USA, 2005; pp. 1–757.
3. Ishikawa, K.; Karahashi, K.; Ishijima, T.; Cho, S.I.; Elliott, S.; Hausmann, D.; Mocuta, D.; Wilson, A.; Kinoshita, K. Progress in nanoscale dry processes for fabrication of high-aspect-ratio features: How can we control critical dimension uniformity at the bottom? *Jpn. J. Appl. Phys.* **2018**, *57*, 06JA01.
4. Vincent, M. Donnelly and Avinoam Kornblit, Plasma etching: Yesterday, today, and tomorrow. *J. Vac. Sci. Technol.* **2013**, *31*, 050825.
5. Seong, I.H.; Lee, J.J.; Cho, C.H.; Lee, Y.S.; Kim, S.J.; You, S.J. Characterization of SiO_2 over Poly-Si Mask Etching in Ar/C4F8 Capacitively Coupled Plasma. *Appl. Sci. Converg. Technol.* **2021**, *30*, 176–182.
6. Lee, Y.; Seong, I.; Lee, J.; Lee, S.; Cho, C.; Kim, S.; You, S. Various evolution trends of sample thickness in fluorocarbon film deposition on SiO_2. *J. Vac. Sci. Technol. A* **2022**, *40*, 013001.
7. Jeon, M.H.; Mishra, A.K.; Kang, S.K.; Kim, K.N.; Kim, I.J.; Lee, S.B.; Sin, T.H.; Yeom, G.Y. Characteristics of SiO_2 etching by using pulse-time modulation in 60 MHz/2 MHz dual-frequency capacitive coupled plasma. *Curr. Appl. Phys.* **2013**, *13*, 1830–1836.
8. Lee, H.-C.; Chung, C.-W. Electron heating and control of electron energy distribution for the enhancement of the plasma ashing processing. *Plasma Sources Sci. Technol.* **2015**, *24*, 024001.
9. Susa, Y.; Ohtake, H.; Jianping, Z.; Chen, L.; Nozawa, T. Characterization of CO_2 plasma ashing for less low-dielectric-constant film damage. *J. Vac. Sci. Technol. A* **2015**, *33*, 061307.
10. Neralla, S. (Ed.) Chemical Vapor Deposition: Where we are and the Outlook for the Future. In *Chemical Vapor Deposition, Recent Advances and Applications in Optical, Solar Cells and Solid State Devices*; IntechOpen: London, UK, 2016; pp. 247–251.
11. Vasudev, M.C.; Anderson, K.D.; Bunning, T.J.; Tsukruk, V.V.; Naik, R.R. Exploration of Plasma Enhanced Chemical Vapor Deposition as a Method for Thin-Film Fabrication with Biological Applications. *ASC Appl. Mater. Interfaces* **2013**, *5*, 3983–3994.
12. Waskow, A.; Avino, F.; Howling, A.; Furno, I. Entering the plasma agriculture field: An attempt to standardize protocols for plasma treatment of seeds. *Plasma Processes Polym.* **2022**, *19*, 2100152.
13. Attri, P.; Ishikawa, K.; Okumura, T.; Koga, K.; Shiratani, M. Plasma Agriculture from Laboratory to Farm: A Review. *Proesses* **2020**, *8*, 1002.
14. Ito, M.; Ohta, T.; Hori, M. Plasma agriculture. *J. Korean Phys. Soc.* **2012**, *60*, 937–943.
15. Fridman, G.; Friedman, G.; Gutsol, A.; Shekhter, A.B.; Vasilets, V.N.; Fridman, A. Applied Plasma Medicine. *Plasma Processes Polym.* **2008**, *5*, 503–533.
16. Schlegel, J.; Köritzer, J.; Boxhammer, V. Plasma in cancer treatment. *Clin. Plasma Med.* **2013**, *1*, 2–7.
17. Weltmann, K.-D.; Polak, M.; Masur, K.; von Woedtke, T.; Winter, J.; Reuter, S. Plasma Processes and Plasma Sources in Medicine. *Contrib. Plasma Phys.* **2012**, *52*, 644–654.
18. Adamovich, I.; Murphy, A.B.; Bruggeman, P.J.; Turner, M.M.; Hamaguchi, S.; Favia, P.; Oehrlein, G.S.; Pitchford, L.C.; Starikovskaia, S.; Bogaerts, A.; et al. The 2017 Plasma Roadmap: Low Temperature Plasma Science and Technology. *J. Phys. D Appl. Phys.* **2017**, *50*, 323001.
19. Godyak, V.A.; Piejak, R.B.; Alexandrovich, B.M. Electron Energy Distribution Function Measurements and Plasma Parameters in Inductively Coupled Argon Plasma. *Plasma Sources Sci. Technol.* **2002**, *11*, 525–543.
20. Godyak, V.A.; Piejak, R.B.; Alexandrovich, B.M. Plasma Sources Science and Technology Measurement of Electron Energy Distribution in Low-Pressure RF Discharges Measurements of Electron Energy Distribution in Low-Pressure R F Discharges. *Plasma Sources Sci. Technol.* **1992**, *18*, 36–58.
21. Kortshagen, U.; Pukropski, I.; Zethoff, M. Spatial Variation of the Electron Distribution Function in a Rf Inductively Coupled Plasma: Experimental and Theoretical Study. *J. Appl. Phys.* **1994**, *76*, 2048–2058.
22. Cherrington, B.E. The Use of Electrostatic Probes for Plasma Diagnostics—A Review. *Plasma Chem. Plasma Process.* **1982**, *2*, 113–140.
23. Kim, S.J.; Lee, J.J.; Lee, Y.S.; Cho, C.H.; You, S.J. Crossing Frequency Method Applicable to Intermediate Pressure Plasma Diagnostics Using the Cutoff Probe. *Sensors* **2022**, *22*, 1291.
24. Welzel, S.; Hempel, F.; Hübner, M.; Lang, N.; Davies, P.B.; Röpcke, J. Quantum Cascade Laser Absorption Spectroscopy as a Plasma Diagnostic Tool: An Overview. *Sensors* **2010**, *10*, 6861–6900. [CrossRef] [PubMed]
25. Conde, L. An Introduction to Langmuir Probe Diagnostics of Plasmas. *Madr. Dept. Física* **2011**, 1–28.

26. Kim, J.H.; Choi, S.C.; Shin, Y.H.; Chung, K.H. Wave Cutoff Method to Measure Absolute Electron Density in Cold Plasma. *Rev. Sci. Instrum.* **2004**, *75*, 2706–2710. [CrossRef]
27. Lopaev, D.V.; Volynets, A.V.; Zyryanov, S.M.; Zotovich, A.I.; Rakhimov, A.T. Actinometry of O, N and Atoms. *J. Phys. D Appl. Phys.* **2017**, *50*, 075202. [CrossRef]
28. Tserepi, A.; Schwarzenbach, W.; Derouard, J.; Sadeghi, N. Kinetics of F atoms and fluorocarbon radicals studies by threshold ionization mass spectrometry in a microwave CF4 plasma. *J. Vac. Sci. Technol. A* **1997**, *15*, 3120. [CrossRef]
29. Singh, H.; Coburn, J.W.; Graves, D.B. Mass spectrometric detection of reactive neutral species: Beam-to-background ratio. *J. Vac. Sci. Technol. A* **1999**, *17*, 2447. [CrossRef]
30. Hecimovic, A.; D'Isa, F.; Carbone, E.; Drenik, A.; Fantz, U. Quantitative gas composition analysis method for a wide pressure range up to atmospheric pressure-CO_2 plasma case study. *Rev. Sci. Instrum.* **2020**, *91*, 113501. [CrossRef]
31. *ISO 14291:2012(en)*; Vacuum Gauges—Definitions and Specifications for Quadrupole Mass Spectrometers. ISO: Geneva, Switzerland, 2012.
32. Data Center for Plasma Properties. Available online: https://dcpp.kfe.re.kr (accessed on 15 June 2022).
33. Singh, H.; Coburn, J.W.; Graves, D.B. Appearance Potential Mass Spectrometry: Discrimination of Dissociative Ionization Products. *J. Vac. Sci. Technol. A Vac. Surf. Film.* **2000**, *18*, 299–305. [CrossRef]
34. Lee, Y.S.; Oh, S.H.; Lee, J.J.; Cho, C.H.; Kim, S.J.; You, S.J. A Quantification Method in Quadrupole Mass Spectrometer Measurement. *Appl. Sci. Converg. Technol.* **2021**, *30*, 50–53. [CrossRef]

Article

Spectroscopy of Laser-Induced Dielectric Breakdown Plasma in Mixtures of Air with Inert Gases Ar, He, Kr, and Xe

Andrew Martusevich [1,2], Roman Kornev [3], Artur Ermakov [3], Igor Gornushkin [4,*], Vladimir Nazarov [1,2], Lyubov Shabarova [3] and Vladimir Shkrunin [3]

[1] Laboratory of Translational Free Radical Biomedicine, Sechenov University, 119991 Moscow, Russia
[2] Laboratory of Medical Biophysics, Privolzhsky Research Medical University, 603005 Nizhny Novgorod, Russia
[3] Institute of Chemistry of High-Purity Substances, 603951 Nizhny Novgorod, Russia
[4] BAM Federal Institute for Materials Research and Testing, 12489 Berlin, Germany
* Correspondence: igor.gornushkin@bam.de

Abstract: The generation of ozone and nitrogen oxides by laser-induced dielectric breakdown (LIDB) in mixtures of air with noble gases Ar, He, Kr, and Xe is investigated using OES and IR spectroscopy, mass spectrometry, and absorption spectrophotometry. It is shown that the formation of NO and NO_2 noticeably depends on the type of inert gas; the more complex electronic configuration and the lower ionization potential of the inert gas led to increased production of NO and NO_2. The formation of ozone occurs mainly due to the photolytic reaction outside the gas discharge zone. Equilibrium thermodynamic analysis showed that the formation of NO in mixtures of air with inert gases does not depend on the choice of an inert gas, while the equilibrium concentration of the NO^+ ion decreases with increasing complexity of the electronic configuration of an inert gas.

Keywords: laser-induced dielectric breakdown (LIDB); nitrogen monoxide; nitrogen dioxide; ozone; emission spectroscopy; inert gases; thermodynamic analysis

1. Introduction

The study of the biological and therapeutic effect of such gas mediators as ozone (O_3) and nitrogen monoxide (NO) is an urgent task; therefore, the issues of personalization of ozone therapy and control of its effectiveness, studies of the biological activity, and sanogenetic effects of NO are relevant.

Despite the fact that ozone, an allotropic form of oxygen, has been widely used in medicine for more than a hundred years, new possibilities for its use are opening up. The biological effect of ozone includes an immunomodulatory property, an increase in cellular energy, and an increase in antioxidant protection [1]. The possibilities of using ozone in dental practice [2], autohemotherapy in the form of an ozonized sorbent [3], and rectal infusion [4] are widely presented.

Another actively developing direction is the use of nitric oxide in medical practice, for example, for the treatment of pulmonary diseases. In addition, several effects of singlet oxygen therapy [5–8] are associated with the presence of NO in the gas flow of the singlet oxygen generator. Among the functional effects, one can note the activation of blood flow through small-diameter vessels (capillary bed) [7], the activation of antioxidant expansion without stimulating free radical oxidation [9,10], the enhancement of energy reduction and detoxification of cells, the enzymes in them [5,6], etc.

As shown in review [11], it is convenient and economically expedient to synthesize NO or O_3 in situ from air using a plasma process. Barrier discharges are often used to generate ozone [12–17]. Garamoon et al. [17] studied the formation of ozone in an oxygen–air barrier discharge and it was found that the concentration of O_3 increases by more than three times when the pressure decreases by two times. It was also found that increasing the length of

the gas discharge tube, the interelectrode distance, and the applied potential also increases the ozone yield.

Moreover, ICP plasma and microwave discharges can be used to produce O_3. Fuller et al. [18] determined the conditions for the formation of metastable active O and O_2 particles responsible for ozone generation, and it was shown that their concentrations depend on the pressure and discharge power, while the amount of dissociated oxygen does not exceed 2%. Cvelbar et al. [19] studied the emission spectra of an oxygen-containing ICP at pressures of 10–300 Pa. The emission spectra contains lines of atomic oxygen at 438.8, 615.6, 645.4, 777.2, and 844.6 nm, and molecular bands of metastable O_2 ($b^1\Sigma_u^+$) at 760.5 nm and O_2^+ ($b^4\Sigma_g^-$—$a^4\Pi_u$); these particles presumably contributed to the formation of ozone. Pulsed and stationary microwave discharges were used in [20] to study the formation of nitrogen oxides; strong NO production and low NO_2 production was demonstrated.

An additional important requirement for medical gases is their purity. At the same time, the gas treated with discharges with metal electrodes can be contaminated with metal particles resulting from spark erosion of the electrodes or cathode sputtering. These impurities are characteristic of barrier and spark discharges, while in ICP impurities can come from the surface of a quartz tube upon contact with the ICP plasma. Therefore, it is important to look for the type of discharge in which these processes are excluded. One of the promising discharges capable of generating ozone and NO without contamination is the optical gas breakdown by a high-power laser, the so-called laser-induced dielectric breakdown (LIDB).

Stricker [21] studied the characteristics of laser breakdown in pure N_2 and O_2 in the pressure range from 1 to 50 atm using a Nd:YAG laser with a wavelength of 1.064 μm and a pulse duration of 10 ns. The emission spectra revealed the presence of O^{++} and O^+ with lifetimes of 15–20 ns and 2 μs, respectively, while no O or O_2 emission was observed. Svatopluk et al. [22] studied the breakdown of $CO-N_2-H_2O$ gas mixtures using a near-IR laser with a power of 85 J and a pulse duration of 450 ps. The various stages of the discharge were studied using time-resolved OES and gas composition using FTIR and GC. The presence of N_2O, CO_2, ethane, acetylene, ethylene, and acetone was detected in the mixture after its laser irradiation. Gornushkin et al. [23] studied the formation of ozone and nitrogen oxides during multiple laser breakdown of oxygen–nitrogen mixtures at atmospheric pressure. About 2000 laser pulses from a Nd:YAG with irradiance of 10^{10} W cm^{-2} were fed into the sealed reaction chamber. The chamber with a long capillary was designed to measure the absorption of O_3, NO, and NO_2 depending on the number of laser pulses. The source of light for measuring the absorption was continuous radiation emitted by the plasma during the first 0.2 μs of its evolution. A kinetic model has been developed that takes into account the chemical reactions between atmospheric components and laser breakdown products. In the model, the laser plasma was considered as a source of nitric oxide and atomic oxygen, the formation rate of which was calculated based on the measured absorption of NO, NO_2, and O_3. The calculated NO, NO_2, and O_3 concentration profiles were in good agreement with the measured profiles on the time scale of 0–200 s; the calculated yield of 2×10^{12} ozone molecules was consistent with the model prediction. This study was important for a general understanding of the chemistry of laser plasmas and for elucidating the nature of spectral interferences and matrix effects important in a spectrochemical analysis.

The aim of this work is to study the spectral features of LIDB plasma in a mixture of air with Ar, He, Kr, and Xe and their influence on the synthesis of NO, NO_2, and O_3. In the future, we plan to use this technology for medical purposes. Air is offered as the most accessible precursor that does not require additional costs from a potential customer (for example, additional gas cylinders), and also reduces the risks associated with the use of gas equipment (working with high-pressure oxygen is dangerous). Inert gases that do not form stable compounds with air components are chosen as electron sources.

2. Materials and Methods

The experimental setup is shown in Figure 1a. This allows a comparative study of the plasma generated by LIDB in various gas mixtures using OES, IR, and mass spectrometry. The plasma is created inside a quartz cylinder 40 mm in diameter, placed in a gas-tight stainless-steel chamber with quartz windows for input and output of laser radiation and registration of radiation spectra (Figure 1b). The window for recording IR spectra is made of zinc selenide (ZnSe). To create plasma, an Nd:YAG laser is used with the parameters shown in Table 1. Before filling the chamber with gases, it is pumped out by fore vacuum and turbomolecular pumps (not shown in Figure 1a). The dosing of inert (Ar, He, Ne, and Kr) and working (air, CO_2) gases is controlled by precision gas flow regulators (RRGs). To study the synthesis of active particles, the UV-Vis, IR, and mass spectrometers were connected to the setup. A high-sensitivity Nanogate 24/3 camera was used to record the plasma light. The camera was located at 30 cm from the reactor wall, perpendicular to its cylindrical axis.

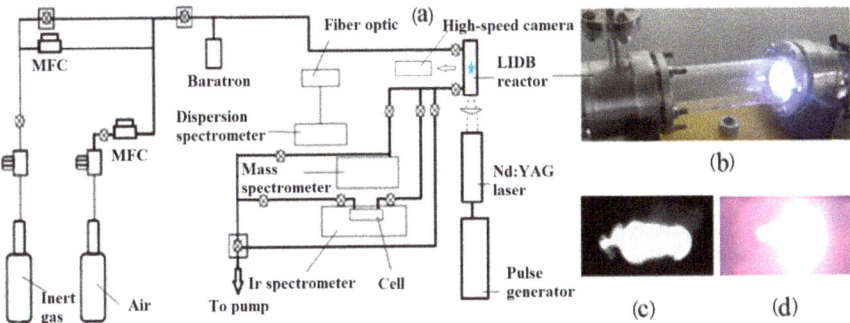

Figure 1. (a) Schematic of experimental setup; (b) LIDB reactor; (c) image of the discharge recorded by camera Nanogate 24/3; (d) photograph of the discharge.

Table 1. Laser parameters.

Wavelength, μm	1.064
Radiation frequency, Hz	10
Energy per pulse, mJ	820
Pulse duration, ns	15
Area of a focal spot, cm^2	10^{-2}
Pulse power, MW	50
Irradiance, GW/cm^2	5
Fluence, J/cm^2	80
Photon flux density, $cm^{-2} s^{-1}$	2.9×10^{28}
Electric field strength, MV/cm	1.4

2.1. Emission Spectroscopy

Optical emission spectroscopy (OES) was used to detect reaction products in plasma. Emission spectra were recorded by an AvaSpec-ULS2048CLEVO-RM-USB3 multichannel 2048-pixel fiber optic spectrometer with ultra-low light scattering. The spectra were recorded in the range 217.5–710 nm with a resolution of 0.17 nm. The spectrometer was activated by a TTL pulse coming from the laser. Time-resolved spectra were obtained by shifting the delay time of the start of recording subsequent measurements by 1 μs, up to 270 μs, when only noise could be recorded. The signal accumulation time varied from 30 to 60 μs.

2.2. IR Spectroscopy

The IR spectra of gas mixtures before and after laser irradiation were recorded in the range 450–7000 cm^{-1} on a BrukerVertex 80v spectrometer with a DTGS detector with a resolution of 1 cm^{-1}. For this, we used an IR gas cell with an optical path length of 10 cm (see Figure 1a) and pressure inside the cell from several tens to several hundreds of Torr. The windows for recording IR spectra in the IR cell are made of zinc selenide (ZnSe).

2.3. Mass Spectrometry of Gases

Mass spectra of gaseous mixtures before and after laser irradiation were recorded on an ExtorrXT300(M) SeriesRGA quadrupole mass spectrometer with a resolution of 1 amu. The residual vacuum in the mass spectrometer chamber was 5×10^{-8} Torr. The working pressure was varied from 1×10^{-6} to 1×10^{-5} Torr to observe the mass spectra of gas mixtures with different concentrations of components. The pressure was controlled by pressure sensors of the mass spectrometer and normalized to the band of the Ar$^+$ carrier gas, which did not take part in the reaction.

2.4. Spectrophotometry of Gases

The absorption spectra were recorded with an SF-2000 spectrophotometer before and after laser irradiation at time intervals of 0 and 10 min in the range of 200–1100 nm with an exposure time of 0.2 s with a resolution of 1 nm. The resulting signal was displayed after fifteen accumulation cycles. The spectrophotometric analysis was carried out by introducing 1 atmosphere of a mixture of air and an inert gas in a ratio of 1:1 into a cylindrical cell-reactor made of fluoroplast with plane-parallel quartz glasses. Before irradiation with the Nd:YAG laser, the spectrum of each of the mixtures was recorded as a blank experiment.

Ozone concentration was calculated using the Hartley band (200–350 nm) with a maximum at $\lambda = 254$ nm. The absorption cross section of the remaining components present in our reaction mixture in this wavelength range was more than two orders of magnitude lower. The NO$_2$ concentration was calculated in the range of 350–430 nm at the maximum $\lambda = 400$ nm. This region of the spectrum is transparent for ozone. The concentrations were calculated based on the Bouguer–Lambert–Beer law. The molar extinction coefficients of 3000 M^{-1}·cm^{-1} for O$_3$ at 254 nm and 157 M^{-1}·cm^{-1} for NO$_2$ at 400 nm were taken from the database [24].

2.5. Thermodynamic Analysis of the System Air + Ar (Kr, Xe)

To estimate the possible yield of ozone and nitrogen oxides in the LIDB plasma, a thermodynamic analysis of gas mixtures was carried out. The calculations were performed out using open-source software [25], which implies local thermodynamic equilibrium (LTE) and is based on the Gibbs free energy minimization algorithm. This model assumes the equation of state for an ideal gas and allows for a small amount of condensed phase due to its negligible volume compared to gaseous particles. The question of whether LTE occurs in LIDB plasma is rather controversial. It is generally accepted that LTE is possible at intermediate and late stages of plasma plume development, somewhere between 1–20 µs, when the characteristic times of the hydrodynamic flow become larger than the characteristic time for chemical reactions. A more complex set of conditions under which a plasma can be considered close to LTE is given in [26,27].

3. Results and Discussion

3.1. Emission Spectra of Ar, He, Kr, Xe in LIDB Plasma

The main factors affecting the plasma lifetime are the electron configuration, ionization potential, and the presence of metastable states of plasma particles. Figure 2 shows the emission spectra of pure He, Ar, Kr, and Xe. The electronic configuration of He is simple compared to other inert gases; it has many electronic levels near a very high (24.6 eV) ionization potential (Table 2) [28]. It was difficult to create a laser breakdown of helium at atmospheric pressure because of its low density; this was possible only at a pressure of

2 atm. The existence of helium plasma is largely due to the presence of a metas $^3S_1$ level with a lifetime of 6×10^5 s (Table 3). In the emission spectrum of He, shown in Figure 2b, only one line is visible at 587 nm, corresponding to the transition between levels with energies of ~23 eV and ~21 eV [29].

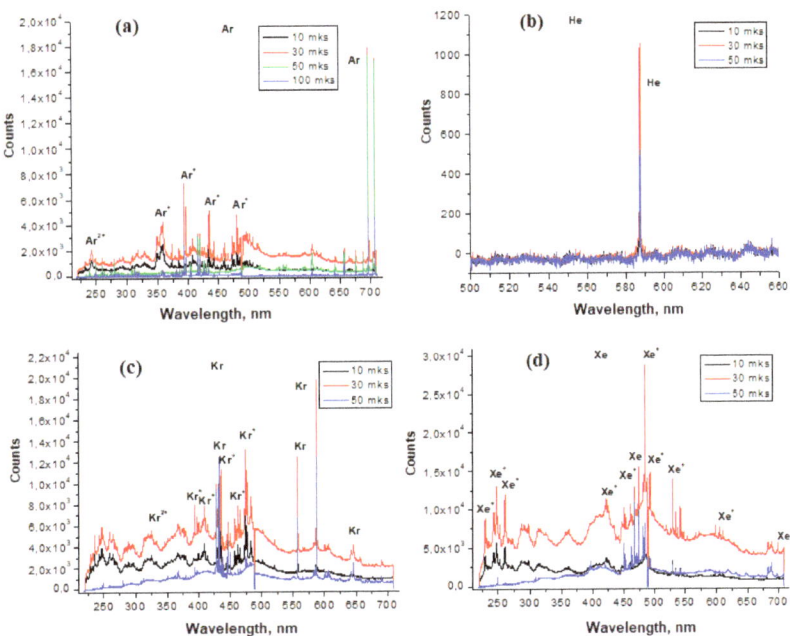

Figure 2. Emission spectra of LIDB plasma in: (**a**) Ar; (**b**) He at 2 atm; (**c**) Kr; (**d**) Xe.

Table 2. Electronic configuration, ionization potentials, and characteristics of metastable states of Ar, He, Kr, and Xe [28].

Inert Gas	Electronic Configuration	Ionization Energy, eV	Metastable States	
			Excitation Energy, eV	Lifetime, s
He	$1s^2$	24.6	19.82 (2^3S_1)	6×10^5
Ar	$1s^2 2s^2 2p^6 3s^2 3p^6$	15.8	11.55 ($4^3P^0_2$) 11.72	>1.3 -
Kr	$1s^2 2s^2 2p^6 3s^2 3p^6 3d^{10} 4s^2 4p^6$	14.0	9.91 10.5	- -
Xe	$1s^2 2s^2 2p^6 3s^2 3p^6 3d^{10} 4s^2 4p^6 4d^{10} 5s^2 5p^6$	12.1	8.32 9.4	- -

The emission spectra of Ar, Kr, and Xe have many lines in the UV, visible, and IR regions (Figure 2a,c,d). This is due to their electronic configuration with many electronic levels (Table 2) distributed between ground state and ionization potential. The intensity of the lines in the emission spectrum of the argon plasma shows that its lifetime is more than 100 μs. Most likely, this is due to the presence of long-lived metastable Ar states with energies of 11.55 and 11.72 eV and a lifetime of more than 1.3 s (Table 2). For He, Kr, and Xe plasmas, the signal drops to the noise level at times longer than 50 μs.

Table 3. Comparative intensities of the lines of excited air particles.

Line (Band), nm	Transition	Air	Air/He = 9	Air/Ar = 1	Air/Kr = 1	Air/Xe = 1
N^+ (399.5)	$^1D_2$-$^1P^o_1$	strong	strong	strong	medium	weak
N^+ (463.0)	$^3P_2$-$^3P^o_2$	strong	strong	-	-	-
N^+ (500.5)	$^3D_3$-$^3F^o_4$	strong	strong	strong	medium	-
N^+ (568.0)	$^3D_3$-$^3P^o_2$	medium	medium	medium	medium	-
N^+ (594.1)	$^3D^o_3$-$^3P_2$	medium	medium	medium	-	-
O (615.8)	$^5D^o_4$-$^5P_3$	weak	weak	weak	weak	-
O^+ (407.6)	$^4F_{9/2}$-$^4D^o_{7/2}$	strong	strong	strong	-	-
O^{2+} (334.0)	$^3S_1$-$^3P^o_2$	medium	medium	medium	-	-

Figure 3 shows the emission spectra of pure air, as well as its mixtures with Ar, He, Kr, and Xe. In the emission spectrum of air (Figure 3a), noticeable lines of a singly-charged nitrogen ion appear; oxygen is represented by both an atomic line and lines of ions O^+ and O_2^+. Most of the spectral features disappear after 40 µs, only weak transitions of atomic oxygen at 615.8 nm and the O^+ ion (640–660 nm) are preserved. After 50 µs, only noise is observed.

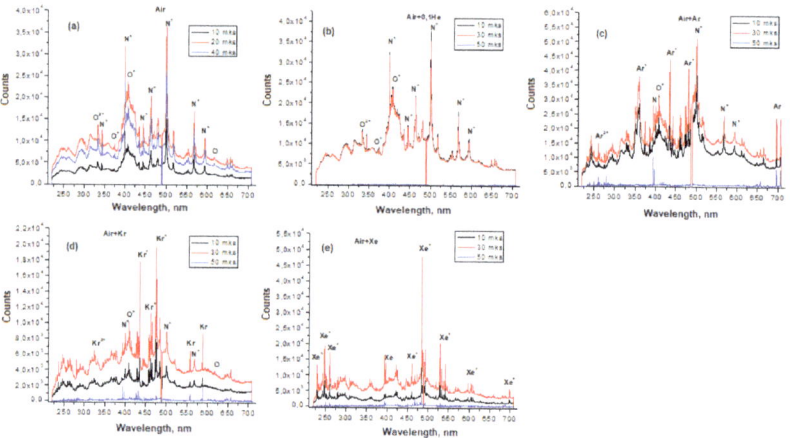

Figure 3. Emission spectra of air and its mixtures with inert gases in LIDB plasma: (**a**) pure Air; (**b**) Air + He; (**c**) Air + Ar; (**d**) Air + Kr; (**e**) Air + Xe.

When analyzing the spectrum of the Air + He mixture (Figure 3b), the overall picture of the spectrum does not noticeably change, the spectral features of pure air also remain. It should be noted that the formation of a gas discharge is observed at a ratio of Air/He ≥ 9. At a higher concentration of He in air, the discharge is not initiated. The line of helium itself is not observed in the emission spectrum. In this regard, it can be assumed that the observed plasma is formed by the air components O_2 and N_2, whose ionization potentials (Table 2) are 12.2 and 15.6 eV, respectively, and the configuration of electronic levels is complex [30]. Thus, He does not affect the composition of the reactive components of the Air + He mixture.

Argon makes it possible to extend the glow time of the discharge of the Air/Ar = 1 mixture up to 80 µs (Figure 4c). The spectrum is rich in lines of atomic and ionized argon. As well as in the case of pure air, N^+ and O^+ transitions were recorded. The intensity of the nitrogen line N^+ at 500.5 nm is noticeably higher than in the previous cases. The intensity of the N^+ line at 399.5 nm is lower, and the N^+ line at 463.0 nm is not observed. The lines of ionized oxygen are intensified.

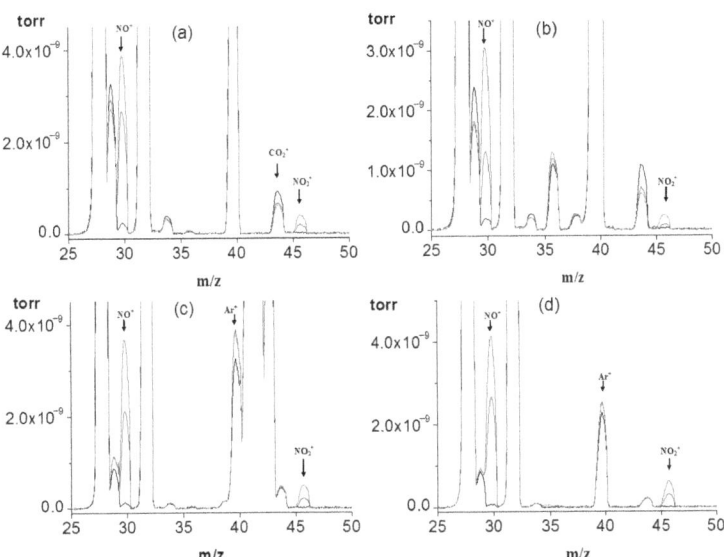

Figure 4. Mass spectra of mixtures of inert gases and air before and after exposure to LIDB plasma: (**a**) Air + He; (**b**) Air + Ar; (**c**) Air + Kr; (**d**) Air + Xe. Black line—before plasma exposure, red line—after 1 h of plasma treatment, green line—after 2 h of plasma treatment.

For the Air/Kr = 1 mixture (Figure 3d), the plasma lifetime is 50 μs, and the spectrum pattern noticeably changes. The most intense are the lines of singly-charged krypton at 473.9 nm and 435.5 nm. The lines of air components noticeably decrease, the N^+ line at 594.1 nm is absent, and the bands of ionized oxygen O_2^+ disappear.

The emission spectrum of the Air/Xe = 1 mixture (Figure 3e) contains mainly xenon lines. Of the air components, only a weak N^+ line at 399.5 nm is observed. The most intense Xe^+ line is observed at 484.3 nm, and the plasma lifetime is 60 μs.

Figure 3b–e shows that only lines of air components are observed in the Air + He mixture. In Air + Ar and Air + Kr mixtures, both lines of air components and lines of inert gases appear. In an Air + Xe mixture, the Xe lines are visible, and the lines of the air components are represented by a weak N^+ line.

It can be assumed that the ionization of helium is very difficult due to the high ionization potential, and the energy is directed to the ionization of air. Xe ionizes more easily than other inert gases; energy is spent on its ionization and excitation of its metastable states. Table 3 presents the comparative intensities of the lines of excited air particles in accordance with the above considerations.

Since the plasma produced by LIDB is close to equilibrium, gas ionization occurs not only due to electron impact, but also due to the impact of heavy particles. Therefore, the following reactions can be written:

$$N_2 \xrightarrow{t} 2N \tag{1}$$

$$O_2 \xrightarrow{t} 2O \tag{2}$$

$$N_2 \xrightarrow{t} 2N^+ + 2e \tag{3}$$

$$O_2 \xrightarrow{t} 2O^+ + 2e \tag{4}$$

$$O_2 \xrightarrow{t} 2O^{2+} + 4e \tag{5}$$

$$N \xrightarrow{t} N^+ + e \tag{6}$$

$$O \xrightarrow{t} O^+ + e \tag{7}$$

$$O \xrightarrow{t} O^{2+} + 2e \qquad (8)$$

$$e + N_2 \rightarrow 2N^+ + 2e \qquad (9)$$

$$e + O_2 \rightarrow 2O^+ + 2e \qquad (10)$$

$$e + O_2 \rightarrow 2O^{2+} + 3e \qquad (11)$$

$$e + N \rightarrow N^+ + 2e \qquad (12)$$

$$e + O \rightarrow O^+ + 2e \qquad (13)$$

$$e + O \rightarrow O^{2+} + 3e \qquad (14)$$

Figure 4 shows the mass spectra of mixtures of air with inert gases before and after exposure to LIDB plasma. In all the presented mass spectra, the bands of N_2 and O_2 included in the air, as well as NO^+ and NO_2^+ ions corresponding to NO and NO_2 molecules, are identified. The ratio of the relative intensity of the NO^+ and NO_2^+ fragments in the mass spectrum of the nitrogen dioxide molecule, according to the reference tables provided by the suppliers of equipment for mass spectrometry, is 100:37. In the mass spectra obtained in this work, this ratio fluctuates within 100:(7÷17), from which it follows that both NO_2 and NO molecules are formed during the reaction according to the following schemes:

$$N + O \rightarrow NO \qquad (15)$$

$$N + 2O \rightarrow NO_2 \qquad (16)$$

$$NO + O \rightarrow NO_2 \qquad (17)$$

$$e + N^+ + O \rightarrow NO \qquad (18)$$

$$e + N + O^+ \rightarrow NO \qquad (19)$$

$$2e + N^+ + O^+ \rightarrow NO \qquad (20)$$

$$e + NO + O^+ \rightarrow NO_2 \qquad (21)$$

The peak of the NO^+ ion in the recorded mass spectra consists of the contribution of the NO^+ ion of the NO molecule present in the air before the plasma-chemical reaction, the contribution of the NO^+ ion of the NO molecule formed during the plasma-chemical reaction, and the contribution of the NO^+ fragment from the NO_2 molecule formed during the plasma-chemical reaction. The relative intensity of the NO^+ ion related to the NO molecule was calculated as the difference between the intensities before and after irradiation and the contribution of the NO^+ fragment from the NO_2 molecule to the intensity of this peak.

The histogram in Figure 5 shows the dependence of the relative intensity of the NO and NO_2 bands after treatment for 1 and 2 h on the composition of the initial mixture. It can be seen that, depending on the complexity of the electronic configuration of an inert gas, as well as a decrease in its ionization potential, there is a tendency for an increase in the NO compound in the reaction products. No pronounced dependence is observed for the NO_2 compound.

Of course, the factors that determine the influence of an inert gas on the chemical reactions occurring in the LIDB plasma are not limited to the complexity of the electronic configuration alone. The density of the gas, which increases with increasing atomic mass, can also have an effect. The processes of Pfennig ionization play a role, which are different for different inert gases, as well as the size of the atom; for xenon it is maximum, which means that the cross section of collisions with atoms and electrons is larger than for other inert gases. However, we can assume that these factors are also determined by the structure of the atom, and their influence can be considered separately in the appropriate statement of the problem.

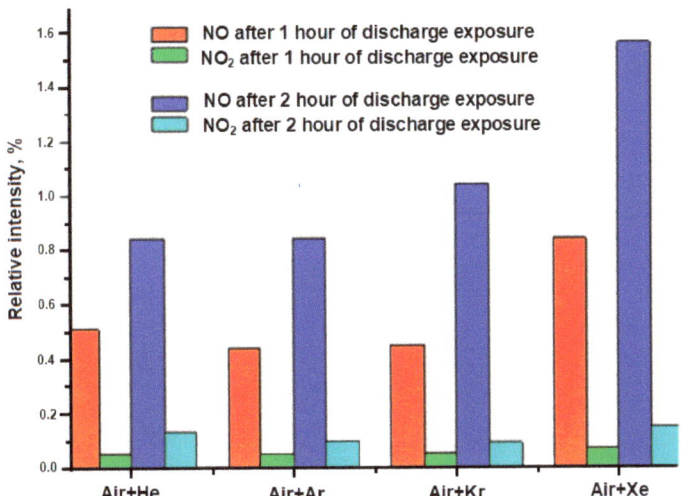

Figure 5. The dependence of the relative intensity of the NO and NO_2 bands after 1 and 2 h of irradiation of the initial mixture.

Figure 6 shows IR spectra of mixtures of air with Ar, He, Kr, and Xe, which confirm the formation of NO_2. For all mixtures, the spectra after gas discharge treatment contain nitrogen dioxide absorption bands in the ranges of 2860–2978 (Figure 6a) and 1564–1650 cm^{-1} (Figure 6b) [25]. The intensity of the NO_2 bands in the spectra increases depending on the complexity of the electronic configuration of the inert gas, which agrees with the mass spectrometry data. The lowest intensity of the NO_2 band is recorded in the Air + 0.1He mixture, the highest in Air + Xe.

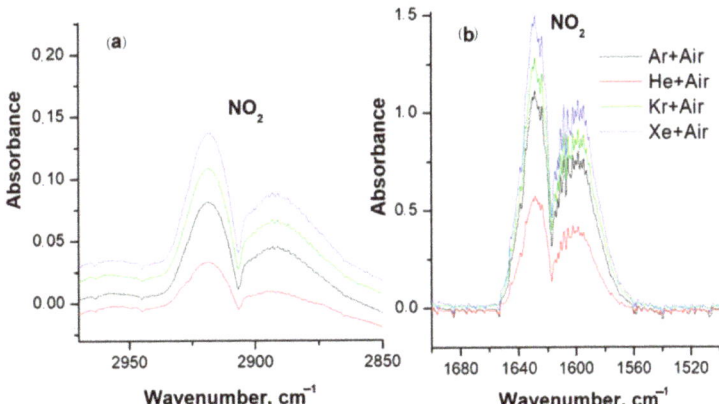

Figure 6. IR spectra of air mixtures with Ar, He, Kr, and Xe. (**a**) nitrogen dioxide absorption bands in the ranges of 2860–2978 cm^{-1}; (**b**) nitrogen dioxide absorption bands in the ranges of 1564–1650 cm^{-1}.

It can be seen that, despite the fact that the lines of excited air components are not resolved in the emission spectrum of the Air + Xe mixture due to their low intensity, as well as the insufficient resolution of the spectrometer, the reaction with formation of NO and NO_2 proceed. We can assume a change in the mechanism of formation of these compounds depending on the gas mixture. The atomic (15)–(17) and ionic (18)–(21) mechanisms of NO and NO$_2$ formation in pure air, air + Ar, and air + Kr change to the molecular mechanism (22), (23) in air + Xe:

$$N_2 + 2O_2 \rightarrow 2NO_2 \quad (22)$$

$$N_2 + O_2 \rightarrow 2NO \quad (23)$$

Figure 7 shows the concentrations of O$_3$ and NO$_2$ as a function of time during LIDB plasma treatment of mixtures of air with argon, xenon, and krypton. In the case of the Air + Xe mixture, there is a jump in the concentrations of O$_3$ and NO$_2$ up to the values $(3.0 \pm 0.2) \times 10^{-4}$ and $(2.8 \pm 0.2 \times 10^{-2}$ vol.%, respectively, in the first ten minutes of irradiation. Further growth of concentrations slows down and by 2 h of irradiation their values are $(3.6 \pm 0.1) \times 10^{-4}$ vol.% for ozone and $(3.1 \pm 0.2) \times 10^{-2}$ vol.% for NO$_2$.

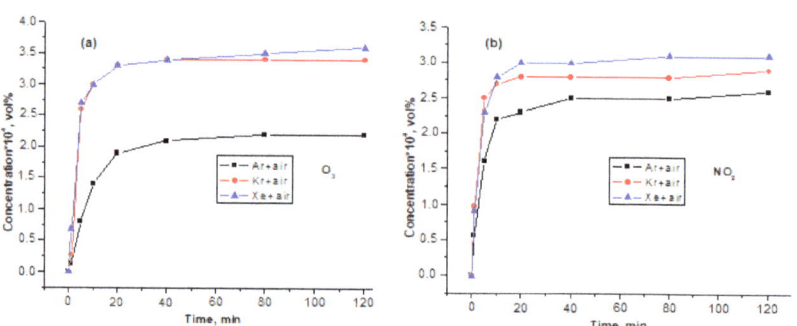

Figure 7. Dependence of O$_3$ (**a**) and NO$_2$ (**b**) concentrations on plasma treatment time of air mixtures with Ar, He, Kr and Xe.

A similar situation is observed for the Air + Kr mixture, a sharp increase in the concentrations of O$_3$ and NO$_2$ within 10 min to the values $(3.0 \pm 0.2) \times 10^{-4}$ vol.% (O$_3$) and $(2.8 \pm 0.2) \times 10^{-2}$ vol.% (NO$_2$). However, after two hours of irradiation, the Air + Kr mixture produces lower amounts of ozone and nitrogen dioxide, $(3.4 \pm 0.2) \times 10^{-4}$ and $(2.9 \pm 0.2) \times 10^{-2}$ vol.%, respectively. A tendency to an even lower yield of O$_3$ and NO$_2$ manifests itself when passing from krypton to argon, where the ozone concentration in the Air + Ar mixture after two hours of irradiation is $(2.2 \pm 0.3) \times 10^{-4}$ vol.%, and the nitric oxide concentration $(2.6 \pm 0.2) \times 10^{-2}$ vol.%.

Possible mechanisms of NO$_2$ formation were discussed above. The formation of ozone, according to [23], cannot occur in the gas discharge zone, since, due to instability, the O$_3$ molecule immediately reacts with N. The most likely mechanism for its formation is the photolysis reaction occurring in the volume of the LIDB reactor according to the following scheme:

$$h\nu + O_2 + O \rightarrow O_3 \quad (24)$$

3.2. Thermodynamics

Considering that the LIDB plasma at atmospheric pressure is close to equilibrium, thermodynamic analysis can be used to consider plasma-chemical reactions. Figure 8 shows the thermodynamically equilibrium composition of air, as well as its mixtures with Ar, He, Kr and Xe. Analyzing the dependences obtained, it can be seen that in pure air, the formation of O$^+$ and N$^+$ ions begins at temperatures above 6000 K. Their presence is seen in the emission spectrum (Figure 3a). For Air + Ar and Air + Kr mixtures, the formation of O$^+$ and N$^+$ ions also begins from a temperature above 6000 K, but at the same time, the Kr$^+$ ion in the Air + Kr mixture is formed at T = 5400 K, while the Ar$^+$ ion, in the Air + Ar mixture, at T = 6000 K. For the Air + Xe mixture, the formation temperature of the Xe$^+$ ion decreases to 4500 K, while the formation of O$^+$ and N$^+$ ions also starts from a temperature above 6000 K.

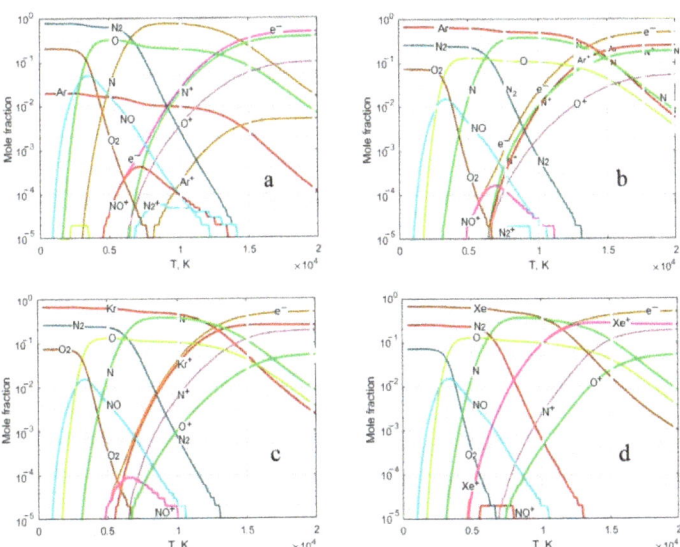

Figure 8. The equilibrium composition of: (**a**) air (N:O:Ar = 78:21:1) and its mixtures; (**b**) N:O:Ar = 39:11:50; (**c**) N:O:Kr = 39:11:50; (**d**) N:O:Xe = 39:11:50. P = 1 bar.

However, while in the emission spectra of Air + Ar and Air + Kr mixtures, one can see O^+ and N^+ ions together with Ar^+ and Kr^+ ions, in the emission spectrum of Air + Xe mixtures, no O^+ and N^+ ions are observed. Thus, depending on the complexity of the electronic configuration of the inert gas, the temperature of the gas in the plasma decreases and in the plasma of the Air + Xe mixture its value is in the range of 4500–6000 K.

It can be seen that the main compounds formed in this system are NO and NO^+. The equilibrium concentration of NO in pure air and in mixtures of air with inert gases is at the same level and does not depend on the choice of inert gas. On the contrary, the equilibrium concentration of the NO^+ ion in pure air is maximum and decreases in mixtures of air with inert gases. Moreover, the concentration of NO^+ is the lower, the more complex the electronic configuration of the inert gas.

4. Conclusions

It was shown that the addition of He to air does not fundamentally change the spectral pattern of air. In contrast, the addition of Ar suppresses the N^+ band at 463.0 nm, while the other bands of nitrogen ions slightly decrease. The addition of Kr leads to even greater suppression of the line intensities of nitrogen ions, as well as oxygen ions. It should be noted that when He, Ar, and Kr are added, the atomic oxygen line retains a low intensity. The addition of Xe results in complete suppression of the air component bands.

The formation of molecules in LIDB plasma can be summarized as follows.

1. The formation of NO and NO_2 in LIDB plasma noticeably depends on the type of inert gas.
2. An increase in the concentration of NO and NO_2 is affected by the complication of the electronic configuration of an inert gas, as well as a decrease in its ionization potential.
3. The complication of the electronic configuration of an inert gas, as well as a decrease in its ionization potential, leads to the suppression of the lines of nitrogen and oxygen ions. Thus, it can be assumed that the main mechanism affecting the chemical transformations in these mixtures and in particular, the formation of NO and NO_2 is atomic.
4. The formation of ozone occurs outside the gas discharge zone by the photolytic reaction.

5. Using the method of thermodynamic analysis, it has been established that the equilibrium concentration of NO in in 1:1 mixtures of air with inert gases does not depend on the choice of an inert gas. On the contrary, the equilibrium concentration of the NO$^+$ ion decreases as the electronic configuration of the inert gas becomes more complex.

In the future, it is planned to use this method on real systems, for example, for the treatment of abiogenic media obtained after irradiation with gas mixtures (for example, saline) with a further transition to blood treatment in order to increase its antioxidant potential.

Author Contributions: Conceptualization, I.G., R.K., A.M. and V.N.; methodology, I.G., R.K., V.S. and A.E.; software, V.S., A.E. and L.S.; validation, V.N. and R.K.; formal analysis, I.G. and R.K.; investigation, I.G., A.E., V.S. and R.K.; resources, A.M., V.N. and L.S. data curation, R.K., writing—original draft preparation, I.G. and R.K.; writing—review and editing, I.G., R.K. and L.S.; visualization, R.K., V.N. and L.S.; supervision, A.M., I.G. and V.N.; project administration, A.M., R.K. and V.N.; funding acquisition, A.M. and V.N. All authors have read and agreed to the published version of the manuscript.

Funding: This research work was supported by the academic leadership program Priority 2030 proposed by Federal State Autonomous Educational Institution of Higher Education I.M. Sechenov First Moscow State Medical University of the Ministry of Health of the Russian Federation (Sechenov University).

Institutional Review Board Statement: Not applicable.

Informed Consent Statement: Not applicable.

Data Availability Statement: Data are available upon request.

Conflicts of Interest: The authors declare no conflict of interest.

References

1. Hoppe, P.; Praml, G.; Rabe, G.; Lindner, J.; Fruhmann, G.; Kessel, R. Environmental Ozone Field Study on Pulmonary and Subjective Responses of Assumed Risk Groups. *Environ. Res.* **1995**, *71*, 109–121. [CrossRef] [PubMed]
2. Baysan, A.; Lynch, E. The Use of Ozone in Dentistry and Medicine. *Prim. Dent. Care* **2005**, *12*, 47–52. [CrossRef] [PubMed]
3. Belianin, I.I.; Shmelev, E.I. The use of an ozonised sorbent in treating patients with progressive pulmonary tuberculosis combined with hepatitis. *Ter. Arkh.* **1994**, *66*, 29–32. [PubMed]
4. Carpendale, M.T.; Freeberg, J.; Griffiss, J.M. Does Ozone Alleviate AIDS Diarrhea? *J. Clin. Gastr.* **1993**, *17*, 142–145. [CrossRef] [PubMed]
5. Martusevich, A.A.; Peretyagin, S.P.; Martusevich, A.K. Molecular and cellular mechanisms of the action of singlet oxygen on biological systems. *M. Tech. Med.* **2012**, *2*, 128–134.
6. Martusevich, A.A.; Solovieva, A.G.; Martusevich, A.K. Influence of singlet oxygen inhalation on the state of blood pro- and antioxidant systems and energy metabolism. *Bull. Exper. Biol. Med.* **2013**, *156*, 41–43. [CrossRef] [PubMed]
7. Martusevich, A.K.; Peretyagin, S.P.; Martusevich, A.A.; Peretyagin, P.V. Effect of ROS inhalations on systemic and local hemodynamics in rats. *Bull. Exp. Boil. Med.* **2016**, *161*, 634–637. [CrossRef] [PubMed]
8. Onyango, A.N. The Contribution of Singlet Oxygen to Insulin Resistance. *Oxid. Med. Cell. Longev.* **2017**, *2017*, 8765972. [CrossRef]
9. Di Mascio, P.; Martinez, G.R.; Miyamoto, S.; Ronsein, G.E.; Medeiros, M.H.G.; Cadet, J. Singlet Molecular Oxygen Reactions with Nucleic Acids, Lipids, and Proteins. *Chem. Rev.* **2019**, *119*, 2043–2086. [CrossRef]
10. Pfitzner, M.; Preuß, A.; Röder, B. A new level of in vivo singlet molecular oxygen luminescence measurements. *Photodiag. Photodyn. Ther.* **2020**, *29*, 101613. [CrossRef]
11. Malik, M.A. Nitric Oxide Production by High Voltage Electrical Discharges for Medical Uses: A Review. *Plasma Chem. Plasma Proc.* **2016**, *36*, 737–766. [CrossRef]
12. Eliasson, B.; Hirth, M.; Kogelschatz, U. Ozone synthesis from oxygen in dielectric barrier discharges. *J. Phys. D* **1987**, *20*, 1421–1437. [CrossRef]
13. Pietsch, G.J.; Gibalov, V.I. Dielectric barrier discharges and ozone synthesis. *Pur. Appl. Chem.* **1998**, *70*, 1169–1174. [CrossRef]
14. Malik, M.A.; Schoenbach, K.H.; Heller, R. Coupled surface dielectric barrier discharge reactor-ozone synthesis and nitric oxide conversion from air. *Chem. Eng. J.* **2014**, *256*, 222–229. [CrossRef]
15. Li, M.; Zhu, B.; Yan, Y.; Li, T.; Zhu, Y. A High-Efficiency Double Surface Discharge and Its Application to Ozone Synthesis. *Plasm. Chem. Plasm. Proc.* **2018**, *38*, 1063–1080. [CrossRef]
16. Pekárek, S. Non-Thermal Plasma Ozone Generation. *Act. Polytech.* **2003**, *43*, 47–51. [CrossRef]
17. Garamoon, A.A.; Elakshar, F.F.; Nossair, A.M.; Kotp, E.F. Experimental study of ozone synthesis. *Plasm. Sour. Sc. Tech.* **2002**, *11*, 254–259. [CrossRef]

18. Fuller, N.C.M.; Malyshev, M.V.; Donnelly, V.M.; Herman, I.P. Characterization of transformer coupled oxygen plasmas by trace rare gases-optical emission spectroscopy and Langmuir probe analysis. *Plasm. Sour. Sc. Tech.* **2000**, *9*, 116–127. [CrossRef]
19. Cvelbar, U.; Krstulović, N.; Milošević, S.; Mozetič, M. Inductively coupled RF oxygen plasma characterization by optical emission spectroscopy. *Vacuum* **2007**, *82*, 224–227. [CrossRef]
20. Babarickij, A.I.; Bibikov, M.B.; Demkin, S.A.; Moskovskij, A.S.; Smirnov, R.V.; Cheban'kov, F.N. Okislenie azota v mikrovolnovyh razryadah atmosfernogo davleniya. *Him. Vys. En.* **2021**, *55*, 487–492. (In Russian)
21. Stricker, J. Experimental investigation of electrical breakdown in nitrogen and oxygen induced by focused laser radiation at 1.064 μ. *J. Appl. Phys.* **1982**, *53*, 851. [CrossRef]
22. Civišs, S.; Babaánkovaá, D.; Cihelka, J.; Sazama, P.; Juha, L. Spectroscopic Investigations of High-Power Laser-Induced Dielectric Breakdown in Gas Mixtures Containing Carbon Monoxide. *J. Phys. Chem. A* **2008**, *112*, 7162–7169. [CrossRef]
23. Gornushkin, I.B.; Stevenson, C.L.; Galbacs, G.; Smith, B.W.; Winefordner, J.D. Measurement and Modeling of Ozone and Nitrogen Oxides Produced by Laser Breakdown in Oxygen–Nitrogen Atmospheres. *Appl. Spectr.* **2003**, *57*, 1442–1450. [CrossRef]
24. Available online: www.uv-vis-spectral-atlas-mainz.org (accessed on 22 November 2022).
25. Available online: https://cearun.grc.nasa.gov (accessed on 22 November 2022).
26. Griem, H.R. *Principles of Plasma Spectroscopy*; Cambridge University Press: Cambridge, UK, 1997; Volume 7.
27. Cristoferetti, G.; Tognoni, E.; Gizzi, L.A. Thermodynamic equilibrium states in laser-induced plasmas: From the general case to laser-induced breakdown spectroscopy plasmas. *Spectr. Act. B* **2013**, *90*, 1–22. [CrossRef]
28. Raizer, Y.P. *Gas Discharge Physics*; Springer: Berlin, Germany, 1991.
29. Available online: https://www.nist.gov/pml/atomic-spectra-database (accessed on 22 November 2022).
30. Petruci, J.F.; Tutuncu, E.; Cardoso, A.A.; Mizaikoff, B. Real-time and simultaneous monitoring of NO, NO2, and N2O using substrate–integrated hollow waveguides coupled to a compact fourier transform Infrared (FT-IR) Spectrometer. *Appl. Spectr.* **2018**, *73*, 1–6. [CrossRef]

Disclaimer/Publisher's Note: The statements, opinions and data contained in all publications are solely those of the individual author(s) and contributor(s) and not of MDPI and/or the editor(s). MDPI and/or the editor(s) disclaim responsibility for any injury to people or property resulting from any ideas, methods, instructions or products referred to in the content.

Article

Low-Temperature Plasma Diagnostics to Investigate the Process Window Shift in Plasma Etching of SiO$_2$

Youngseok Lee [1], Sijun Kim [1], Jangjae Lee [2], Chulhee Cho [1], Inho Seong [1] and Shinjae You [1,3,*]

[1] Department of Physics, Chungnam National University, Daejeon 34134, Korea
[2] Samsung Electronics, Hwaseong-si 18448, Korea
[3] Institute of Quantum Systems (IQS), Department of Physics, Chungnam National University, Daejeon 34134, Korea
* Correspondence: sjyou@cnu.ac.kr

Abstract: As low-temperature plasma plays an important role in semiconductor manufacturing, plasma diagnostics have been widely employed to understand changes in plasma according to external control parameters, which has led to the achievement of appropriate plasma conditions normally termed the process window. During plasma etching, shifts in the plasma conditions both within and outside the process window can be observed; in this work, we utilized various plasma diagnostic tools to investigate the causes of these shifts. Cutoff and emissive probes were used to measure the electron density and plasma potential as indicators of the ion density and energy, respectively, that represent the ion energy flux. Quadrupole mass spectrometry was also used to show real-time changes in plasma chemistry during the etching process, which were in good agreement with the etching trend monitored via in situ ellipsometry. The results show that an increase in the ion energy flux and a decrease in the fluorocarbon radical flux alongside an increase in the input power result in the breaking of the process window, findings that are supported by the reported SiO$_2$ etch model. By extending the SiO$_2$ etch model with rigorous diagnostic measurements (or numerous diagnostic methods), more intricate plasma processing conditions can be characterized, which will be beneficial in applications and industries where different input powers and gas flows can make notable differences to the results.

Keywords: low-temperature plasma; plasma diagnostics; plasma etching; plasma process modeling

Citation: Lee, Y.; Kim, S.; Lee, J.; Cho, C.; Seong, I.; You, S. Low-Temperature Plasma Diagnostics to Investigate the Process Window Shift in Plasma Etching of SiO$_2$. *Sensors* **2022**, *22*, 6029. https://doi.org/10.3390/s22166029

Academic Editors: Bruno Goncalves and Seunghun Hong

Received: 15 June 2022
Accepted: 10 August 2022
Published: 12 August 2022

Publisher's Note: MDPI stays neutral with regard to jurisdictional claims in published maps and institutional affiliations.

Copyright: © 2022 by the authors. Licensee MDPI, Basel, Switzerland. This article is an open access article distributed under the terms and conditions of the Creative Commons Attribution (CC BY) license (https://creativecommons.org/licenses/by/4.0/).

1. Introduction

Plasma is defined as a quasi-neutral gas of charged and neutral particles that exhibits collective behavior [1,2]. The characteristic features that make plasma distinct from other discharge phenomena are utilized in many industrial and research fields in terms of, for instance, controlling the dynamics of the component particles for individual applications [1]. With an enormous range of electron densities and temperatures, which are the most representative parameters, plasma has characteristic physical and chemical properties depending on the electron density and temperature regimes, resulting in a diverse categorization of plasma including material processing plasma and fusion plasma [2].

With the rapid growth of the semiconductor industry in the 20th century, material processing has grown into one of the biggest sub-fields of low-temperature plasma, designated as such by its electron temperature regime [3,4]. The essential processes in semiconductor manufacturing, such as etching [5–9], deposition [10–12], cleaning [13,14], etc., widely employ low-temperature plasma, allowing plasma to play a large role in the microelectronics industry [3,15–17]. Further development of electronic devices, however, requires more advanced plasma techniques to meet market demands, thereby increasing the processing complexity and difficulty. In this circumstance, plasma diagnostics can provide qualitative and quantitative information on plasma parameters for an understanding of the physical and chemical phenomena in the plasma processes, giving rise to the development of plasma technology [18–23].

Obtaining internal plasma parameters via plasma diagnostics can significantly help plasma processing engineers to establish the process window, which can be defined as the condition of the processing equipment or plasma itself that has to be met to realize the purpose of the process. For instance, SiO_2 etching with fluorocarbon (FC) plasma requires that plasma ions be sufficiently strong; otherwise, the FC plasma would form thick FC films on the SiO_2 instead of etching it [24,25]. Another example can be found in a special plasma process called atomic layer deposition (ALD), where a specific temperature window is necessary to realize atomic-scale deposition without defect-producing chemical reactions such as condensation or desorption at temperatures below or above the window, respectively [26]. Similarly, atomic layer etching (ALE), the counterpart of ALD, also has a characteristic process window with respect to the appropriate ion energy range that achieves atomic-scale removal without defect-producing physical reactions such as insufficient removal or sputtering at ion energies below or above the window, respectively [27,28].

There are numerous reports on the demonstration of plasma diagnostics via various methods to achieve the process window [21,23,29–34]. Compared to continuous plasma processes where a single plasma is maintained throughout the processing time, certain plasma processes such as ALE, where two or more kinds of plasma are alternated step by step, may especially benefit from plasma diagnostics. One previous report covers a comprehensive investigation into the discharge physics of ALE plasma, from several fundamental plasma parameters such as electron density and temperature to discharge instability and recovery periods during the ALE process [35].

In this work, a process window shift in SiO_2 etching with FC plasma from a variety of input power is investigated via plasma diagnostics, the tools of which are carefully considered for their appropriateness to the polymeric conditions of FC plasma. Based on a previous report [36] that a steady state of etching conditions is determined by the balance between FC film deposition and SiO_2–FC film removal rates, which are reflected by FC radical and ion energy fluxes, respectively, FC radical density is considered to be the parameter indicating the FC radical flux is this work. The SiO_2 etch model is expressed as follows [36]:

$$\frac{dx_{total}}{dt} = \frac{d(x_{FC} + x_{SiO_2})}{dt} = DR_{FC} - ER_{FC} - ER_{SiO_2}, \quad (1)$$

where x_{total}, x_{FC} and x_{SiO_2} stand for the thickness of the total (FC film + SiO_2), FC film, and SiO_2, while DR_{FC}, ER_{FC} and ER_{SiO_2} stand for the deposition rate of FC films, etch rate of FC films, and etch rates of SiO_2, respectively.

This parameter is diagnosed in real time through quadrupole mass spectrometry with a comparison to etching results also obtained in real time via in situ ellipsometry. Meanwhile, the ion energy flux is estimated from the electron density measured by a cutoff probe, and the plasma potential is found with an emissive probe. These multiple diagnostic results support the understanding of the process window shift occurring from power variation, providing a guideline for external parameter controls for improved processing results. Details of the SiO_2 etching process, as well as the experimental setup and methods for plasma diagnostics, are described below, followed by a discussion on the processing results based on the plasma diagnostics.

2. Experiment

2.1. Plasma Etching

In this work, SiO_2 (iNexus, Inc., Seongnam-si, Korea) etching is conducted not in a continuous manner but rather via ALE. Continuous SiO_2 etching with FC plasma, which normally employs capacitively coupled plasma sources to achieve high-energy ion bombardment for high etch rates, benefits from the synergetic effect of the simultaneous reactions of reactive FC radicals and high-energy bombarding ions on the material surface.

On the other hand, in ALE, the reactions of the radicals and ions are separated to obtain atomic-precision etch control [27,28,37]. A comparison between continuous etching and ALE is illustrated in Figure 1. In continuous SiO_2 etching, both etch gases (e.g., a mixture of Ar and C_4F_8) and radio-frequency (RF) power are employed simultaneously to maintain the processing plasma throughout the etch process.

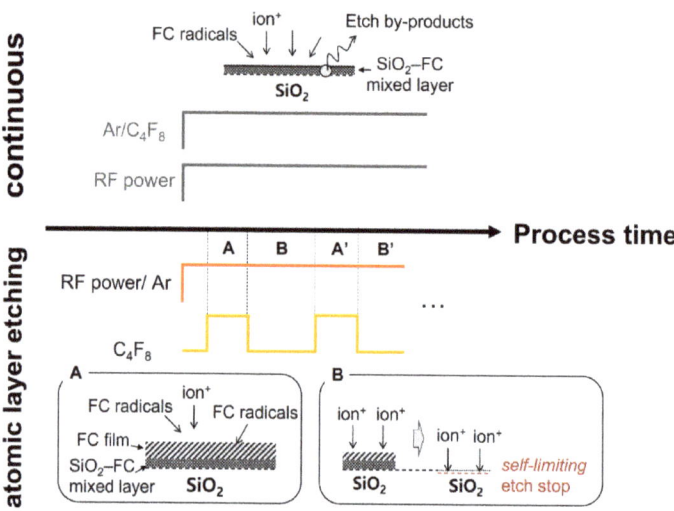

Figure 1. Comparison between continuous etching and ALE.

Alternatively, the ALE process is divided into two steps: Surface modification and removal, as labelled A and B in Figure 1, respectively. In the surface modification step (A), C_4F_8 is injected into a continuous Ar plasma and then dissociates into diverse FC radicals such as C_2F_4 and CF_2 [38,39], allowing an FC film to grow until the C_4F_8 injection is cut off. The following removal step (B) begins with the C_4F_8 cutoff as the Ar ion bombardment physically sputters the deposited FC film. This sputtering continues until both the FC film and SiO_2–FC mixed layer are totally removed and the underlying SiO_2 is exposed. This can only be achieved with well-controlled ion energies that are higher than the sputtering threshold energies of the FC films and the mixed layer but lower than that of SiO_2, a criterion typically called the ion energy window. Satisfying the ion energy window in the removal step of ALE leads to self-limiting etching, meaning that the etching spontaneously stops with the exposure of the new SiO_2 surface (as illustrated in Figure 1), which is the most fundamental characteristic of the ALE process.

An inductively coupled plasma source is employed in the present work, and the plasma chamber has a cylindrical geometry with a diameter of 330 mm and a height of 250 mm. The substrate on which the SiO_2 samples are processed is separated from the ceramic plate by 100 mm. An ellipsometer is mounted to the chamber that allows in situ monitoring of sample thicknesses during the entire etch process. Further details of this ALE chamber setup are described in our previous report [36].

The sequence of the etch process of the present work is illustrated in the lower panel of Figure 1. C_4F_8 is injected in a pulsed manner into continuous Ar plasma for surface modification, followed by the C_4F_8 cutoff that leads to a removal of the modified surface by the Ar ion bombardment. RF power with a frequency of 13.56 MHz is applied, and 44 sccm of Ar is injected into the chamber, resulting in a pressure of 1906 Pa. The flow rate of C_4F_8 is 2 sccm, which barely changes the chamber pressure.

Figure 2a,b plots the results of plasma etching based on the ALE recipe described above at an RF power of 100 and 300 W, respectively. The surface modification and removal steps are clearly separated in Figure 2a, where the thickness increases with the C_4F_8 injection and then decreases and saturates after the C_4F_8 cutoff. Since the FC film deposition during each modification step and the self-limiting etch trend during each removal step are well produced, the ALE condition of Figure 2a is considered to be in the ALE window.

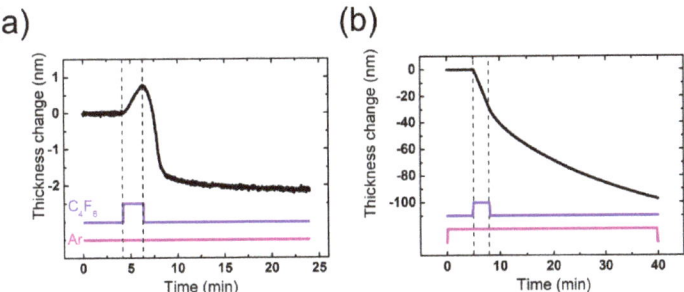

Figure 2. Thickness change in SiO_2 during one cycle of ALE at (**a**) 100 W and (**b**) 300 W.

On the other hand, increasing the RF power from 100 W to 300 W results in significantly different trends, as shown in Figure 2b. Here, C_4F_8 injection into the continuous Ar plasma actually leads to continuous SiO_2 etching with FC plasma rather than the FC film deposition (surface modification) of ALE. This reflects that an increase in RF power shifts the processing condition out of the ALE window, and thus plasma diagnostics should follow to determine the causes of this process window shift.

2.2. Plasma Diagnostics

It is important to determine appropriate diagnostic parameters to rigorously investigate plasma processing. In the present work, the target parameters are chosen based on the SiO_2 FC plasma etch model, explaining that a change in the material thickness is determined by the balance between its increasing and decreasing rates [36]; for an FC film, the increase rate corresponds to the FC film deposition rate and the decrease rate corresponds to the sum between its physical sputtering and chemical etch rates with SiO_2. Whereas, for SiO_2, its increase rate is assumed to be zero since there is no SiO_2 source during the etching, and the decrease rate is set to be the same as the chemical etch rate of the FC film. The dominant parameters related to the deposition and removal rates of the FC film are the FC radical flux and ion energy flux, respectively, which can be described with the specific internal plasma parameters of FC radical density, electron density, and plasma potential. Below, diagnostic data acquisition and processing methods, as well as the geometry of the diagnostic tools, are described in detail.

2.2.1. Electron Density Measurement

Since electron density is one of the most fundamental plasma parameters, there have been numerous studies on the development of electron density diagnostics. Langmuir proposed a historic plasma diagnostic tool, eponymously named the Langmuir probe, that provided not only fundamental parameters such as electron density and temperature but also electron energy distribution functions [40]. This probe has contributed enormously to a deeper understanding of plasma dynamics and characteristics, and to date, still plays a crucial role in plasma research areas including plasma physics [18,41]. However, Langmuir probe diagnostics significantly deteriorate in the harsh environments of plasma in material processing such as etching and deposition and the use of various processing gas mixtures; severe polymer deposition on the probe tip interrupts the probe operation, and the complexity of processing gas mixtures gives rise to considerable theoretical errors during data processing [2].

Microwave plasma diagnostics have emerged as an excellent alternative to the Langmuir probe, especially for processing plasma [42]. The most noticeable feature of microwave diagnostics is that they are barely perturbed by contamination from polymer deposition in processing plasma, as well as by RF noises to which most electrical diagnostic tools such as the Langmuir probe are vulnerable [22]. Among various types of microwave diagnostic tools developed over the years, the cutoff probe was chosen for the present work due to its simplicity in manufacturing and utilization [43–45].

The physics of the cutoff probe measurement is the cut-off phenomenon in plasma [2,46]. Plasma has a characteristic electron oscillation frequency, normally termed plasma frequency, which is expressed in terms of electron density as follows:

$$f_{\text{plasma}} = \frac{1}{2\pi}\sqrt{\frac{n_e e^2}{\epsilon_0 m_e}}, \qquad (2)$$

where n_e is the electron density, e is the elemental charge, ϵ_0 is the vacuum permittivity, and m_e is the electron mass. The higher the electron density, the higher the plasma frequency, enabling the plasma to act similarly to a metal. When an electromagnetic (EM) wave meets plasma, if the wave frequency is higher than the plasma frequency, then the wave will pass through the plasma as if it is a dielectric medium, but if the wave frequency is lower than the plasma frequency, it will not pass through since the plasma, in this case, acts as a conductive medium. This shows that finding the cutoff frequency of plasma will provide its electron density.

The geometry of the cutoff probe used in this work is illustrated in Figure 3a. Except for the vacuum components, a cutoff probe typically consists of two coaxial cables and a signal generator. The two cables play the roles of radiating and detecting antennas. In this work, the antennas are made by stripping one end of two sub-miniature version A (SMA) cables that have no connector by approximately 10 mm and separating them by approximately 3 mm to allow the plasma to fill the space between the antennas. A vector network analyzer (S33601B, SALUKI TECHNOLOGY, Taipei, Taiwan) is used as a microwave signal generator with frequencies ranging from hundreds of kHz to 8.5 GHz.

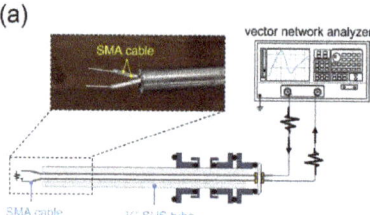

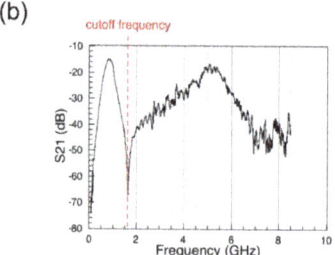

Figure 3. (a) Schematic of the cutoff probe measurement system and (b) characteristic S21 spectrum of the cutoff probe.

With such a cutoff probe measurement system, one can obtain the S21 spectrum that shows the cutoff frequency, the typical form of which is plotted in Figure 3b. The sharp increase and decrease in S21 before the cutoff frequency have been reported to originate from the resonance between the sheath capacitances and plasma inductance. The electron density can then be calculated from the obtained S21 spectrum using Equation (2) [47].

2.2.2. Plasma Potential Measurement

Plasma potential is also one of the most fundamental plasma parameters. Although the plasma potential can theoretically be measured with the Langmuir probe, its use is limited in processing plasma diagnostics due to probe tip contamination, as explained above. As an alternative, an emissive probe was chosen in the present work. The working principle of the emissive probe is similar to that of the Langmuir probe, but the critical difference between them is that the probe tip of the emissive probe emits thermionic electrons by Ohmic heating, as its name implies. Probe tip heating impedes polymer deposition, allowing the emissive probe to endure the processing plasma environment.

The physics of the emissive probe is briefly described as follows [2,48,49]. A W wire immersed in plasma without being electrically connected to its surroundings other than the plasma has a floating potential since the sheath between the W wire and plasma is effectively filled with positive charges that keep the flux of positive ions from the plasma to the W wire equal to that of electrons from the plasma to the wire. As illustrated in Figure 4a, a floating DC power supply connected to the W wire produces a conduction current that gives rise to Ohmic heating in the W wire, leading to thermionic electron emission. The positive and negative potential of the W wire is measured with digital multimeters (101, FLUKE, Washington, DC, USA) to estimate the probe potential at the center where the Ohmic heating is the strongest. Figure 4b shows an example of the emissive probe data of the change in the floating potential of the W wire as a function of the applied DC voltage, or heating voltage. When the heating voltage is significantly low, thermionic emission barely occurs, and the probe potential remains unchanged (see region (i) in Figure 4b). As the heating voltage increases, thermionic electrons start to affect the floating sheath potential, leading to an increase in the probe potential (regime (ii)). When the heating voltage is sufficiently high, the emission current becomes balanced with the plasma electron current and the probe potential levels off at the plasma potential (regime (iii)). In short, with a sufficiently high heating voltage, the plasma potential can be obtained by simply measuring the average of the positive and negative potential of the thermionic electron-emitting W wire.

2.2.3. FC Radical Density Qualitative Measurement

FC radical densities such as CF_2 and CF_3 are measured with a quadrupole mass spectrometer (QMS) (PSM, Hiden Analytic, Warrington, PA, USA). Intensively developed over decades, QMSs have been widely used for gas-phase species diagnostics in plasma [50]. A QMS, also known as a residual gas analyzer, measures the partial pressures of each gas-phase species in a vacuum [51]. It is normally equipped in the main chamber via an orifice with a diameter on the micrometer scale, which allows the QMS chamber to maintain a higher vacuum level than that of the main chamber so that particles transiting in the QMS arrive at the detector with no collision.

The typical components of a QMS are an ionizer, a mass filter, and a detector [50]. In QMSs, gas-phase neutral species should be ionized before entering the mass filter, which is a quadrupole with two pairs of electrodes biased with opposite RF and DC voltages. The applied RF and DC voltages determine which mass will pass through the quadrupole filter. The detector then reads the current of the filtered charged particles with a specific mass, the intensity of which implies the amount of the species with that specific mass in the main chamber. The operation parameters of the QMS in this work are as follows: 70 eV ionization energy, 100 µA emission current, and 1900 V detector multiplying voltage.

Since QMSs only offer the intensities of the detected signals in arbitrary units, plasma diagnostics with QMSs require thorough modeling to obtain quantitative densities of gas-phase species. Nevertheless, they are still powerful in process monitoring since qualitative changes in signals are sufficient to monitor changes in processing plasma with the variation of external parameters such as power or pressure. Another strength of QMS diagnostics is that they are able to monitor plasma processing in real time. Plasma diagnostics inserting probes into the plasma, such as the cutoff and emissive probes, have a limitation for the in situ monitoring of material processing due to the shadowing of the plasma on materials induced by the probe insertion. In situ monitoring plays a particularly essential role in processes where the plasma dynamically changes with time such as in ALE.

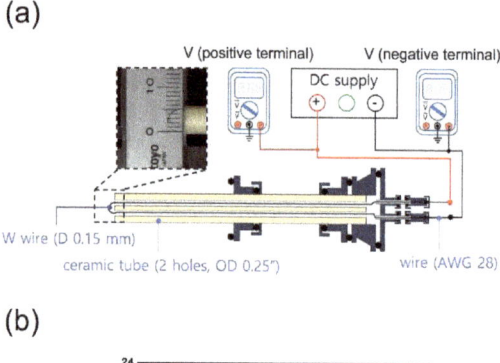

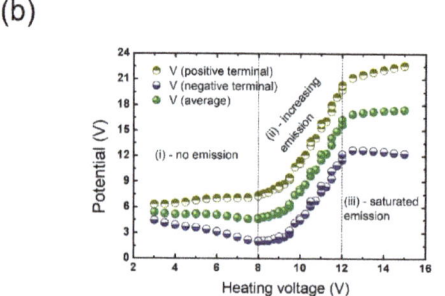

Figure 4. (**a**) Schematic of the emissive probe measurement system and (**b**) characteristic potential vs. heating voltage plot of the emissive probe.

3. Results and Discussion

Figure 5 plots the diagnostic results of the cutoff probe and emissive probe, showing changes in the electron density and plasma potential of Ar and Ar/C_4F_8 plasma with increasing RF power. Remembering that the ALE modification step changes from FC film deposition to continuous SiO_2 etching as the RF power increases from 100 W to 300 W in the modification step (see Figure 2), the diagnostic results here show that the electron density increases from 1.6×10^9 cm^{-3} to 3.71×10^{10} cm^{-3}, approximately 20 times, while the plasma potential decreases from 30 V to 10 V, by two-thirds, which implies that for the ion energy flux, the particle's number of ions significantly increases with slightly decreased energies. This trend of the changes in electron density and plasma potential is also found in Ar plasma in the removal step of the ALE process. The electron density increases from 1.48×10^{10} cm^{-3} to 1.492×10^{11} cm^{-3}, approximately 10 times, while the plasma potential decreases from 15 V to 11 V, by one-third. According to the SiO_2 etch model introduced above, an increase in ion energy flux leads to a higher decrease rate of material thickness [36]. It is thus possible that the different ALE results at the 100 W and 300 W levels of RF power stem from an increase in the ion energy flux in both Ar and Ar/C_4F_8 plasma, which is found via electron density and plasma potential, but only if the FC radical densities remain unchanged with the RF power increase. Therefore, FC radical diagnostic results should follow to evaluate the change in FC film deposition rates for a more rigorous model interpretation.

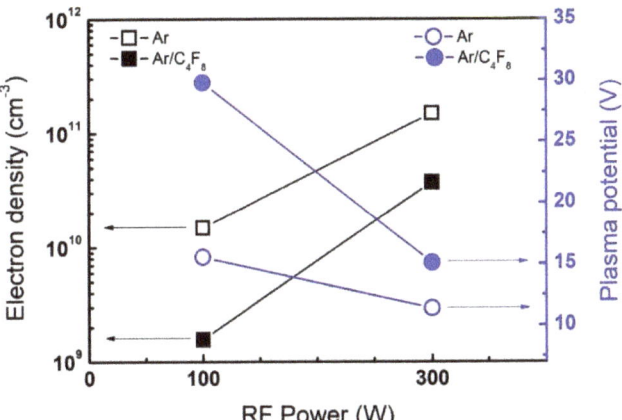

Figure 5. Changes in electron density and plasma potential with an increase in RF power from 100 W to 300 W in Ar and Ar/C_4F_8 plasma.

Figure 6 plots the results of FC radical density measurements using a QMS during one cycle of ALE at different RF powers of 100 W and 300 W. It is shown that during C_4F_8 injection, all of the polymeric radical species' densities, except F, decrease with the increase in RF power, likely indicating that the FC film deposition rate is lower at 300 W than 100 W. Here, the increase in ion energy flux with increasing RF power elucidates the processing regime transition from FC film deposition to continuous etching during the surface modification step.

After the C_4F_8 cutoff, the FC radical densities instantaneously decrease regardless of RF power, as shown in Figure 6a–d. However, it is noticeable that the FC radical densities at 300 W do not fully return to the level before C_4F_8 injection but stop at an intermediate level and then slowly decrease. This may be attributed to the FC radicals absorbed into the chamber wall in the modification step being physically sputtered by ions bombarding the wall. Note that even at 300 W RF power, it is possible for FC films to form on the wall since there is no chemical etching between SiO_2 and FC films, leading to a lower thickness decrease rate in the SiO_2 etch model, unlike in the SiO_2 surface where continuous etching occurs instead of FC film deposition in the modification step. Although the FC films on the wall that act as an FC radical source in the removal step also exist under the RF 100 W condition, the ion energy flux at 100 W may not be sufficient to induce an observable generation of FC radicals from the wall compared to that of the gas phase in the plasma (see Figure 5), leading to the full decrease in the FC radical densities after the C_4F_8 cutoff (note that the ion flux at 300 W is almost 10 times higher than that at 100 W).

We stress that the FC radical supply from the wall is considered to result in a significant difference in the thickness trends between 100 W and 300 W in the removal step shown in Figure 2; the thickness eventually saturates at 100 W, while it continuously decreases at 300 W. An increase in ion energy flux with increasing RF power leads the ion sputtering of FC films on the wall to no longer be negligible, driving it to act as an undesirable etchant source. Meanwhile, the effects of ion sputtering of the FC film deposited on the wall at 100 W give rise to infinitesimal drifts in the etched amount per cycle (EPC) as the ALE process proceeds (see Figure 2a). The increase in the EPC cycle by cycle may be led by an increase in the undesirable FC radical flux from the wall that accumulates cycle by cycle. This result implies that chamber wall conditioning needs to be carefully considered at certain processing conditions; more rigorous investigations into this issue will be conducted in future work.

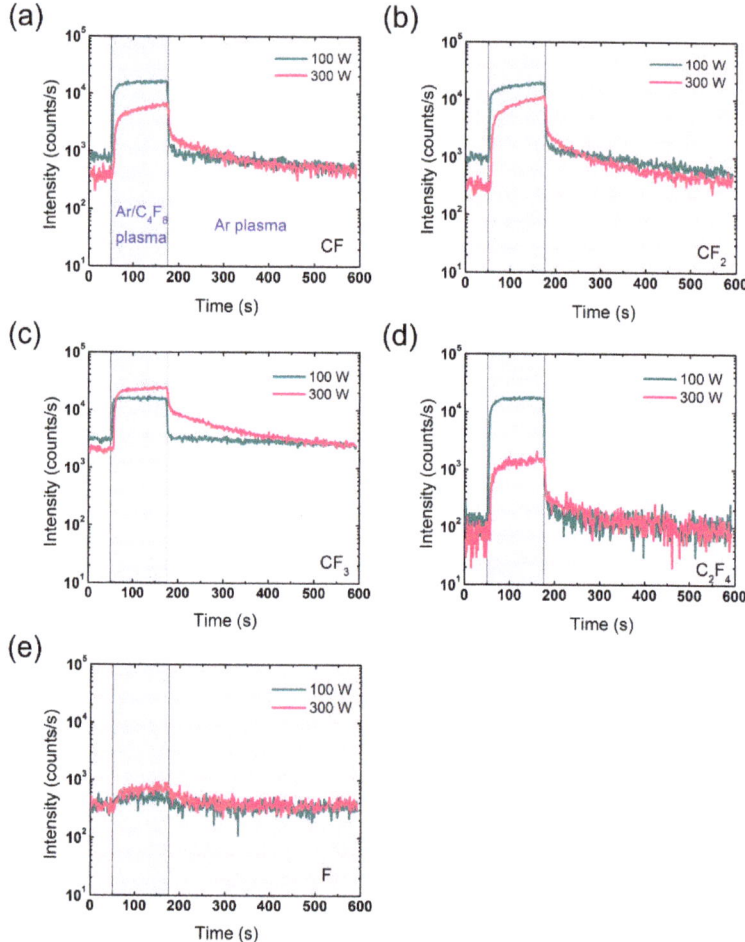

Figure 6. FC radical densities measured using a QMS during one cycle of ALE at 100 W and 300 W of (**a**) CF, (**b**) CF_2, (**c**) CF_3, (**d**) C_2F_4, and (**e**) F.

Synthesizing these multiple plasma diagnostic results, we summarize that the process window shift with increasing RF power is caused by an increase in the electron density, a decrease in plasma potential, and a decrease in polymeric radical densities during the ALE process. The consistency between the changes in the process trend and the plasma parameters is in good agreement with the previously reported SiO_2 etch model [36], and additional analysis considering the FC radical induced by ion sputtering of the deposited FC films on the wall well explains the continuous etching after the C_4F_8 cutoff at 300 W. Accordingly, the multiple plasma diagnostic methods in the present work are expected to be beneficial to establishing finely tuned ALE windows in the future.

4. Conclusions

As plasma processing has become widely employed in material processing, plasma diagnostic techniques play a bigger role in understanding and manipulating processing plasma for better outcomes. In the present work, the process window shift, where an increase in RF power pushes the processing condition out of the window, was investigated via multiple plasma diagnostic methods. Based on the previously reported SiO_2 etch model, target species for the diagnostics of electron density, plasma potential, and FC radical densities were chosen. The obtained diagnostic results were able to sufficiently explain the process window shift, and in addition, were in good agreement with the etch model prediction.

It is worth mentioning that the utilization of multiple diagnostic tools to monitor the same plasma makes it easier to interpret the results of plasma processes, as shown in this work. Ultimately, for some complex plasma processes, such as ALE where plasma dynamically changes during the process, in situ plasma diagnostic methods are expected to offer more informative diagnostic results, allowing more precise and appropriate process controls.

Author Contributions: Conceptualization, Y.L.; validation, Y.L. and C.C.; formal analysis, Y.L. and S.K.; investigation, Y.L.; data curation, Y.L., S.K., J.L., C.C. and I.S.; writing—original draft preparation, Y.L.; writing—review and editing, S.K., C.C. and I.S.; visualization, Y.L. and S.K.; supervision, S.Y. All authors have read and agreed to the published version of the manuscript.

Funding: This research was supported by National Research Council of Science & Technology (NST) grants by the Korean government (MSIP) (No. CAP-17-02-NFRI, CRF-20-01-NFRI); the Next-generation Intelligence semiconductor R&D Program through the Korea Evaluation Institute of Industrial Technology (KEIT) funded by the Korean government (MOTIE); the Korea Institute of Energy Technology Evaluation and Planning (KETEP) and the MOTIE of the Republic of Korea (20202010100020); the MOTIE (20009818, 20010420) and KSRC (Korea Semiconductor Research Consortium) support program for the development of future semiconductor devices; a Korea Institute for Advancement of Technology (KIAT) grant funded by the Korean government (MOTIE) (P0008458, HRD Program for Industrial Innovation); a Basic Science Research Program through the National Research Foundation of Korea (NRF) funded by the Ministry of Education (NRF-2020R1A6A1A03047771); and the KIMM Institutional Program (NK236F) and NST/KIMM.

Institutional Review Board Statement: Not applicable.

Informed Consent Statement: Not applicable.

Data Availability Statement: The data presented in this study are available on request from the corresponding author.

Conflicts of Interest: The authors declare no conflict of interest.

References

1. Chen, F.F. *Introduction to Plasma Physics and Controlled Fusion*, 3rd ed.; Springer: Cham, Switzerland, 2018; ISBN 978-3-319-22308-7.
2. Lieberman, M.A. *Principles of Plasma Discharges and Materials Processing*, 2nd ed.; John Wiley & Sons: Hoboken, NJ, USA, 2015; ISBN 9786468600.
3. Samukawa, S.; Hori, M.; Rauf, S.; Tachibana, K.; Bruggeman, P.; Kroesen, G.; Whitehead, J.C.; Murphy, A.B.; Gutsol, A.F.; Starikovskaia, S.; et al. The 2012 Plasma Roadmap. *J. Phys. D Appl. Phys.* **2012**, *45*, 253001. [CrossRef]
4. Chen, F.F. Industrial Applications of Low-Temperature Plasma Physics. *Phys. Plasmas* **1995**, *2*, 2164–2175. [CrossRef]
5. Donnelly, V.M.; Kornblit, A. Plasma Etching: Yesterday, Today, and Tomorrow. *Vac. Sci. Technol. A* **2013**, *31*, 050825. [CrossRef]
6. Aachboun, S.; Ranson, P.; Hilbert, C.; Boufnichel, M. Cryogenic Etching of Deep Narrow Trenches in Silicon. *J. Vac. Sci. Technol. A Vac. Surf. Film.* **2000**, *18*, 1848–1852. [CrossRef]
7. Abe, H.; Yoneda, M.; Fujiwara, N. Developments of Plasma Etching Technology for Fabricating Semiconductor Devices. *Jpn. J. Appl. Phys.* **2008**, *47*, 1435–1455. [CrossRef]
8. Seong, I.H.; Lee, J.J.; Cho, C.H.; Lee, Y.S.; Kim, S.J.; You, S.J. Characterization of SiO_2 over Poly-Si Mask Etching in Ar/C_4F_8 Capacitively Coupled Plasma. *Appl. Sci. Converg. Technol.* **2021**, *30*, 176–182. [CrossRef]
9. Yoo, S.W.; Cho, C.; Kim, K.; Lee, H.; You, S. Characteristics of SiO_2 Etching by Capacitively Coupled Plasma with Different Fluorocarbon Liquids (C_7F_{14}, C_7F_8) and Fluorocarbon Gas (C_4F_8). *Appl. Sci. Converg. Technol.* **2021**, *30*, 102–106. [CrossRef]
10. Martinu, L.; Poitras, D. Plasma Deposition of Optical Films and Coatings: A Review. *J. Vac. Sci. Technol. A Vac. Surf. Film.* **2000**, *18*, 2619–2645. [CrossRef]
11. Randhawa, H. Review of Plasma-Assisted Deposition Processes. *Thin Solid Film.* **1991**, *196*, 329–349. [CrossRef]
12. Yoo, S.W.; You, S.J.; Kim, J.H.; Seong, D.J.; Seo, B.H.; Hwang, N.M. Effect of Substrate Bias on Deposition Behaviour of Charged Silicon Nanoparticles in ICP-CVD Process. *J. Phys. D Appl. Phys.* **2016**, *50*, 35201. [CrossRef]
13. Isabell, T.C.; Fischione, P.E.; O'Keefe, C.; Guruz, M.U.; Dravid, V.P. Plasma Cleaning and Its Applications for Electron Microscopy. *Microsc. Microanal.* **1999**, *5*, 126–135. [CrossRef]

14. Petasch, W.; Kegel, B.; Schmid, H.; Lendenmann, K.; Keller, H.U. Low-Pressure Plasma Cleaning: A Process for Precision Cleaning Applications. *Surf. Coat. Technol.* **1997**, *97*, 176–181. [CrossRef]
15. Kim, S.J.; Lee, J.J.; Lee, Y.S.; Kim, D.W.; You, S.J. Finding the Optimum Design of the Planar Cutoff Probe through a Computational Study. *AIP Adv.* **2021**, *11*, 025241. [CrossRef]
16. Adamovich, I.; Baalrud, S.D.; Bogaerts, A.; Bruggeman, P.J.; Cappelli, M.; Colombo, V.; Czarnetzki, U.; Ebert, U.; Eden, J.G.; Favia, P.; et al. The 2017 Plasma Roadmap: Low Temperature Plasma Science and Technology. *J. Phys. D. Appl. Phys.* **2017**, *50*, 323001. [CrossRef]
17. Oh, T.; Cho, C.; Ahn, W.; Yook, J.; Lee, J.; You, S.; Yim, J.; Ha, J.; Bae, G.; You, H.; et al. Enhanced RCS Reduction Effect. *Sensors* **2021**, *21*, 8486. [CrossRef]
18. Godyak, V.A.; Piejak, R.B.; Alexandrovich, B.M. Electron Energy Distribution Function Measurements and Plasma Parameters in Inductively Coupled Argon Plasma. *Plasma Sources Sci. Technol.* **2002**, *11*, 525–543. [CrossRef]
19. Godyak, V.A.; Piejak, R.B.; Alexandrovich, B.M. Plasma Sources Science and Technology Measurement of Electron Energy Distribution in Low-Pressure RF Discharges Measurements of Electron Energy Distribution in Low-Pressure R F Discharges. *Plasma Sources Sci. Technol.* **1992**, *18*, 36–58. [CrossRef]
20. Kortshagen, U.; Pukropski, I.; Zethoff, M. Spatial Variation of the Electron Distribution Function in a Rf Inductively Coupled Plasma: Experimental and Theoretical Study. *J. Appl. Phys.* **1994**, *76*, 2048–2058. [CrossRef]
21. Cherrington, B.E. The Use of Electrostatic Probes for Plasma Diagnostics-A Review. *Plasma Chem. Plasma Process.* **1982**, *2*, 113–140. [CrossRef]
22. Kim, S.J.; Lee, J.J.; Lee, Y.S.; Cho, C.H.; You, S.J. Crossing Frequency Method Applicable to Intermediate Pressure Plasma Diagnostics Using the Cutoff Probe. *Sensors* **2022**, *22*, 1291. [CrossRef]
23. Welzel, S.; Hempel, F.; Hübner, M.; Lang, N.; Davies, P.B.; Röpcke, J. Quantum Cascade Laser Absorption Spectroscopy as a Plasma Diagnostic Tool: An Overview. *Sensors* **2010**, *10*, 6861–6900. [CrossRef]
24. Rueger, N.R.; Beulens, J.J.; Schaepkens, M.; Doemling, M.F.; Mirza, J.M.; Standaert, T.E.F.M.; Oehrlein, G.S. Role of Steady State Fluorocarbon Films in the Etching of Silicon Dioxide Using CHF3 in an Inductively Coupled Plasma Reactor. *J. Vac. Sci. Technol. A Vac. Surf. Film.* **1997**, *15*, 1881–1889. [CrossRef]
25. Chang, W.S.; Yook, Y.G.; You, H.S.; Park, J.H.; Kwon, D.C.; Song, M.Y.; Yoon, J.S.; Kim, D.W.; You, S.J.; Yu, D.H.; et al. A Unified Semi-Global Surface Reaction Model of Polymer Deposition and SiO_2 Etching in Fluorocarbon Plasma. *Appl. Surf. Sci.* **2020**, *515*, 145975. [CrossRef]
26. George, S.M. Atomic Layer Deposition: An Overview. *Chem. Rev.* **2010**, *110*, 111–131. [CrossRef]
27. Kanarik, K.J.; Lill, T.; Hudson, E.A.; Sriraman, S.; Tan, S.; Marks, J.; Vahedi, V.; Gottscho, R.A. Overview of Atomic Layer Etching in the Semiconductor Industry. *J. Vac. Sci. Technol. A Vac. Surf. Film.* **2015**, *33*, 020802. [CrossRef]
28. Oehrlein, G.S.; Metzler, D.; Li, C. Atomic Layer Etching at the Tipping Point: An Overview. *ECS J. Solid State Sci. Technol.* **2015**, *4*, N5041–N5053. [CrossRef]
29. Cho, C.; You, K.; Kim, S.; Lee, Y.; Lee, J.; You, S. Characterization of SiO_2 Etching Profiles in Pulse-Modulated Capacitively Coupled Plasmas. *Materials* **2021**, *14*, 5036. [CrossRef]
30. Seo, B.H.; Kim, J.H.; You, S.J.; Seong, D.J. Laser Scattering Diagnostics of an Argon Atmospheric-Pressure Plasma Jet in Contact with Vaporized Water. *Phys. Plasmas* **2015**, *22*, 123502. [CrossRef]
31. Chun, I.; Efremov, A.; Yeom, G.Y.; Kwon, K.H. A Comparative Study of $CF_4/O_2/Ar$ and $C_4F_8/O_2/Ar$ Plasmas for Dry Etching Applications. *Thin Solid Film.* **2015**, *579*, 136–143. [CrossRef]
32. Boris, D.R.; Fernsler, R.F.; Walton, S.G. The Spatial Profile of Density in Electron Beam Generated Plasmas. *Surf. Coat. Technol.* **2014**, *241*, 13–18. [CrossRef]
33. Lee, J.; Efremov, A.; Kwon, K.H. On the Relationships between Plasma Chemistry, Etching Kinetics and Etching Residues in $CF_4+C_4F_8+Ar$ and $CF_4+CH_2F_2+Ar$ Plasmas with Various CF_4/C_4F_8 and CF_4/CH_2F_2 Mixing Ratios. *Vacuum* **2018**, *148*, 214–223. [CrossRef]
34. Gaboriau, F.; Cartry, G.; Peignon, M.C.; Cardinaud, C. Etching Mechanisms of Si and SiO_2 in Fluorocarbon ICP Plasmas: Analysis of the Plasma by Mass Spectrometry, Langmuir Probe and Optical Emission Spectroscopy. *J. Phys. D Appl. Phys.* **2006**, *39*, 1830–1845. [CrossRef]
35. Yoon, M.Y.; Yeom, H.J.; Kim, J.H.; Chegal, W.; Cho, Y.J.; Kwon, D.C.; Jeong, J.R.; Lee, H.C. Discharge Physics and Atomic Layer Etching in Ar/C_4F_6 Inductively Coupled Plasmas with a Radio Frequency Bias. *Phys. Plasmas* **2021**, *28*, 063504. [CrossRef]
36. Lee, Y.; Seong, I.; Lee, J.; Lee, S.; Cho, C.; Kim, S.; You, S. Various Evolution Trends of Sample Thickness in Fluorocarbon Film Deposition on SiO_2. *J. Vac. Sci. Technol. A* **2022**, *40*, 013001. [CrossRef]
37. Faraz, T.; Roozeboom, F.; Knoops, H.C.M.; Kessels, W.M.M. Atomic Layer Etching: What Can We Learn from Atomic Layer Deposition? *ECS J. Solid State Sci. Technol.* **2015**, *4*, N5023–N5032. [CrossRef]
38. Li, X.; Ling, L.; Hua, X.; Oehrlein, G.S.; Wang, Y.; Vasenkov, A.V.; Kushner, M.J. Properties of C_4F_8 Inductively Coupled Plasmas. I. Studies of Ar/c-C_4F_8 Magnetically Confined Plasmas for Etching of SiO_2. *J. Vac. Sci. Technol. A Vac. Surf. Film.* **2004**, *22*, 500. [CrossRef]
39. Vasenkov, A.V.; Li, X.; Oehrlein, G.S.; Kushner, M.J. Properties of C-C_4F_8 Inductively Coupled Plasmas. II. Plasma Chemistry and Reaction Mechanism for Modeling of Ar/c-C_4F_8/O_2 Discharges. *J. Vac. Sci. Technol. A Vac. Surf. Film.* **2004**, *22*, 511. [CrossRef]
40. Conde, L. *An Introduction to Langmuir Probe Diagnostics of Plasmas*; Universidad Politécnica de Madrid: Madrid, Spain, 2011; pp. 1–28.

41. Godyak, V.A. Electron Energy Distribution Function Control in Gas Discharge Plasmas. *Phys. Plasmas* **2013**, *20*, 101611. [CrossRef]
42. Lebedev, Y.A. Microwave Discharges: Generation and Diagnostics. *J. Phys. Conf. Ser.* **2010**, *257*, 012016. [CrossRef]
43. Kim, S.J.; Lee, J.J.; Lee, Y.S.; Kim, D.W.; You, S.J. Effect of an Inhomogeneous Electron Density Profile on the Transmission Microwave Frequency Spectrum of the Cutoff Probe. *Plasma Sources Sci. Technol.* **2020**, *29*, 125014. [CrossRef]
44. Kim, S.J.; Lee, J.J.; Kim, D.W.; Kim, J.H.; You, S.J. A Transmission Line Model of the Cutoff Probe. *Plasma Sources Sci. Technol.* **2019**, *28*, 055014. [CrossRef]
45. You, K.H.; You, S.J.; Na, B.K.; Kim, D.W.; Kim, J.H.; Seong, D.J.; Chang, H.Y. Cutoff Probe Measurement in a Magnetized Plasma. *Phys. Plasmas* **2018**, *25*, 013518. [CrossRef]
46. Kim, J.H.; Choi, S.C.; Shin, Y.H.; Chung, K.H. Wave Cutoff Method to Measure Absolute Electron Density in Cold Plasma. *Rev. Sci. Instrum.* **2004**, *75*, 2706–2710. [CrossRef]
47. Kim, D.W.; You, S.J.; Na, B.K.; Kim, J.H.; Chang, H.Y. An Analysis on Transmission Microwave Frequency Spectrum of Cut-off Probe. *Appl. Phys. Lett.* **2011**, *99*, 18–21. [CrossRef]
48. Sheehan, J.P.; Raitses, Y.; Hershkowitz, N.; Kaganovich, I.; Fisch, N.J. A Comparison of Emissive Probe Techniques for Electric Potential Measurements in a Complex Plasma. *Phys. Plasmas* **2011**, *18*, 073501. [CrossRef]
49. Sheehan, J.P.; Hershkowitz, N. Emissive Probes. *Plasma Sources Sci. Technol.* **2011**, *20*, 063001. [CrossRef]
50. Singh, H.; Coburn, J.W.; Graves, D.B. Appearance Potential Mass Spectrometry: Discrimination of Dissociative Ionization Products. *J. Vac. Sci. Technol. A Vac. Surf. Film.* **2000**, *18*, 299–305. [CrossRef]
51. Lee, Y.S.; Oh, S.H.; Lee, J.J.; Cho, C.H.; Kim, S.J.; You, S.J. A Quantification Method in Quadrupole Mass Spectrometer Measurement. *Appl. Sci. Converg. Technol.* **2021**, *30*, 50–53. [CrossRef]

Article

Uncertainties in Atomic Data for Modeling Astrophysical Charge Exchange Plasmas

Liyi Gu [1,2,*], Chintan Shah [3,4,5] and Ruitian Zhang [6,7]

1. SRON Netherlands Institute for Space Research, Niels Bohrweg 4, 2333 CA Leiden, The Netherlands
2. RIKEN High Energy Astrophysics Laboratory, 2-1 Hirosawa, Wako 351-0198, Saitama, Japan
3. NASA Goddard Space Flight Center, 8800 Greenbelt Rd., Greenbelt, MD 20771, USA; chintan@mpi-hd.mpg.de
4. Max-Planck-Institut für Kernphysik, Saupfercheckweg 1, D-69117 Heidelberg, Germany
5. Lawrence Livermore National Laboratory, 7000 East Avenue, Livermore, CA 94550, USA
6. Institute of Modern Physics, Chinese Academy of Sciences, Lanzhou 730000, China; zhangrt@impcas.ac.cn
7. University of Chinese Academy of Sciences, Beijing 100049, China
* Correspondence: l.gu@sron.nl

Abstract: Relevant uncertainties of theoretical atomic data are vital to determining the accuracy of plasma diagnostics in a number of areas, including, in particular, the astrophysical study. We present a new calculation of the uncertainties on the present theoretical ion-impact charge exchange atomic data and X-ray spectra, based on a set of comparisons with the existing laboratory data obtained in historical merged-beam, cold-target recoil-ion momentum spectroscopy, and electron beam ion traps experiments. The average systematic uncertainties are found to be 35–88% on the total cross sections, and 57–75% on the characteristic line ratios. The model deviation increases as the collision energy decreases. The errors on total cross sections further induce a significant uncertainty to the calculation of ionization balance for low-temperature collisional plasmas. Substantial improvements of the atomic database and dedicated laboratory measurements are needed to obtain the current models, ready for the X-ray spectra from the next X-ray spectroscopic mission.

Keywords: charge exchange; X-ray astrophysics; atomic data; plasma diagnostics

Citation: Gu, L.; Shah, C.; Zhang, R. Uncertainties in Atomic Data for Modeling Astrophysical Charge Exchange Plasmas. *Sensors* **2022**, *22*, 752. https://doi.org/10.3390/s22030752

Academic Editor: Bruno Miguel Soares Gonçalves

Received: 10 December 2021
Accepted: 14 January 2022
Published: 19 January 2022

Publisher's Note: MDPI stays neutral with regard to jurisdictional claims in published maps and institutional affiliations.

Copyright: © 2022 by the authors. Licensee MDPI, Basel, Switzerland. This article is an open access article distributed under the terms and conditions of the Creative Commons Attribution (CC BY) license (https://creativecommons.org/licenses/by/4.0/).

1. Introduction

Charge exchange plasma can be found in a broad range of astrophysical environments, including, in particular, the interfaces where the solar wind ions interact with neutrals in comets and planetary atmospheres [1–4], but potentially also in supernova remnants [5,6], star-forming galaxies [7,8], active galactic nuclei [9], and clusters of galaxies [10,11]. The modeling of the X-ray spectrum of charge exchange has become possible recently thanks to the efforts of Smith et al. [12] and Gu et al. [13]. These models are crucial to interpreting the observations, as well as to understanding the physical sources that power the plasma.

There is an increasing demand from the astronomical community that the plasma model should provide an estimate of the systematic uncertainties for the atomic data used. This is triggered by the accumulating evidence that the uncertainties from the atomic data, which are not accounted for at present, are as significant as the typical errors from instrumental calibration (see [14] for a recent example). So far, there is no systematic estimate of the uncertainties of the existing charge exchange models, making it difficult to assess the accuracy of the scientific results obtained with these models.

Most of the charge exchange reaction rates in existing models are obtained in theoretical calculations, with only a few laboratory benchmarks performed by several groups with various experimental methods (see, e.g., cross-beam/merged-beam neutral setups: [15–17]; tokamak and laser-produced plasmas: [18–20]; cold-target recoil-ion momentum spectroscopy (COLTRIM): [21–25]; electron beam ion trap (EBIT): [26–32]). A recent comparison using the data from the EBIT measurements [33] showed that the model and the laboratory spectra differ significantly in both line energies and strengths, for the L-shell charge exchange between nickel ions and neutral particles. Another recent example is that the COLTRIMS measurement by Xu et al. [34] showed that the model calculations might differ from the measurements by 20–50% for the state-selective cross sections of Ne^{8+} and Ne^{9+} charge exchange. In this work, we compile a sample of existing laboratory measurements on charge exchange total cross sections, and state-selective cross sections, as well as characteristic X-ray line ratios, and put forward a systematic assessment of the model accuracy.

This paper is arranged as follows. In Section 2, we describe the sample and the results of the benchmark, and in Section 3, we discuss the potential improvement with future EBIT and COLTRIMS measurements. The benchmark is directly applied to the charge exchange model and atomic data [13] in the SPEX [35] software. Throughout the paper, the errors are given at a 68% confidence level.

2. Methods and Results
2.1. Total Cross Sections

First, we compare the SPEX calculations with existing laboratory results for a number of ions on their total cross sections for atomic hydrogen targets. The SPEX atomic data do not constitute one uniform set of theoretical calculations, but a mixture of three different types of approaches: (1) the rates derived with the empirical scaling reported in Gu et al. [13] (G16 hereafter), which was based on a numerical approximation to a collection of historical theoretical and experimental rates; (2) the multi-channel Landau-Zener method (hereafter MCLZ) reported in Mullen et al. [36]. The atomic data generated by MCLZ are also publicly available in the Kronos database (https://www.physast.uga.edu/research/stancil-group/atomic-molecular-databases/kronos, accessed on 1 December 2021); and (3) the recommended values (hereafter RCMD), based on dedicated calculations, including, in most cases, the quantum-mechanical and classical molecular-orbital close-coupling methods, and the atomic-orbital close-coupling method. The G16 approach can calculate, for any ions with a given atomic number and charge, the MCLZ data covering most of the H- and He- like ions with atomic number up to 30, and the RCMD rates are available for a small set of key ions, e.g., O VII [37], N VII [38], and C VI [39].

All the three datasets are tested when the corresponding theoretical cross sections (σ_{theo}) and experimental cross sections (σ_{exp}, see Table 1) are available. Examples are shown in Figure 1 for the C VI and O VII data. For C VI, the three calculations converge at the energy range from $\sim$100 eV/amu to 4×10^4 eV/amu, while the MCLZ data do not cover higher energies, and the G16 and RCMD data miss the low energy part. For O VII, the difference between the three calculations becomes more significant than in the case of C VI. The cross section derivatives shown in Figure 1 indicate that the differences in the shapes of the three theoretical calculations become, in general, larger at lower collision energies.

In Figure 2, we plot the distributions of the absolute errors $\sigma_{\text{exp}} - \sigma_{\text{theo}}$ of the theoretical models. The standard deviations of the absolute errors are 1.4×10^{-15} cm^{-2}, 2.0×10^{-15} cm^{-2}, and 1.3×10^{-15} cm^{-2}, for the G16, MCLZ, and RCMD calculations, respectively. As shown in Figure 2b, the absolute errors of G16 become more scattered, and on average larger, at lower collision velocities. The standard deviations of the error distributions are 1.8×10^{-15} cm^{-2} for $v < 600$ km s^{-1}, and 0.9×10^{-15} cm^{-2} for $v \geq 600$ km s^{-1}.

We also summarize the relative deviations ($\sigma_{\text{exp}} - \sigma_{\text{theo}}$) / σ_{theo} of the three calculations in Figure 2. The average absolute values of the fractional deviations are 55%, 88%, and 35% for the G16, MCLZ, and RCMD datasets. Similar to G16, the MCLZ calculation also has larger relative errors for low-velocity collisions, while the RCMD calculation shows fairly constant deviations for the velocity range considered. For high-energy collisions of $v > 3000$ km s^{-1}, the three methods show reasonable agreement with the laboratory results within uncertainties <50%.

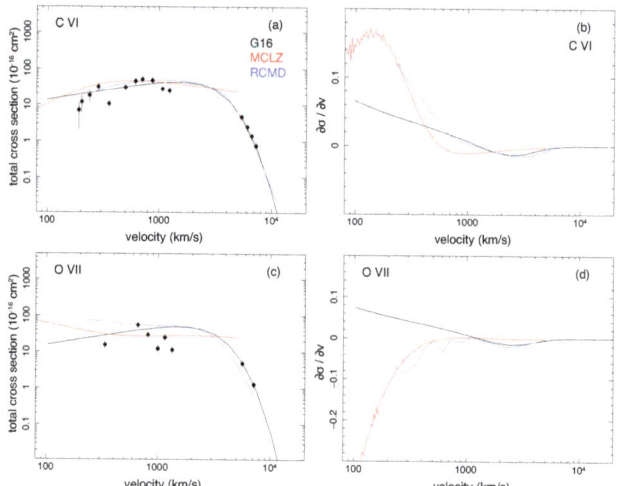

Figure 1. Total cross sections as a function of collision velocity and the cross section derivatives with respect to the velocity for C^{6+} (**a**,**b**) and O^{7+} (**c**,**d**) ions interacting with hydrogen atoms, resulting in C VI and O VIII ions. The data points are experimental results from Goffe et al. [40], Phaneuf et al. [41], Panov et al. [42], and Meyer et al. [43]. Approximate errors of 15% [43] are shown, except for the low energy (<500 km s^{-1}) data of C VI, for which the actual errors were reported in the original paper. The solid lines are the model values from the calculations with the G16 (black), MCLZ (red), and RCMD (blue) methods. The abbreviations are explained in the text.

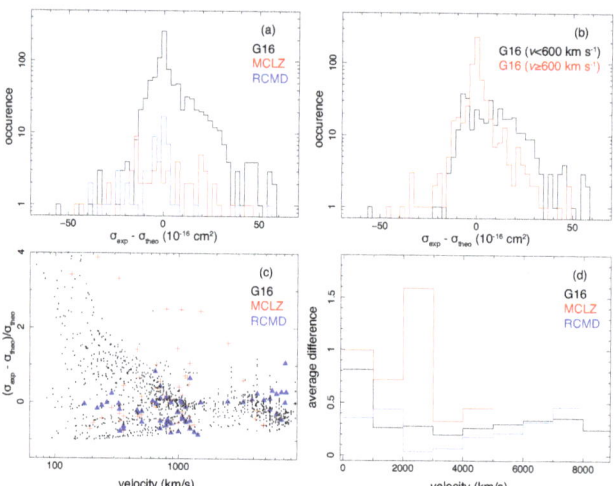

Figure 2. Distributions of the absolute (**upper**) and relative (**lower**) deviations of the theoretical cross sections from the experimental results obtained with the measurements summarized in Table 1. (**a**) Diagrams of the absolute errors for the G16 (black), MCLZ (red), and RCMD (blue) theories. (**b**) Diagrams of the absolute errors for G16 for low-collision velocities (black) and high velocities (red). (**c**) The relative deviations for the G16 (black points), MCLZ (red crosses), and RCMD (blue triangles) calculations. (**d**) The average deviations in absolute values for the three methods in each velocity interval.

Table 1. Experimental cross section data.

Reference	Type [a]	Ion	Theory Data
Shah et al. [44]	total	Li^{q+} (q = 1–3)	G16
Seim et al. [45]	total	Li^{q+} (q = 2–3), N^{q+} (q = 2–5), Ne^{q+} (q = 3–5)	G16
Goffe et al. [40]	total	B^{q+} (q = 1–5), C^{q+} (q = 1–4)	G16
Goffe et al. [40]	total	C^{q+} (q = 5, 6), N^{7+}	G16, MCLZ, RCMD
McCullough et al. [46]	total	B^{2+}, C^+, N^+, Mg^{2+}	G16
Crandall et al. [47]	total	B^{q+} (q = 2–5), C^{q+} (q = 3, 4), N^{q+} (q = 3, 4), O^{q+} (q = 5, 6)	G16
Gardner et al. [48]	total	B^{q+} (q = 2–4), C^{q+} (q = 2–4), N^{q+} (q = 2–5), O^{q+} (q = 2–5)	G16
Phaneuf et al. [49]	total	C^{q+} (q = 1–4), N^{q+} (q = 1–5), O^{q+} (q = 1–5), Si^{q+} (q = 2–7)	G16
Nutt et al. [50]	total	C^{2+}	G16
Phaneuf et al. [41]	total	C^{q+} (q = 3, 4), O^{q+} (q = 2–6)	G16
Phaneuf et al. [41]	total	C^{q+} (q = 5, 6)	G16, MCLZ, RCMD
Sant'Anna et al. [51]	total	C^{3+}	G16
Ciric et al. [52]	total,nl	C^{q+} (q = 3, 4), N^{5+}, O^{6+}	G16
McCullough et al. [53]	total,nl	C^{3+}	G16
Panov et al. [42]	total	C^{4+}, N^{5+}, O^{6+}, Ne^{8+}	G16
Panov et al. [42]	total	C^{q+} (q = 5, 6), N^{q+} (q = 6, 7), O^{q+} (q = 7, 8), Ne^{q+} (q = 9, 10)	G16, MCLZ, RCMD
Dijkkamp et al. [54]	total,nl	C^{q+} (q = 3, 4), N^{5+}, O^{6+}	G16
Fritsch & Lin [55]	total,nl	C^{4+}	G16
Hoekstra et al. [56]	total,nl	C^{4+}	G16
Stebbings et al. [57]	total	N^+, O^+	G16
Fite et al. [58]	total	O^+	G16
Meyer et al. [43]	total	B^{q+} (q = 2–5), C^{q+} (q = 3, 4), N^{q+} (q = 3, 4)	G16
Meyer et al. [43]	total	O^{q+} (q = 3–6), Si^{q+} (q = 4–9), Fe^{q+} (q = 4–15)	G16
Meyer et al. [43]	total	O^{q+} (q = 7, 8)	G16, MCLZ, RCMD
Havener et al. [59]	total	O^{5+}	G16
Huber [60]	total	Ne^{q+} (q = 2–4), Ar^{q+} (q = 2–4, 6)	G16
Kim et al. [61]	total	Si^{q+} (q = 2–7)	G16
Beijers et al. [62]	nl	O^{3+}	G16
Rejoub et al. [63]	total	Ne^{3+}	G16
Havener et al. [64]	total	Ne^{4+}	G16
Bruhns et al. [65]	total	Si^{3+}	G16
Havener et al. [66]	total	C^{3+}	G16
Mroczkowski et al. [67]	total	Ne^{2+}	G16
Pieksma & Havener [68]	total	B^{4+}	G16
Folkerts et al. [69]	total	N^{4+}	G16

[a]: total = total cross section, nl = nl-resolved cross section.

The laboratory results should have their own uncertainties; however, these values are available for only a part of the measurements. Here, we provide a rough estimate of the combined measurement uncertainty. The mean systematic uncertainties on the cross sections measured in, e.g., Meyer et al. [43], Draganić et al. [70], Cabrera-Trujillo et al. [71], are approximately 15% for the energy range considered. Assuming that this value can be applied to the other laboratory results, the measurement uncertainties are about 1% for the sample used in testing the G16 calculation, and ∼4% for the MCLZ and RCMD results. These relatively minor uncertainties can be accepted as the errors of the theoretical deviations obtained above (e.g., 55%, 88%, and 35% for the G16, MCLZ, and RCMD approaches).

The total charge exchange cross section is needed not only for calculating the charge exchange emission, but also to derive the ionization concentration for general cosmic plasmas in collisional ionization or photoionization equilibrium. The uncertainties in the theoretical calculation would introduce systematic uncertainties to the charge state distribution for the low-temperature plasmas where ions and neutral atoms coexist. As shown in Figure 3, we present two test cases on the concentration calculations of N and O ions in collisional ionization equilibrium (CIE). Here, we assume uncertainties of 50% on the charge exchange recombination rates. The induced errors on the charge distributions of N I and O I would become 10% and 60% at an equilibrium temperature of 1.2 eV. The difference between N I and O I errors reflects the different relative contribution of charge exchange to the total recombination in the concentration calculation. This result suggests that the charge exchange atomic data are vital to the modeling accuracy of lowly ionized species for collisional plasmas. It is expected that similar uncertainties would also apply to photoionization modeling, which includes the charge exchange component in the same way.

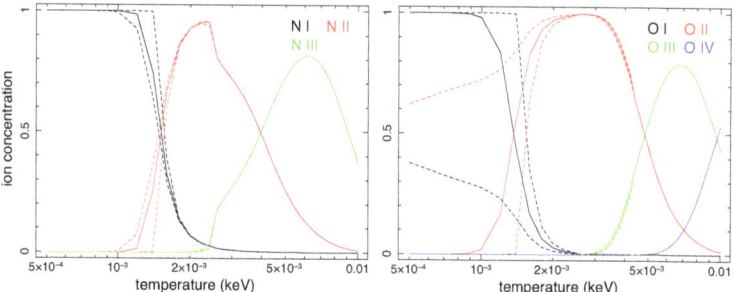

Figure 3. Charge state distributions of N (**left**) and O (**right**) as a function of equilibrium temperature for the CIE plasma, calculated with SPEX version 3.06.01. The dashed lines show the calculations when the charge exchange recombination rates are changed by 50%, while the other ionization and recombination data are kept intact.

2.2. Cross Sections for the Peak nl Shells

Next, we examine the state-resolved cross sections. The selective population of high-n levels of the recombining ions is known to be a characteristic property of the charge exchange reaction. The distribution functions on the quantum numbers n and l are key to the calculation of the spectrum, though the present theory still cannot fully reproduce the nl distributions measured in the laboratory [27,33,72].

As shown in Figure 4, we compare the laboratory measurements of four reactions with theoretical calculations, using the G16 method. G16 is the only calculation available in SPEX for the ions tested. It defines empirically n of the most populated levels as functions of the collision velocity, charge, and ionization potential. For the four test cases, G16 successfully predicts the peak n: $n = 3$ for C IV and O III, $n = 4$ for N V and O VI. The cross sections of the peak n levels, however, show deviations from the G16 values at the low energies. For C IV and O III, the measured values for $v = 100$ km s^{-1} are higher by a factor of ∼2.5 than the theoretical ones. This is probably because the G16 method underestimates the total cross sections at low energies, as already shown in Figure 2. For $v > 500$ km s^{-1}, the G16 calculations become consistent, with the measurements within 40% for the peak n.

To assess the l-distribution function, in Figure 4, we also compare the cross sections of the np subshells. The l-distribution defined in G16 is a smooth function that switches as a function of velocity between the different empirical l distributions introduced in Janev & Winter [73] (see also Equations (4)–(8) and Appendix B in [13]). The G16 cross sections on the np shells are lower, by a factor of 2–5, than the experimental values for $v < 500$ km s^{-1}. The deviations again become much smaller at higher velocities. To summarize above, the G16 method could reproduce the nl-resolved cross sections for the test cases with an accuracy of ∼40% for $v > 500$ km s^{-1}, while for the low-velocity collision, the G16 cross sections, as well as the line intensities calculated based on the atomic data, are much less reliable.

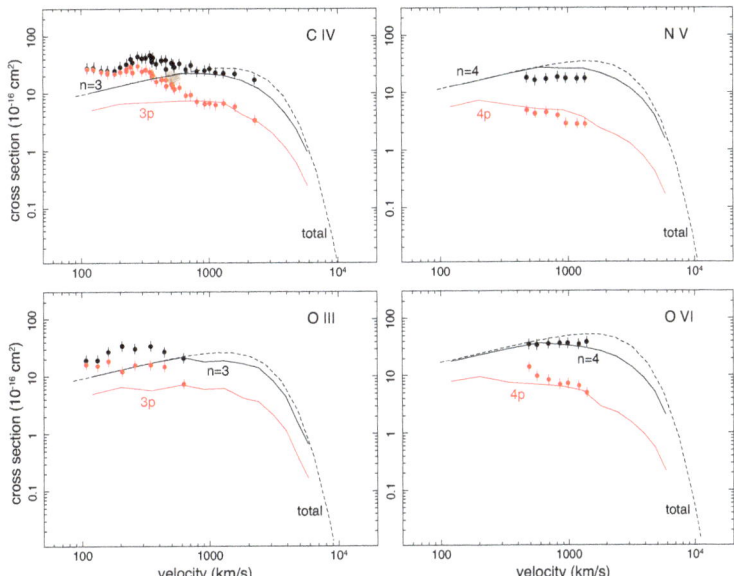

Figure 4. State-selective cross sections as a function of collision velocity for C IV, N V, O III, and O VI. The data points are taken from merged-beam experiments (see Table 1 for detail), for the peak n shells (black) and the np subshells (red). The approximate errors of 15% [43] are shown. The black solid lines are the G16 calculations of the peak n shells, and the red lines are the G16 data for the np subshells. The dash lines are the G16 calculations of the total cross sections.

2.3. Line Ratios

The large ratios between $1s - np$ ($n > 2$) and $1s - 2p$ lines are often used as characteristic diagnostics of the highly charged charge exchange plasma [10,11]. It is known that the line ratios would decrease with increasing collision velocity, because a high-speed collision might yield captures on high angular momentum states, producing more $1s - 2p$ transitions through cascade. So, the line ratios can often be utilized as a probe of collision velocity [27,74]. The accuracy of the velocity measurement is therefore determined by the quality of the atomic data.

In Figure 5, we plot the comparison of the line ratio calculations and experiments for C VI and O VIII. The experimental data are taken from the beam-gas measurements by Andrianarijaona et al. [75] for C VI and Seely et al. [76] for O VIII. A caveat of the comparison is that these experiments used the Kr atom as a target, while the original theoretical calculations are based on capture from H atom. As reported in Leung & Kirchner [77], the line ratios from Kr and H collisions are somewhat different, in particular for the low-energy regime, even though the ionization potentials of Kr and H atoms are nearly the same. To compensate this discrepancy, we calculate the H-to-Kr scalings as a function of velocities on both C VI and O VIII line ratios, using the theoretical results reported in Leung & Kirchner [77] (in their Figures 3 and 6), and apply the scalings to the G16, MCLZ, and RCMD line ratios. The scaled line ratios should represent a better approximation to the collisions with the Kr target.

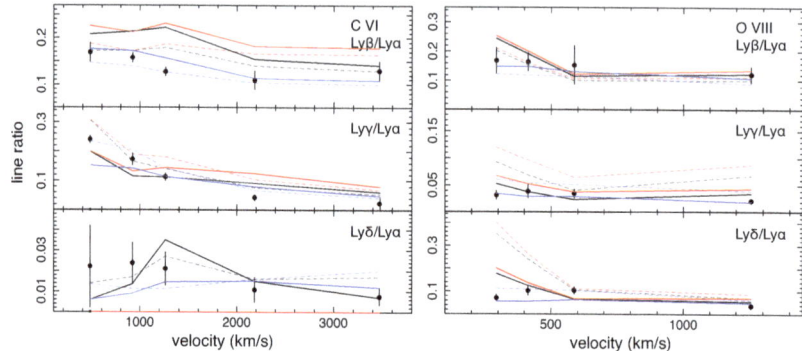

Figure 5. Comparison of experimental and theoretical line ratios for the C^{6+} (**left**) and O^{8+} (**right**) reactions. The experimental data from Andrianarijaona et al. [75] and Seely et al. [76] are plotted as data points, and the predictions from G16, MCLZ, and RCMD are shown in black, red, and blue curves. The dashed curves show the calculations of C^{6+} and O^{8+} collisions with H atoms, and the solid lines show the collisions with Kr atoms obtained using the scaling of Leung & Kirchner [77].

As shown in Figure 5, the experiments and calculations yield the same peak n, $n = 4$ for C VI and $n = 5$ for O VIII, though the line ratios still differ at several velocities. One of the main discrepancies occurs between the scaled G16/MCLZ and the lab data for the C VI Lyβ/Lyα line ratio, where the two theoretical values exceed the measured one by about 70% at $v = 1000$ km s^{-1}. The RCMD calculation shows better agreement with the lab values on this line ratio.

A more extensive comparison can be seen in Table 2. It is a compilation of several laboratory efforts, including the recent electron beam ion trap devices with X-ray spectral analysis carried out at both low and high resolutions. The EBIT devices simulate charge exchange reactions only at low collision energies. The average relative discrepancies (experiment-theory/theory) of the line ratios are 0.63, 0.77, and 0.54 for the G16, MCLZ, and RCMD calculations, respectively. For the peak n shell, the average discrepancies are 0.61, 0.81, and 0.56 for the three models. These differences are significantly larger than those on the modeling of collisional ionization equilibrium plasmas (∼10–40%, [14,78,79]), suggesting that the state-of-the-art charge exchange spectral models, even with dedicated theoretical calculations, are still less reliable than those for the CIE plasma.

Table 2. Experimental and theoretical Line ratios.

Ion	v (km s^{-1})	Ratio	Experiment	G16	MCLZ	RCMD	Reference [a]
N VII	794	Lyβ/Lyα	0.76	0.12	0.13	0.10	1
		(Lyγ + Lyδ)/Lyα	0.62	0.18	0.20	0.29	
O VII	724	Lyβ/Lyα	0.19	0.07	0.09	0.07	
		(Lyγ + Lyδ)/Lyα	0.24	0.11	0.47	0.07	
O VIII	774	Lyβ/Lyα	0.13	0.11	0.10	0.11	
		(Lyγ + Lyδ)/Lyα	0.17	0.15	0.18	0.14	
Ne IX	743	Lyβ/Lyα	0.04	0.04	0.12	—	
		(Lyγ + Lyδ)/Lyα	0.05	0.05	0.18	—	
Ne X	783	Lyβ/Lyα	0.12	0.08	0.08	0.08	
		(Lyγ + Lyδ)/Lyα	0.11	0.06	0.08	0.04	
O VIII	293	Lyβ/Lyα	0.169 ± 0.044	0.244 [b]	0.254	0.149	2
		Lyγ/Lyα	0.032 ± 0.008	0.053	0.068	0.035	
		Lyδ/Lyα	0.071 ± 0.014	0.177	0.201	0.057	
		Lyϵ/Lyα	0.0065 ± 0.003	0.054	0.0061	0.027	
O VIII	414	Lyβ/Lyα	0.165 ± 0.030	0.192	0.202	0.149	
		Lyγ/Lyα	0.039 ± 0.012	0.038	0.053	0.030	
		Lyδ/Lyα	0.103 ± 0.02	0.125	0.138	0.057	
		Lyϵ/Lyα	0.005 ± 0.0076	0.031	0.0024	0.019	
O VIII	586	Lyβ/Lyα	0.154 ± 0.006	0.115	0.123	0.132	
		Lyγ/Lyα	0.035 ± 0.008	0.024	0.038	0.030	
		Lyδ/Lyα	0.104 ± 0.015	0.066	0.068	0.064	
		Lyϵ/Lyα	0.0048 ± 0.0061	0.015	0.00086	0.014	

Table 2. Cont.

Ion	v (km s^{-1})	Ratio	Experiment	G16	MCLZ	RCMD	Reference [a]
O VIII	1256	Lyβ/Lyα	0.121 ± 0.027	0.122	0.135	0.108	
		Lyγ/Lyα	0.022 ± 0.004	0.035	0.044	0.020	
		Lyδ/Lyα	0.037 ± 0.011	0.055	0.071	0.050	
		Lyϵ/Lyα	0.0048 ± 0.0028	0.023	0.00045	0.0090	
C VI	477	Lyβ/Lyα	0.169 ± 0.023	0.208 [b]	0.226	0.177	3
		Lyγ/Lyα	0.240 ± 0.012	0.198	0.199	0.152	
		Lyδ/Lyα	0.022 ± 0.020	0.0062	2.8×10^{-6}	0.0061	
C VI	924	Lyβ/Lyα	0.157 ± 0.012	0.214	0.213	0.173	
		Lyγ/Lyα	0.173 ± 0.023	0.115	0.132	0.142	
		Lyδ/Lyα	0.024 ± 0.009	0.014	2.7×10^{-6}	0.0091	
C VI	1262	Lyβ/Lyα	0.128 ± 0.009	0.222	0.231	0.157	
		Lyγ/Lyα	0.113 ± 0.012	0.112	0.144	0.114	
		Lyδ/Lyα	0.021 ± 0.008	0.035	2.2×10^{-6}	0.015	
C VI	2185	Lyβ/Lyα	0.109 ± 0.019	0.154	0.182	0.114	
		Lyγ/Lyα	0.043 ± 0.011	0.091	0.125	0.080	
		Lyδ/Lyα	0.011 ± 0.006	0.015	6.7×10^{-7}	0.015	
C VI	3466	Lyβ/Lyα	0.130 ± 0.021	0.141	0.178	0.108	
		Lyγ/Lyα	0.024 ± 0.018	0.061	0.080	0.048	
		Lyδ/Lyα	0.0076 ± 0.004	0.007	5.0×10^{-7}	0.012	
O VII	low	He$_{high}$/Heα	0.167	0.168	0.152	0.058	4
Ne IX	low	He$_{high}$/Heα	0.162	0.161	0.133	—	
Ar XVII	low	He$_{high}$/Heα	0.191	0.133	—	—	
Fe XXV	low	He$_{high}$/Heα	0.267	0.156	0.079	—	
O VIII	low	Ly$_{high}$/Lyα	1.006	0.786	0.887	0.366	
Ne X	low	Ly$_{high}$/Lyα	1.207	0.690	0.865	0.210	
Mg XII	low	Lyβ/Lyα	0.227 ± 0.040	0.179	0.205	—	5
Mg XII	low	Lyγ/Lyα	0.133 ± 0.022	0.070	0.083	—	
Mg XII	low	Lyδ/Lyα	0.044 ± 0.015	0.038	0.046	—	
Mg XII	low	Lyϵ/Lyα	0.095 ± 0.015	0.028	0.030	—	
Mg XII	low	Lyζ/Lyα	0.030 ± 0.018	0.221	0.120	—	
Mg XII	low	Lyη/Lyα	0.080 ± 0.014	0.091	0.287	—	
S XVI	low	Lyβ/Lyα	0.203 ± 0.070	0.153	0.171	—	5
S XVI	low	Lyγ/Lyα	0.082 ± 0.016	0.055	0.064	—	
S XVI	low	Lyδ/Lyα	0.053 ± 0.011	0.028	0.033	—	
S XVI	low	Lyϵ/Lyα	0.053 ± 0.008	0.017	0.020	—	
S XVI	low	Lyζ/Lyα	0.016 ± 0.005	0.012	0.014	—	
S XVI	low	Lyη/Lyα	0.029 ± 0.008	0.024	0.014	—	
S XVI	low	Lyθ/Lyα	0.111 ± 0.019	0.149	0.101	—	
S XVI	low	Lyι/Lyα	0.165 ± 0.024	0.058	0.165	—	

[a]: references 1 = Greenwood et al. [72]; 2 = Seely et al. [76]; 3 = Andrianarijaona et al. [75]; 4 = Wargelin et al. [80]; 5 = Betancourt-Martinez [81]. [b]: H-to-Kr scaling has been applied to the theoretical line ratios for O VIII and C VI; see text for details.

3. Discussion and Ending Remarks

Based on a large sample of laboratory measurements, we have systematically compared the commonly used charge exchange atomic data to the experimental results. The G16, MCLZ, and RCMD calculations utilized in the SPEX code do not fully reproduce the measurements, with notable, and likely velocity-dependent discrepancies in both total cross sections, state-resolved cross sections, and line ratios in the X-ray spectra. While the ease of the use of the present CX model is beneficial for the X-ray astronomical community, it should be used with caution, in particular for non-charge-exchange experts. The unresolvable disagreements call for advanced theoretical calculations for especially the low collision energy regime, in combination with more laboratory measurements with, in particular, EBIT and COLTRIMS facilities.

The previous EBIT experiments have provided relevant benchmarks to the predicted cross sections for electron capture into specific principal quantum number states n. However, a comparison with the angular-momentum l-resolved cross sections is challenging, as they depend on the collision energy; and the EBIT measurements are limited to low collision energies (<10 eV/u) [27]. Besides, the charge exchange process not only produces X-ray lines, but also generates lines in the ultraviolet and optical band as the Rydberg levels populated by charge exchange relax through radiative cascades to the ground state of the ion. Thus, the simultaneous measurements of EUV and optical charge exchange cascade photons at the EBIT would be of interest, and they could provide additional information on the population of nl-states for plasma modeling [29]. Furthermore, possible multi-electron capture contributions from the molecular targets used in the EBIT measurements can also be avoided by using an atomic hydrogen target, where only single-electron capture can occur [82]. Atomic hydrogen is of particular interest as it is also the most abundant neutral element in the universe, and it makes a comparison between laboratory measurements and astrophysical observations more reliable.

Besides EBIT, the COLTRIMS and beam-gas experiments have been providing reliable measurements on velocity-dependent total and state-resolved cross sections. The improvement in the momentum measurement technique allows nl selectivity, and for a few cases, it might even be able to resolve the spin state. The state-of-the-art measurement accuracy is about 11% for both the total and nl-resolved cross sections [83].

A systematic measurement of the cosmic abundant ions with the COLTRIMS facilities, in combination with simultaneous EBIT X-ray spectroscopy, is desirable for the astronomical community. A consistent and continuous effort will be needed to ensure that the charge exchange atomic data will be ready for the high-resolution X-ray spectra taken with next-generation missions, XRISM (launch due in 2023, [84]) and Athena (early 2030s, [85]).

Assessing uncertainties carried out by the theoretical atomic data is also vital to the success of the upcoming missions. The atomic physics and plasma code community has already begun this work, with a persistent effort on the evaluation of the errors on electron impact excitation and transition probability data [14,86–90], as well as errors on photon impact data and modeling [91]. One implication from the aforementioned works, including the present work on the charge exchange modeling, is that the classical assumption of constant model uncertainty (e.g., 20% on line emissivity) is no longer valid, since the uncertainties are proven to vary significantly with the underlying model and its key parameters.

Author Contributions: Conceptualization, L.G., C.S. and R.Z.; Data curation, L.G.; Formal analysis, L.G.; Investigation, L.G.; Methodology, L.G.; Project administration, L.G.; Writing—original draft, L.G., C.S. and R.Z.; Writing—review & editing, L.G., C.S. and R.Z. All authors have read and agreed to the published version of the manuscript.

Funding: This research was funded by European Union's Horizon 2020 Programme under the AHEAD2020 project grant agreement n. 871158.

Acknowledgments: SRON is supported financially by NWO, the Netherlands Organization for Scientific Research. C.S. acknowledge support from an appointment to the NASA Postdoctoral Program at the NASA Goddard Space Flight Center, administered by the Universities Space Research Association, under contract with NASA, by the Lawrence Livermore National Laboratory (LLNL) Visiting Scientist and Professional Program Agreement, and by Max-Planck-Gesellschaft (MPG).

Conflicts of Interest: The authors declare no conflict of interest.

References

1. Lisse, C.M.; Denner, l.K.; Englhauser, J.; Harden, M.; Marshall, F.E.; Mumma, M.J.; Petre, R.; Pye, J.P.; Ricketts, M.J.; Schmitt, J.; et al. Discovery of X-ray and Extreme Ultraviolet Emission from Comet C/Hyakutake 1996 B2. *Science* **1996**, *274*, 205–209. [CrossRef]
2. Cravens, T.E. Comet Hyakutake X-ray source: Charge transfer of solar wind heavy ions. *Geophys. Res. Lett.* **1997**, *24*, 105–108. [CrossRef]
3. Bodewits, D.; Christian, D.J.; Torney, M.; Dryer, M.; Lisse, C.M.; Dennerl, K.; Zurbuchen, T.H.; Wolk, S.J.; Tielens, A.G.G.M.; Hoekstra, R. Spectral analysis of the Chandracomet survey. *Astron. Astrophys.* **2007**, *469*, 1183–1195. [CrossRef]
4. Branduardi-Raymont, G.; Bhardwaj, A.; Elsner, R.F.; Gladstone, G.R.; Ramsay, G.; Rodriguez, P.; Soria, R.; Waite, J.H., Jr.; Cravens, T.E. A study of Jupiter's aurorae with XMM-Newton. *Astron. Astrophys.* **2006**, *463*, 761–774. [CrossRef]
5. Katsuda, S.; Tsunemi, H.; Mori, K.; Uchida, H.; Kosugi, H.; Kimura, M.; Nakajima, H.; Takakura, S.; Petre, R.; Hewitt, J.W.; et al. Possible charge-exchange X-ray emission in the cygnus loop detected withsuzaku. *Astrophys. J.* **2011**, *730*, 24. [CrossRef]
6. Cumbee, R.S.; Henley, D.B.; Stancil, P.C.; Shelton, R.L.; Nolte, J.L.; Wu, Y.; Schultz, D.R. Can charge exchange explain anomalous soft X-ray emission in the cygnus loop? *Astrophys. J.* **2014**, *787*, L31. [CrossRef]

7. Liu, J.; Wang, Q.D.; Li, Z.; Peterson, J.R. X-ray spectroscopy of the hot gas in the M31 bulge. *Mon. Not. R. Astron. Soc.* **2010**, *404*, 1879. [CrossRef]
8. Zhang, S.; Wang, Q.D.; Ji, L.; Smith, R.K.; Foster, A.R.; Zhou, X. Spectral modeling of the charge-exchange X-ray emission from M82. *Astrophys. J.* **2014**, *794*, 61. [CrossRef]
9. Gu, L.; Mao, J.; O'Dea, C.P.; Baum, S.A.; Mehdipour, M.; Kaastra, J.S. Charge exchange in the ultraviolet: Implication for interacting clouds in the core of NGC 1275. *Astron. Astrophys.* **2017**, *601*, A45. [CrossRef]
10. Gu, L.; Kaastra, J.; Raassen, A.J.J.; Mullen, P.D.; Cumbee, R.; Lyons, D.; Stancil, P.C. A novel scenario for the possible X-ray line feature at ∼3.5 keV. *Astron. Astrophys.* **2015**, *584*, L11. [CrossRef]
11. Gu, L.; Mao, J.; de Plaa, J.; Raassen, A.J.J.; Shah, C.; Kaastra, J. Charge exchange in galaxy clusters. *Astron. Astrophys.* **2018**, *611*, A26. [CrossRef]
12. Smith, R.; Foster, A.; Brickhouse, N. Approximating the X-ray spectrum emitted from astrophysical charge exchange. *Astron. Nachrichten* **2012**, *333*, 301–304. [CrossRef]
13. Gu, L.; Kaastra, J.; Raassen, A.J.J. Plasma code for astrophysical charge exchange emission at X-ray wavelengths. *Astron. Astrophys.* **2016**, *588*, A52. [CrossRef]
14. Akamatsu, H.; Akimoto, F.; Allen, S.W.; Angelini, L.; Audard, M.; Awaki, H.; Axelsson, M.; Bamba, A.; Bautz, M.W.; Blandford, R.; et al. Atomic data and spectral modeling constraints from high-resolution X-ray observations of the Perseus cluster with Hitomi. *Publ. Astron. Soc. Jpn.* **2018**, *70*, 12. [CrossRef]
15. Bodewits, D.; Juhász, Z.; Hoekstra, R.; Tielens, A.G.G.M. Catching Some Sun: Probing the Solar Wind with Cometary X-ray and Far-Ultraviolet Emission. *Astrophys. J.* **2004**, *606*, L81–L84. [CrossRef]
16. Dijkkamp, D.; Gordeev, Y.S.; Brazuk, A.; Drentje, A.G.; De Heer, F.J. Selective single-electron capture into (n, l) subshells in slow collisions of C^{6+}, N^{6+}, O^{6+} and Ne^{6+} with He, H_2 and Ar. *J. Phys. B At. Mol. Phys.* **1985**, *18*, 737–756. [CrossRef]
17. Trassinelli, M.; Prigent, C.; Lamour, E.; Mezdari, F.; Mérot, J.; Reuschl, R.; Rozet, J.-P.; Steydli, S.; Vernhet, D. Investigation of slow collisions for (quasi) symmetric heavy systems: What can be extracted from high resolution X-ray spectra. *J. Phys. B At. Mol. Opt. Phys.* **2012**, *45*, 085202. [CrossRef]
18. Beiersdorfer, P. Highly charged ions in magnetic fusion plasmas: Research opportunities and diagnostic necessities. *J. Phys. B At. Mol. Opt. Phys.* **2015**, *48*, 144017. [CrossRef]
19. Lepson, J.K.; Beiersdorfer, P.; Bitter, M.; Roquemore, A.L.; Hill, K.; Kaita, R. Charge exchange produced emission of carbon in the extreme ultraviolet spectral region. *J. Phys. Conf. Ser.* **2015**, *583*, 012012. [CrossRef]
20. Rosmej, F.B.; Lisitsa, V.S.; Schott, R.; Dalimier, E.; Riley, D.; Delserieys, A.; Renner, O.; Krousky, E. Charge-exchange-driven X-ray emission from highly ionized plasma jets. *EPL Europhys. Lett.* **2006**, *76*, 815–821. [CrossRef]
21. Ali, R.; Beiersdorfer, P.; Harris, C.L.; Neill, P.A. Charge-exchange X-ray spectra: Evidence for significant contributions from radiative decays of doubly excited states. *Phys. Rev. A* **2016**, *93*, 012711. [CrossRef]
22. Ali, R.; Neill, P.A.; Beiersdorfer, P.; Harris, C.L.; Raković, M.J.; Wang, J.G.; Schultz, D.R.; Stancil, P.C. On the Significance of the Contribution of Multiple-Electron Capture Processes to Cometary X-ray Emission. *Astrophys. J.* **2005**, *629*, L125–L128. [CrossRef]
23. Ali, R.; Neill, P.A.; Beiersdorfer, P.; Harris, C.L.; Schultz, D.R.; Stancil, P.C. Critical test of simulations of charge-exchange-induced X-ray emission in the solar system. *Astrophys. J.* **2010**, *716*, L95–L98. [CrossRef]
24. Fischer, D.; Feuerstein, B.; Dubois, R.D.; Moshammer, R.; López-Urrutia, J.C.; Draganic, I.; Lörch, H.; Perumal, A.N.; Ullrich, J. State-resolved measurements of single-electron capture in slow Ne^{7+}-and Ne^{8+}-helium collisions. *J. Phys. B At. Mol. Opt. Phys.* **2002**, *35*, 1369. [CrossRef]
25. Xue, Y.; Ginzel, R.; Krauß, A.; Bernitt, S.; Schöffler, M.; Kühnel, K.U.; López-Urrutia, J.R.C.; Moshammer, R.; Cai, X.; Ullrich, J.; et al. Kinematically complete study of electron transfer and rearrangement processes in slow Ar16+-Ne collisions. *Phys. Rev. A* **2014**, *90*, 052720. [CrossRef]
26. Allen, F.I.; Biedermann, C.; Radtke, R.; Fussmann, G.; Fritzsche, S. Energy dependence of angular momentum capture states in charge exchange collisions between slow highly charged argon ions and argon neutrals. *Phys. Rev. A* **2008**, *78*, 032705. [CrossRef]
27. Beiersdorfer, P.; Olson, R.E.; Brown, G.V.; Chen, H.; Harris, C.L.; Neill, P.A.; Schweikhard, L.; Utter, S.B.; Widmann, K. X-ray Emission Following Low-Energy Charge Exchange Collisions of Highly Charged Ions. *Phys. Rev. Lett.* **2000**, *85*, 5090–5093. [CrossRef]
28. Betancourt-Martinez, G.L.; Beiersdorfer, P.; Brown, G.; Kelley, R.L.; Kilbourne, C.A.; Koutroumpa, D.; Leutenegger, M.; Porter, F. Observation of highly disparate K-shell X-ray spectra produced by charge exchange with bare mid-Zions. *Phys. Rev. A* **2014**, *90*, 052723. [CrossRef]
29. Dobrodey, S. The Faculty of Physics and Astronomy. Ph.D. Thesis, Heidelberg University, Heidelberg, Germany, 2019.
30. Leutenegger, M.A.; Beiersdorfer, P.; Brown, G.; Kelley, R.L.; Kilbourne, C.A.; Porter, F.S. Measurement of Anomalously Strong Emission from the 1s−9p Transition in the Spectrum of H-Like Phosphorus Following Charge Exchange with Molecular Hydrogen. *Phys. Rev. Lett.* **2010**, *105*, 063201. [CrossRef] [PubMed]
31. Shah, C.; Dobrodey, S.; Bernitt, S.; Steinbrügge, R.; López-Urrutia, J.R.C.; Gu, L.; Kaastra, J. Laboratory measurements compellingly support a charge-exchange mechanism for the "dark matter" ∼3.5 kev X-ray line. *Astrophys. J.* **2016**, *833*, 52. [CrossRef]
32. Wargelin, B.J.; Beiersdorfer, P.; Neill, P.A.; Olson, R.E.; Scofield, J.H. Charge-Exchange Spectra of Hydrogenic and He-like Iron. *Astrophys. J.* **2005**, *634*, 687–697. [CrossRef]

33. Betancourt-Martinez, G.L.; Beiersdorfer, P.; Brown, G.V.; Cumbee, R.S.; Hell, N.; Kelley, R.L.; Kilbourne, C.A.; Leutenegger, M.A.; Lockard, T.E.; Porter, F.S. High-resolution Charge Exchange Spectra with L-shell Nickel Show Striking Differences from Models. *Astrophys. J. Lett.* **2018**, *868*, L17. [CrossRef]
34. Xu, J.W.; Xu, C.X.; Zhang, R.T.; Zhu, X.L.; Feng, W.T.; Gu, L.; Liang, G.Y.; Guo, D.L.; Gao, Y.; Zhao, D.M.; et al. Measurement of n-resolved State-selective Charge Exchange in Ne(8,9)+ Collision with He and H2. *Astrophys. J. Suppl. Ser.* **2021**, *253*, 13. [CrossRef]
35. Kaastra, J.S.; Mewe, R.; Nieuwenhuijzen, H. SPEX: a new code for spectral analysis of X & UV spectra. In *UV and X-ray Spectroscopy of Astrophysical and Laboratory Plasmas*; University of California: Berkeley, CA, USA, 1996; pp. 411–414.
36. Mullen, P.D.; Cumbee, R.S.; Lyons, D.; Stancil, P.C. Charge exchange-induced X-ray emission of fe xxv and fe xxvi via a streamlined model. *Astrophys. J. Suppl. Ser.* **2016**, *224*, 31. [CrossRef]
37. Wu, Y.; Stancil, P.C.; Schultz, D.R.; Hui, Y.; Liebermann, H.P.; Buenker, R.J. Theoretical investigation of total and state-dependent charge exchange in O6+ collisions with atomic hydrogen. *J. Phys. B At. Mol. Opt. Phys.* **2012**, *45*, 235201. [CrossRef]
38. Wu, Y.; Stancil, P.C.; Liebermann, H.P.; Funke, P.; Rai, S.N.; Buenker, R.J.; Schultz, D.R.; Hui, Y.; Draganic, I.N.; Havener, C.C. Theoretical investigation of charge transfer between N6+ and atomic hydrogen. *Phys. Rev. A* **2011**, *84*, 022711. [CrossRef]
39. Nolte, J.L.; Stancil, P.C.; Liebermann, H.P.; Buenker, R.J.; Hui, Y.; Schultz, D.R. Final-state-resolved charge exchange in C5+ collisions with H. *J. Phys. B At. Mol. Opt. Phys.* **2012**, *45*, 245202. [CrossRef]
40. Goffe, T.V.; Shah, M.B.; Gilbody, H.B. One-electron capture and loss by fast multiply charged boron and carbon ions in H and H2. *J. Phys. B At. Mol. Phys.* **1979**, *12*, 3763–3773. [CrossRef]
41. Phaneuf, R.A.; Alvarez, I.; Meyer, F.W.; Crandall, D.H. Electron capture in low-energy collisions of Cq+ and Oq+ with H and H2. *Phys. Rev. A* **1982**, *26*, 1892–1906. [CrossRef]
42. Panov, M.N.; A Basalaev, A.; O Lozhkin, K. Interaction of Fully Stripped, Hydrogenlike and Heliumlike C, N, O, Ne and Ar Ions with H and He Atoms and H2 Molecules. *Phys. Scr.* **1983**, *T3*, 124–130. [CrossRef]
43. Meyer, F.W.; Phaneuf, R.A.; Kim, H.J.; Hvelplund, P.; Stelson, P.H. Single-electron-capture cross sections for multiply charged O, Fe, Mo, Ta, W, and Au ions incident on H and H2 at intermediate velocities. *Phys. Rev. A* **1979**, *19*, 515–525. [CrossRef]
44. Shah, M.B.; Goffe, T.V.; Gilbody, H.B. Electron capture and loss by fast lithium ions in H and H2. *J. Phys. B At. Mol. Phys.* **1978**, *11*, L233. [CrossRef]
45. Seim, W.; Muller, A.; Wirkner-Bott, I.; Salzborn, E. Electron capture by Lii+ (i = 2,3), Ni+ and Nei+ (i = 2, 3, 4, 5) ions from atomic hydrogen. *J. Phys. B At. Mol. Phys.* **1981**, *14*, 3475–3491. [CrossRef]
46. McCullough, R.W.; Nutt, W.L.; Gilbody, H.B. One-electron capture by slow doubly charged ions in h and H2. *J. Phys. B At. Mol. Phys.* **1979**, *12*, 4159–4169. [CrossRef]
47. Crandall, D.H.; Phaneuf, R.A.; Meyer, F.W. Electron capture by slow multicharged ions in atomic and molecular hydrogen. *Phys. Rev. A* **1979**, *19*, 504–514. [CrossRef]
48. Gardner, L.D.; Bayfield, J.E.; Koch, P.M.; Sellin, I.A.; Pegg, D.J.; Peterson, R.S.; Crandall, D.H. Electron-capture collisions at keV energies of boron and other multiply charged ions with atoms and molecules. II. Atomic hydrogen. *Phys. Rev. A* **1980**, *21*, 1397–1402. [CrossRef]
49. Phaneuf, R.A.; Meyer, F.W.; McKnight, R.H. Single-electron capture by multiply charged ions of carbon, nitrogen, and oxygen in atomic and molecular hydrogen. *Phys. Rev. A* **1978**, *17*, 534–545. [CrossRef]
50. Nutt, W.L.; McCullough, R.W.; Gilbody, H.B. Electron capture by C^{2+} and Ti^{2+} ions in H and H2. *J. Phys. B At. Mol. Phys.* **1978**, *11*, L181–L184. [CrossRef]
51. Sant'Anna, M.; Melo, W.S.; Santos, A.; Shah, M.B.; Sigaud, G.M.; Montenegro, E.C. Absolute measurements of electron capture cross sections of C3+ from atomic and molecular hydrogen. *J. Phys. B At. Mol. Opt. Phys.* **2000**, *33*, 353–364. [CrossRef]
52. Ciric, ; D.; Brazuk, A.; Dijkkamp, D.; De Heers, F.J.; Winter, H. State-selective electron capture in C^{3+}-H, H2 collisions (0.7–4.6 keV amu^{-1}) studied by photon spectroscopy. *J. Phys. B At. Mol. Phys.* **1985**, *18*, 3629–3639. [CrossRef]
53. McCullough, R.W.; Wilkie, F.G.; Gilbody, H.B. State-selective electron capture by slow C2+ and C3+ ions in atomic hydrogen. *J. Phys. B At. Mol. Phys.* **1984**, *17*, 1373–1382. [CrossRef]
54. Dijkkamp, D.; Ciric, D.; Vileg, E.; De Boer, A.; De Heer, F.J. Subshell-selective electron capture in collisions of C4+, N5+, O6+ with H, H2 and He. *J. Phys. B At. Mol. Phys.* **1985**, *18*, 4763–4793. [CrossRef]
55. Fritsch, W.; Lin, C.D. Atomic-basis study of electron transfer into C3+(nl) orbitals in C4++H and C4++Li collisions. *J. Phys. B At. Mol. Phys.* **1984**, *17*, 3271–3278. [CrossRef]
56. Hoekstra, R.; Beijers, J.P.M.; Schlatmann, A.R.; Morgenstern, R.; de Heer, F.J. State-selective charge transfer in slow collisions ofC4+with H and H2. *Phys. Rev. A* **1990**, *41*, 4800–4808. [CrossRef]
57. Stebbings, R.F.; Fite, W.L.; Hummer, D.G. Charge Transfer between Atomic Hydrogen and N+ and O+. *J. Chem. Phys.* **1960**, *33*, 1226. [CrossRef]
58. Fite, W.L.; Smith, A.C.H.; Stebbings, R.F. Charge transfer in collisions involving symmetric and asymmetric resonance. *Proc. R. Soc. Lond. Ser. A Math. Phys. Sci.* **1962**, *268*, 527–536. [CrossRef]
59. Havener, C.C.; Huq, M.S.; Krause, H.F.; Schulz, P.A.; Phaneuf, R.A. Merged-beams measurements of electron-capture cross sections for O5++H at electron-volt energies. *Phys. Rev. A* **1989**, *39*, 1725–1740. [CrossRef] [PubMed]
60. Huber, B.A. Electron capture by slow multiply charged Ar and Ne ions from atomic hydrogen. *Eur. Phys. J. A* **1981**, *299*, 307–309. [CrossRef]

61. Kim, H.J.; Phaneuf, R.A.; Meyer, F.W.; Stelson, P.H. Single electron capture by multiply charged ^{28}Si ions in atomic and molecular hydrogen. *Phys. Rev. A* **1978**, *17*, 854–858. [CrossRef]
62. Beijers, J.P.M.; Hoekstra, R.; Morgenstern, R. State-selective charge transfer in slow collisions of with H and. *J. Phys. B At. Mol. Opt. Phys.* **1996**, *29*, 1397–1408. [CrossRef]
63. Rejoub, R.; Bannister, M.E.; Havener, C.C.; Savin, D.W.; Verzani, C.J.; Wang, J.G.; Stancil, P.C. Electron capture by Ne3+ ions from atomic hydrogen. *Phys. Rev. A* **2004**, *69*, 052704. [CrossRef]
64. Havener, C.C.; Rejoub, R.; Vane, C.R.; Krause, H.F.; Savin, D.W.; Schnell, M.; Wang, J.G.; Stancil, P.C. Electron capture by Ne4+ ions from atomic hydrogen. *Phys. Rev. A* **2005**, *71*, 034702. [CrossRef]
65. Bruhns, H.; Kreckel, H.; Savin, D.W.; Seely, D.G.; Havener, C.C. Low-energy charge transfer for collisions of Si3+ with atomic hydrogen. *Phys. Rev. A* **2008**, *77*, 064702. [CrossRef]
66. Havener, C.C.; Muller, A.; Van Emmichoven, P.A.Z.; Phaneuf, R.A. Low-energy electron capture by C3+ from hydrogen using merged beams. *Phys. Rev. A* **1995**, *51*, 2982–2988. [CrossRef]
67. Mroczkowski, T.; Savin, D.W.; Rejoub, R.; Krstić, P.S.; Havener, C.C. Electron capture by Ne2+ ions from atomic hydrogen. *Phys. Rev. A* **2003**, *68*, 032721. [CrossRef]
68. Pieksma, M.; Havener, C.C. Low-energy electron capture by B4+ ions from hydrogen atoms. *Phys. Rev. A* **1998**, *57*, 1892–1894. [CrossRef]
69. Folkerts, L.; Haque, M.A.; Havener, C.C.; Shimakura, N.; Kimura, M. Low-energy electron capture by N4+ ions from H atoms: Experimental study using merged beams and theoretical analysis by molecular representation. *Phys. Rev. A* **1995**, *51*, 3685–3692. [CrossRef] [PubMed]
70. Draganić, I.N.; Seely, D.G.; Havener, C.C. Low-energy charge transfer between C^{5+} and atomic hydrogen. *Phys. Rev. A* **2011**, *83*, 054701. [CrossRef]
71. Cabrera-Trujillo, R.; Bruhns, H.; Savin, D.W. Acceptance-angle effects on the charge transfer and energy-loss cross sections for collisions of C4+ with atomic hydrogen. *Phys. Rev. A* **2020**, *101*, 052708. [CrossRef]
72. Greenwood, J.B.; Williams, I.D.; Smith, S.J.; Chutjian, A. Measurement of Charge Exchange and X-ray Emission Cross Sections for Solar Wind–Comet Interactions. *Astrophys. J.* **2000**, *533*, L175–L178. [CrossRef] [PubMed]
73. Janev, R.; Winter, H. State-selective electron capture in atom-highly charged ion collisions. *Phys. Rep.* **1985**, *117*, 265–387. [CrossRef]
74. Otranto, S.; Olson, R.E.; Beiersdorfer, P. X-ray emission cross sections following charge exchange by multiply charged ions of astrophysical interest. *Phys. Rev. A* **2006**, *73*, 022723. [CrossRef]
75. Andrianarijaona, V.M.; Wulf, D.; McCammon, D.; Seely, D. G.; Havener, C. C. Radiance line ratios Ly-β/Ly-α, Ly-γ/Ly-α, Ly-δ/Ly-α, and Ly-ϵ/Ly-α for soft X-ray emissions following charge exchange between C^{6+} and Kr. *Nucl. Instrum, Methods Phys. Res. B* **2015**, *350*, 122. [CrossRef]
76. Seely, D.G.; Andrianarijaona, V.M.; Wulf, D.; Morgan, K.; McCammon, D.; Fogle, M.; Stancil, P.C.; Zhang, R.T.; Havener, C.C. Line ratios for soft-X-ray emission following charge exchange between O8+ and Kr. *Phys. Rev. A* **2017**, *95*, 052704. [CrossRef]
77. Leung, A.C.K.; Kirchner, T. Lyman line ratios in charge-exchange collisions of C6+ and O8+ ions with hydrogen and krypton atoms. *Phys. Rev. A* **2018**, *97*, 062705. [CrossRef]
78. Gu, L.; Raassen, A.J.J.; Mao, J.; de Plaa, J.; Shah, C.; Pinto, C.; Werner, N.; Simionescu, A.; Mernier, F.; Kaastra, J. X-ray spectra of the Fe-L complex. *Astron. Astrophys.* **2019**, *627*, A51. [CrossRef]
79. Gu, L.; Shah, C.; Mao, J.; Raassen, A.; De Plaa, J.; Pinto, C.; Akamatsu, H.; Werner, N.; Simionescu, A.; Mernier, F.; et al. X-ray spectra of the Fe-L complex. II. Atomic data constraints from the EBIT experiment and X-ray grating observations of Capella. *Astron. Astrophys.* **2020**, *641*, 93. [CrossRef]
80. Wargelin, B.J.; Beiersdorfer, P.; Brown, G.V. EBIT charge-exchange measurements and astrophysical applications. *Can. J. Phys.* **2008**, *86*, 151–169. [CrossRef]
81. Betancourt-Martinez, G. Benchmarking Charge Exchange Theory in the Dawning Era of Space-Born High-Resolution X-ray Spectrometers. Ph.D. Thesis, University of Maryland, College Park, MD, USA, 2017.
82. Leutenegger, M.; Beiersdorfer, P.; Betancourt-Martinez, G.L.; Brown, G.; Hell, N.; Kelley, R.L.; Kilbourne, C.A.; Magee, E.W.; Porter, F.S. Characterization of an atomic hydrogen source for charge exchange experiments. *Rev. Sci. Instrum.* **2016**, *87*, 11E516. [CrossRef] [PubMed]
83. Han, J.; Wei, L.; Wang, B.; Ren, B.; Yu, W.; Zhang, Y.; Zou, Y.; Chen, L.; Xiao, J.; Wei, B. Measurement of Absolute Single and Double Electron Capture Cross Sections for O6+ Ion Collisions with CO_2, CH_4, H_2, and N_2. *Astrophys. J. Suppl. Ser.* **2021**, *253*, 6. [CrossRef]
84. Maejima, H.; Angelini, L.; Costantini, E.; Edison, M.R.; Herder, J.-W.D.; Ishisaki, Y.; Matsushita, K.; Mori, K.; Guainazzi, M.; Kelley, R.L.; et al. Concept of the X-ray Astronomy Recovery Mission. *Space Telesc. Instrum. 2018 Ultrav. Gamma Ray* **2018**, *10699*, 1069922.
85. N.; ra, K.; Barret, D.; Barcons, X.; Fabian, A.; Herder, J.W.D.; Piro, L.; Watson, M.; Adami, C.; Aird, J.; Afonso, J.M.; et al. The Hot and Energetic Universe: A White Paper presenting the science theme motivating the Athena+ mission. *arXiv* **2013**, arXiv:1306.2307.
86. Bautista, M.; Fivet, V.; Quinet, P.; Dunn, J.P.; Gull, T.R.; Kallman, T.; Mendoza, C. Uncertainties in atomic data and their propagation through spectral models. I. *Astrophys. J.* **2013**, *770*, 15. [CrossRef]

87. Loch, S.; Pindzola, M.; Ballance, C.; Witthoeft, M.; Foster, A.; Smith, R.; O'Mullane, M. The propagation of uncertainties in atomic data through collisional-radiative models. In Proceedings of the Eighth International Conference on Atomic and Molecular Data and Their Applications: ICAMDATA-2012, Gaithersburg, MD, USA, 30 September–4 October 2012; Volume 1545, p. 242. [CrossRef]
88. Yu, X.; Del Zanna, G.; Stenning, D.C.; Cisewski-Kehe, J.; Kashyap, V.L.; Stein, N.; Van Dyk, D.A.; Warren, H.P.; Weber, M.A. Incorporating Uncertainties in Atomic Data into the Analysis of Solar and Stellar Observations: A Case Study in Fe xiii. *Astrophys. J.* **2018**, *866*, 146. [CrossRef]
89. Foster, A.; Heuer, K. PyAtomDB: Extending the AtomDB Atomic Database to Model New Plasma Processes and Uncertainties. *Atoms* **2020**, *8*, 49. [CrossRef]
90. Morisset, C.; Luridiana, V.; García-Rojas, J.; Gómez-Llanos, V.; Bautista, M.; Mendoza, A.C. Atomic Data Assessment with PyNeb. *Atoms* **2020**, *8*, 66. [CrossRef]
91. Mehdipour, M.; Kaastra, J.; Kallman, T. Systematic comparison of photoionised plasma codes with application to spectroscopic studies of AGN in X-rays. *Astron. Astrophys.* **2016**, *596*, A65. [CrossRef]

MDPI AG
Grosspeteranlage 5
4052 Basel
Switzerland
Tel.: +41 61 683 77 34

Sensors Editorial Office
E-mail: sensors@mdpi.com
www.mdpi.com/journal/sensors

Disclaimer/Publisher's Note: The statements, opinions and data contained in all publications are solely those of the individual author(s) and contributor(s) and not of MDPI and/or the editor(s). MDPI and/or the editor(s) disclaim responsibility for any injury to people or property resulting from any ideas, methods, instructions or products referred to in the content.

www.ingramcontent.com/pod-product-compliance
Lightning Source LLC
LaVergne TN
LVHW070218100526
838202LV00015B/2060